The Principles and Practice
of Human Physiology

The Principles and Practice
of Human Physiology

Editors
O.G. EDHOLM
University College, London

J.S. WEINER
London School of Hygiene
and Tropical Medicine

1981

ACADEMIC PRESS
A Subsidiary of Harcourt Brace Jovanovich, Publishers
London New York Toronto Sydney San Francisco

ACADEMIC PRESS INC. (LONDON) LTD.
24/28 Oval Road,
London NWI

United States Edition published by
ACADEMIC PRESS INC.
111 Fifth Avenue
New York, New York 10003

British Library Cataloguing in Publication Data

The principles and practice of human
physiology.

1. Human physiology
I. Edholm, Otto Gustaf   II. Weiner,
Joseph Sidney
612   AP34.5   80-40831

ISBN 0-12-231650-9

Filmset by Asco Trade Typesetting Ltd.,
Hong Kong

**Printed in Great Britain by
Whitstable Litho Ltd., Whitstable, Kent**

Contributors

E. JOAN BASSEY, Department of Physiology and Pharmacology, University Hospital and Medical School, University of Nottingham, Nottingham NG72UH.

DAVID DENISON, Lung Function Unit, Brompton Hospital, London SW3.

O.G. EDHOLM, Bartlett School of Architecture and Planning, University College London, Wates House, 22 Gordon Street, London WC1-OQB.

DAVID H. ELLIOTT, OBE, Shell (UK) Limited, P.O. Box 148, Shell-Mex House, Strand, London WC2R ODX.

PETER FENTEM, Department of Physiology and Pharmacology, University Hospital and Medical School, University of Nottingham, Nottingham NG7 2UH.

DENNIS HILL, North-east Thames Regional Health Authority, 40 Eastbourne Terrace, London W2 3QR.

S.M. HILTON, Department of Physiology, The Medical School, University of Birmingham, Birmingham B15 2TJ.

PETER HOWARD, OBE, Institute of Aviation Medicine, Farnborough, Hampshire GU14 6SZ.

F.J. IMMS, Medical Research Council Environmental Physiological Unit, 242 Pentonville Road, London N1 9LB.

M.W. JOHNS, 187 Rathmines Road, Hawthorn East, Victoria 3123, Australia.

IVOR H. MILLS, Department of Medicine, Addenbrooke's Hospital, Cambridge CB2 2QQ.

J.N. MILLS, formerly of the Department of Physiology, Stopford Building, University of Manchester, Manchester M13 9PD. (Professor Mills is now deceased.)

D.S. MINORS, Department of Physiology, Stopford Building, University of Manchester, Manchester M13 9PD.

J. TINKER, Intensive Therapy Unit, Middlesex Hospital, Mortimer Street, London W1N 8AA.

J.M. WATERHOUSE, Department of Physiology, Stopford Building, University of Manchester, Manchester M13 9PT.

J.S. WEINER, MRC Environmental Physiology Unit, London School of Hygiene and Tropical Medicine, Keppel Street (Gower Street), London WC1E 7HT.

Preface

It is customary to give some reason or make some excuse for having prepared, edited or written a book. In this case, there were discussions between the two of us concerning a new course in Applied Human Physiology that J.S.W. was planning, and we began to think of a possible textbook. At that time there had been an enquiry about the feasibility of a second edition of "The Physiology of Human Survival". Some of the authors concerned were enthusiastic but others were not, and it was agreed that there would be no point in going ahead. However, we felt that a new book could be related to the old one; one title we considered was "The Physiology of Human Endeavour". Such a title should help to explain the contents of the present book, as well as the omissions. Two of the original authors, Professor Sidney Hilton and Air Commodore Peter Howard, are contributors with the same titles as before. Most of the other chapters contain material previously included but some which were in "The Physiology of Human Survival" have been omitted. The emphasis, we felt, should be on "Human Endeavour" and so chapters on nutrition, pregnancy and childbirth, and irreversible changes, have not been repeated. On the other hand, Dr Dennis Hill has contributed an important chapter on instrumentation, and Dr Tinker has described the techniques used in the intensive care units. The progress of human physiology in recent years owes much to the development of instrumentation, especially non-invasive techniques. As a result, it has become possible to carry out work in the field with a degree of sophistication close to that commonly achieved in the laboratory. It is relevant to stress this aspect since so many of the contents are concerned with the results of field work. We would have liked to have included more, principally accounts of the extensive studies carried out under the auspices of the Human Adaptability Section of the International Biological Programme. However, the cost of books is related to their length and we have been conscious of the needs of students. It is appropriate, therefore, to begin our thanks with the name of Academic Press which has patiently and courteously, helped and encouraged us and advised us on the arcane problems of publishing today, including costs.

There are many others to thank, including of course the authors who have had to endure nagging and then irritating editorial changes but, more seriously, have had their manuscripts trimmed to that we could keep down the overall length. We also record our thanks to the numerous colleagues and publishers who have generously given permission for the reproduction of figures, diagrams and tables. Any omission is due to our editorial incompetence, and is not the fault of individual authors.

The checking and final preparation of the manuscript have been done by Mrs Marie Strong and Miss Winifred Bailey, and we are greatly indebted to them.

Amongst the particulars often included in the Preface is the choice of units. Originally we asked our contributors to use the SI system, but then Air Commodore Peter Howard sent us a short but telling comment, which follows this preface.

O.G. EDHOLM
J.S. WEINER

Meretricious Metrication— A Physiologist's Lament

Peter Howard

Physiology has a long and happy history of promiscuous intercourse with other sciences and, as befits a discipline, one the major preoccupations of which has been with adaptation of the organism, it has adopted and adapted the best features of its consorts. This is nowhere more true than in the field of physiological "quantity surveying", and the measurement of pressure supplies a good example of miscegeny. Arterial blood pressure is described in the unit of millimetres of mercury, and to use other standards would be almost unthinkable. Venous pressures, on the other hand, are normally so low that a more "sensitive" scale must be found; water replaces mercury and inches often take the place of millimetres. The environmental physiologist must ape his engineering colleagues by quoting pressures in pounds per square inch or in atmospheres (which may themselves be absolute or relative to the outside world). For the diving fraternity, pressure must be expressed in feet of sea water; for aviators, millibars are *de rigeur*, but some American altimeters are still scaled in inches of mercury.

The notion that science is somehow better if it makes measurements in metric terms will not be argued here. The adoption of "a rationalized and coherent system of units"—the Système Internationale d'Unités, or SI—provided, for physics, engineering, chemistry and industry, a consistent structure of fundamental and derived quantities that could be used by all. Once its language had been learned, it offered a scientific and technical Esperanto that deposed Angstrom and Gauss but commemorated Telsa and Hertz. In the SI dictionary there is a place for only one unit of pressure. The newton is the measure of force and the square metre that of area; pressures must accordingly have the dimension of newtons per square metre. (Pascal had the same meaning but, after a brief popularity, accompanied Maxwell to oblivion.)

Consistency and Christianity may both be very laudable, but the indiscriminate imposition of either by international decree cannot be defended. Indeed, the missionaries of the Système Internationale have been obliged to recognize the cumbrance of expressing a pressure of 120 atmospheres as 12 375 000 N m^{-2}, although they cannot bring themselves to accept the more manageable approximation of 1800 psi. The attempt to smelt the life sciences and to re-cast them into a common scientific mould is condoned by some biologists and many editors who prize orderliness and uniformity above comfort and convenience. Statements such as "Inflation of an anti-*g* suit to 5 psi increases the intragastric pressure by 140 mm Hg and the central venous pressure by 2 cm water" are readily understood by the audience for whom they are intended; translating the three parameters to 34 473.790 N m^{-2}, 18 665.136 N m^{-2} and 196.128 N m^{-2} adds nothing but confusion and frustration.

An even more serious consequence of the mindless insistence upon metrication is that it invests numbers with a specious accuracy. For example, a recent edition of an authoritative textbook proclaims that "enrichment of the inspired air with oxygen is required at altitudes in excess of 3048 metres". A more exact traduction of the original text would have yielded a figure of 3047.995 m, for that is the accepted equivalent of the 10 000 feet cited by the author. The use of round numbers implies the approximation that must be inherent in all quantitative statements about physiological systems, few of which have uncertainties of less than $\pm$ 5 per cent.

Pharmacists have long-since lost their scruples, but physiologists should cleave unto their time-honoured parlance.

Punctualists who are unhappy with the physical inaccuracy and the unitary admixture of "At a height of 63 000 feet the barometric pressure is 47 mm Hg, and water boils at 98.4 °F" should be obliged to perform the conversions for themselves. They may well find that the savour and the sense are lost in the translation.

Contents

CHAPTER 1

Introduction to The History of Human Physiology

O.G. Edholm and J.S. Weiner

The study of human physiology is a recent development; the greater part of our physiological knowledge is based on animal experimentation. Nevertheless, today a detailed account can be given of the physiological functions and control of the human body, based on the systematic use of man as an experimental animal. Observation of the sick or diseased man has been, and continues to be, the duty of the physician, but, as Douglas and Priestley wrote in 1924, "it is only by studying the *normal* subject that we can appreciate correctly the exquisite co-ordination of the different functions". The very difficulties of human experimentation, the need to work without anaesthesia, the care required to avoid any harm, have provided the opportunities for learning about essential aspects of physiology which cannot be done by the most careful observation of the anaesthetized animal.

It may be asked, all the same, "Why this insistence on man? Why not rely on animal work?". Physiologists studying man depend greatly on progress made with animal models and have often used these as preliminary or complementary to human experiments (e.g. the goat in underwater physiology). However, man has many distinct biological characteristics, and much human activity takes place in situations unknown in the animal world. Moreover, man improves on his physiological adaptivity by recourse to technology; success in such technology depends on applying human biology.

The history of physiology based on human experiments may be described as occurring in three phases. The long period from classical times through the Renaissance to the age of Enlightenment at the end of the eighteenth century comprises the first phase. Spectacular but sporadic discoveries were overshadowed by the progressive development of animal experimentation. The second phase, covering the nineteenth century and the period up to World War I, was characterized by the gradual emergence of systematic studies of human physiology. In the third phase, from World War I to the present day, the study of man has been strengthened and indeed has become institutionalized.

The First Phase

It will not be possible to do more than mention a few of those who made impressive and lasting contributions during the centuries included in the first phase. The overthrow of the Galenical system of medicine, which was regarded as sacrosanct for more than a millenium, was begun by the anatomists in the sixteenth and seventeenth centuries. Sanctorius introduced the technique of careful and meticulous study (of himself) and recorded changes of body weight due to sweating as well as the effects of eating and digestion. William Harvey, as part of his demonstration of the circulation of the blood, showed how the veins of the forearm swell when the blood flow is obstructed. He also made the role of the venous valves clear. Other notable achievements included Robert Hooke's use (1664) of a crude decompression chamber and Blagden's (1775) spectacular experiments on exposure to extreme heat. The work of these men and other talented pioneers such as Borelli (blood clotting) Hales (blood pressure) and Haller (muscle action) marks the initiation of our understanding of the main systems of the human body.

The Second Phase

There was a rapid increase of physiological investigation starting in Germany during the early years of the nineteenth century. The great German physiologists such as Ludwig, Müller and Liebig attracted students from all over the world. The French physiologists, also influenced by the Germans, soon developed their own characteristic schools, culminating in the work of Claude Bernard. But none of these studied man; indeed, it seems probable that human physiology was regarded as unnecessary, or that it was impossible to carry out scientifically rigorous work on man. As a result, the monumental work by J. Müller (1840) entitled *"Handbuch der Physiologie des Mensch"* has only a few references to the results of measurements or experiments on man. And the "Textbook of Human Physiology" by Landois and Stirling (1884) contains virtually no description of studies of man. This textbook, originally published in German, went through many editions in English, and was widely read by medical students. Physiology as a science flourished and grew, nurtured by medicine or, rather, by the new professional medical schools. Within the hospitals, research advanced the knowledge of medicine and some of this knowledge was of direct interest to the physiologist. Towards the end of the nineteenth century, systematic clinical observation began to provide substantial physiological information. The functions of the thyroid, adrenal and pituitary glands were elucidated by the study of patients with disorders of these glands. And neurologists, examining patients suffering from diseases such as multiple sclerosis or the effects of cerebral haemorrhage or cerebral tumours, from their clinical and pathological studies contributed substantially to an understanding of the central nervous system.

In one field work on man dominated even in the nineteenth century. The science of nutrition and metabolism may be said to stem from the work of Lavoisier. His colleague, Seguin, acted as a human guinea-pig and on him was demonstrated that oxygen consumption was lowest during fasting and resting in comfortable conditions—the first measurement of the "basal metabolism". The oxygen consumption increased when he was cooled and when he ate food, and very substantially when he did physical work. Lavoisier laid the foundations of respiratory physiology, of the study of energy balance, and the metabolic aspects of nutrition and also of work physiology, and all this 200 years ago.

Work continued, in Germany, during the nineteenth century on many aspects of nutrition and of energy balance. Pettenkoffer and Voigt developed accurate calorimeters in which human subjects could live and be studied. One of the triumphs associated with human calorimetry was the demonstration by Rubner that the law of the conservation of energy applied to man. Rubner and his colleagues carried out extensive work on human physiology, showing the distinction between "physical" and "chemical" modes of heat production and regulation, and also developing practical application as in their studies of ventilation and of clothing. One consequence was the foundation in 1913 of the *Institut für Arbeitsphysiologie* at Dortmund. Rubner was the first director, followed by Atzler. Before his death in 1932, Rubner had lived to see his Institut develop into a foremost research centre for applied human physiology.

Another distinguished German physiologist, Züntz, worked in the field of respiration over the years 1882–1912. He introduced the concept of the "dead space" (1882) and, from the results of his Alpine expedition in 1906, was able to disprove the acapnia theory of Mosso. He studied the role of haemoglobin in both O_2 and CO_2 transport, and introduced the nitrous oxide method for the measurement of cardiac output in 1911, just before Krogh and Lindhard. But it was his systematic studies of the energy cost of various activities which were to the programme of work at the Dortmund Institut für Arbeitsphysiologie. The O_2 consumption involved in walking and climbing, with and without loads, was measured and it was shown that energy cost increased with fatigue and the recruitment of additional muscles (1891).

Parallel with these nineteenth century developments in Germany, Paul Bert was exploring the problems of altitude. In 1862, James Glaisher made a balloon ascent to a height, it was claimed, of 29 000 ft, and gave a detailed account of the effects of a reduced barometric pressure. He also described motion sickness and barotrauma. However, it was Bert who systematically investigated the precise effects, by the use of a decompression chamber, and showed that the physiological actions of O_2 and CO_2 in the inspired air depend on their partial pressures. His famous book *"La Pression Barométrique"* was published in 1878. French physiologists continued to make many notable contributions in this field; Hallion and Tissot studied the effects of altitude on the circulation and respiration in man; Viault demonstrated that red cells are increased in number in high mountain dwellers.

Paul Bert also examined the effects of a raised atmospheric pressure; he discovered the cause of "bends" in workers in compressed air and he also showed that oxygen is poisonous when breathed at high atmospheric pressure.

In the UK, physiology developed slowly. The first Chair (at University College) was only established in 1870, to be followed closely by Oxford, Edinburgh and Cambridge, but it was animal physiology which dominated. The first professor of physiology at University College, was J. Burdon Sanderson, who shortly moved to the new Chair at Oxford. He invited his nephew, J.S. Haldane, to join him in 1887 and so ushered in the first great productive era of human and applied physiology in this country. Before joining his uncle's department, Haldane had studied the composition of air in schools, houses and also sewers. This led him to examine ventilation and the problem of "vitiated" air. Brown Sequard and d'Arsonval (1888) had claimed that a toxic substance was given off in expired air. Haldane and Lorraine Smith (1893) were able to disprove this view. However, the idea persisted in the form that CO_2 was responsible for the sensation of stuffiness in overcrowded rooms, and it was this belief which stimulated Haldane and his colleagues to investigate the role of carbon dioxide in the control of respiration, which culminated in the classical paper (with Lorraine Smith and J.G. Priestley) on the role of CO_2 in the regulation of pulmonary ventilation. Haldane, through his work on respiration and blood gas composition, attracted many distinguished collaborators to Oxford. They made many contributions to the understanding of human responses to work, to changes in air temperature, humidity and pressure. Haldane demonstrated throughout his career the value of applied work and of the interplay between the laboratory and the practical problems of industry. He was asked to advise about the conditions which might cause colliery accidents, and the visits he had to pay to the site of explosions and fires in mines made him resolve so to improve conditions that such disasters could be prevented. This led him to investigate fire-fighting, diving, tunnelling and rescue operations. Just as his experiments with "vitiated" air, a typical "applied" problem, led to the evaluation of the role of CO_2 in the control of respiration, a typical "basic" problem, so the work on fire-damp or carbon monoxide (applied) led to his fundamental work on the combination of haemoglobin and carbon monoxide. As a further development, Haldane designed a clinical haemoglobinometer using carbon monoxide, and a method for estimating blood volume.

Haldane is famous for his work on respiration, but he could also be credited with advances in knowledge of temperature regulation, and this was based on his observations in hot and humid mines. He gave the first precise account of the phenomenon of acclimatization to heat in man (1905) based on his studies of sweat rate in mines. His interest in diving stemmed from observations on the effects on man of working in high pressures during the construction of tunnels. Haldane, Damar and Boycott (1905–07) worked out the technique of stage decompression, halving the pressure at each stage, on which the British diving tables were based. Stage decompression superseded the technique of slow uniform decompression of divers which had been introduced by Paul Bert in 1878. The remarkable success achieved in this field of underwater physiology was vividly described by K. Donald at the J.S. Haldane Centenary Meeting held in Oxford in 1974.

The effects of high altitudes were another topic of interest to Haldane, particularly the physiology of acclimatization. In 1894 Haldane and Lorraine Smith visited Christan Bohr at Copenhagen and learnt the methods he used in blood gas analysis. They also discussed the question of oxygen secretion in the lungs. Bohr was convinced that such secretion took place and Haldane soon became equally convinced. However, others were not so certain, and the controversy over O_2 secretion continued for many decades.

Haldane showed throughout his career how successfully experimentation on man could be used, not only to solve practical problems affecting human performance but also to advance basic knowledge of physiology. Haldane was a pioneer whose influence became worldwide, and this tradition continues still, specifically in respiratory physiology at the University of Oxford.

Joseph Barcroft began his career at Cambridge after deciding reluctantly not to complete his medical training at St George's Hospital. His first research, suggested by Langley (1895) was on the metabolism of the salivary gland, but Barcroft soon realized he needed an understanding of oxygen transport to solve his problems, and began his study of haemoglobin. He learned the ferricyanide method from Haldane and they published a joint paper. Barcroft continued to work on haemoglobin and his results were eventually summarized in "The Respiratory Function of Blood". He met Christian Bohr in 1906 during the meeting of the British Association for the Advancement of Science in South Africa. Some time after that he met August Krogh, who was an assistant in Bohr's department in Copenhagen, and he and Krogh became firm friends; they shared a common interest as they were both critical of the theory of oxygen secretion in the lungs. Barcroft first went on a high altitude expedition in 1910, organized by Züntz on an international basis, to Trinidad. On this expedition Barcroft measured the effect of high altitude on the O_2 dissociation curve of haemoglobin, and in the following year he was able to get confirmation of the shift in the curve at the high altitude laboratory which had recently been opened on Monte Rosa in the Italian alps. In the same year, Haldane organized a high altitude physiological expedition to Pike's Peak, Colorado; the other members included Yandell Henderson and Schneider from the USA, and C.G. Douglas from Oxford (of Douglas bag fame). Haldane believed that the results obtained on this expedition supported the view that there was oxygen secretion by the lungs at altitude. (Further details of Barcroft's contributions, especially at high altitudes, will be given below.)

The third leading UK figure in human physiology at the turn of the century was Leonard Hill. He studied cerebral circulation and the effects of gravity first on animals and then on man (1897), and some of his discoveries of the effects of increased gravity on the circulation in man, were to become important for aviation in the Second World War. Hill and Barnard developed a method similar to that of Riva Rocci for measuring blood pressure in man, but their paper appeared a few months later. Using the cuff method, Hill was able to show the influence of gravity on the circulation in man when changing from a horizontal to a vertical position (1897). He also examined the effects of exercise, oxygen lack and CO_2 excess as well as hot baths on blood pressure. His colleague, Martin Flack, had discovered (with Arthur Keith) the auriculo-ventricular node, and later was to become an aviation physiologist in the First World War.

Hill became interested in the effects of high atmospheric pressure. Together with J.J.R. Macleod and Major Greenwood, they extended Paul Bert's observations on the increased solution of nitrogen in the blood in hyperbaric conditions, by showing that there was an increased nitrogen concentration in the urine. They also used Bert's uniform decompression method as Hill was initially critical of Haldane's stage decompression; later he supported this method and was to improve it. In experiments with Flack he confirmed (1906) that the Haldane–Priestley principle (1905) of constancy in alveolar CO_2 partial pressure held true up to barometric pressures of six atmospheres.

At the turn of the century the study of the physiological effects of exercise began in Denmark. Compulsory physical education in schools was started there as early as 1814, but for the greater part of the nineteenth century such education was empirical. Ling's Swedish drill, introduced about the middle of the century, was (erroneously) claimed to be based on "the laws that govern the human body" and had been enthusiastically taken up in many countries, especially by the military who had conscripts to train. As a result of long agitation by educationists in Denmark, the government was eventually persuaded to support proper study, and in 1909 J. Lindhard was appointed Lecturer in the theory of gymnastics at the University of Copenhagen.

August Krogh and Lindhard were close friends and they worked together on the physiology of exercise. Krogh designed a bicycle ergometer which was used inside the respiration chamber he had built, and, with Lindhard, was able to make accurate measurements of the net efficiency of mechanical work and to show the effects of different diets. The nitrous oxide method for measuring cardiac output was also developed by them in 1912. Krogh and Lindhard were interested in the effects of high altitude, and they built a low pressure chamber which is still in use in Copenhagen.

In America the first physiological laboratory was founded by H.P. Bowditch at Harvard in 1872. Although his main interest was in isolated tissues, Bowditch also devoted much time to anthropometric studies. However, earlier in the century one of the most dramatic episodes in human physiology was provided by William Beaumont's study (published in 1833) of human gastric function. This he achieved by examining Alexis St Martin, who had a permanent fistulous opening into the stomach due to a gunshot wound. Beaumont observed, through this opening, the secretion of gastric juice and the presence of HCl in it, the movements of the stomach and the time needed to digest various foods.

The first sustained research in human physiology in the USA was in the field of energy balance by Atwater and Benedict, who used an elaborate human calorimeter to study the effects of various diets and the calorie cost of different activities. Nutritional and metabolic studies of man continued to be an important area of investigation in the USA. Prominent workers included Chittenden, Lusk and du Bois who published the surface area formula based on height and weight in 1915. Yandell Henderson, who was a member of Haldane's Pike's Peak high altitude expedition in 1911, made important contributions on the role of carbon dioxide in maintaining physiological equilibria. Problems of heating and ventilation were also of interest in the United States, and C.E.A Winslow had begun the investigations which were to become of both applied and theoretical importance after the First World War. In 1913, Winslow was a member of the New York State commission on ventilation. There is one distinctive American contribution to the study of industrial work and fatigue; the movement for the scientific management of work or "work study" was inaugurated by Taylor and Gilbreth at the turn of the century—a reflection of the growing American promotion of mass production.

The major advances made by W.B. Cannon at Harvard in the understanding of homeostasis were based on animal experimentation. In his famous book "Bodily changes in Pain, Hunger, Fear and Rage" published in 1915, he closely and convincingly argues for the applicability of his findings to man. Much subsequent work has confirmed Cannon's arguments and he may indeed be claimed as one who has made outstanding contributions to human physiology.

In the period of the second phase, roughly 1815–1914, discoveries based on human experimentation increased, with greater attention being paid to practical problems of human adaptation and survival. It was the period of the long post–Napoleonic, Victorian–Edwardian era, and one which, by comparison with the twentieth century, appears as a time of tranquility; yet wars occurred and may well have stimulated research. The coincidence between the frequent use of balloons in the Paris siege of 1870 and Paul Bert's interests does not seem fortuitous.

One area of physiological activity may serve as an illustration of the influence that the two World Wars were to exert on human physiology, as on all sciences and technology. It is no accident that J.S. Haldane and Leonard Hill were so deeply involved in submarine and diving research, for the early years of the twentieth century witnessed the sharpening of Anglo-German competition in the ominous form of naval rivalry.

The Third Phase

The modern phase has been dated from the First World War. This is arbitrary, but convenient, as the War affected research in so many ways. Many of the physiologists whose work has been briefly described lived through and continued work after the War. In the UK, Haldane, Barcroft and Leonard Hill were all involved in the new problems associated with gas warfare and the ventilation of dug-outs. Barcroft was sent to France by the War Office to study victims of the first poison gas attacks, with the objective of improving methods of treatment. It was then that his interest in the possibility of oxygen secretion was rekindled as it was evident that such a process might play an important part in patients with seriously damaged lungs. After the War, he continued his investigations of O_2 secretion, and organized and led an expedition to the Andes in 1922. This was a great physiological success, with experiments and measurements made at higher altitudes than ever before. Further evidence was obtained of the shift in the oxyhaemoglobin dissociation curve, and also further convincing evidence against the theory of oxygen secretion. From observations on the journey, Barcroft subsequently showed the reservoir function of the spleen and made the important finding during the tropical part of the journey of an increased blood volume. Such an effect was subsequently shown to be a factor in the circulatory adaptation during acclimatization to heat. (The actions of antidiuretic hormone and of aldosterone in this circulatory adjustment were eventually elucidated by human physiologists in Oxford in 1960.)

Amongst the members of the expedition was A.V. Bock from Harvard, who spent 1920–21 in Barcroft's laboratory. During this time L.J. Henderson (of Harvard) visited Cambridge and he was able to provide research funds for the expedition. This link between Barcroft and Henderson was one factor in the genesis of the Harvard Fatigue Laboratory where further high-altitude expeditions were to be organized.

High-altitude research continued at Cambridge, using a low pressure chamber in which Barcroft carried out many experiments on himself. A colleague of Barcroft was Bryan Matthews who combined great skill in using modern technology for the study of neurophysiology (in association with E.D. Adrian) with an interest in high altitudes.

In his later years, Barcroft worked in another major field of human physiology, that of the pre-neonate and the neonate. His animal experiments paved the way for the direct study of the human newborn's respiratory, circulatory and metabolic functions.

Haldane continued his respiratory studies after the war but also spent much time on mining problems. He attracted many to Oxford to work on respiration, principally on man; Priestley, Douglas and Boycott were his colleagues for many years. The work of the Oxford school was described in "Respiration" by Haldane and Priestley, published in 1922, with a last edition in 1935.

Leonard Hill became Director of applied physiology at the newly formed National Institute for Medical Research in 1914. After the war he worked with Martin until his retirement in 1930, on the effects of environmental temperature, humidity and air movement on human comfort, health and work capacity. He was primarily concerned with practical aspects and was not specifically interested in temperature regulation or the physiology of sweating. An important contribution was Hill's development of the kata-thermometer with the assistance of O.W. Griffith and Martin Flack (1916–19). Interest in thermal comfort, heating and ventilating was developing rapidly in the USA where the American Society of Heating and Ventilation Engineers (ASHVE) had established a research laboratory. Amongst the many distinguished workers in the ASHVE laboratories was Yaglou who devised the Effective Temperature Scale. The use of this scale makes it possible to compare, physiologically, environments with varying combinations of temperature, humidity and air movement. Later, T. Bedford in the UK modified the scale to take account of radiation, and the corrected effective temperature scale is still used today.

Winslow and Herrington developed a wide range of studies on temperature sensation, temperature regulation, thermal comfort and particularly in relation to heating and ventilation. In 1933, the John B. Pierce Foundation Laboratory was established at Yale University, New Haven. Pharo Gagge joined Winslow and Herrington; they developed the technique of partitional calorimetry. At the Pierce Laboratories, the Russell Sage Foundation, the ASHVE Research Laboratories, and in a number of university departments of physiology, work continued in the 1920s and 1930s on various aspects of temperature regulation, and the effects of the environment on man. Amongst those who were prominent was Cuthbert Bazett at Philadelphia, E. DuBois and J.D. Hardy at the Russell Sage, Murlin and Alan Burton at Michigan.

One of the most important developments in the USA was the establishment of the Harvard Fatigue Laboratory, which was founded by L.J. Henderson in 1926 with the support of President Lovell of Harvard and the Dean of the Harvard School of Business Administration. "The foundation of the Laboratory" (later wrote L.J. Henderson) "resulted from considerations ... of physiology, applied physiology and sociology. The work of the fatigue laboratory may be described as an attempt to make contributions towards the establishment of what may be called Human Biology".

The Fatigue Laboratory had 20 years of successful work to its credit when it was closed in 1946 and its members were dispersed over the United States. In those 20 years, almost everyone who contributed to human physiology in the mid-twentieth century worked there or visited the laboratory. A leading member was Bruce Dill, who also acted as Head of the Laboratory after L.J. Henderson died in 1942. Some of the results of the extensive field studies were described by Bruce Dill in his delightful book "Heat, Life and Altitude". He also wrote an account of the Laboratory and of L.J. Henderson. In these accounts are given many of the names of those who worked in the Laboratory, too many to be listed here. What needs to be emphasised is the great influence exerted by the Fatigue Laboratory, not only in the USA but also in many other countries. There were particularly close links between Denmark and Harvard; E. Asmussen and Marius Nielsen, and H. Christensen, were frequent visitors. They had continued the work begun in Copenhagen by Lindhard and Krogh on the physiology of muscular exercise, including the study of training and sport. They and their colleagues expanded their experiments; under the aegis of the League of Nations they examined all aspects of severe exercise, including heavy manual work.

In the UK, A.V. Hill had been working on the aerobic and anaerobic components of muscular contraction before he went in 1920–21 to Cornell University in the USA. He had been a considerable athlete and spent his year studying the track athletes of the University; his publications and books set the foundations for future work on sports physiology.

Study of the thermal environment, and of thermal comfort, developed in the UK as a result of work during the First World War. The health of munition workers, many of them women, caused concern, and this led to a series of investigations of industrial physiology under the auspices of the Industrial Fatigue Research Board, later named the Industrial Health Research Board. Vernon was the principal investigator; his team included T. Bedford, Warner and Weston. The reports of the Industrial Health Research Board covered a variety of industrial problems and may be regarded as providing the foundation in the UK of what was to be called "ergonomics" after the Second World War.

Another centre for Human Physiology was developing in Japan, where Yas Kuno had begun his studies of temperature regulation in man just before the First World War. For several years his work did not make much impact in the West, since he published in Japanese. He spent some time in University College, London and subsequently published in English "The Physiology of Human Perspiration" in 1934. This book, which may properly be regarded as a classic, established Kuno's international reputation. Many of Kuno's pupils have continued with studies of temperature regulation and climatic physiology, including Ito, Ogata and Yoshimura.

Applied physiology has also been an active field in Japan, stimulated by the foundation in 1921 of the Institute of the Science of Labour in Tokyo.

Clinical Physiology

An important aspect of human physiology is its relation to clinical medicine, and there has developed in the twentieth century a rapidly growing field of clinical physiology. A leader and initiator of this development was Sir Thomas Lewis, who founded in 1917 the Journal *Heart*, later renamed *Clinical Science*. Sir Thomas Lewis had wide interests in medicine, although he was principally concerned with the circulation and especially cardiology. He laid the foundations of the scientific study of medicine by observations and measurements made in controlled laboratory conditions, and was himself a brilliant research man. Among his many contributions may be cited the first demonstration of cold-induced vasodilatation and the hunting response, the first accurate measurement of reactive hyperaemia and the effects of sympathectomy on the peripheral circulation in man. In his department at University College Hospital he trained many who have since become famous in clinical medicine and clinical research, and who have in their turn established departments of clinical physiology.

At Queen's University, Belfast, Henry Barcroft (son of Sir Joseph) established a centre for human physiology which may be regarded as an offshoot of Sir Thomas Lewis's laboratory, where R.T. Grant had developed a plethysmograph for measuring the blood flow in the human forearm. The venous occlusion technique which he employed had fallen into disuse, but Grant soon showed what an effective research tool it was. Henry Barcroft and his colleagues, in their turn, improved the plethysmograph and, with the advent of war, carried out detailed studies of the effects of temperature, and then the effects of blood loss and of fainting (in collaboration with John McMichael and Sharpey–Schafer at the Hammersmith Postgraduate Medical School). The studies on the peripheral circulation continued at

Belfast under Henry Barcroft's successor, A.D.H. Greenfield, and in turn his successor Professor I. Roddie.

Second World War

The problems with which military commanders had to contend in World War II included many which required the work and expertise of human physiologists. In turn, this led to a great expansion of studies in many aspects of human physiology.

In the UK, the problems of the RAF were urgent and demanding, including the effects of increased gravity, of oxygen lack, as well as complex problems involving the design of aircraft, of clothing and providing an adequate thermal environment. This lead to the establishment of a research group under the direction of Professor Bryan Matthews (now Sir Bryan) at the Royal Aircraft Establishment, Farnborough. Eventually this developed into the Institute of Aviation Medicine, where work both of applied and fundamental importance continues to be done. The problems of the Army and Navy were dealt with by the relevant personnel research committees established by the Medical Research Council in collaboration with the Medical Directors General of the Army and Navy. These committees set up specialized working groups to deal with particular problems. The necessary field and laboratory work was carried out by MRC research workers together with service personnel and service facilities. The Royal Naval Physiological Laboratory was established to deal with diving problems, and at the Army Operational Research Group, set up later in the war, a number of physiological or quasi physiological studies were carried out, including investigations of personnel problems.

Apart from those working in these major centres, many physiologists were involved in applied human physiology, e.g. examining the effects of haemorrhage. By the end of the War, the experience and knowledge of human physiology had been greatly expanded, and this was to be continued after the War by the MRC, both through its personnel committees and by the establishment of an applied physiology unit at Oxford and the Division of Human Physiology (1949–74) at the National Institute for Medical Research, in effect a revival of Leonard Hill's department (1913–33). The service laboratories continued to expand, and today the Institute of Aviation Medicine, the Army Personnel Research Establishment and the Institute of Naval Medicine employ a large number of human physiologists.

Across the Atlantic there was a remarkable development of similar work in Canada, mainly in Toronto, where laboratories were built with special facilities such as climatic chambers, large human centrifuges, and decompression chambers. Here, outstanding work was done in which the physiological principles of clothing were investigated and the problems of protection against low temperature extensively studied.

There was close collaboration between the Canadians and the Americans, where the Harvard Fatigue Laboratory was the key establishment. The Climatic Research Laboratory, as conceived by W.H. Forbes, was developed by Paul Siple and was later under the direction of H.S. Belding. R.E. Johnson began his extensive studies of climate and nutrition for the Army. At the Harvard Fatigue Laboratory itself, major work was done on the effects of heat on man, and on acclimatization. The Aeromedical Laboratory at Wright Field was expanded and the staff was strengthened by the secondment there of Bruce Dill, G. Hall and O. Benson. At Fort Knox, the Army Medical Research Laboratory was established, and here Steve Horvath worked. The Naval Medical Research Institute was organized at Bethesda by Al Behnke.

German physiology in the Second World War was affected by the pressures exerted by military and political demands. Possibly as a result, the Germans did not have the advantage provided to the RAF by the Institute of Aviation Medicine in the form of superb oxygen equipment. However, after the War, the Institut für Arbeitsphysiologie at Dortmund recovered and, for example, important work was soon started there by E.A. Muller on muscle training. Industrial physiology was studied, and energy expenditure measured using the Kofranyi–Michaelis respirometer.

Applied Physiology

The Second World War provided many problems concerning the health and efficiency of the fighting man. These included questions about the effects of hot or cold climates, fatigue, physical fitness and also more unusual ones of high altitude, acceleration and the design of equipment. There were the problems of survival at sea, and on land, and the effects of hunger and thirst. Since all these were of considerable military importance the resources to study them were made available. In a sense, applied human physiology came of age during and immediately following the war. Many biologists came into the field, since the number of physiologists with experience of human physiology was too small to deal with the multitude of urgent demands.

Inevitably, the war-time level of work was not maintained in peace-time; although some remained, the majority of biologists who had been recruited returned to their original fields. There is no longer the pressure of military urgency forcing the need for *ad hoc* work, but the problems remain; they have not all been solved. In addition to military needs, physiology since the War has furnished extensive and often dramatic applications to industry, sport, life in the tropics and deserts, polar expeditions, deep sea exploration, mountaineering and the conquest of Everest, space travel and the landing on the Moon.

But there has also been a comparable level of achievement in basic and theoretical physiology which has taken two forms: the acquisition of a very large body of new knowledge and original discovery, and, as in science generally, an elaboration and more precise reaffirmation of previously known phenomena. For all these developments, applied and basic, the growth of the technology of instrumentation has expanded the possibility of exploring human body function, and this process is still accelerating at the present time. Nowhere is this more evident than in clinical practice, now a predominant area of applied physiology. The extent to which the physiology of homeostatic control is involved in the management of patients requiring life support systems provides an illustration. Such systems include the monitoring and regulation of cardiovascular function, of acid–base equilibria, of respiration, oxygen and CO_2 transport, kidney function, endocrine secretion and of energy balance.

One of the most intriguing features of human physiological experimentation in this third and present phase has been the continuous development and interplay of "invasive" and "non-invasive" techniques. Catheterization and ultrasonics, for example, have made it possible to study the physiology of the heart and circulation and to explore them to a degree comparable to but without the trauma involved in animal experiments.

In this period the enhanced application of human physiology has called forth and been based on a corresponding increase in fundamental knowledge. Only a few examples can be cited here.

The physiology of muscular exercise, following A.V. Hill's classic studies of the 1920s on the aerobic and anaerobic components of muscular contraction, has been extended to the elucidation of the detailed energetics of muscular work, including negative work, of the endocrine control of metabolic demands of muscle activity, the applications of the electromyogram and, particularly striking, the development of the biopsy technique (in Sweden) to reveal the enzyme systems at the cellular level. "Respiration," the classic book by Haldane and Priestley, was first published in 1922. Since then the understanding of the physiological mechanisms involved in respiration have developed largely under the impact of control theory and cybernetics. For an impressive illustration of the present highly sophisticated analysis of respiratory function, the reader should consult the J.S. Haldane Centenary Volume (1974) edited by Cunningham and Lloyd.

The biology and biophysics of heat regulation owes its present advanced status to workers in many countries. In the years following Kuno's pioneer monograph of 1934, the physiology of the sweat gland has become better understood, including sweat gland efficiency during acclimatization to heat, liability to fatigue and details of the osmotic work and energy sources involved in its ionic exchanges.

Indices of heat stress have been devised, empirical such as the P4SR scale, or rational such as the Belding and Hatch Index. At the Pierce Laboratory, analogue modelling of the temperature regulating system has been developed, inspired by J.D. Hardy. A war-time development should be mentioned; C. Bazett (Philadelphia), Pharo Gagge (Pierce Laboratory), and Alan Burton (Canada) invented the "clo" unit of insulation for use in the assessment of clothing assemblies, a step of fundamental importance in the scientific study of clothing.

Many other examples could be given of advances in human physiology; some will be found in succeeding chapters of this book.

This historical introduction has necessarily been terse and, inevitably, there have been many omissions. It is a common convention that only those who are dead should be mentioned in any historical review; some, who fortunately are still living and active, have been cited as it seemed to us important to stress the continuity in the development of human physiology. Short though this account may be, it should be clear how the subject has been taken up and developed in different countries, and how it may decline at one stage in one country and begin to grow and develop in another.

Many of the physiological problems which have been mentioned are described in detail in this book. It is hoped that some who read it will accept the challenge such problems represent. Research in the field of human physiology is not easy; compared with animal physiology, the proper restrictions on experiment and the sheer difficulty of obtaining suitable subjects mean that research output for the human physiologist can be disappointingly slow. However, when results are obtained they are frequently of crucial importance. Human physiology is an essential branch of physiology in its own right, and is also the unquestioned basis of human medicine. It will be seen from the following chapters that the application of human physiology is not only to clinical medicine but is also involved in a wide range of non-clinical situations. The effects of heat and cold on human performance as well as the effects of extreme conditions on survival, the importance and meaning of the term "physical fitness", the mechanisms involved in rehabilitation, are all topics which affect everyday life even though they are of clinical interest as well. The problems of weightlessness, high acceleration and hypoxia are of particular importance in aviation; mountaineers are interested in hypoxia, and the adaptations displayed by permanent high altitude inhabitants are of interest to all biologists. North Sea Oil has emphasized the dangers of deep sea diving, and the peculiar physiological problems associated with the breathing of oxygen–helium mixtures at high pressure. A knowledge of the physiology of emotion and its relationship to "stress" is required for an informed appraisal of its possible role in disease. Such knowledge is essential also for an understanding of human behaviour. The chapters on techniques and methods used in physiological measurement illustrate the dramatic growth in the application of technology in modern medicine, especially in the intensive care wards. Such developments have made possible the equally dramatic increase in research in human physiology. For many years the problems of human experimentation undoubtedly have made it difficult to reach the same standards as in animal work. Increasingly, during and since the Second World War, by the use of equipment of greater sophistication, human physiological studies in the laboratory have approached and then reached the levels of achievement of animal experimentation, at least in most fields. The triumph of recent years is that such high standards can now be attained in field work, outside the laboratory.

Not only human physiology, but applied human physiology has come of age.

References

Adolph, E.F. (1947). "Physiology of Man in the Desert", Interscience, New York.

Asmussen, E. (1972). August Krogh—Physiologist. In "Respiratory Adaptation and Capillary Exchange", (Ed. A.S. Paintal), Patel Chest Institute, Delhi.

Barcroft, J. (1926). "The Respiratory Function of the Blood", Cambridge University Press, London.

Bert, P. (1943). "Barometric Pressure", Translated by Mary Hitchcock and Fred Hitchcock. College Book Company, Columbus, Ohio.

Boylan, J.W. (Ed.) (1956). "Some Founders of Physiology", American Physiological Society, Washington, D.C.

Cunningham, D.J.C. and Lloyd, B.B. (Ed.) (1963). "The Regulation of Human Respiration", J.B.S. Haldane Centenary Symposium. Blackwell, Oxford.

Dill, D.B. (1938). "Life, Heat and Altitude", Harvard University Press, Cambridge, MA.

Dill, D.B. (1967). "The Harvard Fatigue Laboratory", Suppllement to Circ. Res. Vol. XX and XXI.

Dill, D.B. (1977). L.J. Henderson: his transition from physical chemist to physiologist. Physiologist 20, 1-15.

Douglas, C.G. (1936). J.S. Haldane, 1860–1936. *Obituary Notices of Fellows of the Royal Society*, **2**, 115–139.

Douglas, C.G. (1953). L.E. Hill, 1886–1952. *Obituary Notices of Fellows of the Royal Society*, **8**, 431–444.

Douglas, C.G. and Priestley, J.G. (1924). "Human Physiology", Oxford University Press, Oxford.

Green, F.H.K. (1947). "Medical Research in War", *Report of the M.R.C. 1939–1945*. H.M.S.O. London.

Haldane, J.S. and Priestley, J.G. (1922). "Respiration", Clarendon Press, Oxford. (New edn, 1933.)

Hill, A.V. (1927). "Muscular Movement in Man; The Factors Governing Speed and Recovery from Fatigue", McGraw Hill, New York.

Kuno, Yas (1934). "The Physiology of Human Perspiration", Churchill, London.

Pledge, H.T. (1939). "Science Since 1500", H.M.S.O. London.

Winslow, C.E.A. and Herrington, L.P. (1949). "Temperature and Human Life", Princeton University Press, New Jersey.

CHAPTER 2
Work Physiology

E.J. Bassey and P.H. Fentem

Introduction

Over the centuries human achievement has come to depend more and more on inventiveness and less and less on motor skills and physical exertion, until today we have two sharply contrasting situations. Heavy manual work, sometimes under adverse environmental conditions, is still the daily round for some, and this is particularly so in developing countries. However, in affluent societies technology has been exploited so successfully that physical labour is almost unnecessary and life for many people is now possible without it, while at the same time, for a few, the peaks of man's endeavour are ever more demanding of man himself. The off-shore search for oil has posed problems of work in cold water at great depth. Space exploration has required physical work in the absence of gravity, of atmospheric pressure and in extremes of environmental temperature. The quest for better and better athletic performance has led to greater demands on élite sportsmen and even to attempts to manipulate their physiology in artificial ways.

Physical performance continues to depend upon energy output, neuro-muscular function and those psychological factors which affect skill and performance. These elements continue to be revealed in more interesting detail, and there are new dimensions which require discussion. Is physical work a worthwhile human activity? Is it beneficial to health? Is it important for the quality of life for normal people and must they shun the sedentary life which technology offers after all? Exercise has always been of interest because it can test physiological control mechanisms to their limit. The value of observing the disturbances caused to these mechanisms by disease has proved worthwhile in the laboratory, using formal exercise testing, and in the field, using human monitoring.

Understanding of the physiology of exercise requires a good working knowledge of the cardiovascular and respiratory systems, some knowledge of the locomotor and nervous systems and also of renal function and the control of body fluids. It has been assumed that the reader already possesses the basic information in these areas. Simple descriptions have been omitted and the emphasis of this chapter is on the elucidation of the control systems involved in exercise and their interplay, and also on controversial aspects and their recent developments. Much remains to be satisfactorily explained, both about the normal response to exercise and also about the links between exercise and health.

The Cellular Basis of Muscular Work

Mechanical work

When a muscle shortens whilst performing work against a constant load the contraction is described as *isotonic*. It is a straightforward matter to calculate the external work performed and the power exerted (Table 1).

When a muscle shortens as it develops tension the work is said to be *concentric*. When a muscle lengthens at a controlled rate actively resisting the lengthening tension this is *eccentric* work. Lowering a weight with the forearm at a slow controlled speed is an example of eccentric work of biceps allowing extension of the elbow. The work performed in this case is called

TABLE 1   Units of work

The unit of WORK (force × distance) is the joule
    1 joule = 1 newton metre

The unit of FORCE (mass × acceleration) is the newton
    1 newton = 1 kg metre s^{-2}

The force acting upon 1 kg at the acceleration of gravity (at the surface of the earth)
= 9.8 newtons

The unit of POWER (work per unit time, i.e. rate of performing work; force × velocity)
    1 watt = 1 joule s^{-1}

---

negative work, whereas lifting against gravity is positive work. Figure 1 shows the classical relationship between force and velocity of the change in length of a muscle. It demonstrates that the greatest force develops with eccentric work. This may explain the increased risk of muscle and tendon injuries when running downhill.

When the muscle contracts and generates a force which does not produce any shortening of the muscle or movement at the joint involved, the contraction is described as isometric (same length). In this case the work performed is the development of force and its maintenance for the duration of the contraction; no external physical work is performed and the energy balance is impossible to study. Since the muscles never perform work at a mechanical efficiency better than about 21%, four-fifths of this energy is released internally as heat.

Detailed knowledge of the magnitude of the body's heat loads was required for the development and design of the environmental controlled suit used by astronauts during extra-vehicular work in space and on the surface of the moon. Complete insulation from environmental extremes was combined with water-filled cooling tubes in the suit.

The contractile process

Various hypotheses exist regarding the changes in the ultrastructure of the myofibrils which occur during the contraction of a muscle cell and they have been intensively studied (Huxley, 1974). When a nerve impulse has depolarized the myoneural junction it spreads to depolarize the surrounding membrane and the tubules of the sarcoplasmic reticulum of the cell. As a consequence calcium ions are released from storage vesicles in the sarco-plasmic reticulum. The calcium combines with troponin, a part of the actin

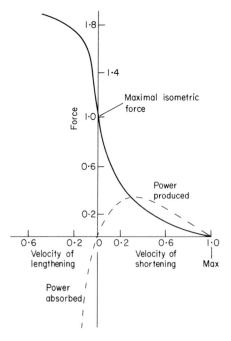

FIG. 1 Relationship between the maximal force developed and the rate of change of length of a muscle (solid line). The maximal power developed (dotted line) is the product of the force × the velocity. (From Åstrand and Rodahl, 1977.)

molecule, and this complex inactivates a mechanism, as yet undefined, which inhibits the interaction of actin and myosin filaments. The formation of the calcium–troponin complex causes a change in the alignment of tropomyosin strands which are linked to the actin molecule. This frees the active sites on the actin molecule and makes them available for bridges with adjacent myosin molecules, the mechanism upon which cell contraction depends. Adenosine triphosphate is believed to take part in the formation of these intermolecular bridges providing the energy for movement. Myosin takes up ATP and then behaves as an ATPase. Huxley (1974) provides a detailed account of the likely physical events, which explains the formation of cross-bridges and the relative movement between muscle filaments. When the muscle length changes, by stretching or contraction, the extent of the overlap between the two kinds of filament changes and so does the extent of the molecular surfaces available for bridging. It is not surprising that there is an optimal sarcomere length at which the muscle can exert its greatest tension. At a shorter length, overlap of two sets of actin filaments interferes

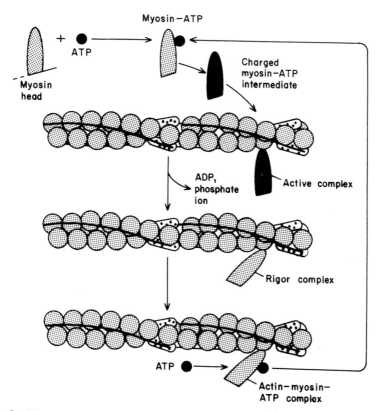

Myosin–ATP

Myosin head

ATP

Charged myosin–ATP intermediate

ADP, phosphate ion

Active complex

Rigor complex

ATP

Actin–myosin– ATP complex

FIG. 2 Diagram of the molecular events thought to take place at the cross-bridge during a single contraction of a muscle cell, showing the changes involving actin, myosin and ATP. (Modified from Murray and Weber, "The cooperative action of muscle proteins". © *Scientific American*, Inc., 1974. All rights reserved 1974.)

with bridging, whilst at longer lengths the molecular surface available for the making of bridges is reduced.

Energy transformations

Movement and physical work require large quantities of fuel and oxygen for the release of energy in the muscles. The extent and scale of the change in rate of energy transformation which takes place with heavy physical exertion are quite remarkable. The rate of energy transformation of a man asleep, the resting metabolic rate (RMR), expressed in kilocalories min^{-1}

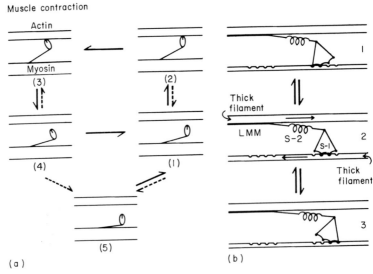

FIG. 3 Diagrams of the physical events thought to take place during contraction of a striated muscle. (a) (1) Actin and myosin separated; (2) cross-bridge formation; (3) shortening; (4) re-dissociation; (5) well-ordered cross-bridge array giving rise to the pattern seen in resting muscle. (*In* Lymn and H.E. Huxley, 1972.) (b) Development of (a) to incorporate elastic and step-wise shortening elements. The strength of binding increases from 1 to 3. The myosin head can be detached from position (3) with the utilization of a molecule of ATP, this is the predominant process during shortening. During stretch the head can dissociate from position 1 without using ATP (redrawn from Huxley and Simmons, 1971). (From Åstrand and Rodahl, 1977.)

(or kilojoules), is about one kcal min^{-1} (4.2 kJ), if he weighs 70 kg. This may rise on heavy exertion to 20 kcal min^{-1} (84 kJ), which is 1.3 kilowatts, with peaks as high as 3 kilowatts.

The physiological challenge of exertion involves the provision of fuel (see Table 2) and oxygen to maintain these rates of energy transformation and the removal of waste products of metabolism, carbon dioxide and heat from the muscles and from the body.

The cellular chemical reactions which sustain the transfer of energy within the skeletal muscle cell have been described in detail (Åstrand and Rodahl, 1977). Figure 4 summarizes the various reaction chains. Their final common pathway to the contractile proteins is through the ATP/ADP system. Adenosine triphosphate can be regenerated from ADP using energy from various sources. In the absence of oxygen there is a small store of phosphocreatine which can supply energy for high intensity work but only for a matter of seconds. Also, the initial steps in the breakdown of glycogen or glucose as far as pyruvic acid can supply more energy very rapidly, using

TABLE 2   Energy content of a man
(body weight 75 kg, 20% fat)

	kJ	kcal	
ATP	4	1	
Phosphocreatine	15	3.6	
Glycogen[a] Liver	4 600	1 100	200
Muscle			400
Glucose	160	40	
Fat	420 000	105 000	
Protein	100 000	25 000	

[a] Muscles and liver replete with glycogen
From Astrand and Rodahl, 1977.

cytoplasmic enzymes, but without oxygen; for the next step the build-up of pyruvic acid leads to the formation of large quantities of lactic acid, which, because of its acidity, inhibits further breakdown of glucose within a few minutes. In the presence of oxygen both the pyruvic acid and derivatives of free fatty acids are oxidized in the Krebs cycle in mitochondria, releasing further energy for rephosphorylation of ADP. Remarkably low tensions of oxygen (1 or 2 mm Hg) seem sufficient for the mitochondrial enzymes to function (see Fig. 4).

Red and white muscle fibres

Since muscles are required to perform work of different kinds and at different speeds it is not surprising that there are several kinds of muscle fibres and that muscles differ in the proportion of each which they contain.

It was noted during the nineteenth century that in animals some muscles appeared to be dark or red compared with others which were light or white. Muscles which are constantly active, such as the diaphragm, the oribularis oculi and the soleus in some species, are dark. It has been shown subsequently that dark muscle has a higher blood flow, more numerous mitochondria, and high concentrations of oxidative enzymes (Dubowitz and Brooke, 1973). Furthermore, the speed of contraction and relaxation of red muscle is slower than that of white. Red muscle is therefore suited to aerobic exercise of moderate intensity and long duration, whereas white muscle is more suited to anaerobic exercise of high intensity and short duration. The characteristics of these two types of muscle are summarized in Table 3.

Histochemical staining of transverse sections of muscle can be used to demonstrate the mitochondrial oxidative enzymes such as succinic de-

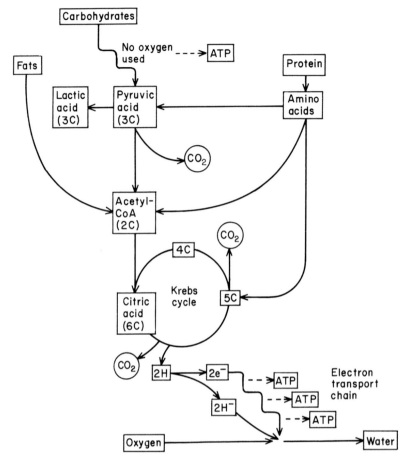

FIG. 4 Diagrammatic summary of reaction chains leading to energy release through ATP in the cell. For some of the steps the number of carbon atoms is given. (From Åstrand and Rodahl, 1978.)

hydrogenase, cytochrome oxidase or NADH tetrazolium reductase which are involved in the aerobic metabolism of glucose and free fatty acids. They show high concentrations of these enzymes in a large proportion of *fibres* in red muscle. These are usually designated "slow twitch" or "Type I" fibres, relatively few of which are found in white muscle. Fibres which do not stain positively for oxidative enzymes contain them in lower concentrations and usually have a positive reaction for myosin ATPase; they are known as "fast twitch" or "Type II" (Table 4). They have a high capacity for fast energy release and a well-developed anaerobic enzyme system.

TABLE 3   Characteristics of slow and fast muscles

	Slow	Fast
Macroscopic appearance	Red	White
Twitch characteristics	Slow	Fast
Fibre types	Mainly Type I	Mainly Type II
Function	Mainly postural and dynamic endurance exercise	Mainly isometric work or high-intensity, dynamic exercise
Anaerobic capacity	Moderate	High
Capillary supply	High	Low
Tolerance of sustained work (Endurance)	High	Low

TABLE 4   Characteristics of Type I and Type II muscle fibres

	Type I	Type II
Concentrations of oxidative enzymes	High	Low
Concentration of myosin ATPase	Low	High
Concentration of lactate dehydrogenase	Low	High
Recruitment in dynamic exercise:		
low or moderate intensity	From the onset	At exhaustion
severe intensity	From the onset	From the onset
Recruitment in isometric exercise	$< 20\%$ maximum	$> 20\%$ maximum

In man no muscle appears to be composed entirely of either Type I or Type II. Generally, muscles which have a postural function have a predominance of Type I. In an autopsy study of 36 muscles in six subjects, soleus and tibialis anterior had 87% and 73% Type I whereas triceps and biceps, which play little part in posture, have less than 50% Type I (Polgar et al., 1973). Human muscle biopsies have revealed the same pattern.

It is likely that all muscle fibres innervated by a single motor nerve (i.e. one motor unit) will be of a single type. There is evidence both from animal experiments and from studies of the effects of disease of the nervous system in man, that motor nerves may play an important role in determining both the biochemical and physiological properties of muscle. Thus Buller et al. (1960) showed by cross-innervation experiments that transference of the nerve innervating a fast muscle (extensor digitorum longus) to a slow muscle (soleus) and vice versa led to a reversal of the twitch characteristics of the muscles. More recently it has been shown that cross-innervation causes a partial reversal of the histochemical characteristics of the muscle fibres (Dubowitz, 1967). It has been argued that this is a direct trophic effect of the nerves but it may well be due to the tonic frequency of impulses.

In animal experiments tenotomy (tendon section) of the soleus quickens the twitch speed of the muscle. The likely mechanism of this change is that muscle in its normal situation in the body is slightly stretched, and is therefore subjected to a steady low frequency stream of motor nerve impulses resulting reflexly from stimulation of stretch receptors. Following tenotomy, the muscle is no longer stretched and therefore not subject to reflex motor activity. This hypothesis has been tested by implanting electrodes around the nerve to soleus and showing that stimulation of the tenotomized soleus muscle at five or ten impulses a second prevents shortening of the twitch time. Furthermore, stimulation of intact fast muscles at these frequencies converts these into slow muscles (Salmons and Vrbova, 1967).

In patients who have suffered strokes or traumatic section of the spinal cord, the percentage of fast twitch fibres in leg muscles is increased. Some of these patients exhibit the condition of spasticity in the lower limbs which is characterized by increased muscle tone and hyperactive tendon reflexes (Grimby et al., 1976; Landin et al., 1977). The increase in percentage of Type II fibres is possibly due to the frequency of impulses in motor nerves being too high to retain slow twitch characteristics in fibres (Salmons and Vrbova, 1967).

These findings support the view that slow muscles which contain a predominance of Type I fibres are involved in the maintenance of posture, in which situation they will be in a state of partial contraction or "tone" for long periods. It has been shown in man that Type I fibres are recruited during dynamic exercise of low or moderate intensity; with their high oxidative capacity they are well suited to this role. At the onset of exhaustion, Type II fibres are recruited. During isometric exercise up to 20% of maximum, Type I fibres are recruited and contractions of this intensity may be held for long periods, probably because an adequate blood supply is maintained. When tensions greater than 20% of maximum are maintained, Type II fibres are recruited from the onset. These stronger contractions of muscle interrupt the blood flow and therefore energy needs to be provided anaerobically (Gollnick, 1974).

Further support for these hypotheses of the function of Type I and Type II muscle fibres comes from studies of athletes. A group of international and Olympic standard distance runners had a much higher percentage of slow twitch fibres in the gastrocnemius muscle than either untrained subjects or lesser athletes (Costill et al., 1976).

The concentrations of oxidative enzymes in muscle may be increased by endurance training; the highest levels in athletes are found in those muscles used for endurance activity, e.g. legs in distance runners, arms in rowers. Athletes competing in strength or explosive events such as weight-lifting and sprinting have a higher than normal proportion of Type II fibres (Gollnick *et al.*, 1972). Since it appears that neither endurance training nor sprint training affects the proportion of Type I or Type II fibres respectively, then top-class athletes may be born, not made. However, recent evidence suggests that within the Type II group there are subdivisions and that some Type II fibres have a greater capacity for oxidative energy release than others (Jansson *et al.*, 1978). Moreover it has been shown that fibres can move from one subdivision to another, increasing their oxidative capacity with training and vice versa (Andersen and Henriksson, 1977a). So a suspicion remains that a small amount of transformation from Type II to Type I may be possible. In the main the effects of training are hypertrophy of the respective fibre type and an increase of concentrations of the appropriate enzymes.

Heat production

The production of heat which occurs when muscles perform mechanical work means that there is an increase in local muscle temperature, and heat is transferred to the blood flowing through the tissues and conducted to the surrounding tissues. Muscle temperature in the limbs can vary widely. It may be as low as 20 °C in resting muscle in a cool environment and working muscle may reach 2 °C above core temperature. Rapid energy release is favoured by the increased temperature. For each degree rise, there is a 13% increase in the metabolic rate of the cell. In prolonged exercise core temperature also rises to several degrees above normal and is then controlled about a new set point. Rectal temperatures as high as 41 °C have been recorded in marathon runners (Rowell, 1977); such rectal temperatures are considerably influenced by the elevated temperature of the venous blood returning from the working muscles of the legs. The temperature of the blood perfusing the temperature regulating centres in the hypothalamus may be lower since it will be cooled by the high air flow over the face (Cabanac and Caputa, 1979).

Thermoregulation in prolonged strenuous exercise makes demands on the cardiovascular system that conflict with the demands of working muscle and the maintainance of water balance. This is discussed in another chapter.

Cardiovascular Responses

Changes in cardiovascular function and oxygen transport systems can achieve as much as a twenty-fold increase in oxygen uptake during strenuous exercise. In order to achieve this there is a five- or six-fold increase in the cardiac output, whilst the remaining increase can be attributed to increased oxygen extraction from arterial blood and changes in the properties of haemoglobin. The capacity for aerobic work is largely determined by the maximum cardiac output which can be achieved.

Cardiac output

The autonomic nervous system is concerned in many of the adjustments which are normally made during exercise.

The cardiac output is increased by a reduced parasympathetic and increased sympathetic drive which has inotropic as well as chronotropic effects. A general sympathetic α-noradrenergic vasoconstrictor outflow ensures the maintenance of blood pressure and redistributes the cardiac output to the dilated beds of the working muscles. The stimulation of the adrenal medulla to release adrenaline contributes to this.

Studies of greyhounds whilst they were racing on the track (Donald *et al.*,

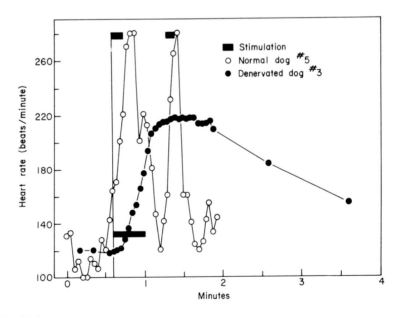

FIG. 5(a)

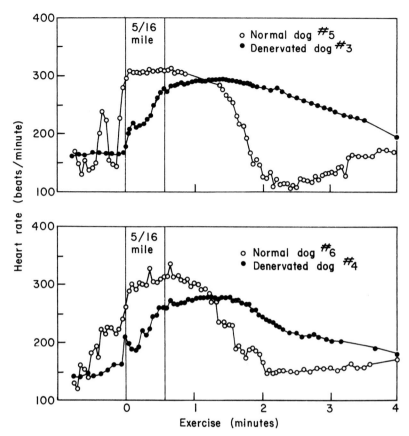

FIG. 5 Response of the heart rate in normal and denervated greyhounds (a) to the sight of a rabbit (stimulation); and (b) to a 5/16 mile race. The normal dogs had themselves undergone a thoracotomy and sham denervation. (From Donald *et al.*, 1964.)

1964) showed that these animals increased their cardiac output principally by raising their heart rate (Fig. 5). Decreased parasympathetic and increased sympathetic drives were responsible. However, the same investigators went on to demonstrate that after surgical denervation of the heart the dogs had almost unchanged track times. Complementary experiments on the treadmill showed that the manner in which the heart responded to the requirements of exercise was quite different (Fig. 6). The necessary increase in cardiac output was achieved entirely by an increase in stroke volume from an increased end-diastolic volume. It was clear that after cardiac denervation the animals were relying heavily on heterometric autoregulation (the Starling mechanism) as a reserve mechanism which was only involved to a small extent when autonomic control was available.

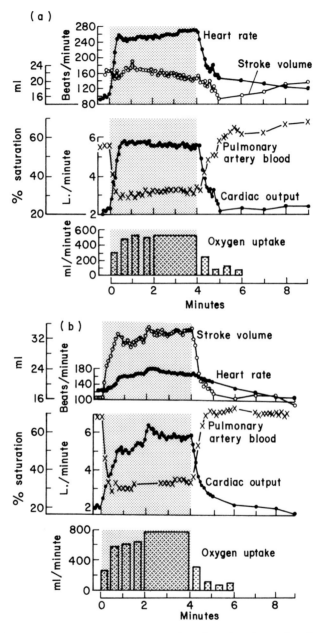

FIG. 6 Heart-rate response to severe exercise after chronic cardiac denervation. Dogs ran on a treadmill for 4 min at 5.5 km h^{-1} (21% gradient). A, normal dog. B, dog with heart previously denervated. Note that the cardiac output and oxygen uptake increase to a similar extent. (From Donald *et al.*, 1964.)

Similar observations have been made in man. With submaximal work in the upright position the increase in cardiac output is attributable mainly to an increase in heart rate and to some extent stroke volume. When the effect of the autonomic nervous system on the heart is blocked by atropine and by propranolol the cardiac response changes (Fig. 7). Since the normal rate changes are now impossible, a partially compensated increase in cardiac output is achieved by a bigger stroke volume occurring because of increased filling time and the Starling mechanism. However, Ekblom et al., (1972) have shown that the maximum cardiac output is somewhat curtailed and so is the maximum work time, despite increased oxygen extraction by the muscles which maintains maximum oxygen uptake.

Maximum heart rates tend to be lower with advancing age but to a very variable extent. Shephard (1978) quotes a man of 64 years with a maximum heart rate of 192 b min^{-1}. This is much higher than would be predicted by Astrand's formula (220 − age in years). Heart rate, stroke volume, and cardiac output at a given submaximal work level change little with age (Bassey, 1978). This is because an age-based decline (4 b min^{-1} per decade found by Cotes et al., 1973) appears to be offset by a rise due to reduced activity levels.

The cardiac output at a given submaximal oxgyen uptake is similar in many types of exercise, e.g. arms and/or leg work, cycling, walking, running, swimming. In the erect posture it is consistently 1–2 litre min^{-1} less than in the recumbent posture. During arm work the cardiac output will be achieved with a higher heart rate and smaller stroke volume than during leg work. This seems to be a general effect on heart rate drive of small working muscle groups.

Autonomic control of heart rate

Heart rate is linearly related to oxygen uptake over the whole range of work with the same muscle group but the way in which this is achieved is not well understood. There are major central and reflex influences as well as interactions with blood pressure controlling systems. The direct action of the higher motor centres on the cardiovascular centre has been demonstrated in a variety of different experiments. If a subject attempts to obey a command to perform muscular work and if the muscle response is prevented or attenuated, the cardiovascular response is appropriate to the central command and not to the work performed. For example, several investigators (Asmussen et al., 1965) find that heart rate and blood pressure responses are greater at a given work load when muscle strength is reduced by partial curarization. Thus the cardiovascular response is related to the

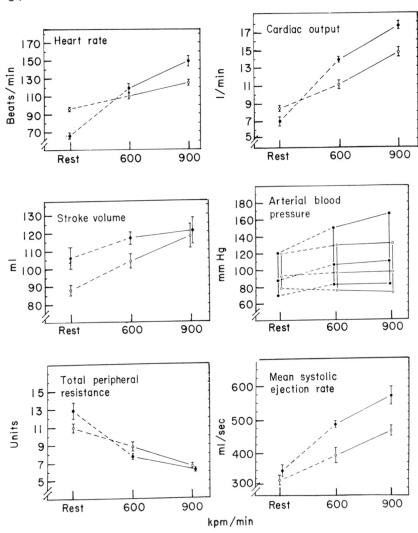

FIG. 7   The cardiovascular changes with exercise after treatment with atropine and propranolol to block the parasympathetic and sympathetic drive to the heart. Eight healthy subjects undertook two levels of moderate exercise before (solid circles) and after (open circles) the two drugs. Mean values ±1 SEM are presented except for arterial blood pressure where means alone have been plotted. (From Nordenfelt, 1971.)

greater motor command required by a given level of muscular work when the subject is weaker. Goodwin *et al.* (1972) investigated the cardiovascular response of human subjects to isometric work in experiments in which the central motor command needed to achieve a given tension was varied. The afferent nerve fibres arising from muscle spindles are excited by high frequency vibration of the muscle. If these afferents are activated in a contracting muscle they reflexly cause an involuntary increase in motor activation of the muscle, so that less central command is required to maintain the same tension. The hypertension and tachycardia in response to isometric exercise is decreased. On the other hand, if the afferents from the antagonist muscle are excited they cause reflex inhibition of the contracting muscle, so that a greater central command is required and the subject perceives that the conscious effort has increased. The resultant hypertension and tachycardia are greater.

Evidence for a muscle–heart reflex comes from experiments in which direct electrical stimulation of the skeletal muscle caused an increase in heart rate identical in its time course and magnitude to one obtained during voluntary contraction of the same muscle (Hollander and Bouman, 1975). On the basis of animal experiments (Mitchell *et al.*, 1977) it appears that the afferent nerve fibres of this reflex are small type III or type IV fibres. The fibres principally originate in free nerve endings.

The factors responsible for excitation of these endings during muscle contraction have not been identified with certainty. The increase in local interstitial potassium concentration and in osmolarity may have a role, and so may mechanoreceptors which would provide for the rapid early phase of the increase in heart rate at the onset of work. The relative importance of central command and this peripheral control mechanism remains to be elucidated. It seems likely that, as with many control systems in the body, there is a fail-safe design involving several systems for achieving the same ends. If both operate simultaneously then their effects may be gated if they impinge on the same central motorneurones.

The response of the vasomotor centres to changes in blood pressure may also change during exercise. The response of the systemic arterial pressure receptors may be tested during exercise by observing the degree of brady-cardia which occurs in response to a rise in arterial pressure. Smyth *et al.* (1969) have used an intravenous solution of phenylephrine to produce a transient rise in pressure. There is a proportional relationship, both at rest and during exercise, between the increment in systolic arterial pressure and the prolongation of the pulse interval (R–R interval). However, the slope of R–R interval on pressure is very much less steep during exercise and it appears that the reflex mechanisms respond less powerfully, thus allowing blood pressure to rise.

Systemic arterial blood pressure

Systemic arterial blood pressure rises with exercise but the increase may be small depending upon how and where it is measured. The position of a catheter introduced into the arterial tree proves to be of more importance than had previously been realised. In exercise, lateral pressures in brachial and femoral arteries are significantly higher than aortic lateral pressure, which is reported not to differ from rest pressures. This may be due to resonance in the peripheral arteries. Moreover, if values for aortic pressure are recorded with the catheter tip pointing upstream then a component due to impact (kinetic energy) is measured (Jorgenson et al., 1977). Indeed with the catheter tip pointing downstream changes in aortic pressure occurring with exercise are very small. The faster flow will exaggerate the differences due to catheter position. However, since the maintenance of driving pressure in peripheral arteries is what is important for the perfusion of tissues, the measurement of upstream pressures in such arteries seems valid. The baroreceptors in the walls of the carotid sinus and aortic arch that are sensitive to stretch or mechanical deformation of the wall will be sensitive to lateral pressure. They also respond to rate of change of pressure as well as absolute levels. It is not known whether their discharge increases during exercise.

Using direct methods, brachial arterial blood pressure is raised during exercise in linear proportion to the increase in oxygen uptake. The rise in systolic pressure is marked and reaches a maximum of about 200 mm Hg in young healthy subjects. The diastolic pressure does not change, so the rise in mean pressure is modest. The rise is greater with rhythmic arm work than with leg work and very dramatically greater with sustained contractions (e.g. isometric work at more than 20% of maximum). The interaction between normal baroreceptor control of blood pressure and the needs of working muscle during exercise is not well understood. There is some evidence for a decreased baroreceptor sensitivity during exercise as described above (Bristow et al., 1971). This would be brought about by the contraction of smooth muscle in the walls and there may also be central inhibition of their influence.

The rise with sustained contractions is noteworthy because it is observed even when a small mass of muscle is involved in the contraction. A finger flexor, for example, working at 50% of its own maximum voluntary contraction produces as big a rise in blood pressure after 1 min as a large muscle like quadriceps working at the same percentage of its maximum (Fig. 8). The absolute tensions may be very different but, if the relative tensions are the same, the time course of the rise in blood pressure is similar.

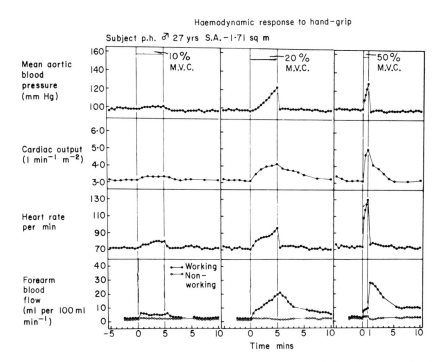

Haemodynamic response to hand-grip

Subject p.h. ♂ 27 yrs S.A. – 1·71 sq m

FIG. 8   Cardiovascular responses in one subject during and after three sustained hand-grip contractions. The contractions were held for 5 mins at 10% and 20% of maximum voluntary contraction and to fatigue at 50%. The data obtained in two separate experiments. (From Lind *et al.*, 1964.)

The muscle, by contracting, has reduced its own blood supply by compression of the blood vessels, and the accumulation of metabolities of anaerobic energy release probably gives rise to increased afferent drive to medullary cardiovascular control centres.

Arm work has greater pressor effects for this reason. When work is performed with the arms lifted above the head the rise in pressure is considerable. Astrand *et al.* (1968) studied this effect by comparing the arterial blood pressures recorded whilst men were hammering nails into the ceiling with the pressures of the same man hammering nails into wood at bench height. Older men have higher resting blood pressures and also greater increases in arterial pressure with physical exertion than do younger men. Hypertensive individuals are found to have greater increments than those who are normotensive.

Central venous pressure

The pressure in the great veins close to the heart fluctuates with each cardiac cycle and with each breath within the range 0–5 cm of blood. It is an important determinant of cardiac filling and is itself dependent upon venous return. Venous return is maintained by the pumping action of working leg muscles as they compress the deep veins and upon the respiratory fluctuations in both intrathoracic and intraabdominal pressure.

During prolonged exercise at a constant work load the heart rate rises steadily and there is a reciprocal fall in stroke volume. The cardiac output is maintained. The oxygen uptake also increases by as much as 5–10% without increase in work load, perhaps due to a temperature dependent increase in the metabolic rate or to a decrease in efficiency of movement. It is possible that the fall in stroke volume is the primary change because there is evidence that cardiac filling pressures fall with increasing duration of work. Eklund and Holmgren (1964) describe a group of patients in whom the right ventricular end-diastolic pressure fell from 2.7 mm Hg to 0 mm Hg between the tenth and fiftieth minutes of exercise. The diastolic pulmonary artery pressure also fell from 12.5 mm Hg over the same time. Thus both left and right ventricular filling pressures may fall. Radiographically measured heart volumes are reduced at the time that ventricular filling pressures are observed to fall, but the reduction is not explained by the small reduction in plasma volume caused by sweating and increased formation of interstitial fluid. It is likely that it is the result of the peripheral redistribution of the circulating blood volume in order to increase heat dissipation, which will be discussed below.

Veins and venous return

During heavy exercise the volume of blood in the veins will be actively reduced by sympathetic venoconstriction, and it will also be reduced passively in the veins of those parts in which the tissue blood flow is diminished. With running or walking the return of blood to the heart is greatly facilitated by contractions of the leg muscles. The pumping action of the calf and thigh muscles is more effective than any other single factor in assisting venous return and keeping the volume of blood in the veins small (Fig. 9).

The increase in ventilation also improves the effectiveness with which the abdomino-thoracic pump draws blood back to the heart.

The blood flow to the working muscles

A man weighing 70 kg may have as much as 20 kg of skeletal muscle. With many forms of exercise a large proportion of this muscle mass will be used in the work, and muscle blood flow may rise as high as 1 litre kg^{-1} min^{-1}, a twelve-fold increase. Since the resting cardiac output is 5 litre min^{-1} there is clearly no way in which this increase in muscle blood flow could be achieved by redistribution alone and an increase of nearly five-fold in cardiac output is required. This increase in blood flow to exercising muscle is achieved by the interaction of a number of mechanisms and regulated by the local concentration of metabolities. The substances released from the working cells are vasodilator; they include potassium, carbon dioxide, hydrogen ions, ATP, adenosine and phosphate and perhaps

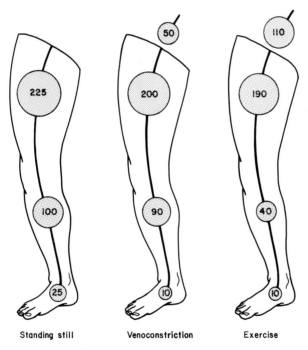

FIG. 9 Schematic representation of the incremental volumes of blood which enter the thigh, calf, and foot with change in posture from lying to standing still, and the effects of maximal venoconstriction, and of muscular exercise, on these volumes. (From Ludbrook, 1966.) (Courtesy of Charles C. Thomas, Springfield, III.)

bradykinin and lactic acid. A fall in the tissue tension of oxygen is also dilator; so is the increased heat and the increase in local osmolarity of tissue fluid resulting from the splitting of large molecules into several smaller molecules with the release of energy. The metabolites effectively override what would otherwise be a vasoconstrictor action of noradrenaline released at sympathetic nerve endings (Fewings *et al.*, 1965, see Fig. 10). This is seen in muscle tissue not being used in the exercise.

The small rise in mean arterial blood pressure which accompanies exercise must also increase muscle blood flow a little. The contribution from this rise in perfusion pressure varies with the severity of the exertion and the type of exercise; isometric exercise may be accompanied by a dramatic increase in pressure (see Fig. 8). Blood flow into the muscle may be seriously limited during contraction by high intramuscular pressures. The insufficient blood flow terminates strong sustained contractions in less than a minute. In rapid intermittent work the highest blood flow occurs between contractions and after exercise.

Many authors have tried to implicate the sympathetic cholinergic vasodilator fibres in the genesis of a vasodilatation preparatory for exercise. There

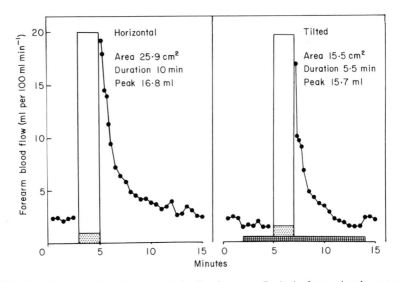

FIG. 10   Post-exercise hyperaemia in the forearm. Period of sustained contraction shown by stippled bar. Period of 45° feet down tilt shown by hatched bar. The area under the curve is the total extra volume of blood flow attributable to the exercise. The duration of increased flow and the peak increment in flow are also given. (From Fewings *et al.*, 1965.)

is experimental evidence for such a pathway and it appears to be involved in the acute and sudden dilatation of skeletal muscle resistance vessels in emotional stress, fainting and also in the chronic vasodilatation associated with prolonged anxiety. There is no evidence that it is useful even at the onset of exercise. Work increases as a step function, heart rate rises rapidly and yet V_{O_2} rise only slowly, taking about one minute to plateau. The fast anaerobic pathways have to be used and there is some initial production of lactate. This would not be expected if sympathetic cholinergic vasodilatation were happening.

Blood flow to splanchnic areas

When an individual exerts himself at a work rate or oxygen consumption that is greater than a certain threshold, which is about 30% of his maximum, the blood flow to the kidneys and splanchnic tissues becomes reduced by the action of the sympathetic nervous system. With an increasing work rate above this threshold the renal and splanchnic blood supplies are cut progressively to a value 20–30% of the flow at rest, allowing up to 2 litre min^{-1} to be redistributed to the working muscles.

Coronary blood flow

This will increase in proportion to the myocardial oxygen uptake, and is achieved by the parallel increase in cardiac output and blood pressure as well as by the autoregulatory effects of pO_2 on the coronary resistance vessels. The oxygen consumption of left ventricular myocardium may increase from 10 to 60 ml 100 g^{-1}. The requirement for oxygen is met mainly by an increase in blood flow and to some extent by an increase in oxygen extraction. The normal human heart has the capacity to increase oxygen extraction by as much as 50%. A number of factors have been identified as major influences on myocardial oxygen consumption, (a) heart rate, (b) internal work (i.e. the generation of pressure and intramyocardial tension), (c) external work (i.e. the ejection of the stroke volume against the aortic pressure), (d) the contractile state of the heart. For many years the external stroke work (c), the product of stroke volume and arterial pressure, was considered the major source of work although it had been recognized that pressure generation is more expensive than volume work, and that pressure changes reflect changes in wall tension only if ventricular volume is not altered. The tachycardia of exercise is probably responsible for 30% of the increment in coronary oxygen consumption; when external cardiac

TABLE 5   An estimate of the major components of cardiac energy expenditure during heavy work

(a)	Increase in heart rate	15%
(b)	Internal work (pressure generation)	45%
(c)	Resting metabolism and improved contractility	30%
(d)	External work (forward flow of stroke volume)	10%

work is high, because cardiac output is high, then internal work of pressure generation accounts for about 50% of the oxygen consumption. It has never been possible to draw up the energy balance sheet by measuring all the components in the same experiment. Table 5 has been prepared from available estimates of the energy cost of each function at high levels of work.

The myocardial pO_2 is the major determinant of the calibre of coronary resistance vessels. Excesses of CO_2, hydrogen ions, lactic acid and adenosine have a mild vasodilator action. Any rise in diastolic aortic pressure will also increase coronary flow.

Skin blood flow

The heat load of exercise is dissipated by increased evaporative loss, as well as by convective and radiant losses, the magnitude of which depends upon the temperature and humidity of the environment. The dilatation of skin blood vessels permits the transfer of heat from the body core to the shell and thence to the surroundings. It takes 3–5 min from the start of exercise before thermoregulatory vasodilatation in the skin occurs. The initial sympathetic vasoconstriction of skin vessels mediated by sympathetic nervous activity gives way to dilatation of vessels, including distal arteriovenous anastomoses and a shift of blood volume into cutaneous veins. The mechanism of this vasodilatation is not clearly understood. It may be due to increased activity of a neurogenic active vasodilator system, it has been linked by some investigators to sweating and local bradykinin formation. The resolution of this problem awaits better methods for measuring skin blood flow. Skin vasoconstriction occurs with high work loads and there is a generalized increase in sympathetic nervous outflow when heat stress is superimposed on exercise. This pattern of sympathetic nervous activity ensures the optimal distribution of cardiac output and checks the peripheral displacement of blood volume. However, severe heat stress can lower maximum oxygen uptake by reducing and diverting the cardiac output away from muscle (Fig. 11). There is a point at which vasodilator effects overcome sympathetic vasoconstriction, and circulatory collapse due to heat stress may ensue.

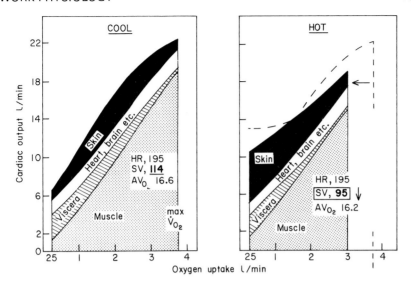

FIG. 11   Estimated distribution of cardiac output in response to graded upright exercise in cool (25.6 °C) and hot (43.3 °C) environments. First part of dashed line shows average peak cardiac output during heat stress at rest and second part shows curve obtained in cool conditions during exercise. The reduction in the heat is due to a reduced stroke volume. SV = stroke volume; AV_{O_2} = arteriovenous difference of O_2 content of blood (ml dl^{-1}). (From Rowell, 1974.)

Cerebral blood flow

Blood flow to the brain has been shown to remain unchanged or to increase slightly during submaximal exercise. It is likely that during exercise at exhausting levels when the arterial pCO_2 falls to 30 mm Hg (4 kPa) the blood flow will fall, though the measurements have not been made. Certainly alkalaemia causes cerebral vasoconstriction in other circumstances, for example voluntary hyperventilation at rest or during submaximal work.

Tissue fluid movement

The formation of tissue fluid is increased in exercising muscle. More capillaries are open and there is an increase in capillary area as well as in capillary pressure. The mean capillary pressure increases from 15–20 to 25–30 mm Hg. The local increase in osmolarity may also contribute to an increased rate of formation, large energy rich molecules breaking into smaller units. However, there is also local production of metabolic water.

The contracting muscles facilitate the movement of fluid along the lymphatics and its return to the circulation. Some tissue fluid will be absorbed into the capillaries in those tissues with reduced perfusion. The net effect of these transfers of fluid between the vessels and the tissues is to produce some degree of haemoconcentration and a fall in blood volume. The fall in blood volume is small unless the rate of sweat production is high and prolonged. The fall in total blood volume is without effect on maximal oxygen uptake.

Oxygen extraction

The increasing oxygen transport to the working muscles is not only provided by the increased blood flow but also by greater oxygen extraction. With increasing work load this is reflected in a steady fall in the oxygen content of venous blood, for example the content of femoral vein blood may fall to 14 ml dl^{-1} with leg work, whilst the content of mixed venous blood may fall to 20 ml dl^{-1} (Fig. 12). The high overall extraction, seen as a low mixed venous oxygen content, is in part explained by the reduction in blood flow to non-exercising muscles and to regions which normally receive a high blood flow but extract little O_2 per litre of blood. During strenuous

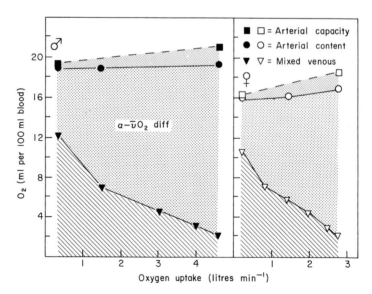

FIG. 12 Oxygen-binding capacity and measured oxygen content of arterial blood: calculated oxygen content of mixed venous blood at rest and during work up to maximum on a bicycle ergometer. Mean values for five male and five female subjects, 20–30 years of age. (From Astrand and Rodahl, 1977.)

exercise oxygen delivery to the tissues will also be improved by a shift of the oxygen dissociation curve to the right as a result of acidaemia, an increased venous CO_2 tension and an increase in local temperature.

Pulmonary Ventilation

Pulmonary ventilation increases with exercise and so provides the gaseous exchange required for aerobic energy metabolism. It keeps in step or ahead of the circulating transport system and does not limit maximal oxygen uptake in young normal subjects.

From a value of 6 litre min^{-1} at rest when oxygen uptake is about 0.3 litre min^{-1}, minute ventilation increases to 80–100 litre min^{-1} at an oxygen uptake of about 3 litre min^{-1}. It may reach 150 litre min^{-1} and even 200 litre min^{-1} with maximal physical effort by élite athletes. As Fig. 13 shows

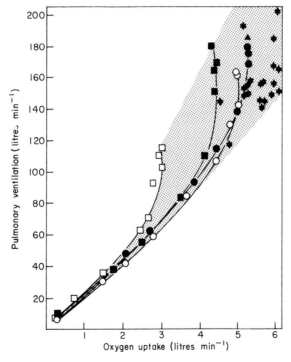

FIG. 13   Pulmonary ventilation at rest and during exercise (running or cycling). Ranges of values for four individuals (represented by circles and squares) and single values (stars) for top athletes working at maximal rates. (From Åstrand and Rodahl, 1977.)

ventilation increases during muscular work in proportion to the increase in oxygen uptake up to a certain level (Owles' point), after which the increase in ventilation becomes steeper. The cause of this disproportionate change is not understood. Increased arterial hydrogen ion concentration [H$^+$] due to anaerobic energy release, increased reflex drive from exercising muscles and thoracic muscle spindle behaviour have been suggested as possible causes.

Pulmonary ventilation changes with prolonged work at constant moderate intensity. In individuals who are used to exercise but not highly trained there is a progressive time-dependent tachypnoea and hyperventilation, arterial carbon dioxide tension (p_{a,CO_2}) falls but regulation of arterial [H$^+$] is almost perfect. This "ventilatory drift" is analogous to the progressive tachycardia or "cardiovascular drift" which also occurs during prolonged work at a constant workload. Since it is accompanied by a gradually rising intravascular temperature, the explanation of this "ventilatory drift" may lie with the increase in thermal drive to respiration or in the effect of increasing concentrations of circulating noradrenaline acting through the carotid body chemoreceptors. However, it is difficult to see how arterial [H$^+$] is maintained constant in the face of hyperventilation unless [H$^+$] production is increasing to drive the ventilation. The raised temperature may be linked to an increasing [H$^+$] perhaps through changes in dissociation constants.

Maximum exercise ventilation falls with age, probably as a consequence of the concomitant reduction in vital capacity.

The ventilatory response to arm work is greater than the response to leg work. This difference is perhaps analogous to the effects on arterial blood pressure and may reflect the apparently greater reflex drive of a given rate of work performed by a smaller muscle group.

During physical work of low-intensity ventilation is increased mainly by increase in tidal volume. With heavy work the frequency of respiration also increases. The frequency at which an individual breathes appears to be one at which the energy cost of the respiratory movements themselves are minimal (Milic–Emili et al., 1960). The same principle seems to apply during exercise; except that respiratory movements often become entrained with the rhythmic movements of running or cycling. The energy required for respiratory movements is small at rest (0.5–1.0 ml O$_2$ per litre of ventilation). This rises steadily with increasing exertion and may amount to 10% of the total oxygen uptake during heavy work. There is a theoretical limiting factor in maximal exercise which is the point at which any further increase in oxygen uptake will be consumed in achieving the increased ventilation by which it will be provided. In practice, maximal ventilation rates are well below this point.

Pulmonary gas exchange

Gaseous exchange between pulmonary capillary blood and alveolar air is less complete during exercise than at rest. The alveolar to arterial p_{O_2} difference, $(A - a) d_{O_2}$, becomes slightly greater during exercise even when it is mild. This is a little unexpected, since this gradient for oxygen is greatly dependent upon ventilation–perfusion relationships and the inequalities in ventilation and perfusion, particularly at the apex of the lung at rest, should diminish with exercise. Overall ventilation increases more than cardiac output and with the rise in pulmonary artery pressure exercise should ensure more even perfusion of the different parts of the lung. This effect may be offset by reason of the shorter time spent by erythrocytes in

TABLE 6   Pulmonary ventilation to perfusion ratio ($\dot{V}_A : \dot{Q}$) during exercise: results of multiple inert gas analysis

	Rest ($\dot{V}_{O_2} = 0.27$ litre min^{-1})	Mild work ($\dot{V}_{O_2} = 1.1$ litre min^{-1})
Mixed–venous blood gas values (mm Hg)		
p_{O_2}	39	31
p_{CO_2}	45	52
$\dot{V}_A$ and $\dot{Q}$ (litre min^{-1})		
$\dot{V}_A$	5.4	25.6
$\dot{Q}$	5.8	11.7
$\dot{V}_A : \dot{Q}$		
Mean	0.93	2.3
Dispersion[a]	± 0.34	± 0.46
Range	0.49–2.1	0.93–6.6
p_{O_2} (mm Hg)		
Alveolar	100) 7[b]	108) 9[b]
End pulmonary capillary	93) 3[c]	99) 8[c]
Arterial	90	91
Alveolar to arterial difference, $(A - a) d_{O_2}$	—	—
	10	17

[a]Log of standard deviation (SD).
[b]Portion of the $(A - a) d_{O_2}$ due to $\dot{V}_A : \dot{Q}$ inhomogeneity.
[c]Portion of the $(A - a) d_{O_2}$ due to anatomical shunt. We assume that limitations to alveolar capillary diffusion contribute nothing to the $(A - a) d_{O_2}$.
From Dempsey et al. (1977).

traversing the pulmonary capillary; the transit time falls from 0.75 to 0.25 s in severe exercise, or by venous shunting from the right to left sides of the heart. As Dempsey *et al.* (1977) have shown (see Table 6) when the gradient is subdivided into the alveolar to pulmonary capillary component and the pulmonary end-capillary to arterial difference (venous shunting) the component due to ventilation–perfusion inequalities remains the same whilst there is a pulmonary end-capillary to arterial difference of 8 mm Hg representing a venous shunt of about 1%. This could be due to an increased flow through the bronchial circulation which is not exposed to gaseous exchange.

The regulation of breathing

The control of ventilation during exercise is still not understood. There are only minor changes in p_{a,O_2}, p_{a,O_2} and arterial blood pH and these are quite insufficient to explain the increase in ventilation. Impulses originating centrally in parallel with the motor command to the working muscles may be responsible, as is thought to happen in the control of heart rate during exercise. Alternatively, a hypothesis which fits the various findings well is that impulses carried in afferent nerves from working muscles are sufficient to drive respiration roughly in step with the exercise; the role of the central respiratory neurones remains unchanged, that is, a negative feedback mechanism operates based upon the p_{a,CO_2} which adjusts the drive to respiratory muscles so as to maintain the p_{a,CO_2} constant. This is over-ridden during vigorous exercise when the p_{a,CO_2} may fall, because acidaemia mainly due to lactic acid, a fall in arterial p_{O_2} and a rise in body temperature produce additional respiratory stimulation.

$$\overset{\displaystyle OH}{\underset{\displaystyle |}{}}$$

$$H^+ + CH_3 \cdot CH \cdot COO^-$$

$$P_A CO_2 \rightleftharpoons CO_2 + H_2O \rightleftharpoons H_2CO_3 \rightleftharpoons H^+ + HCO_3^-$$

$$Na^+ + HCO_3^-$$

Increase in ventilation

If the exercise is severe enough for the rate of production of lactate to exceed its rate of breakdown or its rate of escape from the muscle cells, then the exercise intensity is said to be above the anaerobic threshold and it will not be possible for the exercise to be continued indefinitely. Muscle cells are more tolerant of acidity than other tissues but when the pH has fallen to a critically low value glycolysis is inhibited, energy release stops and contraction fails. In exhausted muscle cells pH values of 6.3 have been recorded. Voluntary contraction usually ceases long before this stage is reached due to local ischaemic pain produced by metabolites.

The lactic acid is buffered and in prolonged exercise the steady production of increasing quantities of lactate is offset by hyperventilation and the fall in p_{a,CO_2} as well as by metabolism of lactate in the heart, liver and non-working skeletal muscle, so that the pH of arterial blood is maintained fairly constant (see Fig. 14) but the concentrations of lactic acid are much higher in the working muscle cells.

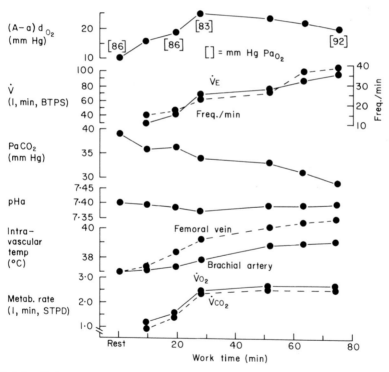

FIG. 14   Pulmonary response to prolonged exercise (66% of maximum oxygen uptake). Mean values from six untrained subjects working at sea level. Blood gases corrected to observed vascular temperatures. (From Dempsey *et al.*, 1977.)

Breathlessness

There is a feeling of breathlessness or laboured and uncomfortable respiration at high levels of exertion and even at modest levels in people unused to physical effort whose anaerobic threshold is very low. The conscious awareness that ventilation is increased depends upon afferent information from the muscle spindles in respiratory muscles and from thoracic joint receptors. Whether the effort of the respiratory muscles is appropriate to their length and tension during respiratory movement appears to determine the intensity of the distress. During severe exercise in normals or in some patients with lung disease the effort is much greater for any given length of the respiratory muscles.

Hormonal and Metabolic Responses

Prolonged physical activity produces marked changes in metabolic rate. It must be possible for the flux of metabolites through the energy producing pathways in muscle to increase and for the fuels from the storage tissues of the body, namely liver and adipose tissue, to be mobilized in sufficient quantities. These processes appear to depend both on biochemical regulation by metabolic processes and on the levels of circulating hormones, chiefly insulin, glucagon and the catecholamines.

In adipose tissue there is a dynamic equilibrium between lipolysis and esterification. Tissue triglyceride is broken down to fatty acids and those which do not enter the blood are reactivated and re-esterified to form triglyceride. This equilibrium is sensitive to feedback control from blood levels of free fatty acids and so responds to the rate at which they are utilized by the muscles. Their release from adipose tissue is also sensitive to catecholamines which may increase the rate of turnover in the cells or influence the situation by increasing blood flow through adipose tissue. β-Blockers such as propranolol reduce the availability of free fatty acids and shift the R.Q. nearer to 1.

In muscle the importance of fatty acid oxidation is that it spares the glycogen stores to some extent, although it can never completely replace glucose as an energy source. The Krebs cycle requires some glycolytic replenishment. Free fatty acids can provide energy at the rate of only up to about 6 kcal min^{-1}. The rest of the energy must come from glucose stores or glucose uptake from the blood stream. During moderate exercise this presents no problem but when work intensity increases acutely the release of free fatty acids can only supply about 65% of an individual's aerobic capacity. Under these conditions of high intensity work glycogen stores clearly set the limit on the duration of the work. In the case of prolonged running there will eventually be a fall in blood glucose sufficient to explain fatigue and exhaustion. The degree to which fatty acids can augment energy release depends upon physical training which increases their contribution: starvation also increases their availability.

During prolonged exercise various mechanisms operate to establish a hormonal response which favours glucose production, gluconeogenesis in the liver and lipid mobilization. There is an increased release of catecholamines, glucagon, cortisol and growth hormones. Insulin levels fall. The response seems to involve both glucose sensitive cells in the CNS and pancreas and may involve metabolic sensors in the exercising muscles themselves. There also appears to be a release of certain amino acids from exercising muscles which could be taken up by the liver and used for gluconeogenesis.

After exercise muscle glycogen is restored from dietary sources under the influences of insulin or from liver gluconeogenesis. This can take many hours.

Renal function

During severe exercise, renal blood flow decreases to allow maximal redistribution of cardiac output to the exercising muscles. Nevertheless differential constriction of the afferent and efferent glomerular arterioles, the efferent arteriole constricting the more, preserves glomerular filtration and a greater proportion of the remaining diminished renal plasma flow is filtered. Inulin clearance, the best measure of GFR, may fall by 30% in a burst of severe exertion sufficient to drop renal blood flow to 50% of the flow at rest. Urine flow also shows a decrease of 30%. Levels of antidiuretic hormone increase during exercise resulting in a fall in free water clearance. Renin and aldosterone levels also rise in the blood and there is a decrease in urinary sodium excretion and an increased potassium excretion. These responses help to maintain blood volume in the face of increased water loss through sweating.

Although not of functional significance there is an increased excretion of albumen in urine formed during exercise. This probably occurs because of an increase in glomerular permeability; several factors contribute, the rise in body temperature and the slowing of glomerular blood flow due to efferent arteriolar constriction being the most important.

Some subjects may have haemoglobinuria and others myoglobinuria. March haemoglobinuria is rare; it occurs in strenuous upright exercise including running on hard surfaces. It is thought to be caused by mechanical damage to red blood cells caught beneath the bones of the foot during impact of heel strike. It was first observed in soldiers after a long march, hence its name. Myoglobin can appear in the urine 24–48 hours after exercise severe enough to cause damage to muscle fibres. It is a small molecule, 17 000 molec. wt, so that it easily passes the glomerular membrane.

Disturbances of water and electrolytes are much more severe during exercise in a hot environment and this will be discussed in Chapter 3.

The Effects of Repetitive Exertion (Training)

When exercise is performed regularly and with sufficient vigour there are marked physiological improvements in the function of several body systems. The nature of these adaptive changes depends upon the muscles used, the type of exercise, and the previous degree of training. Different types of exercise will, to different extents, develop motor skill, increase strength or improve capacity for rhythmic exercise. Research over the last ten years and particularly the analysis of muscle biopsies has considerably advanced our understanding of the effects of training. Emphasis has shifted peripherally from the heart and central circulation, where the changes are more apparent than real, to the working muscles now known to be the major site of the changes induced by training and which reflexly produce some of the central changes. This is why training is largely specific to the muscles used. The type of exercise produces differential effects because there are two types of muscle cell. Exercise which is predominantly aerobic will produce its main improvements in Type I muscle cells, and anaerobic work in the Type II cells. High intensity exercise of moderate duration such as middle distance running will train both systems. The development of motor skills depends upon adaptive changes in the central nervous system and will not be considered further. Increased muscle strength results from relatively few, forceful contractions as in weight lifting (e.g. more than 80% of maximum once a day). Muscle cells hypertrophy in response. The increase in muscle cell proteins is due both to an increase in protein synthesis and to a decrease in the rate of protein degradation. There is also an increase in the concentration of aerobic enzymes responsible for rapid energy release.

The adaptive changes which result in an improved aerobic exercise capacity occur primarily in the skeletal muscles fibres and also, to some extent, in the cardiovascular system and the autonomic nervous system. These are the long-term changes associated with rhythmic endurance training. They do not necessarily involve muscle hypertrophy or increase in strength unless the exercise intensity is high, but always include an increase in the capacity of the muscles for aerobic metabolism and sometimes improvements in the oxygen transport system. Considerable information is now available regarding these changes. Training effects are cumulative and empirically it is well established that capacity for aerobic exercise requires a minimum of 20 min of extra exercise three times a week. More frequent or longer sessions do not seem to change the rate of improvement in relatively sedentary individuals. Moreover, it is becoming apparent that many athletes over-train. They need more than 60 min training per week but Sebastian Coe and others have demonstrated that runners do not need to run 200 miles a week, to reach a peak performance; 50 miles or even less may be enough.

The intensity of the exercise must also be above a critical minimum, but there is no maximum. The more intense it is within the limits of damage or excessive stiffness the faster the improvements. This is the principle of overload. The minimum is often specified as 60% of the individual's maximum oxygen uptake or 60% of the heart rate range from resting to maximum. The practical rule of thumb is that the exercise intensity must be above that to which the individual is accustomed.

A large number of different adaptive changes are induced by training and their time course varies. Some of the changes in concentration of key oxidative enzymes are rapid, taking place within a few days; hypertrophy of muscle cells takes weeks and maximum oxygen uptake can continue to increase slowly after the initial rapid raise for many months (see Fig. 15). The reason for this slow increase is not understood. None of the changes induced by training are permanent and the various improvements revert with similarly differing time courses.

Improvements in the metabolic capacity of skeletal muscle fibres

These changes have been observed in experiments with animals and confirmed by human muscle biopsies taken from individuals at various times during endurance training programmes. Holloszy (1976) has reviewed them in detail and the principle changes are included in Table 7. Skeletal muscle which has adapted to endurance exercise may have up to twice as many mitochondrial cristae per gram as untrained muscle. The mitochondria contain the oxidative enzymes. (Figure 16 shows the internal structure of a mitochondrion.) The increase in cellular myoglobin (Table 8) is also important. Myoglobin is not only an oxygen store but also seems able to facilitate the diffusion of oxygen through the cytoplasm to the mitochondria. The changes are not entirely restricted to one fibre type, at least in

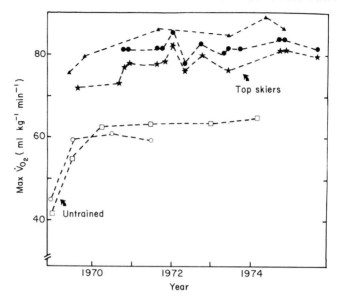

FIG. 15   Data on maximal oxygen uptake from cross-country skiers and two normal subjects who started intensive training in 1969. One (open squares) shows a progressive increase over several years. (From Åstrand and Rodahl, 1978.)

TABLE 7   Adaptive responses in skeletal muscle cells during training

Response of mitochondria
 Increase in size and number
 Increase in mitochondrial ATPase
 Increase in enzymes of citric acid cycle
 Increase in enzymes of fatty acid catabolism
 Increase in enzymes of ketone utilization

Capacity to oxidize substrates
 Increased capacity to oxidize
  Carbohydrate
  Fat/fatty acids
  Ketones
  Pyruvate

Other metabolic changes
 Consequent reduction in quantities of lactate produced at comparable rates of
  glycolysis
 Increase in hexokinase activity (other glycolytic enzymes unchanged)
 Increase in myoglobin content

From Holloszy (1976).

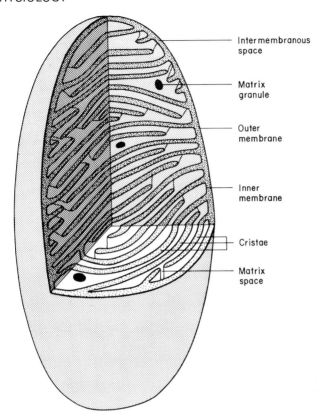

Intermembranous space

Matrix granule

Outer membrane

Inner membrane

Cristae

Matrix space

FIG. 16 Mitochondrion shown as a three-dimensional structure cut away in three planes to show the cristae. (From Burkitt *et al.*, 1979.) (Reproduced with permission from Churchill, Livingstone.)

man. Biopsy specimens always contain a mixture of the two types of fibre. Animal studies show that in general the changes are greatest in the slow twitch red fibres.

Because of the greater oxidative capacity of the cell more fat can be used as an energy source. This is reflected in a lower R.Q. Muscle stores of glycogen are depleted more slowly and so capacity for prolonged endurance exercise improves. In addition anaerobic work is required later in exercise and at higher work loads.

Training does not produce any change in metabolic efficiency of the muscles. The pattern of energy transformations does not change. There is no greater output of mechanical work for the fuel consumed, contrary to the common misconception that training means improved biochemical/ mechanical efficiency.

Changes in blood flow and total peripheral resistance

Capillary density has been observed to increase gradually during a period of endurance training. Andersen and Henriksson (1977b) observed an increase from 1.36 to 2.00 capillaries per fibre with such a training programme. This increases the availability of oxygen to the muscle cells by shortening the diffusion pathway. Blood flow to working muscle is not increased after training but femoral venous p_{O_2} is reduced.

The mean arterial blood pressure during exercise is no higher after training, even though the maximum cardiac output increases along with maximum oxygen consumption. Systolic pressure is sometimes even found to drop by about 10 mm Hg. Thus the well-trained individual appears to have an extreme ability to reduce resistance to flow through muscle tissue; he achieves very high systemic blood flow with very little rise in pressure. The total peripheral resistance drops despite vasoconstriction of skin and splanchnic areas.

In submaximal exercise the blood flow through the working muscles is probably unchanged or reduced by training. At maximal levels of work, the flow capacity is probably improved. The circumstantial evidence for this is strong but has not been confirmed by direct measurement. When maximal oxygen uptake increases, 10–20% of the increase with an accompanying increase in blood flow will need to go to the respiratory muscles which can provide more ventilation if the oxygen can be transported.

The reduction in heart rate

After training the heart rate is reduced. The rate is lower at rest and during submaximal exercise; maximal heart rates remain the same or drop very slightly. If the heart rate response to exercise at a given oxygen uptake is compared with the response before training in the same subject performing the same type of work, then a decrease of up to 30 beats min^{-1} may be observed depending upon the intensity of the training and the pre-training baseline (Saltin et al., 1968). Muller, even in 1943, before the biochemistry of muscle had been much studied, suggested that the reduction observed with exercise might be secondary to peripheral changes in the trained muscle. He tested a young man, 18 years of age, who undertook a ten-week training programme on a bicycle ergometer. His heart rate at a work load corresponding to an oxygen uptake of 1.6 litre min^{-1} fell from 129 beats min^{-1} to 95 beats min^{-1}. Three weeks after the training programme had been completed his heart rate for the same amount of work had returned to 126 beats min^{-1}. However, the previous week he had begun a strenuous arm training programme which involved hand cranking in the standing position. When tested by hand cranking his heart rate for work at 1.41 litres oxygen min^{-1} had fallen from 140 to 116 beats min^{-1} in a week. Evidently the heart rate reduction resulting from training with the arms did not carry over to exercise performed with leg muscles.

Clausen (1977) has been able to confirm this observation and to investigate the mechanisms in a series of elegant longitudinal studies of the training of young healthy subjects (see Fig. 17). He too has compared in the same individual the responses to exercise of limb muscles which have been trained with those which have not been trained. Compared with measurements made before training he found substantial reductions in submaximal

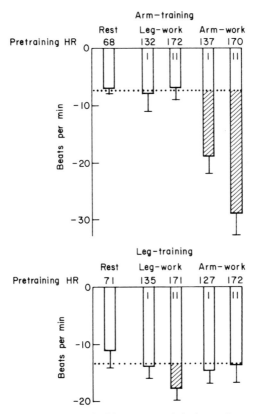

FIG. 17 Reduction in heart rate (HR) at rest and during submaximal exercise, at two levels of intensity I and II, with the arms or legs, after arm-training (above) and leg-training (below). Mean values from 13 subjects. Matched areas indicate the fall in HR in excess of the fall obtained at rest and during exercise with non-trained extremities. (From Clausen, J.P. (1973). *In* "Das Chronisch Kranhe Herz", (Eds H. Roskamm and H. Rendall), F.K. Schattaner, Stuttgart. Reproduced with permission.)

heart rates recorded during exercise with the trained muscle, just as Muller did, but also smaller reductions of similar size in both resting heart rates and submaximal heart rates recorded during exercise with untrained muscles. These latter reductions appear to depend upon a shift in the balance between parasympathetic and sympathetic control towards increased vagal tone. Thus the effect is central. The larger reduction, found using trained muscles, also includes a peripheral effect not primarily affecting the heart or autonomic nervous system. Saltin *et al.* (1976) have reported experiments performed with a similar objective which suggest that in some circumstances only the peripheral effect is apparent. In their experiments only one leg was trained and a work test was performed with one leg, either the trained leg or the untrained leg. Whereas the heart rate was significantly lower after endurance exercise when testing the trained leg (13 beats min^{-1} less) there was no significant change when testing the untrained leg (7 beats min^{-1} less) (only three subjects were tested). He did not measure heart rates but other studies consistently report a reduction in resting heart rate after training. Several different patterns of response may occur, depending upon the balance before training in a particular individual between sympathetic and parasympathetic tone, pretraining values for various other central and peripheral factors and the way they change with exercise. The peripheral response is always apparent.

The explanation of the peripheral effect appears to be found in the reduced needs of the exercising muscle for arterial blood arising from improved local oxygen transport and muscle performance. The muscle cells after training can work at a lower tissue p_{O_2} because of their improved aerobic capacity. It is probable that this also leads to lower concentrations of anaerobic metabolites at any given work rate and so to a reduced peripheral drive from the working muscles to the cardiac and vasomotor centres. This leads to lower heart rates, reduced sympathetic vasoconstrictor outflow, less severe splanchnic redistribution and a reduction in perceived exertion (Ekblom and Goldbarg, 1971).

Changes in the central circulation of the heart

A small increase in blood volume is sometimes seen with training. Maximum cardiac output increases substantially thus improving maximal oxygen uptake, but submaximal cardiac output at a given oxygen uptake does not change in normal subjects. The reduction in heart rate is compensated by an increase in stroke volume due partly to the increased filling time and partly to more powerful contraction of the heart muscle. There is a net gain because myocardial work for a given cardiac output is consistently reduced (see p. 41).

In young subjects oxygen transport capacity is also increased by better extraction of oxygen from the muscle capillaries, judging by the reduced femoral venous p_{O_2} during maximal leg exercise. The observed reduction in systemic mixed venous p_{O_2} is also explained by this improved oxygen extraction in working muscles. In older men only stroke volume increases.

The improvement in myocardial performance has been studied by Clausen et al. (1973). The subjects undertook five weeks' training with leg exercise. Thereafter heavy submaximal exercise with the arms produced a greater cardiac output, stroke volume, and arterial blood pressure. The total peripheral resistance remained unchanged. The improvement in stroke volume in the face of the same peripheral resistance suggests an enhancement of myocardial contractility. Comparisons between well-trained subjects and sedentery subjects, including autopsy findings, X-ray measurements and echocardiographic measurements suggest that a year of severe exertion increases cardiac ventricular volume and wall thickness. Comparisons of 20 world-class distance runners, eight University class distance runners and 10 lean non-athletes from a University population were undertaken by Underwood and Schwade (1977) using echocardiography and electrocardiography. The athletes had large hearts, they tended to cluster towards the upper limit of normal and 30% had left ventricular hypertrophy by ECG voltage criteria. Longitudinal studies of athletes suggest that there is a modest increase in heart mass occurring over years of training but that this regresses again slowly once competitive training ceases.

There do not seem to be any adaptive changes in the mitochondria or respiratory enzymes. However, the cardiac muscle cells do hypertrophy in response to strenuous endurance exercise as the studies mentioned above indicate. In animals actomyosin and myosin ATPase increase with repeated exercise, which may explain the improvement in myocardial contractility. There is no good evidence that exercise increases the coronary capillary network, the calibre of coronary arteries, or the formation of collaterals.

Fatigue

Fatigue may be defined in physiological terms as the failure to generate power in the muscles. It can be due to failure at any one of the links in the chain from the central command originating in cortical or cerebellar brain centres to the sliding of the actin and myosin filaments. Some of the links in the chain can be studied in isolated tissue preparations. Nerve conduction, neuromuscular transmission and muscular contraction can continue in response to repetitive stimulation for many hours in these experiments, but the neuromuscular junction appears to fatigue first when supplies of acetylcholine no longer meet the demand.

In practice, fatigue is a subjective symptom. The phenomenon of exhaustion in man is notoriously difficult to study because of the complicated interplay of psychological factors including motivation, sensitivity to discomfort and responses to environmental conditions. Respiratory and muscular discomfort in strenuous exercise are caused by the build up of the acidic metabolic products of anaerobic metabolism. Fatigue in a maintained isometric contraction is muscular rather than respiratory and is due to the rapid accumulation of these metabolites in the absence of an adequate local blood flow.

Intramuscular tension depends upon the force of contraction and, if it is high, impedes the blood flow. Thus endurance depends upon the intramuscular pressure so generated remaining below the level which compromises the circulation, whereas, with dynamic contractions, the frequency of contractions needs to be sufficiently low relative to the force to allow an adequate supply between contractions. With repetitive movements under heavy load pain from intramuscular pressure and tendon tension will be a subjective factor.

The concept that the accumulation of lactate results in fatigue has had considerable appeal since the hypothesis was proposed by Hill and Kupalov in 1929. It is attractive because intracellular pH can interfere with contractile function (see p. 49). Training has a protective effect against both lactate accumulation and the development of fatigue during exercise because it improves aerobic capacity; anaerobic sources are not required until the intensity of exercise is much higher, and also the metabolism of lactate by tissues other than working muscle is improved.

Fatigue is characterized by changes in neuromuscular coordination as well as in muscle cell metabolism. If an individual performs repetitive work with one limb to the point of exhaustion, there is a progressive change in the pattern of motor unit activity. The total contractile force as measured by the integrated electromyogram gets progressively bigger. The contractions of the muscles involved become less smooth and more clonic. This appears to be caused by the increased synchronization of the contracting motor units.

During prolonged, strenuous exercise depletion of body carbohydrate stores results in the development of physical exhaustion and symptoms of hypoglycaemia. Replacement of glucose stores and water loss during marathon running does not enable the athlete to run for ever. Exhaustion may then be due to loss of electrolytes and ionic imbalance.

Maximum oxygen uptake is traditionally defined in terms of a short bout of exercise but work capacity depends upon the time for which it must be sustained (see Fig. 18). If exercise is to be sustained for hours then the anaerobic contribution to energy release must be small enough for the lactate production to be metabolized at the rate at which it is produced.

The maximum oxygen uptake of record-breaking endurance athletes has not increased over the years although their performance has improved. This must be due to a greater capacity for an aerobic energy release which extends the peak in Fig. 18 or to psychological factors. Athletes rarely break their own records, which suggests that the second factor is important. The limits to athletic performance are still thought to be psychological rather than physiological, although clearly there are physiological limits which may be the muscle's strength, its capacity to take up and use oxygen, or the capacity of the cardiovascular system to deliver oxygen depending on the event in question.

Fatigue after walking for many hours in normal people is different since aerobic capacity is adequate and homeostatic mechanisms are not disturbed. Minor trauma to joints and muscles may be responsible both for the fatigue and the ensuing stiffness in these circumstances.

Summary of endurance training effects

The scale of the improvement in exercise performance which can be achieved with training is considerable. The classical study undertaken by Saltin *et al.* (1968) in which they determined over 70 days the extent of the changes from the deconditioned state after three weeks' bed-rest through to the well-trained state after weeks of strenuous intermittent running at about 90% of maximum V_{O_2} are shown in Fig. 19. The maximal oxygen uptake

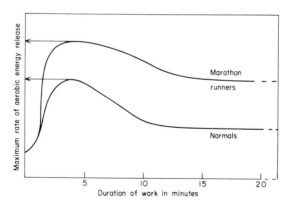

FIG. 18   Diagram of the relation between endurance time and the rate of energy release from aerobic sources.

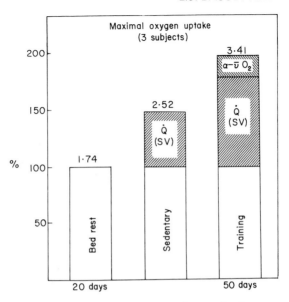

FIG. 19   Mean values for three male subjects for maximal oxygen uptake after bed-rest for 20 days, a habitually sedentary life and intensive training for 50 days. The increase is due to an increased cardiac output ($\dot{Q}$) due to an increased stroke volume (SV) since maximal heart rates were unchanged. After training there is also an increase in the $a - \bar{v} - O_2$ difference. (*In* Saltin *et al.*, 1968. From Åstrand and Rodahl, 1977.)

after six weeks was twice that found after bed-rest. Significant differences were also apparent at submaximal levels of work, such as a 30 beats min^{-1} reduction in heart rate.

In summary, at maximal levels of exercise of relatively short duration (three or four minutes) the rate of working, the oxygen uptake, cardiac output and ventilation are known to increase, and muscle blood flow and muscle oxygen extraction probably also increase. There is no increase in muscle strength unless the training has been at intensities greater than 80% of maximum. At submaximal levels the most easily measured change is the reduction in heart rate. This is compensated by a rise in stroke volume so that there is no change in cardiac output. In the working muscles the capacity for extracting oxygen from the blood flow improves because of increased density of the capillary network, and increased concentration of aerobic enzymes. The blood flow may therefore decrease, giving lower femoral p_{O_2} levels and sparing splanchnic blood flow. The perception of effort at any given work intensity is also less perhaps because of the decreased splanchnic vasoconstriction or the lower concentration of anaerobic metabolites in the working muscle, or perhaps because fewer motor units need to be recruited. The changes are mainly a consequence of improvements in the working muscle, although there are central changes as well.

It is often forgotten that there can also be remarkable improvements in capacity for prolonged endurance exercise. This should be measured more frequently than it is. There are patients in several centres in North America who, a year or two following recovery from myocardial infarction, have trained to the point that they can run the marathon distance in three to four hours. Such patients have been observed to have a 55% increase in maximum oxygen uptake, which is considerable and achieved largely because of the biochemical and cardiovascular changes which have been described.

Levels of Physical Exertion at Work

For most people in this country, earning a living does not require much muscular work. An individual at work sits at a desk or stands at a bench, and the energy spent on the job during an eight-hour shift may amount to no more than 800 kcals (about 30% of a total daily energy expenditure). In sharp contrast many developing countries still need to rely on heavy manual work, and workers such as the sugar cane cutters in Latin America or the Sudan may expend nearly 4000 kcals (68% of their much larger daily total) (see Table 8).

If man is affluent enough to choose his life-style he emerges as a lazy animal who avoids strenuous physical effort when possible, seeking to replace his own muscular effort with machines, whether for performing his daily task, for his means of transport or for his entertainment. He has a

TABLE 8   Daily energy expenditure in kcal in various kinds of employment

	Sedentary (clerical managerial)	Industrial (skilled, semiskilled)	Manual (3rd world)
At work	800	1200	2600
At home	1000	800	800
In bed	500	500	400[a]
Total	2300	2500	3800
% used on the job	35%	48%	68%

[a] Lighter body weight and shorter sleeping night

tendency to natural indolence, and powerful motivating drives such as hunger and thirst are required to stir him into vigorous activity. His level of daily energy expenditure is the minimum compatible with his continued survival in his present state. For other peoples (living in subsistence economies) survival depends upon obtaining a job, any job, however physically demanding, and holding it down in the face of competition. In this situation the level of activity may still be the minimum compatible with survival but that minimum may also be near to the maximum level of activity of which the man is capable.

Levels of daily activity and rates of work are determined not only by a man's physiology but also by social expectations and economic rewards. In this country fear of starvation has been eliminated but in accordance with the biological principles shaped by natural selection man will in general adopt levels of activity which give the maximum reward for the minimum of effort. Observed levels of work owe as much to modern technology and trade union negotiation as to physiological constraints.

Nevertheless there are important physiological considerations. The total possible amount of energy expended daily is limited, and there are also limits to the intensity and duration of the bouts of physical work contained within that total. Moreover, in many rhythmic activities such as walking or pedalling, there are optimum rates of work associated with optimum frequencies of movement. These rates, which are found to be the preferred ones, enable the total job to be done with the minimum internal energy expenditure.

An upper limit will be set to daily energy expediture by a number of factors. Dietary intake of calories and essential nutrients such as protein must in the long run keep pace with expenditure, or there will be a gradual loss of weight leading eventually to muscle weakness and loss of working capacity. The subject will complain with perfect accuracy of lack of energy.

Body size and physical condition will also set an upper limit. A large well-trained man whose weight is due to muscle rather than to fat, will have a higher work capacity and will be capable of a much higher daily expenditure of energy than a small fat man who is not used to exercise. Fully trained, well nourished, well built individuals will be able to expend energy at rates of about 4000 kcals (16.5 MJ) day after day. Values of this order are found amongst lumberjacks, miners and sugar cane cutters.

Daily energy expenditure is distributed between periods of exercise and periods of rest. The contributions made by resting metabolism and various activities is given in Table 9. It is clear that high daily expenditure must entail either many short periods of very intense exercise (i.e. intermittent work) or prolonged periods of more moderate exercise (i.e. steady-state work). The limiting factors are different in each case.

TABLE 9   Typical day's energy expenditure of reference man
weighing 65 kg and 25 years old

Time spent	Activity	Energy Expediture	
		Mean Rate (kcal min^{-1})	Total (kcal)
8 h	at work, mostly standing	2.5	1200
8 h	leisure time		
1 h	pottering	2.0	120
$\frac{1}{2}$ h	walking	5.0	150
4 h	sitting	1.4	320
$1\frac{1}{2}$ h	domestic work	5.0	450 1040
8 h	rest in bed	1.1	530
		Day's Total	2770

Maximum values for working capacity and oxygen uptake, whether gross or per unit body weight, have little meaning unless the duration of the work is specified. Duration can range from the seconds taken by dockers to manipulate cargo, to the hours in the case of the miner or foot soldier. For any one person the maximum values for rate of performing work will be inversely proportional to its duration.

High intensity work of short duration depends to a large extent on anaerobic energy sources. Lactic acid and other metabolites accumulate rapidly and within a few minutes they contribute to the production of muscular pain and breathlessness. The maximum rate of work is determined by the man's muscle strength, the proportion of aerobic/anaerobic energy release and his tolerance of the discomfort produced in the muscles. It is possible to work at rates of over 400 watts for a few minutes at a time. There are optimum patterns for intermittent work which are related to the speed at which the lactic acid produced can be metabolized (this happens in cardiac muscle, other skeletal muscle and the liver) during the rest periods. Alternate periods of two minutes' work followed by one minute's rest have been found to give the highest mean rate of work (see Fig. 20) and can be sustained for long periods. In some industries this is a common pattern of work, for instance, loading concrete pipes onto lorries (see Fig. 21). Repeated bouts of sustained isometric work (see p. 36) can be considered an extreme form of intermittent work.

There is a gradation between intermittent and steady work; if the intensity of the work falls then the work can be continued for longer before

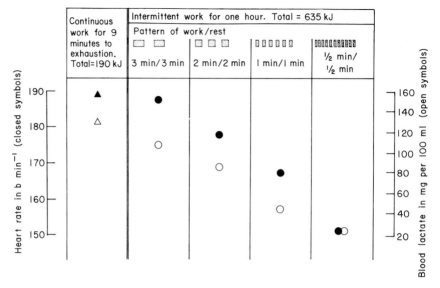

FIG. 20   Optimum patterns of work. Alternating periods of $\frac{1}{2}$ min of work and rest allow the same amount of work to be done in a given time with a much lower heart rate and lactate level than if the work and rest periods are longer. (Based on data from Astrand and Rohahl, 1977.)

fatigue sets in. If the maximum work capacity is defined as the maximum rate which can be maintained for three or four min (also measured as maximum oxygen uptake), then at intensities below about 50% of that maximum the duration becomes indefinitely long. It stretches to hours and a maximum value for duration is difficult to measure precisely. The causes of fatigue are different and depletion of energy stores becomes important. The small amount of lactic acid can be metabolized as fast as it is produced (see p. 49).

The change from intermittent to steady work can therefore be defined as the level of exertion at which the lactic acid metabolism can keep pace with its production. This is known as the lactic acid or anaerobic threshold. As a result of training, this point is raised, occurring at a higher percentage of the short-term working capacity because of the increased concentration of mitochondria and oxidative enzymes in the muscles (see p. 53). The well-trained compared with the unfit or inactive have a capacity for steady prolonged work which is more than proportionately higher than any increase in V_{O_2} max or ability to sustain isometric work. The Olympic-standard marathon runner can expend energy for many hours at 70–80% of his maximum short-term level, but a normal sedentary individual can only

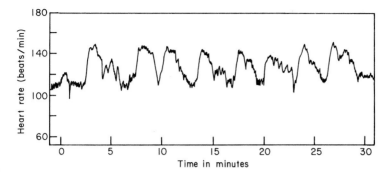

FIG. 21   Heart rate recorded from a man aged 64 years loading concrete pipes on to a lorry by hand.

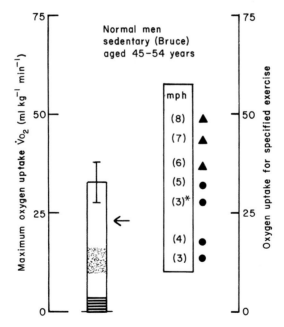

FIG. 22   The mean maximum oxygen uptake ± I.S.D. (left-hand scale) is shown for a group of sedentary normal men (Bruce *et al.*, 1973) along with the oxygen uptake required (right-hand scale) for walking (●) or running (▲) at various speeds (m.p.h.) and uphill* or against the wind. The range of resting metabolic rates for these subjects is indicated by the barred area. The level of exercise at which symptoms usually develop and the range of lactic acid thresholds expected for this group of subjects is indicated by the stippled area.

work below 50% of maximum and remain in a steady state. A coronary patient may be in a worse state.

These considerations of the physiological basis of intermittent and steady state energy expenditure provide some rational basis for determining work rates in industry and agriculture, but do not provide a complete prescription. Body size (and male gender which relates to size) confer inevitable advantages and so do well trained muscles. It can be argued that the training factor will resolve itself, since the job will train (or decondition) the man. In the end, the prescription has to be an individual one; the preferred rates of work will probably be beween 30 and 60% of short-term maximum depending upon how well-trained the man is, and there will be wide variation in short-term maximum.

Top-class endurance athletes can expend energy at a rate which uses up oxygen at up to 80 ml kg^{-1} min^{-1} (3.47 mmol kg^{-1} min^{-1}), but a deconditioned middle-aged man may have a maximum oxygen uptake of only 25 ml kg^{-1} min^{-1} (1.12 mmol kg^{-1} min^{-1}). These values are observed to drop with age but this may be mainly because of reduced levels of activity with increasing age.

The perception of effort during work is related to the intensity of the work, expressed as a percentage of the individual's own short-term maximum (see p. 60). Borg (1973) has devised several descriptive scales consisting of subjective verbal ratings of perceived exertion (see Fig. 23). The middle of the scale, corresponding to preferred rates of steady-work (about 30 or 40% of short-term maximum), contains the neutral words "neither

6	
7	Very, very light
8	
9	Very light
10	
11	Fairly light
12	
13	Somewhat hard
14	
15	Hard
16	
17	Very hard
18	
19	Very, very hard
20	

FIG. 23   Borg scale for rating perceived exertion during exercise. The numbers on the scale correspond roughly to a tenth of the heart rate expected for the exertion perceived. (From Borg, 1973.)

light nor hard". The scale is also well correlated with percent maximum heart rate, so that the middle of the scale corresponds to heart rates of 100–120 beats min^{-1}, which are associated with moderate levels of oxygen uptake (1.0–1.5 litre min^{-1}), which can be maintained for long periods. Once subjects have become familiar with it in relation to a particular activity the Borg scale is reliable and can be a useful means of assessing the severity of a particular task for an individual (see Fig. 24). It holds good for a wide variety of subjects of differing fitness, body composition and state of motivation. The association with heart rate is not a causal one, since its linear relation with perceived exertion can be dissociated by drugs, such as propranolol (Ekblom and Goldbarg, 1971). The lower heart rates for the same work rate after β-adrenoceptor blockade are associated with the perception of greater effort.

Preferred rates of work usually correspond to optimum frequencies of rhythmic movement. Walking, pedalling a bicycle, stairclimbing and lift work with the arms have optimum frequencies for which efficiency is at its maximum (see Fig. 25). The graph is a shallow one with a blunt minimum of internal energy expenditure. There is little change in efficiency over quite a wide range of frequencies but at the extremes efficiency falls. This is because at low frequencies an excessive amount of postural stabilization is required and the energy stored during movement in stretched muscles and tendons or in the pendulum swings of a limb is not being as fully recovered. At high frequencies the energy stored in the pendulum swing is also partly lost and an increased amount of energy is used in overcoming friction and inertia in the muscles and joints.

The preferred pace frequencies of normal walking are influenced by leg

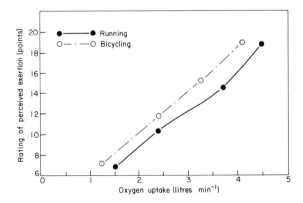

FIG. 24   The relation between perceived exertion and oxygen uptake during work on a bicycle ergometer and on a treadmill. (From Ekblom and Goldbarg, 1971.)

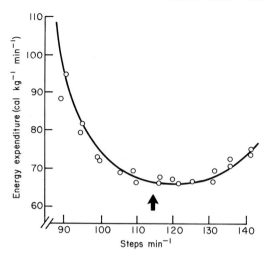

FIG. 25   Relation between energy expenditure and stepping frequency for walking at 90 m min^{-1}. The arrow indicates the freely chosen frequency. (From Zarragh and Radcliffe, 1978.)

length and found to fall within the optimal range; much of the stored energy is used in the next step (Cavagna and Kaneko, 1977). This is not so in running, for which very little of the stored energy is re-used and there is no optimal frequency. At about 7 km h^{-1} walking becomes very inefficient and it is more economical to run. It is likely that there is a broad band of efficient frequencies for many activities and that preferred rates of work fall within them.

Physical Work and Health

In the past the working environment has posed serious risks to health, and physical effort has often been blamed for producing the symptoms of disease. It is therefore not surprising that the idea that high levels of physical exertion at work might be beneficial to health has met with opposition. However, recent concern to identify the causes of the rising incidence of coronary heart disease has provided the impetus for a reassessment of the role of exercise and there is good evidence that health is improved by physical effort either at work or in leisure time. Moreover, it is increasingly clear that the benefits of exercise extend beyond any role it may have in modifying the natural history of coronary heart disease. The worthwhile improvements in physiological function which occur with regular physical work constitute the promotion of positive health rather than a prevention of disease.

Some of the confusion arises from the breadth of meaning which is invested in the term "health" and from the restrictions which are imposed when it is measured. Health may reasonably be defined as the possession of a full capacity to meet the various physical, psychological and social challenges of life. Man can adapt successfully to the challenge of his environment by making use of his various physiological reserves or by modifying his behaviour. Traditionally, health has been regarded as an absence of disease. This leads to a study of the incidence of disease and to questions about the effects of levels of exercise on morbidity and mortality. The difficulties inherent in this approach can be illustrated by supposing that measles is influenced by levels of daily physical activity. What is the incidence of measles in a population of children? How many children die from the disease or its complications? Is the disease least severe in the most active children? Are the most active children least likely to catch the disease? The first two questions are easily answered provided the disease is notifiable and has a clear diagnosis, but the last two questions are impossible to answer with certainty. There are too many uncontrolled variables to enable firm conclusions to be drawn.

Absence of disease does not necessarily imply the presence of good health. There is an intermediate state in which people are neither well nor ill (Fig. 26). The addition of this category poses different questions but they are just as difficult to answer. An association between high levels of exercise and low incidence of minor ailments has been suggested, but there is no evidence that a change in exercise levels can move people from one category to another. Those who suffer most from minor ailments and indulge in self-medication to the greatest degree may also be those who least enjoy exercise.

It is possible to think of a spectrum stretching continuously from poor to good health. The measurement of health as a whole rather than some of its contributory factors is probably impossible. Subjective estimates of general health are coloured by expectations, for example elderly people expect to deteriorate. They will report that they enjoy good health, forgetting to add in parentheses "for my age", which in their minds accounts for their troubles with their joints, lungs, hearts and blood vessels. Moreover, subjective reports of the amount of exercise taken are notorious over-estimates.

Yet despite these and other difficulties, it is useful to consider the relationship between exercise and two aspects of health, namely the effect of exercise, first upon the incidence and progress of disease or disability and secondly, on the ability to meet the challenge of physical work.

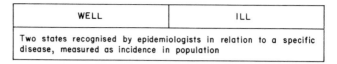

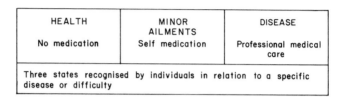

GOOD HEALTH	POOR HEALTH	ILL HEALTH

A continuum which can be measured in an individual, eg. as the capacity to meet the challenge of physical work

FIG. 26   Categories of health.

Exercise and the challenge of physical work in normal people

If one aspect of positive health is the ability to respond to the challenge of physical work, then health may be defined in terms of working capacity. When normal individuals who have previously led sedentary or inactive lives assume a more active life or take part in a regular exercise programme, they immediately become aware of their limited tolerance for physical work and the discomfort it provokes. However, in a matter of only a week or two of regular exercise they are aware of feeling better (Borg, 1973). They do so in part because they experience an improvement in their physical working capacity and a reduced sense of effort for any given task. Harder physical work can be tolerated for longer without fatigue and all daily physical tasks are accomplished more easily, and the range of physical activities in which they can participate is extended.

The link between health and physiological improvements affected by training is important in a society where labour-saving technology has resulted in a widespread reduction in the duration and intensity of obligatory exercise at work. The reduction in manual labour is welcome in many ways. Manual work has been dangerous, unpleasant and even part of our penal system (see Fig. 27).

FIG. 27 Brixton Treadmill (*c.* 1830), a standard nineteenth century punitive treadmill. (From Brunner and Major, 1972.)

Chronic disability and the challenge of physical work

In the United States one in seven of the working population are reported to suffer some limitation of their working capacity through chronic disability. The commonest problems are heart disease, cerebrovascular disease and arthritis. The incidence is probably similar in Britain.

Coronary heart disease may lead to a limitation of working capacity in a number of ways. Infarction may cause sufficient damage to cause heart failure, serious cardiac arrhythmias or anginal pain from myocardial ischaemia. The occurrence of a heart attack may also destroy a patient's confidence in his ability to perform physical work, even though the residual cardiac damage is slight and physical recovery apparently complete.

Training programmes will help the patient with angina by raising the ceiling level of work which causes pain, since one effect of training is to reduce cardiac work (see Fig. 28 (a)), and in many cardiac patients the cardiac output is lower for the same exercise intensity after training.

The patient who has recovered from a heart attack may have a low working capacity (see Fig. 28 (b)) so that walking at a normal pace can provoke symptoms of breathlessness or angina. Considerable improvements in exercise tolerance can be achieved by training as well as the recovery of self-confidence in the ability to work safely and without discomfort.

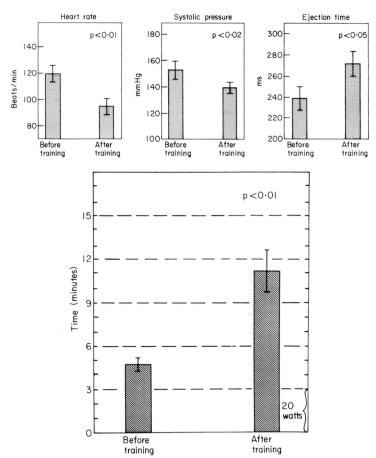

FIG. 28   (a) The effect of training on heart rate, systolic aortic pressure and ejection time. Mean values ± SEM are given for seven patients with angina. The exercise level in both cases was that which just caused the onset of angina before training. (b) The effect of training on exercise tolerance in the same group of patients as in (a). The exercise intensity was increased by 20 watts every 3 min until the onset of angina. (From Epstein *et al.*, 1971.)

The prescription of exercise for rehabilitation programmes should not be restricted to patients with cardiac disease. In chronic obstructive lung disease the patient enters a vicious cycle of breathlessness on exertion, inactivity, weakness, and further breathlessness. (The patient's ability to ventilate his lungs and achieve an adequate oxygen uptake is often severely limited.) Physical activity will not replace destroyed lung tissue but it can help the patient make better use of his existing capacity. Training can ease the situation because the improvements in peripheral muscle function allow a more complete extraction of the oxygen which can be delivered; endurance training, undertaken with the help of bronchodilator drugs has been shown to ameliorate chronic obstructive lung disease (McGavin *et al.*, 1977). Patients with asthma are also helped by training and the frequency of attacks is usually reduced. Exercise can also induce bronchospasm in asthmatics and potential asthmatics who are otherwise symptom-free. This phenomenon is not well understood. It is less likely to occur with swimming and sprinting than with endurance exercise, such as fast walking or running. It can cause modest disability for individuals who do not experience bronchospasm in any other circumstances.

Musculoskeletal impairment due to arthritis, neurological disease, or injury produces a range of limitations from slight to total and the limitation is specific to the damage. Nevertheless, in almost all cases, the challenge of physical work or even of minimal movement will best be met by training and maintaining the power of the remaining healthy muscles. This requires brief near maximum contractions for each muscle at least once a day (Thorstensson, 1977). Similarly, mobility of arthritic joints may be maintained by allowing them to move through their full range at least once a day. The stability of joints depends upon the strength of the muscles surrounding them so the sudden stresses which occur frequently in normal active life are less likely to cause damage when muscles are well trained. The severe stresses of contact sports such as football are a different matter and can give rise to injuries despite well-trained muscles.

There are advantages for sedentary individuals in their maintaining, and even increasing the strength and bulk of trunk and abdominal muscles. It is likely that this reduces the liability to hernias and back pain due to prolapsed intravertebral discs.

Diabetes mellitus can be helped by regular exercise. Glucose tolerance improves and the insulin requirements are reduced. The effect of exercise appears to be an increase in the passage of glucose into muscle cells. Impaired movement of glucose across the cell membranes is one of the major biochemical lesions in diabetes. During exercise and for several hours afterwards insulin becomes more effective in stimulating the movement of glucose. It follows that in diabetes the daily insulin dosage must be matched to the activity levels as well as the dietary intake of carbohydrate. If this match is not achieved there is a risk of hypoglycaemia during exertion.

Obesity is a common condition in Western Europe and North America. In Britain 40% of the adults over the age of 30 years are obese, which is to say that they are more than 10% over the desirable weight for their height, as defined by the actuarial statistics of the insurance companies. The obese may find exercise exhausting because of their weight, but this in turn may keep their muscles and cardiovascular system in good condition.

Exercise has long been neglected as a possible means of controlling or reducing obesity because the amounts of exercise required to burn up a significant amount of fat are very large. A half-hour walk uses about 150 kcals; this amount of energy is contained in one piece of bread or 16 g of fat. However, if a long view of the problem is taken then one kilogram of fat can be lost in just over 60 days provided the extra daily walk is continued and the calorie intake does not change. Similarly 10 kg can be lost in less than two years but two other factors may operate to potentiate the weight-reducing effect. Light exercise for half an hour after a meal has been shown to increase the calorie expenditure by an amount equal to twice the resting thermic effect of the food eaten, in addition to the calorie expenditure required for the external work involved in the exercise. Resting metabolic rate has been found to be raised 25% above the basal level for 15 hours after strenuous exercise (Allen and Quigley, 1977). Thus, although half an hour of squash may expend only 450 kcal, the increase in metabolic rate over the ensuing two days could account for possibly another 550 kcal, making a total of 1000 kcals. A game of squash every two days would therefore produce a sustained weight loss of up to 1 kg fat per week. On the other hand, the rate of weight loss will usually fall slightly over the months because the energy expenditure of walking or running at a particular speed will go down with loss of weight, but speed of movement may increase. Exercise does not stimulate appetite, as is commonly believed, in fact strenuous exercise inhibits it in the short term, so efforts to reduce body weight are likely to be successful if based upon a permanent shift to a more active life style coupled with a modest reduction in energy intake.

Mental ill-health does not give rise to any obvious limitations of physical working capacity, but poor motivation, introversion and depression may lead to poor work performance, unwillingness to work, and a reduction in spontaneous activity levels. Exercise programmes have produced psychological benefits over and above the physical improvements, including improvements in overt behaviour and capacity for recall in geriatric mental patients. Brain-damaged children may be limited in their physical capacities and fail to develop to their optimum levels both physically and mentally unless physical activities are deliberately organized for them.

Exercise and prevention of disease

Health is often thought of as resistance to disease. Those who enjoy good health rarely succumb to the inevitable invasions of bacteria and viruses and recover rapidly if they do. However, there is no good evidence to suggest that health as defined in this way owes anything to exercise or a good capacity for work. The common belief that exercise is good for you can be substantiated by the arguments already put forward, but the benefits do not include a reduced incidence of disease except in a few specific instances which will be described later.

There may be an association between the reported incidence of minor ailments and low levels of physical activity (see Fig. 26, p. 72) but it is likely that those who are free of such ailments or choose to discount them are also the ones who enjoy physical exercise. No evidence is yet available that increased physical exercise can cure or improve minor ailments. Some minor ailments are quite major health problems when judged by the loss of working days to industry. Low back pain is an example of such a condition. Reference has already been made to the alleviation possible by exercise directed to strengthening the appropriate muscles of the abdomen and back.

There are other more insidious diseases caused, not by microorganisms, but by gradual degenerative or otherwise deleterious change. These include arthritis, hypertension and coronary heart disease.

There are three views about the role of exercise in the development of osteoarthritis. Moderate rhythmic exercise such as walking or running may have a beneficial effect by increasing the natural lubrication of the joints and so protecting them against degeneration (Puranen et al., 1975). Those who are already suffering from the condition are usually advised to take exercise in order to prevent the situation from becoming worse (Ekblom et al., 1974). Others take the view that exercise is irrelevant and that osteoarthritis is an inevitable consequence of minor damage to the joint (Solomon, 1978). The third group believe that osteoarthritis is a "wear-and-tear" phenomenon and that exercise only hastens an inevitable process (Murray and Duncan, 1971).

An increase in exercise levels has a beneficial effect in mild hypertension, specifically those with so-called labile hypertension or with diastolic pressure between 100 and 110 mm Hg systolic pressure between 150 and 170 mm Hg (Sannerstedt et al., 1973). The elevation of blood pressure, albeit small, is a significant risk to health. The effect of exercise is to reduce the systolic and diastolic pressure by about 10 mm Hg.

Coronary heart disease (CHD) has doubled in its incidence (in the UK) among men between 40 and 50 years of age in the last 20 years. There are probably a number of factors which interact with each other (see Fig. 29). It is suspected that lack of exercise may be only one of them. Several studies have shown that groups who regularly take strenuous exercise have an incidence which is about half that of less active groups. Naturally there has been interest shown in men with physically active occupations. Studies of London busmen more than 20 years ago produced the first suggestion that physical activity was associated with a low risk of CHD. Although conduc-

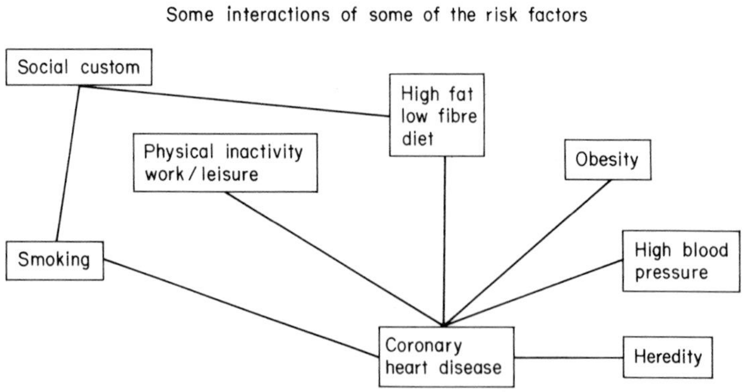

Some interactions of some of the risk factors

FIG. 29   Factors thought to contribute to coronary heart disease.

tors had less disease than drivers, the drivers were obese as well as less active. The most dramatic demonstration has been seen in the study of San Francisco's longshoremen who were followed for 22 years (or to death or to age 75 years) (Paffenbarger and Hale, 1975). Those men who were classified as being in jobs requiring repeated bursts of high energy output had a death rate from CHD about half those found in men performing work which required a medium or low category of energy output. For sudden death, the rate for the men performing the heaviest work was one-third of the rates for the men performing medium or light work. There was no difference between the groups in medium and low categories of work. The state of the coronary arteries of these men at death is not described. It should be realized that the level of energy output of the most active category of longshoremen was very high compared to that of the general population.

Those in sedentary jobs may increase their energy output by having active leisure pursuits. A recent study of middle-aged sedentary male civil servants has shown that those reporting vigorous exercise had an incidence of CHD which was one-third of the incidence in matched inactive controls. In this study vigorous exercise was taken to include contrived exercise and sport, as well as heavy work in the garden, house or garage (Morris *et al.*, 1973; Chave *et al.*, 1978).

A possible mechanism for the role of exercise is emerging which concerns the handling of plasma cholesterol. High levels of cholesterol in the blood are associated with coronary heart disease. Some of this cholesterol is carried by high density lipoproteins (HDLs) and some by low density lipoproteins (LDLs). The LDL tends to lose its cholesterol; some is taken up by cells lining the walls of blood vessels, where it may play a part in the formation of fatty atheromatous plaques which eventually block the vessel or give rise to a thrombus. The HDL molecule has a high affinity for circulating cholesterol which is released in the liver to be eliminated in bile. It has been found that groups with a high HDL:LDL ratio have a reduced incidence of coronary heart disease. Moreover, groups who are active also have a high HDL:LDL ratio, and better still, increased levels of activity have been found to increase that ratio (Lewis *et al.*, 1976). It remains to be seen whether this affects the subsequent course of latent coronary heart disease.

The Investigation of Man's Response to Work

Introduction

The measurement of man's ability to perform muscular work at various degrees of intensity and under a variety of conditions is important in many areas of research as well as in the clinical investigation of disease. Exercise can involve the functional reserves of the lungs and circulation to their full extent so that exercise is, and hence can be, the basis for an excellent test of the functional capability of an individual. The formal exercise test is now used extensively as an investigative tool in many different disciplines; among these are physiology, physical education, sports science, ergonomics, cardiorespiratory medicine, employment medical services and epidemiology (Table 10).

The choice of test

Because the response of muscles to training is so specific, the testing of exercise capacity must be equally specific. Muscle strength and flexibility are specific to muscles and capacity for rhythmic exercise is also specific to a large extent to the muscle groups used. The results of different tests will not necessarily be well correlated. If capacity for exercise is related to a particular job it might be measured in kg or kg m min^{-1}, but if it is a question of an individual's potential then a measure of maximum oxygen uptake per kg of lean body mass would be a more appropriate measure of ability for moderate exercise of moderate duration. Such a test needs to involve dynamic exercise with rhythmic contractions of major muscle groups, usually flexors and extensor muscles contracting alternately. The muscle bulk involved needs to be large so that general cardiorespiratory adaptations are required and the response is not dependent on the local

TABLE 10   Uses for exercise tests

---

1. Estimating fitness for work, sport and other activities

2. Prediction of disease
   e.g. heart disease

3. Cardiological diagnostic investigation
   e.g. angina
        lack of physical exercise
        peripheral vascular disease

4. Assessing prognosis
   e.g. heart disease

5. In monitoring drug therapy
   e.g. antihypertensive treatment

6. Therapeutic
   e.g. to restore confidence in ability to work after myocardial infarction

7. In rehabilitation
   e.g. physiological appraisal of training procedures; effects of medication
                                                        surgery
                                                        physical training

8. In investigation of respiratory disease
   e.g. assessing exercise capacity
        investigating dyspnoea or causes of hypoxia
        diagnosing exercise induced primary alveolar
        hypoventilation

9. Assessment of muscular disease

10. Calibration
    e.g. indirect measures of levels of habitual activity

11. Epidemiological investigations of cardiorespiratory fitness
    e.g. relationship to coronary heart disease

---

response of one or two small muscles. In such a test the maximum oxygen uptake can be determined or the changes in cardiac and respiratory function needed to exchange and transport a known volume of oxygen can be measured. The test is usually performed at several intensities of work spaced over the range of a subject's capacity. It is easier to control the protocol of a test if work load is known as well as oxygen uptake.

Submaximal tests may involve measuring the cardiovascular response in terms of the increase in heart rate to a target work load or conversely measuring the work load required to raise the heart rate to a previously designated level (see Fig. 30). More commonly the relation between heart rate and oxygen uptake is established over a range of work loads (Cotes, 1971). This procedure has a number of advantages over a single load test. Oxygen uptake and heart rate are linearly related during exercise except possibly at low work loads in subjects unaccustomed to testing. The linearity of the result obtained is some measure of the competence of the observers and the quality of the equipment. The V_{O_2} maximum can be predicted from this line using a knowledge of the maximum heart rate for the subject's age. This is not entirely satisfactory because maximum heart rates are variable and also drop with training. Since the relationship between heart rate and oxygen consumption may change in a parallel fashion (see Fig. 31), standardization procedures may be conveniently applied. Determination of the heart rate required to distribute a standard volume of oxygen to the body tissues, either 1 litre or 1.5 litres, has become usual practice. In the case of clinical testing such standardization procedures allow comparison with tables of normal values for age and sex. Low values for standardized heart rate indicate good physical condition and well trained muscles. The disadvantage is that there may be changes in the slope of this relationship as well.

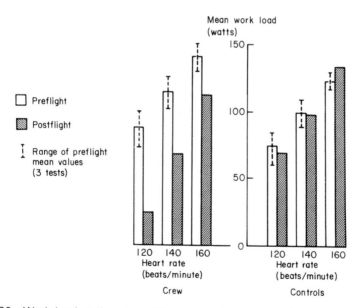

FIG. 30 Work load at three target heart rates in a group of astronauts before and after space flight and in a group of controls. (From C.A. Berry, 1969.)

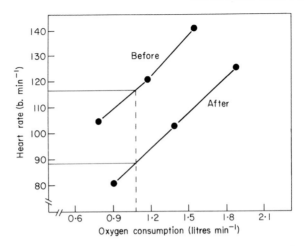

FIG. 31 Heart rate and oxygen consumption measured during work on a bicycle ergometer at three different loads in one subject before and after training showing the standardization procedure and the reduction after training in the interpolated heart rates at 1.1 litre min^{-1} oxygen consumption.

The source of work may be stepping, walking on an inclined treadmill or cycling on an ergometer. The availability of reliable portable monitoring equipment has also made it possible to base tests upon walking and running in field situations (Bassey *et al.*, 1976).

A useful tabulation of the relative merits of exercise tests appears in the WHO Technical reports (see Table 11). Submaximal target tests differ in their aerobic cost (Fig. 32). They also vary a little in the mechanical efficiency with which the subject performs the work and in the value for V_{O_2} (maximum working capacity) (see Fig. 33). The bicycle ergometer elicits the lowest V_{O_2} maximum, probably because the work is concentrated on thigh muscles.

Knowledge of the subject's height, weight, skinfolds and customary level of physical activity, and the temperature, humidity and altitude will usually be required. Monitoring of a subject's progress through the test is helped by the use of a Borg scale (see p. 68).

Steady state

The measurements may be made during a "steady state" when heart rate and oxygen uptake have reached a plateau appropriate to the work. If a continuous series of increasing loads is used, five minutes will be required at each load in order to achieve an almost steady state during the last few

TABLE 11   Relative merit of exercise tests[a]

Criterion	Steps	Type of test Upright bicycle	Supine bicycle	Treadmill
A. Ease of performance				
Familiarity with task required?	+ + +	+ +	−	+ + +[b]
Ease of obtaining high V_{O_2}.	+ +	+ +	±	+ + +
Subject's performance to (V_{O_2}) max	+	+ +	+	+ + +
Ease of instrument calibration	[c]	+ +[d], − −[e]	+ +[d], − −[e]	+ or ±[f]
Ease of measuring applied power	+ +[g]	+ + +	+ + +[g]	[h]
Ease of recording or obtaining the following during maximum test:				
ECG	±	+ +	+ +	±
blood pressure	− −	+ +	+ + +	−
blood samples	− − −	+ +	+ + +	±
respiratory volume and oxygen	±	+ +	+ +	+
Need for providing for emergency care[h]	+	−	+ + +	− − −
Ease of breathing	+ + +	+ +	+	+ + +
Ease of obtaining a nearly continuous increase of effort[h]	±	+ +[d], + + +[e]	+ +[d], + + +[e]	+ + or ±[f]
B. Freedom from undesirable features				
Hazards	+ + + or ±[j]	+	+ + +	− −
Need for skill	+	+	−	+ +[j]
Occurrence of local muscle fatigue at high exercise levels	+	−	− −	+ +
Need for trained personnel	+ +	+ +	+ +	±
Cost of equipment	+ + +	+ +[d], − −[e]	+[d], − −[e]	− − −
Ease of maintenance (including need for constant calibration)	+ + +	+ +[d], ±[e]	+ +[d], ±[e]	±
Freedom from noise	+ + +	±	±	− −
Bulk of equipment[h]	+ + +	+	−	− − −

(Table continued overleaf)

Table 11 Continued

Criterion	Steps	Type of test Upright bicycle	Supine bicycle	Treadmill
B. Freedom from undesirable features				
Ease of transporting equipment[h]	+ + +	+ +[d], ±[e]	±[d], − −[e]	− − −
Need for electricity[h]	c	+ +[d], − −[e]	+ +[d], − −[e]	− − −
Need for neuromuscular− skeletal coordination	−	−	+	− −
Ease of rate control[h]	− −	−[d], + +[e]	−[d], + +[e]	+ + +

[a] This table evaluates each of the four types of test according to the criteria listed in the first column. A grading of + + + indicates easiest, greatest freedom from undesirable features, most advantageous, etc.; a grading of − − − indicates most difficult, least freedom from undesirable features, least advantageous, etc. The intermediate point is represented by a grading of ±. Throughout the table, therefore, the greater the number of plus signs (or the fewer the number of minus signs), the fewer the problems presented by the test concerned.

[b] More difficult when the rate and slope are high.

[c] Unnecessary.

[d] Friction type.

[e] Electric type.

[f] Calibration easy for angle, less easy for rate.

[g] Less easy at maximum power.

[h] Can be estimated only.

[i] Less important factor.

[j] Less at low stepping rate, greater at high rate.

From Shephard *et al.* (1968).

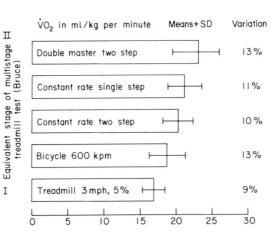

FIG. 32 Aerobic costs of five commonly used submaximal target tests expressed as rate of oxygen uptake per kg body weight. (From Bruce, 1974.)

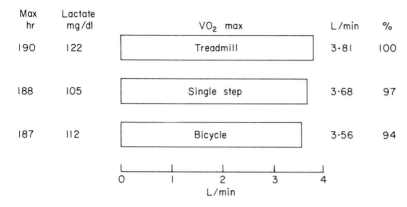

FIG. 33   Mean maximal values for oxygen uptake, heart rate min^{-1} and lactate level in a group of normal men of mean age 26 years. Three different discontinuous methods are compared. (From Shephard *et al.*, 1968.)

minutes. However, it has been satisfactorily demonstrated that comparable results can be obtained from a progressive test with a nearly continuous increase in load (see Fig. 34).

Safety

Safety affects the choice of test. There is a greater risk of falling with a step test and with a treadmill, particularly at high speed, and of cardiac failure with recumbent cycling. When the subjects are old, the choice of test is more difficult. Ideally, the test should not create anxiety; new skills should not have to be learned and the physical activity should be a familiar one, so many investigators favour walking.

Stepping

Shephard based his choice on the comparison (see Table 4) of stepping, treadmill exercise and the bicycle ergometer. He concluded that stepping was the best test for field studies. It is cheap, portable, the subjects show relatively little anxiety or habituation. Good quality electrocardiograms can be obtained, but because the subject is moving the arterial blood pressure is difficult to measure by sphygmomanometry. There are a large number of possible designs for a step test. The range of work load can be varied by changing stepping frequency, 50–150 steps min^{-1} are possible, or by changing the height of each tread. A further departure from the clinical laboratory scene is a step test involving the use of the bottom stairs in the

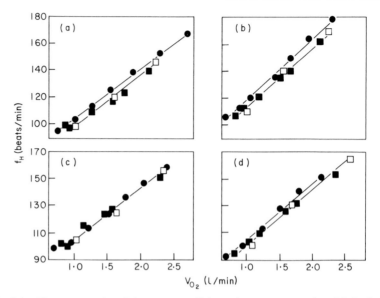

FIG. 34   Measurements of heart rate (f_H) and oxygen uptake (V_{O_2}) during bicycle exercise. "Steady state" values from a discontinuous test (square symbols); values from continuous test (solid circles), for four subjects (a) to (d). (From Bennett and Morgan, 1975.)

home. This is the basis of a home-based fitness test devised for the Canadian Fitness Programme (Cumming and Glenn, 1977). The "Fit-Kit" contains a long-playing record to set the stepping frequency. The subject measures his own pulse rate immediately the test finishes. The possibility of serious errors in the assessment of apparent fitness by such a method has been the cause of some controversy.

The three-minute two-step test ("double Master's") has been widely used for testing the electrocardiographic response to exercise as part of clinical diagnostic procedures. But even in clinical practice the advantages of a multilevel test are apparent, so that any electrocardiographic abnormalities may be related to the patient's own scale of exercise capacity.

The Harvard Pack test is a variation of the Master's test but involves the subject carrying a back-pack containing 22 kg of weight. This test is used as a screening procedure when selecting men for very heavy work, for instance, miners who volunteer to serve in mine rescue teams. It is suitable because it mimics the job.

Bicycle ergometer.

This is the most suitable apparatus if close control of the work rate is required and is the most convenient for a hospital laboratory. The body weight is supported by the saddle and does not determine the work rate in contrast to stepping and walking on a treadmill. Cotes *et al.*, (1973) have analysed the factors which influence this test. The choice of tool is between the more expensive electrically braked ergometer and the cheaper model in which the fly wheel is braked by a friction-belt. With the former the work rate selected remains constant over a wide-range of pedalling frequencies (45–75 rpm). Using the friction-braked ergometer a continuous count of pedal revolutions is needed. The only electrically braked machine which can be calibrated easily is the Ergometre Electronique Ray Thomas (details from Messrs P.K. Morgan, Chatham, Kent, UK). Calibration of the friction-braked machine is straightforward.

Most measurements have been made at a pedalling rate of 50 min^{-1}; rates of pedalling of 60 min^{-1} are recommended for measurements of maximum oxygen uptake. This is within the optimum range; rates of 40 or 70 rpm are less efficient.

Treadmill

The speed and slope of the treadmill can be varied to change the work load. Treadmills which carry a tall gantry around the platform require a room with a high ceiling if full elevation of 40% is to be possible. The length of the running surface needs to be at least two metres if athletes are to run freely enough to be tested to the limit of their ability. The wider the belt the sooner the subject will feel at home on the treadmill. Treadmill walking and running is subjectively very different from the free situation to begin with and several trials are often necessary to accustom the subject.

Treadmills have improved in design and can now be used by frail subjects such as cardiac patients. With a suitable motor they can start with the subject already standing on the track and the test can be performed with the subject wearing a safety harness suspended directly from a "stop" switch. They can be programmed or controlled by the subject to provide a nearly continuous increase in work load or a continuous series of loads with an almost steady state at each level.

Tests in the field

Body-borne tape recorders for continuous unobtrusive recording of heart rate, blood pressure and footfall have made field tests and the monitoring of real-life situations much more practical. More recently still a new portable device (Oxylog) for measuring oxygen uptake and ventilation has become available (see Chapter 12) in addition to the older heavier IMP and K–M respirometers.

On a course out of doors the oxygen required for walking or running at a given rate will be affected by the surface of the course and by the speed and direction of the wind. Pugh (1971) found that a wind speed of 22.5 km h^{-1} increased oxygen consumption by 0.4 litre min^{-1} in runners. Walking is not significantly affected below speeds of 1.5 km h^{-1} (head wind + walking velocity). Body weight affects the oxygen uptake in walking and running both on a treadmill and in the free situation. In a test in which only heart rate and speed can be measured, there is a linear relation between heart rate and the square of the walking speed (up to about 7.5 km h^{-1} which is about the speed at which running becomes more efficient). The heart rate will be influenced by the amount of carried fat but not by body size because heart size increases with total lean body mass.

Results of uphill running and walking outdoors are difficult to interpret. In uphill running, the oxygen requirement seems to be greater than that of the same slope and distance on the treadmill, presumably because during free running the body is lifted through a greater distance.

Preparation for a test

The test should be carried out at a fixed time of day to take account of circadian changes in response. Exhausting physical exercise should be avoided during the preceding day and any exercise comparable with the test during the preceding hour. Stimulants, tobacco or medicaments should not be taken on the day of the test, which should not be performed until at least three hours after a large or fatty meal. There should be a prior medical examination of the respiratory, cardiovascular and locomotor systems including an enquiry regarding symptoms of disease which would alert the investigator to any special risks or difficulties, any absolute contraindications and any condition which may influence the interpretation of the result. A resting 12-lead electrocardiogram should be examined by someone competent to interpret any changes.

Contraindications and precautions

An exercise test should not be performed if the subject has

(i)   symptoms or signs of cardiac failure or myocarditis;

(ii) unexpected symptoms, signs or ECG changes of recent myocardial infarction or ischaemia;

(iii) an acute infection, a metabolic disturbance of uncertain magnitude or a possible pulmonary embolus.

Special precautions are needed in heart disease which involves dysrhythmia or valvular obstruction. A physician needs to be nearby, irrespective of the health of the subject, if high work loads are involved.

There is a significant risk, albeit very small. Rochmiss and Blackburn (1971) surveyed the experience gained in 170 000 exercise tests and estimated morbidity at 2.4 per 10 000 and mortality at 1 per 10 000. Ventricular fibrillation has occurred during the testing of perfectly fit young men. For this reason the personnel conducting the test must agree the symptoms, signs and ECG changes which will be indications for stopping it and must be able to recognize these changes should they occur. Emergency equipment should be available for cardiorespiratory resuscitation; this equipment should include drugs and a defibrillator. The continuous display of the electrocardiogram is desirable and obligatory if patients with heart disease and a risk of dysrhythmia are to be tested.

Observation of the subject must not be relaxed for several minutes following completion of the test. The risk of a serious dysrhythmia seems greatest about three minutes after heavy work has finished. It is possibly of significance that blood catecholamine levels are usually higher at this time than during the exercise itself. The risk is increased rather than decreased by lying down and yet hypotension may occur if the subject remains motionless in the upright posture. Thus it is desirable for the exercise to be tapered off gradually.

The indications for stopping the test include hypotension during the exertion, ventricular dysrhythmias and ECG changes of progressive myocardial ischaemia (see Table 12), all serious signs.

The measurement of cardiorespiratory changes

Detailed study of cardiovascular function requires the measurement of cardiac output, heart rate and systemic arterial pressure together with observation of the ECG (see Chapter 11, by D. Hill). Non-invasive methods are used for research which involves the testing of normal subjects. The use of the indirect Fick method with estimation of the mixed venous p_{CO_2} by a rebreathing method has become attractive since rapid gas analysers have become more generally available. Heart rate is often the only measurement made of cardiac performance; this can be defended by reference to the linear relationship between heart rate and cardiac output (see p. 33). Other

TABLE 12   Possible indications for stopping an exercise test

Symptoms:	chest pain, with or without ECG changes of ischaemia
	fainting
	severe breathlessness
	severe fatigue
	intermittent claudication
Clinical signs:	hypotension
	distress: pallor, cyanosis, deteriorating performance, e.g. tripping or
	staggering
	confusion
	fatigue
	hypertension: systolic BP $\pm$ 250 mm Hg
	failure of BP to rise normally
ECG changes:	ventricular tachycardia or fibrillation
	more than 6 ventricular ectopic beats or a run of 3 or more
	ventricular ectopic beats of the R or T pattern
	a conduction defect other than slight atrioventricular block
	R–ST depression of > 0.2 mV (2 mm) (horizontal or descending
	ST segment)

measurements which may be required include those of pulmonary arterial pressure, systolic ejection time and organ and tissue blood flow.

Heart rate should be counted directly from the ECG or pulsimeter trace for an adequate length of time. Those ratemeters which provide a mean heart rate by electronic methods are useful for monitoring but not for accurate measurement.

Systolic blood pressure can be measured by sphygmomanometry during exercise. Accurate measurements of diastolic pressure are only possible using intra-arterial catheters.

The recording of a noise-free electrocardiogram during exercise requires care in the placement and attachment of the chest electrodes. Measurement of changes in the displacement and slope of the ST segment is particularly dependent upon a "clean" recording. One electrode is usually placed on the manubrium sterni or forehead, the other in the V_4 position clear of the pectoral muscles. On the other hand, the positions which display ST depression most reliably are V_4 together with a right infrascapular electrode position. There have been recent improvements in electrode design but it is still important to obtain a low skin impedance when ECG recordings are to be made during exercise. Some workers find it useful to measure the impedance with a special meter designed to draw very low current and to reapply the electrodes with additional abrading of the skin until the resistance is less than 5 k ohms. Leads also need to be prevented from moving excessively.

The respiratory measurements required will usually include minute ventilation (1 min^{-1}), respiratory frequency and the O_2 and CO_2 content of expired air.

The respiratory exchange ratio, oxygen extraction and tidal volume can be calculated from these measurements. For specific applications, additional measurements may be required such as diffusing capacity, heart volumes, blood lactate levels, arterial O_2 and CO_2 tensions and pH and aspects of muscle function such as the integrated electromyogram. Muscle biopsy is becoming a more widespread technique which yields information about the histology and biochemistry of the muscle.

It is necessary to know at least the height and weight of the subject so that results may be related to the body weight or build; fat-free mass may also be required and this may be deduced from the relation between height and weight, measurements of skinfold thickness, underwater weighing or volumetric measurement of soft-tissue radiography of the thigh.

Exercise testing in clinical medicine

The large functional reserves of a normal individual's lungs and circulation means there may be a considerable loss of function before the reduction in reserves are apparent. Exercise testing will usually demonstrate the loss of these reserves. Equally exercise will often exaggerate any minor defect of function only just evident at rest, making a functional diagnosis so much easier.

Lung disease

A patient's own assessment of his work capacity is poor even when special steps are taken to categorize symptoms and to grade them by matching them to carefully prepared descriptions. The patient makes his judgement only on the basis of recent experience. Exercise testing allows an objective measurement of exercise tolerance and breathlessness.

It is possible to measure the change in the alveolar–arterial gradient for oxygen if a steady state test is used and in this way some of the various causes of arterial hypoxaemia at rest may be distinguished. Thus exercise has a role in diagnosis of restrictive lung disease and the various syndromes associated with primary alveolar hypoventilation. Exercise-induced bronchospasm is a common disability which may not necessarily be associated with frank asthma. Variable breathlessness occurring during exertion and persisting afterwards may be diagnosed by the measurement of maximum force expiratory air flow (forced expiratory volume in the first second, FEV_1 or peak flow rate) after running for 6–10 min.

Cardiovascular disease

An exercise test allows an objective assessment of the severity of exercise induced chest pain and of myocardial ischaemia with or without anginal pain. It is valuable to relate the ECG change which occurs to the level of exercise. A test may establish the cause of chest pain and it may indicate that a lack of customary exercise is the cause of effort intolerance.

Exercise testing has a place in investigative cardiology. In the investigation of disorders of the pulmonary circulation some care is needed that recumbent exercise does not precipitate excessive pulmonary hypertension and right heart failure.

In Seattle the fight against heart disease has become the concern of the whole community and a prospective community study "Seattle Heart Watch" is in progress. Exercise testing is being used (Bruce, 1974) to examine the hypothesis that persons at risk for future ischaemic heart disease or sudden death can be identified by an impairment of their functional aerobic power.

Treadmill testing provides an objective assessment of the severity of obliterative arterial disease of the legs by defining the exercise conditions leading to claudication. In proximal arterial disease, using a sphygmomanometer and doppler flow probe, measurements can be made of the local perfusion pressure in the post-tibial, dorsalis pedis or politeal arteries at rest and after exercise (see Fig. 35).

Exercise may also be useful in the diagnosis of deep vein thrombosis. Thermography of the calf after exercise shows, in the presence of deep vein thrombosis, a patchy rather than uniform distribution of skin radiant temperature.

Measuring the Physiological Response to Physical Activity away from the Laboratory

Thus exercise tests in the laboratory can provide a detailed physiological description of a man's response to physical work. It has become possible to make a similar range of measurements away from the laboratory. This arises from the recent development of reliable, light-weight, battery-driven measuring and recording equipment. The developments have gained im-

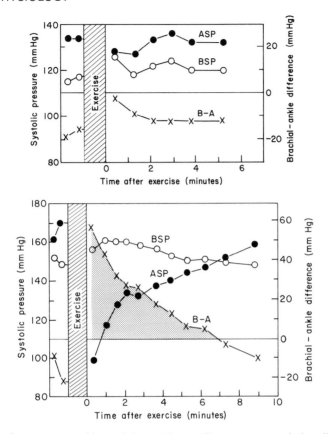

FIG. 35 Changes in ankle and brachial systolic pressure and the difference between these pressures (B − A) following leg exercise (a) in a healthy subject, (b) in a patient with questionable arterial disease. The scale on the left applies to ankle and brachial systolic pressure (ASP and BSP) and the one on the right to brachial–ankle differences (B − A). The pressures were measured using a sphygmomanometer cuff around the calf and upper arm and flow was detected with a doppler bloodflow velocity meter. (From Carter, 1972. Reprinted, by permission, from *New Engl. J. Med.* **287**, 578–582.

petus from the renewed interest in methods of non-invasive measurement in human physiology and clinical measurement, from technical advances in electronics and from anticipation of the rewards which could arise from long-term observation of cardiorespiratory function in everyday life. It has become possible to re-examine questions about the physical activity involved in various jobs and occupations, about the total daily energy expenditure and about man's response to his work. Growing interest in the possible beneficial relationship between daily physical activity and health (see p. 70) requires that these measurements can be made repeatedly over periods of considerable duration (Andersen *et al.*, 1978).

Of the various portable recording devices the Medilog recorder (Goulding, 1976) is particularly convenient and versatile. It can be used to record four different digital or analogue signals continuously for 24 hours. It is suitable for the study of subjects engaged in strenuous work, the elderly, the sick and the disabled. The equipment is unobtrusive, the record is objective and discriminatory. The cost of collecting data is low when compared with the cost of direct observation. The method is particularly well suited to investigations in which the duration and intensity of bouts of physical activity must be defined. The large amounts of data collected can be overwhelming without the use of computer techniques during analysis.

Method of measurement

The value of subjective assessment of physical activity or direct observation has not been entirely displaced by objective methods. Direct observation can be used when the nature of the work to be measured involves many different movements and postures and when the cost of instrumentation is unjustified. Time-lapse cinematography allows the same method to be used off-line. If the subject is restricted in his movements, for example if he is confined to a room, then some help can be obtained from movement sensors, radar surveillance of the rooms or the use of pressure sensitive chairs and floors.

Subjective assessment of physical activity over long or short periods is unreliable. New technology has at least helped by allowing the validation of various procedures. Of the various methods, the interview, questionnaire, recall diaries, contemporary diaries, self-rating scales, and activity scores, the interview is considered to be the best.

Table 13 lists various means of obtaining an objective assessment of physical activity over periods of several hours or days duration. The devices vary according to whether they provide continuous or cumulative recordings of activity. Those which record physical movement will of course only provide information about the use of the limb or limbs to which they are attached.

Several of these methods involve counting heart beats and measuring mean or instantaneous heart rate. This is criticised because exercise is not the only cause of an elevated heart rate. Arousal, mental stress and other environmental stresses can all evoke a tachycardia. In practice the significance of this reservation needs to be reviewed in the light of the objective conditions of a particular study. For example it is of dubious value to

TABLE 13   Methods for the long-term measurement of physical activity

Mechanical pedometers
Tape recording of footfall, accelerations
Actimeters, posture switches, electromagnetic devices
Tape recording of heart rate, cumulative pulse rate counting (SAMI), telemetry of
    heart rate
Tape recording of integrated EMG
Long-term energy expenditure by measurement of oxygen consumption

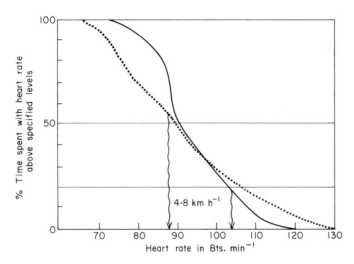

FIG. 36   Two cumulative frequency distribution curves of heart rate against time derived from eight-hour shift records from two men showing for each man the heart rate required for walking at 4.8 km h^{-1} (arrows). (From Bassey *et al.*, 1978.)

record heart rate as an indirect measurement of energy expenditure if the intention is to decide whether an obese subject is in negative or positive energy balance (see Fig. 36). The value of the data is more readily interpreted if the subject is "calibrated" by taking part in a formal exercise test in association with the field measurements (see Fig. 3, p. 82). A self-paced walking test (Bassey *et al.*, 1976) can be used to find the heart rate associated with a standard walking speed of 3 mph (4.8 km h^{-1}) for instance.

Portable methods of measuring oxygen consumption have long been available. Although the latest addition to this range of equipment (the Oxylog) still, as ever, requires the use of a face mask, a continuous record of V_{O_2} can be recorded using a Medilog tape-recorder (see Fig. 37). There is

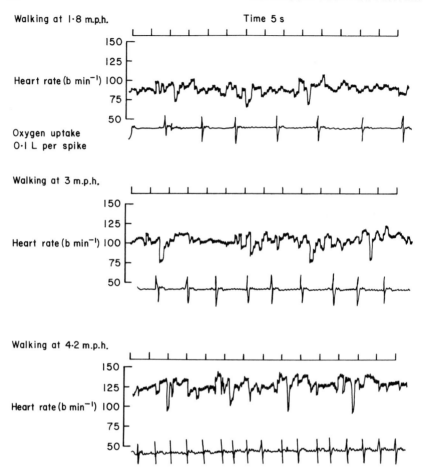

FIG. 37 Continuous recording of oxygen consumption (in 0.1 litre units) and heart rate made in the field with a portable "Oxylog" and Oxford Instruments tape recorder. The heart rate is obtained via a ratemeter from the recorded ECG.

an ever increasing range of situations in which these measurements become informative. They range from studies in which an attempt was made to test the hypothesis that physiological monitoring could be used to equalize the physical strain on workers on a production line, to studies designed to compare the physiological strain for diplegic children with spina bifida of using wheelchairs and prosthetic walking devices (Zakai *et al.*, 1977).

Applications

Knowledge of the physiology of exercise finds application in many fields including ergonomics, the training of athletes, rehabilitation after injury and the provision of assistance for the disabled.

Ergonomics

Ergonomics is "a science devoted to the search for and creation of the best possible conditions of employment so as to eliminate the harm, the boredom, the fatigue of labour" (International Labour Office, 1970).

Ergonomics seeks to achieve optimum relationships between man and his work. It is the natural extension of man's use of his intelligence, from prehistoric times, in the shaping of tools to suit his hands and the task to be performed. It is not a single discipline and seeks to integrate contributions from biological sciences, psychology and engineering.

The principal contribution made by physiologists in this field is in the identification of the conditions under which the body will function best in order to complete the task required. There are optimal ranges of temperature, noise and lighting which are associated with optimal work output and minimal accident rates. The worker's performance and his perception of the effort required are influenced by circadian rhythms of metabolic rate, deep body temperature and exercise responses. There are optimal patterns of work which allow the highest output of energy over long periods for the minimum of fatigue.

Noise, air pollution

The external environment can influence work performances in a number of other ways. Noise of pleasant nature and of moderate intensity may enhance work performance. On the other hand, high noise intensity may exaggerate fatigue and may cause hearing loss and ill-health. Vibrating tools, for example pneumatic drills, produce specific local vascular damage. Air pollution can cause changes in pulmonary function and gas exchange and represents a serious potential health hazard. The special difficulties encountered under water will be described elsewhere in the book.

Matching the man to the job

There are still a few jobs in heavy industries which depend for their performance upon muscular effort and a high energy expenditure. Since physical working capacity depends upon body size as well as training, the worker must be chosen because his genetic endowment of physical strength and skills suit him for a particular job. Therefore, physically demanding jobs, which cannot be scaled down in intensity or timing, like the handling of containers of molten metal in the foundries, may always have to be performed by large individuals with powerful muscles and high maximum working capacities.

On the other hand, many individual tasks and, in theory, jobs on a production line, can be adjusted to take account of the worker's capacity. O'Brien et al. (1979) have investigated the possibility that knowledge of physiological responses to exercise and optimal patterns of energy expenditure can be exploited to achieve an equitable and appropriate distribution of the physical exertion encountered on a production line.

Detailed consideration of biomechanics and physiology allows adjustments to be made in the pattern or mode of completion of the task. When the job consists mainly of lifting, loading and carrying, isometric muscle contractions are involved. The force or tension required and the duration for which it must be sustained will determine the pattern of muscle contraction and the general cardiovascular response. Isometric contraction of muscles can be sustained for only short periods at tensions which are 15% of that muscle's maximum tension or greater. Thus, if sustained contractions are required, for example in load carriage, then the load must be distributed among as many muscle groups as possible. Ambulance personnel can carry loaded stretchers for longer and with less effort (a lower heart rate and arterial blood pressure) if the load is carried on a shoulder harness rather than the hands. The small muscles of the hand and finger flexors will fatigue relatively quickly. Similar weights can be carried for longer in a rucksack then a shopping basket. Draymen can carry their beer barrels, often weighing 70 kg, up narrow staircases by making effective use of the large muscles of the trunk and legs. For lifting heavy weights muscles of the abdomen can be used to support the muscles of the back; if used in concert they allow the creation of a rigid cylinder. Use of the back muscles alone produces a rigid column which is much less strong. If the legs are splayed they provide a stable base and avoid the need for strenuous postural adjustments against sway under load.

The posture in which a man must work has a large influence on the energy expenditure of the job and the stress involved. Hammering nails into the ceiling is much harder than hammering nails into the floor. The limbs have to be used at a greater mechanical disadvantage and the blood supply to the arm muscles is less good with the arms elevated despite some reflex increase in arterial blood pressure. Bench work is easiest if the bench height is optimal and requires little or no bending. The extra energy expended in maintaining awkward postures is high. Quite apart from fatigue and poor work performance it can lead to pain arising from muscles, ligaments and joints and perhaps eventually to permanent articular damage.

Other jobs involve rhythmic repetitive movements similar to the activities of pedalling and walking. Then the amount of work which is possible depends upon the intensity of the work in terms of power output, the frequency of the repetitive movements and the pattern of the work in terms of the ratio of the durations of rest to that of periods of work. Earlier in the chapter consideration has been given to the factors which determine the onset of fatigue during repetitive movement (p. 66) and optimum repetition frequencies for movement (p. 70).

Athletic training

Physiologists are as interested as the athletes themselves in the limits of human physical performance, in its achievement and in the factors which are concerned. World records continue to be broken in track and field events (see Fig. 38). The limits of performance do not appear to have been reached. In the past athletes have been trained by the experience and intuition of coaches and other athletes.

In an earlier section (p. 52) the physiological changes resulting from training, the improvements in muscular and cardiovascular function with regular repetitive muscular work, have been described and shown to be dramatic. Relatively unfit individuals can expect considerable improvement at the beginning of training. Much more effort is needed to achieve the extra performances required for national or international athletic competititon (see Fig. 39).

Athletes are to a large extent born not made. Unless they start with a propitious genetic endowment, no amount of training will bring them into the top rank. Certain body dimensions equip an athlete for effective performance in some athletic events and not others. The sprinter normally has long powerful legs; the marathon runner is often small but with a proportionately large heart; the weight-lifter and shot-putter are large and their muscles are bulky. The ratio of Type I to Type II muscle fibres also seem to be genetically determined (p. 29). In Eastern Europe attempts have been made to select athletes by assessing, using muscle biopsy, those with optimal histochemistry of their muscles. A high proportion of Type I fibres is an advantage to the long distance runner who must depend on a high aerobic capacity. Sprinting and jumping, which require a high power output for a short time, are best served by muscles containing a high proportion of the fast Type II fibres whose metabolism is primarily anaerobic.

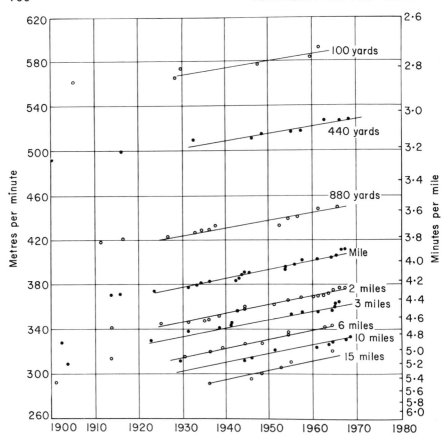

FIG. 38 World speed records for running for the last 50 years. Each circle represents a new record at each distance. (From Ryder *et al.*, "Future performance in foot racing", © 1976 by *Scientific American*, Inc. All rights reserved.)

The specificity of the training process to the muscles used and to the kind of physical activity for which the muscles have to be trained has been mentioned earlier (p. 52). Åstrand's own observations (Table 14) reinforce this point.

If endurance capacity for middle-distance events is required, then the best way to train is to run at speeds a little faster than those of the event but for slightly shorter periods. This is the principle of overload. The taxing of a slightly greater proportion of the aerobic capacity during training increases the number of mitochondria and the concentration of oxidative enzymes in the muscle cells of both types, as discussed previously.

Endurance training is not useful to a sprinter. He may run his race

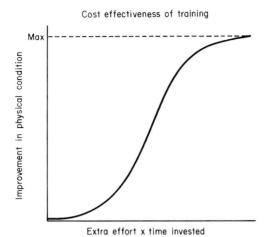

FIG. 39   Cost effectiveness of training. Theoretical relation between training activity and improvement in physical condition.

TABLE 14   Identical twins: maximum working capacity (max V_{O_2} litre min^{-1})

Twin	(1)[a]	(2)	% Difference
Running	3.61	3.56	1
Breast stroke	3.63	2.81	29
Crawl	3.36	2.71	24
Arm stroke	2.74	1.84	49
Leg kick	3.37	2.72	24
Arm cycling	2.16	1.96	10

[a] Olympic swimming competitor
Both twins were physical education students, they had been preparing for Olympic competition. Twin (2) stopped training for competitive swimming one year before these measurements were made but had continued general endurance training. (Astrand, 1978; lecture, University College, London).

without drawing breath since he must run on the rapidly released anaerobic sources of energy. He needs to improve the power output of his leg muscles by increasing the amount of contractile protein and the contraction of anaerobic enzymes in his muscle fibres. This is best done by intermittent running at high speeds, sometimes called interval training. Over short periods the energy expenditure will be higher than that achieved in a test of maximum oxygen uptake which lasts for several minutes and depends upon a steady supply of energy from aerobic as well as anaerobic metabolism.

The principle of training which constantly emerges is that of its specificity. If you want to be a good swimmer, then swim, using a consistent style and aiming at competitive events of a similar duration.

The nutritional state of the athlete can be important. The availability of glycogen stores in the muscle can affect endurance performance. It is now known that a low carbohydrate diet for a few days followed by a very high carbohydrate diet for the 48 hours preceding a race produces the largest possible glycogen stores.

Preparation for competition through attention to the energy stores of the athlete seems an acceptable application of physiological knowledge. The search for a marginal physiological advantage continues. Anabolic steroids have been used for increasing muscle bulk. Expansion of the blood volume by the reinfusion of stored blood has been used in the hope of increasing the athletes' maximal cardiac output. These procedures are regarded as unethical. Certainly they are a risk to health and represent practices, which if allowed in one individual, will tempt others to risk their own health so that they can remain in the competition.

The question arises whether there is any risk for athletes previously in normal health seeking to exercise to the limit of their physiological capacity, particularly at altitude. Certainly they often possess the motivation and willpower to drive themselves beyond the point when respiratory distress and muscular pain will usually stop less motivated individuals. It is unclear whether the discomfort is an important protective mechanism in fit athletes. Those athletes who have died during competition have either had diagnosed diseases, taken sympathomimetic stimulants or allowed themselves to become seriously depleted of body fluids. However, high levels of circulating endogenous catecholamines, high heart rates, and the possibility of sudden changes in cardiac filling may increase the risk of abnormal cardiac rhythms.

Assisting the disabled

There is now a greater willingness to assist the disabled and, because of technological advances, a greater possibility of doing so. The physiologist and biomedical engineer have roles to play in advancing our understanding of locomotion and manipulative movements and of identifying those residual physiological functions and mechanisms which can be exploited to restore some degree of useful movement to injured or maldeveloped limbs or digits.

Those who have lost the use of limbs or muscles will learn to make use of their remaining abilities in new and sometimes very ingenious ways. Restoration depends upon simulating and replacing the function of lost limbs or transducing the energy provided by remaining muscles into new forms which can replace the loss of function.

A central problem is whether to try to replace the lost function with man-made equipment, or to encourage the patient to develop compensatory use of his own residual ability. Those who are born disabled can compensate best. For instance the child born without hands will make use of his feet instead. His hips will retain into adult life the extended range of flexion normally seen only in very young children. The muscle strength, manipulative skill and range of movement of the toes will approach, though never reach, that of the hands of normal individuals.

For those who have lost ability in later life replacement is often more acceptable because they are less capable of the necessary physical and psychological adaptation. The disadvantage of artificial devices is that they still fall short of replacing lost function. The variety of abilities afforded by the human hand, for instance, is as yet impossible to emulate; the best attempts have to rely on complicated systems, but these break down and cause great frustration.

Besides being less effective, an artificial limb is an additional physical burden because it has to be carried. On the other hand artificial legs do allow for some flexibility of movement and for movement in three dimensions. A wheelchair allows independent mobility, but movement against gravity is largely impossible. Stairclimbing on artificial legs is dependent upon the patient retaining the power of hip flexion. The chief problems are the load imposed on stumps and soft tissues by the body weight and the serious lack of postural feedback information necessary for balance. Pain and tissue damage at pressure points can be ameliorated by improved surgical technique at the time of amputation. The ends of the bones can be padded with connective tissue or fat. Part of the body weight can be redistributed to the rim of the stump socket around the upper part of the thigh. The problem of poor balance can be overcome to a remarkable degree by practice, especially in young and very determined patients; meagre postural information can become adequate.

The swivel-walker represents an imaginative attempt to transduce the power output of trunk muscles performing rotational movement into the forward movement of well-braced paralysed legs (see Fig. 40). Like the wheelchair the swivel-walker is only effective on flat surfaces and it has the added disadvantage of requiring a very high energy output for every modest speeds of movement. Its advantage is that it provides for pottering movements yet leaves the hands free. These movements in the home are necessary for cooking, personal hygiene and other household tasks. Crutches and

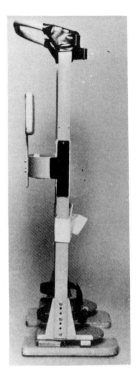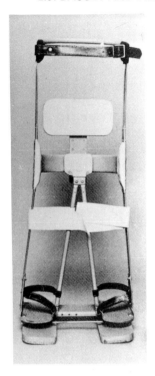

FIG. 40   The Salford swivel-walker. (Zakai *et al.*, 1977.)

calipers allow faster movement over a variety of surfaces, inclines and small steps for a lower energy expenditure but the arms are occupied with the crutches.

The use of the patient's own muscle power, where possible, rather than external sources such as the battery-powered wheelchair, is an advantage for two reasons. The patient has more control over and more feedback information about his situation. He is independent of battery failure and his well-developed muscles extend his abilities in many directions. Moreover, since he is expending energy he is less likely to become overweight. Obesity is a serious problem for the chair-bound which exacerbates the difficulties they already have in moving about. The transfer from wheelchair to car, to bath or to bed is made more difficult if the patient is heavy and the arms are weak; whereas those who make the fullest use of arm and trunk muscles find themselves participating in the Stoke-Mandeville wheelchair Olympics.

The weakness of certain muscle groups may involve a disability which is incapacitating to an extent which is out of step with the progress of the disease or injury. This occurs in patients with myopathies and partial loss of muscle power particularly of proximal limb muscles when they may be unable to rise from a chair. This movement requires powerful but brief contractions of hip and knee extensors. Though the muscle power may not be adequate to lift the body, the patient may still walk. Patients who receive artificial legs face this as an acute problem. Thus, loss of one movement can seriously limit physical activity. The problem has been solved by the use of chairs which provide for power assistance and push the occupant of the chair into the standing position. Loss of hands or of hand movements results in the loss of manipulative skills including feeding, and writing become impossible. These abilities can be restored to a limited extent by man-made replacements which range from the simplicity of the split-hook to the complexity of the myoelectric arm.

The split-hook was developed before the First World War and remains a highly effective tool (see Fig. 41). The halves of the hook are held together by rubber bands which provide a variable tension. It can be opened by a combination of arm extension and shoulder movement which pulls on a cord running from the centre back of the shoulder harness to the side arm

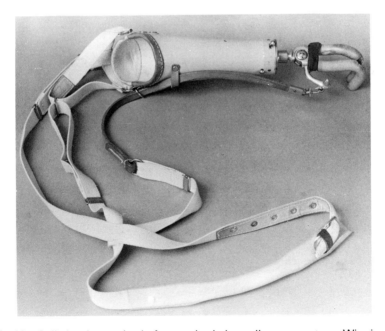

FIG. 41   Split-hook prosthesis for use by below-elbow amputees. Winging of the shoulder blades or extension of the elbow exerts tension on a cord which opens the two halves of the rubber-coated hook. Counter-tension is provided by rubber bands.

of the hook. There is therefore considerable power available, mechanically transduced through a simple system from the patient's own muscles over which he can exert a fine control.

After nearly 20 years the myoelectric arm is still in the development stage. It depends upon a battery-driven power, activated and controlled by the motor nerve signals in the patient's stump. The advantage of this, in theory at least, is that for an adult amputee there is no need to relearn patterns of movement. He only has to send the normal message from the motor cortex down the arm to the phantom hand and the artificial hand will make the appropriate movement in both direction and intensity. In practice, it is difficult to make a hand with an adequately sophisticated response which is also light in weight and robust enough to withstand daily wear and tear. It is necessary to select from the variety of movements of which the human hand is capable, those actions which are most vital for independent daily living. To achieve a hand which has a power grip for holding a gardening trowel, a precision grip (finger tip – thumb tip) for picking up papers or small objects, a suitable grip for a pen, a knife, fork and so on, is a formidable challenge.

Basic problems which remain for replacement of hand function arise in finding materials which are strong, durable and light, and in providing adequate power sources which are light and neat. The problem facing the amputee is whether to sacrifice the afferent information afforded by skin contact for a limited improvement in manipulative ability or aesthetic appearance.

Basic rules for the patient are that his own muscle power should be used where possible, not only to activate and control but also as the source of power, keeping the three functions unified as in normal individuals, and that restoration should be made as young as possible in the congenitally disabled so as to make the fullest use of adaptive capacity. Blind babies of about six months of age can benefit from "sonar spectacles" which give them information about their surroundings by bouncing an audible beam off solid surfaces within a few feet. The returning beam varies in pitch according to the distance of the reflecting surface. These babies begin to reach out for objects very much earlier then they do without the aid, and are therefore less delayed in learning to relate to the world.

Those who are so severely disabled that they are bedridden, such as victims of poliomyelitis, muscular dystrophy or high spinal transections can still use small residual movements to activate electronic systems such as POSSUM which will enable them to type and achieve some control over their immediate environment. Since electronic systems work on binary switching principles, the minimum input required from the patient is one unidirectional movement which need not be strong.

Rehabilitation

The general principles of rehabilitation are the same as those for training and have been described earlier in this chapter. There are a few extra considerations to meet the special needs of patients.

Prolonged bed-rest and disuse of muscles leaves the patient with wasted muscles, partially decalcified bones, stiff joints and a reduced plasma volume. The return to an upright posture restores plasma volume within hours; complete remineralization takes many weeks but the stimulus of gravity is adequate for this.

Mobility must be restored to joints first by passive movement. This is followed by graded exercise designed to restore muscle bulk and strength. The length of muscle will adjust by physiological mechanisms to the optimum for the movements with which it is involved.

The practice of rehabilitating an individual energetically with the needs of his occupation and the tasks of daily living in mind, has at last gained acceptance. Such a programme requires a full understanding of the physiology of exercise.

References

Allen, D.W. and Quigley, B.M. (1977). *Med. J. Aust.* **2**, 434–438.

Andersen, P. and Henriksson, J. (1977a). *Acta Physiol. Scand.* **99**, 123–125.

Andersen, P. and Henriksson, J. (1977b). *J. Physiol.* **270**, 677–690.

Andersen, K.L., Masironi, R., Rutenfranz and Seliger, V. (1978). "Habitual Physical Activity and Health". WHO Regional Publications European Series No. 6. WHO, Copenhagen.

Asmussen, E., Johansen, S.H., Jorgensen, M. and Nielsen, M. (1965). *Acta Physiol. Scand.* **63**, 343–350.

Åstrand, I., Guharay, A. and Wahren, J. (1968). *J. Appl. Physiol.* **25**, 528–532.

Åstrand, P.O. and Rodahl, K. (1977). "Textbook of Work Physiology", 2nd edn McGraw Hill, New York and London.

Bassey, E.J. (1978). *Gerontology* **24**, 66–77.

Bassey, E.J., Fentem, P.H., Macdonald, I.C. and Scriven, P.M. (1976). *Clin. Sci. Molec. Med.* **51**, 609–612.

Bassey, E.J., Fentem, P.H., Fitton, D.L., Macdonald, I.C., Patrick, J.M. and Scriven, P.M. (1978). *In* "Proceedings of the Second International Symposium on Ambulatory Monitoring" (Ed. J.D. Stott), Academic Press, London and New York.

Bennett, T. and Morgan, D.B. (1976). *J. Sports Med.* **16**, 38–44.

Berry, C.A. (1969). *Aerosp. Med.* **40**, 245–254.

Borg, G.A. (1973). *Med. Sci. in Sports* **5**, 90–93.

Bristow, J.D., Brown, E.B., Cunningham, D.J.C., Howson, M.G., Petersen, E.S., Pickering, T.G. and Sleight, P. (1971). *Circulation Res.* **28**, 582–592.

Bruce, R.A. (1974). *Am J. Cardiology* **33**, 715–720.

Bruce, R.A. Kusumi, F. and Hosmer, D. (1973). *Am. Heart J.* **85**, 546–562.

Brunner, H. and Major, J.K. (1972). *J. Hist Ind. Technol.* **9**, 117–151.

Buller, A.J., Eccles, J.C. and Eccles, R.M. (1960). *J. Physiol.* **150**, 399–439.

Burkitt, H., Wheater, P.R. and Daniels, V.G. (1979). *In* "Functional Histology", Churchill Livingstone, London.

Cabanac, M. and Caputa, M. (1979). *J. Physiol.* **289**, 163–174.

Carter, S.A. (1972). *New England J. Med.* **287**, 578–582.

Cavagna, G.A. and Kaneko, M. (1977). *J. Physiol.* **268**, 467–481.

Chave, S.P.W., Morris, J.N., Moss, S. and Semmence, A.M. (1978). *J. Epidemiol. Commun. Health* **32**, 239–243.

Clausen, J.P. (1977). *Physiolog. Rev.* **57**, 779–815.

Clausen, J.P., Klausen, K., Rasmussen, B., and Trapensen, J. (1973). *Am. J. Physiol.* **225**, 675–682.

Costill, D.L., Fink, W.J. and Pollock, M.L. (1976). *Med. Sci. Sports* **8**, 96–100.

Cotes, J.E. (1971). *Scand. J. Resp. Dis. Suppl.* **77**, 123–127.

Cotes, J.E., Berry, G., Burkinshaw, L., Davies, C.T.M., Hall, A.M., Jones, P.R.M. and Knibbs, A.V. (1973). *J. Exp. Physiol.* **58**, 239–50.

Cotes, J.E., Hall, A.M., Johnson, G.R., Jones, P.R.M. and Knibbs, A.V. (1973). *J.Physiol.* 24–25P.

Cumming, G.R. and Glenn, J. (1977). *Canad. Med. Assoc. J.* **117**, 346–349.

Dempsey, J.A., Gledhill, N., Reddan, W.G., Forster, H.V., Hanson, P.G. and Claremont, A.D. (1977). *Ann. N.Y. Acad. Sci.* **301**, 243–261.

Donald, D.E. and Shepherd, J.T. (1964). *Am. J. Physiol.* **207**, 1325–1329.

Donald, D.E., Milburn, S.E. and Shepherd, J.T. (1964). *J. Appl. Physiol.* **19**, 849–852.

Dubowitz, V. (1967). *J. Physiol. (London)* **193**, 481–496.

Dubowitz, V. and Brooke, M.H. (1973). *In* "Muscle Biopsy: a Modern Approach", Saunders, London.

Ekblom, B. and Goldbarg, A.N. (1971). *Acta Physiol. Scand.* **83**, 399–406.

Ekblom, B., Goldbarg, A.N., Kilbom, A. and Astrand, P.-O. (1972). *Scand. J. Clin. Lab. Invest.* **30**, 35–42.

Ekblom, B., Lovgren, D., Alderin, M., Fridstrom, M. and Satterstrom, G. (1975). *Scand. J. Rheumatol.* **4**, 80–86.

Eklund, L.G. and Holmgren, A. (1964). *Acta Physiol. Scand.* **62**, 240–255.

Epstein, S.E., Redwood, D.R., Goldstein, R.E., Beiser, G.D., Rosing, D.R., Glancey, D.L., Reis, R.L. and Stinson, E.B. (1971). *Ann. Intern. Med.* **75**, 263–296.

Fewings, J.D., Roberts, M.L., Stepanas, A.V. and Whelan, R.F. (1965). *Austr. J. Exp. Biol. Med. Sci.* **43**, 547–552.

Gollnick, P.D., Armstrong, R.B., Saubert, C.W., Piehl, K., and Saltin, B., (1972). *J. Appl. Physiol.* **33**, 312–319.

Gollnick, P.D., Armstrong, R.B., Saubert, C.W., et al. (1973). *Pflugers Arch.* **344**, 1–12.

Goodwin, G.M., McCloskey, D.I. and Mitchell, J.H. (1972). *J. Physiol. (London)* **266**, 173–190.

Goulding, L. (1976). *Postgrad. Med. J.* **52**, (Suppl. 7), 14–18.

Grimby, G., Brobert, C., Krottkiewska, I. and Krotkiewski, M., (1976). *Scand. J. Rehabil. Med.* **8**, 37–42.

Hill, A.V. and Kupalov, P. (1929). *Proc. R. Soc. Series B.* **105**, 313–328.

Hollander, A.P. and Bouman, L.N. (1975). *J. Appl. Physiol.* **38**, 272–281.

Holloszy, J.O. (1976). *Prog. Cardiovasc. Dis.* **18**, 445–458.

Huxley, A.F. (1974). *J. Physiol.* **243**, 1–43.

International Labour Office. (1970). "Ergonomics and physical environment factors". *In* "Occupational Safety and Health, Series 21". Geneva.

Jansson, E., Sjödin, B. and Tesch, P. (1978). *Acta Physiol. Scand.* **104**, 235–237.

Jorgensen, C.R., Gobel, F.L., Taylor, H.L. and Wang, Y. (1977). *Ann. N.Y. Acad. Sci.* **301**, 213–223.

Landin, S., Hagenfeldt, L., Saltin, B. and Wahren, J. (1977). *Clin. Sci. Molec. Med.* **53**, 257–269.

Lewis, S., Haskell, W.L., Wood, P.D. Manoogian, N., Bailey, J.E. and Pereira, M. (1976). *Am. J. Clin. Nut.* **29**, 151–156.

Lind, A.R., Taylor, S.H., Humphreys, D.W., Kenelly, B.M. and Donald, K.W. (1964). *Clin. Sci.* **27**, 229–244.

Ludbrook, J. (1966). "Aspects of Venous Function in the Lower Limbs", Thomas, Springfield, Illinois.

Lymn, R.W. and Huxley, H.E. (1972). Cold Spring Harbour Symposium, *Quart. Biol.* **37**, 449.

McGavin, C.R., Gupta, S.P., Lloyd, E.L. and McHardy, G.J.R. (1977). *Thorax 32*, 307–311.

Milic-Emili, G., Petit, J.M. and Deroanne, R. (1960). *Intern. Z. Angew. Physiol.* **18**, 330–340.

Mitchell, J.H., Reardon, W.C., McCloskey, D.I. and Wildenthal, K. (1977). *Ann. N.Y. Acad. Sci.* **301**, 232–242.

Morris, J.N., Adam, C., Chave, S.P.W., Sirey, C., Epstein, L. and Sheehan, D.J. (1973). *Lancet*, **i** 333–339.

Muller, E.A. (1943). *Arbeits Physiol.* **12**, 92.

Murray, R.O. and Duncan, K. (1971). *J. Bone and Joint Surg.* **53B**, 406–419.

Murray, T.M. and Weber, A. (1974). *Sci. Am.* **230**, 58–71.

Nordenfelt, I. (1971). *Cardiovasc. Res.* **5**, 215–222.

O'Brien, C., Smith, W.S., Goldsmith, R., Fordham, M. and Tan, G.L.E. (1979). *In* "Response to Strain. Occupational Aspects," (Eds. C. McKay and T. Cox), 52–60. I.P.C., London.

Paffenbarger, Jr., R.S. and Hale, W.E. (1975). *New Eng. J. Med.* **292**, 545–550.

Polgar, J., Johnson, M.A., Weightman, D. and Appleton, D., (1973). *J. Neurol. Sci.* **19**, 307–318.

Pugh, L.G.C.E. (1971). *J. Physiol.* **213**, 255–276.

Puranen, J., Ala-Ketola, L., Peltokallio, P. and Saarela, J. (1975). *Br. Med. J.* **2**, 424–425.

Rochmis, P. and Blackburn, H. (1971). *J. Am. Med. Ass.* **217**, 1061–1066.

Rowell, L.B. (1974). *Physiolog. Rev.* **54**, 75–159.

Rowell, L.B. (1977). *In* "Problems with Temperature Regulation During Exercise", (Ed. E.R. Nadel), Academic Press, London and New York.

Ryder, H.W., Carr, H.J. and Herget, P. (1976). *Sci. Am.* **234**, 109–114, 118–119.

Salmons, S. and Vbrova, G. (1967). *J. Physiol.* **192**, 39–40P.

Saltin, B., Blomqvist, B., Mitchell, J.H., Johnson, R.L. Jr. Wildenthal, K. and Chapman, C.B. (1968). *Circulation* **38**, (Suppl. 7).

Saltin, B., Nazar, K., Costill, D.L., Stein, E., Jansson, B. and Gollnick, P.D. (1976). *Acta Physiol. Scand.* **96**, 289–305.

Sannerstedt, R., Wasir, H., Henning, R. and Werko, L. (1973). *Clin. Sci. Molec. Med.* **45**, 145s–149s.

Shephard, R.J. (1978). "Physical Activity and Aging", Croom Helm, London.

Shephard, R.J., Allen, C. Benade, A.J.S., Davies, C.T.M., Di Prampero, P.E., Hedman, R., Merriman, J.E., Myhre, K. and Simmons, R. (1968). *Bull. Wld Hlth. Org.* **38**, 765–775.

Smyth, H.S., Sleight, P. and Pickering, G.W. (1969). *Circ. Res.* **24**, 109–121.

Solomon, L.P. (1976). *J. Bone Joint Surg.* **58B**, 176–183.

Thorstensson, A. (1977). *Acta Physiol. Scand.* **100**, 491–493.

Underwood, R.H. and Schwade, J.L. (1977). *Ann. N.Y. Acad. Sci.* **301**, 297–309.

Zakai, Z.A., Griffiths, J.C. and Heywood, O.B. (1977). Proceedings of the 2nd Int. Symp. on Ambulatory Monitoring. (Ed. F.D. Stott). Academic Press, London and New York.

Zarrugh, M.Y. and Radcliffe, C.W. (1978). *Eur. J. Appl. Physiol.* **38**, 215–223.

CHAPTER 3
Thermal Physiology

O.G. Edholm and J.S. Weiner

A fundamental property of man in common with mammals generally is his homeothermy—the ability and necessity to maintain internal body temperature within narrow limits. The survival value of a constantly maintained high temperature to the homeotherm is very evident; it imparts freedom from the dominance of external temperature and permits a high level of muscular, neuronal and cellular activity. Analysis of this property entails the understanding of (a) the nature, significance and limits of the homeothermic state, (b) the functioning of the homeothermic control system and its components, and (c) the thermal exchanges involved in the maintenance of the homeothermic state. From a knowledge of the control system and the associated thermal exchanges it is possible to specify the thermal behaviour of the human body under a wide variety of environmental conditions taking into account such factors as work and clothing. As will become apparent, specification or prediction of the thermal state can be quantitatively fairly precise but there are limitations imposed by the complexities of bodily configuration as well as factors affecting the sensitivity and rate responses of the control system and its components.

Body Temperatures and Gradients

Homeothermy is the capacity of the body to maintain the deep body or "core" temperature (ideally, the temperature of the *milieu intérieur*), within a relatively narrow range under the impact of factors acting to increase the heat gain to or the heat loss from the body. The sites at which the "core" temperature is conveniently sampled are in the rectum, the mouth, the oesophagus, in the ear canal near the tympanum, and in the urine stream. By "normal" temperature is meant the values within the regulated homeothermic range recorded at these sites. There is therefore no single fixed "normal" internal temperature. The conventional "normal" mouth temperature of 98.4 °F (36.9 °C) is best regarded as a useful reference value appropriate to the subject in the post-absorptive "basal" comfortable state.

Thermal gradients within the body are determined by the rate of local heat production and the rate of local heat gain or removal which in turn is a function of local blood flow and tissue conductance. The thermal gradients across the superficial tissues are parabolic in contour as are those along the long axis of the limbs. Environmental conditions, at a constant work load, greatly alter these gradients. Exposure to cold increases the steepness of the gradient; exposure to heat reduces it. As work intensity goes up, the gradient across the skin increases, mainly due to a temperature rise in the deepest layers (Nielsen, 1969). These variations arise because the thermostatic control of the core is brought about by using the "shell" as a variable heat sink and a variable conductor (or insulator) of heat flow from the skin to the outside air. Temperatures at the skin and within the shell will vary with the influx of heat from the environment. To counter heat added in this way or generated by increased heat production within the core, the shell functions as a heat exchanger and cooling system, by providing that the net flow of heat is outward from both core and skin.

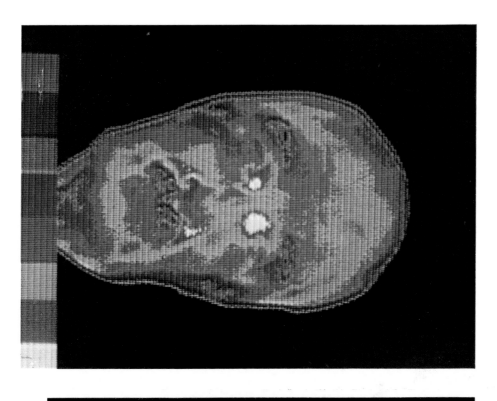

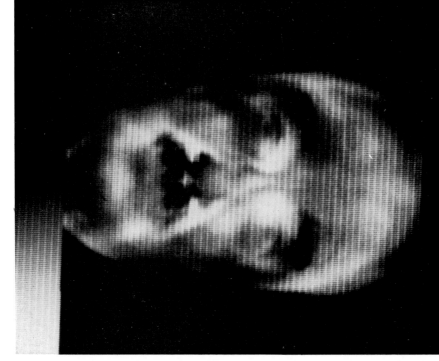

The axial gradient down the extremities is important functionally. It allows heat to flow from the warmer arterial blood across to the closely continguous veins so that some body heat is returned centrally before it reaches the periphery and thereby uses the length of the limb for insulation.

The coefficient of thermal conductance (in kcal m^{-2} h^{-1}) is given by the expression $H_c/A(t_c - t_s)$ where H_c (kcal) is the heat supplied by conduction and by convection due to blood flow, from core to surface, $t_c - t_s$ is the difference between core and mean skin temperature (°C) and A is the duBois surface (m^2) area. This coefficient has been extensively employed. It increases with both work and environmental temperature.

It is not surprising that no single "internal" temperature can be taken to represent the temperature of the core as a whole. The oesophagus is regarded as probably the most representative of t_c (Bligh, 1973) since it provides an indirect measure of the temperature of the arterial blood leaving the heart (Cooper and Kenyon, 1957). Much argument has raged on the value of the rectal temperature as an index of the regulated body temperature. Rectal temperature, in the resting state, is higher than that taken in the mouth, oesophagus, auditory canal, or urine, but is close to that of the liver and brain (Eichna et al., 1951). During leg work t_r is raised by heating of the pelvic cavity by venous blood from the working muscles and is higher by 0.2–0.4 °C than deep oesophagus temperature (t_{oe}), whereas with arm work the temperatures at the two sites are similar (Nielsen, 1969). Piironen (1970) has shown that differences exist between the oesophagel, oral, aural and rectal sites both in time and in the amplitude responses to thermal transients. For measurement of rapid changes the rectal site is unsuitable because of its considerable thermal inertia. Benzinger (1960) claimed that tympanic temperature (t_y) best mirrored the hypothalamic temperature though during body heating t_y was found to be lower than t_r and t_{oe}. All three core temperatures behave in parallel during work independent of environmental conditions over a wide range (Nielsen, 1969). On the whole, the tympanic temperature mirrors transient changes in oesophageal temperature more accurately and predictably than any of the other three sites (Edwards et al., 1978).

Skin temperatures vary over the body surface; the colder the environment the lower the surface temperatures of distal parts compared to the trunk and head. These variations can be beautifully visualized and recorded by the technique of thermography (Figs. 1(a) and 1(b)). Average values for

---

FIG. 1   Thermoscan of head and face. (Left) In colour. The code shows the range from 31 to 35 °C, with 0.5 °C difference between colours. The diversity of skin temperature is demonstrated. (Right) In black and white. The lighter the shade the higher the skin temperature.

skin temperature are obtained from measurements taken at a number of locations (15 sites according to Winslow *et al.* (1936) or 10 according to Nadel *et al.* (1971)), and using weighting factors for the different regions they represent (Hardy and DuBois, 1938).

Estimates of changes in heat content or storage cannot be made directly and depend on estimates of changes of the average temperature of the body as a whole according to the formula $S = \Delta t_b \times W \times s$ where $S =$ the heat debt or loss of stored heat (in kcal), $W =$ body mass (in kg), and $s =$ average specific heat of the tissues (usually taken as 0.83) The change in average body temperature (Δt_b) is obtained by using a formula by Burton (1935): $t_b = 0.65t_r + 0.35t_s$. This formula has been widely used for conditions both of heating and cooling.

Under extreme conditions of heat exposure the endurance time before collapse is related to the rate of heat storage. The upper average body temperature compatible with survival is of the order of 42 °C (rectal) representing a heat storage of about 360 kcal.

Under cooling conditions (depending on the subject's physique and clothing) the average body temperature compatible with survival is surprisingly low, of the order of 25 °C, representing a heat debt of about 700 kcal. There is clearly a greater degree of reversibility from extreme chilling than from extreme over-heating.

As distinct from survival, what are the limits of the homeothermic capacity of the body which depend on the self-regulatory mechanisms whereby steady state core temperatures can be achieved?

As a generalization it can be said that the body can accept or handle, at a steady state, a heat load of a magnitude no greater than that which raises the core temperature to about 41 °C. Conversely, under a cooling load the core temperature can be maintained at a steady state not much lower than about 35–36 °C. Thus homeostatic control under heat load is more efficient than under cooling load.

Components of the Control System (Shivering, Sweating, Circulation)

The responsiveness of the body temperature control system rests on its capacity to register and to respond to changing heating or cooling loads in the shell or the core, and to correct the resulting deviation in the core temperature and restore it to a new steady state. The components of this feedback regulatory system comprise in effect a series of reflex channels. They are shown in the diagrammatic model in Fig. 2. A good deal is now

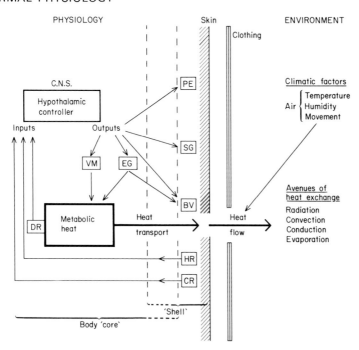

FIG. 2 A simple model to illustrate the more important features of human thermoregulation. PE, pilo-erection; SG, sweat gland; BV, blood vessel; HR, hot receptor; CR, cold receptor; VM, voluntary muscle; EG, endocrine gland; DR, deep receptor. (From Fox, 1974.)

known about the functional features of the elements and linkages making up the system and particularly about the thermosensitive elements which register changes in the periphery and in the core, and the effector components which bring about increase or decrease in heat loss or increase in heat production as required. What still remains to be elucidated is the nature of the integrative thermostatic mechanisms in the central nervous system itself.

In man there is, in fact, a double homeothermic control system as indicated in Fig. 2. One is involuntary and autonomic by which information from thermoreceptors is transmitted and processed in the specific regions of the hypothalamus, the output from which then acts to bring about appropriate changes in blood flow, sweat gland activity or calorigenesis. There is also the perceptual–behavioural system, whereby conscious appreciation of warming or cooling is registered in higher nervous centres, so that voluntary action is taken to reduce or eliminate thermal disturbance as well as to monitor the performance of the involuntary response system.

Thermosensors

Temperature-sensitive receptors are present both in the periphery and in the hypothalamic area. The former signal temperature changes in the skin; the latter, changes in temperature of the local blood supply. The existence of specific cutaneous end-organs has been clearly established in various mammalian species, including primates and in man. Less certain are the claims for receptors in veins, mucous membrane, viscera and the spinal cord. The existence of specific muscle receptors in man has also been postulated. In man, it is believed that the same cutaneous sensors subserve both autonomic and perceptual responses. Animal experiments have revealed the presence of specific thermoreceptor sensors in the hypothalamus (Hellon, 1971).

The morphology of skin thermoreceptors is not yet clearly established. Earlier studies indicated that the receptors subserving thermal sensation and thermoregulation were structures in the "warm" and "cold" spots lying quite superficially in the skin. That they correspond to morphologically specialized receptors such as the Ruffini and Krause end-organs is not now accepted; they appear to be free nerve endings widely distributed over and within the epidermis (Weddell and Miller, 1962). Their signals are carried by either non-myelinated C fibres or small myelinated A fibres and the main afferent spinal pathway is in the lateral spinothalamic tract.

Thermoreceptors both peripheral and central fall into two groups of "warm" and "cold"—they signal both rate of change and static temperature. How these receptors act as thermal transducers is unclear (Iggo, 1970). In the skin, the stimulus is the temperature itself and not the thermal gradient across the receptor terminals (Hensel and Witt, 1959). The receptors display characteristic impulse frequencies. A "warm" receptor shows an increase in impulse signals on sudden warming and a decrease on sudden cooling. A "cold" receptor responds in the opposite manner. The receptors also show a static discharge at a constant skin temperature; the peak discharge for the cold receptor is at a lower temperature than for the warm. (Fig. 3(a), (b)). Dynamic sensitivity is greatest at or near the peak maximum static sensitivity.

That the hypothalamic region is responsive to local changes in its temperature was made clear by the experiments of Magoun et al. (1938). The anterior hypothalamic heat centre can also be activated by direct electrical stimulation (Hess and Stoll, 1944).

Direct evidence for hypothalamic thermoreceptive sensors has come from using a microelectrode to detect the electrical activity of single neurones during stimulation by local cooling or warming by thermode, as well as by warming the carotid blood or by heating or cooling the skin. Only a

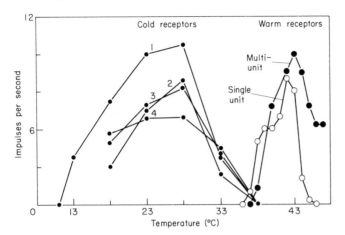

FIG. 3 Static sensitivity curves for four "cold" receptors, ●————● 1 to 4; one single warm receptor, ○————○; and a multi-unit "warm" receptor, ●————●. (From Iggo, 1969.)

proportion of cells in the preoptic anterior hypothalamic region show an appropriate response to local neuronal heating or cooling—so that neurones may be classified as warm-sensitive, cool-sensitive or insensible. A warm sensor gives a positive temperature coefficient for the increase in firing rate; cold sensors give a negative temperature coefficient.

An important finding which bears on the interaction between skin and central temperatures in activating the central mechanisms is that units in the hypothalamus increase their firing rate when the body surface is warmed (in cats) (Wit and Wang, 1968). Hellon (1970) extended this to show that hypothalamic cells increased or reduced their firing rates in response to peripheral warming or cooling (in rabbits.)

Central components

Peripheral and hypothalamic sensors are axonally connected to a neuronal network in the anterior hypothalamus and the preoptic region (Fig. 4). The existence of two linked hypothalamic centres was established by placing neurological lesions at different levels of the brain and by the effects of warming and cooling and by electrical stimulation of different areas. Ablation and local stimulation have localized the control of heat loss to the anterior hypothalamus and preoptic region, levels 1–3 (Fig. 4). Heating the anterior hypothalamic region elicits responses which counteract and inhibit the effects of cooling, while cooling exerts the opposite re-

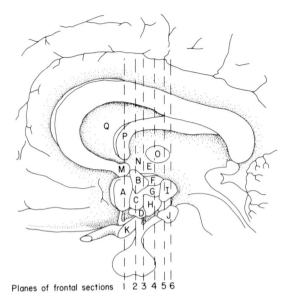

Planes of frontal sections | 2 3 4 5 6

FIG. 4 Sagittal section of human brain with levels (caudal to rostral) for projected localization of areas associated with temperature regulation. A, preoptic nuclei; B, paraventricular nucleus; C, anterior hypothalamic area; D, supraoptic nucleus; E, lateral hypothalamic area; F, dorsal hypothalamic area; G, dorsomedial nucleus; H, ventromedial nucleus; I, posterior hypothalamic area; J, mamillary body; K, optic chiasma; L, lamina terminalis; M, anterior commissure; N, hypothalamic sulcus; O, intermediate mass of thalamus; P, fornix; Q, septum pellucidum. (From Carlson, 1973)

ciprocal effects (Hammel, 1968). The second more posterior centre was identified by Isenschmidt and Krehl (1912); lesions at levels 4–6 (Fig. 4) produce an animal that fails to vasoconstrict or shiver when exposed to cold. Levels 3–6 are involved in the shivering response.

The functional organization of the hypothalamic control system has generated many hypotheses in the form of thermoregulatory system models. As these are, in effect, closed loop negative feedback control systems, the role of the hypothalamic centres cannot be divorced from the linked transfer functions on both the input and output side. These regulatory models are considered below.

A second parallel system exists in which central integration of incoming thermal information manifests itself as perceptual knowledge of changes in the environment and in the body, which leads to behavioural responses. The afferent pathways to the central receptor areas are via the thalamus (postero–lateral and ventro–medial nuclei) and limbic regions to the postcentral frontal area.

Effector pathways

Some six effector nervous pathways can be identified whereby the hypothalamic controller can induce responses to correct departure from homeothermy. Of these pathways, the sympathetic adrenomedullary system (SAM) provides a widely distributed network of noradrenergic terminals activating metabolic and vascular responses, and a motor supply to the adrenal medulla for the release of adrenaline into the circulation.

The separate pathways for particular homeothermic responses are the following.

1. Adrenergic non-medullated nerve fibres of the sympathetic system bring about vasoconstriction of skin vessels. There is probably at the same time inhibition of sympathetic vasodilator nerves. The pilomotor reaction ("goose-flesh") is also mediated by adrenergic sympathetic nerves.

2. The nervous response to heating involves active cutaneous vascular dilatation and inhibition of vasoconstrictor tone. The vasodilatation is exerted through sympathetic nerves. The presence of active vasodilator nerves is evidenced by the fact that skin flows attained after loss of vasoconstriction by sympathectomy are not as high as when the innervation is intact.

3. Shivering is a special pattern of skeletal muscle activity induced by the thermal control centre in the posterior hypothalamus. The ordinary skeletal nerve supply, not the sympathetic innervation, is involved. Shivering consists of synchronous contraction of small groups of motor units, which contract out of phase with other groups and alternate with units of antagonistic muscles so that gross movements are avoided.

4. The sympathetic supply to the adrenal medulla is reflexly stimulated by cold exposure. The adrenaline secreted increases heat production, cardiac output and enhances skin vasoconstriction while tending to dilate blood vessels in the muscles. Metabolic responses to cold involve increased glucogenesis. Glucose and free fatty acid utilization also depend on adrenal medullary stimulation and release of catecholamines.

It is likely that all these metabolic and vascular effects of catecholamine release take place during shivering thermogenesis, and indeed these amines facilitate shivering.

5. Animal experiments indicate that a non-shivering phase of thermogenesis results from the development of enhanced tissue sensitivity to the calorigenic action of noradrenaline, a phenomenon which has also been claimed to occur in man (Joy, 1963; Kang et al., 1970). In animals, chemical blockage of sympathetic ganglia or of adrenergic receptors in tissues have revealed very clearly the mediation of the sympathetic adreno–medullary system in responding to peripheral body cooling or to cold stimulation of the preoptic anterior hypothalamic centres.

6. Sweating is activated through non-medullated sympathetic nerves which are undoubtedly cholinergic. The possibility that eccrine sweat glands are also innervated by adrenergic fibres (since they respond to direct application of adrenaline and noradrenaline) seems unlikely (Foster *et al.*, 1970).

Changes in cutaneous blood flow while neuronally controlled are also affected by direct heating or cooling. Perfusion of the limbs (of cats or dogs) with warmed blood causes dilatation of skin vessels, including the arterio–venous anastomoses; cooling the incoming blood causes intense vasoconstriction of the cutaneous vessels with a pronounced vasodilatation of deeper vessels (Pappenheimer *et al.*, 1948).

With prolonged exposure and acclimatization to heat or cold the effector mechanisms which come into play involve the endocrine systems. Thermal stimuli under cool conditions at the hypothalamic level result in an output of thyrotropic releasing factor (TRF) and hence activation of the thyroid gland. In addition, the cortico-releasing factor (CRF) acting via ACTH on the adrenal gland releases cortisol. Under hot conditions the adrenal cortex releases aldosterone. Hypothalamic heating also stimulates the release of ADH.

Thermoregulatory Functioning

Each of the basic thermoregulatory processes—vasomotor, sweating, calorigenic and behavioural—need to be understood in terms of the functioning of the control elements described above. In each case the rate and intensity of the response depends on the characteristics of the stimuli to the peripheral and central thermoreceptors, the modulation and integration exercised by the hypothalamic controller over incoming and outgoing signals and the response characteristics of the particular effector mechanism.

Vascular responses

Both central and peripheral thermosensors are involved in eliciting vasoconstriction or vasodilatation; the vasomotor drive results from the interaction between central and cutaneous stimuli.

On exposure to cold the vasoconstriction in the skin is at first due to stimulation of cutaneous thermoreceptors. This reflex response is then reinforced and maintained by the activation of the central receptors by a subsequent fall in core temperature.

As shown by Edholm *et al.* (1957) the cutaneous blood flow is regulatored by a balance between vasoconstrictor and vasodilator tone. The threshold oesophageal temperature for eliciting vasodilatation is 37.30–37.35 °C whether the core temperature is falling or rising (Cabanac and Massonnet, 1977). The temperature is also very close to the thresholds for shivering or sweating.

The vasomotor responsiveness of the central receptors is very high. Vasodilatation ensues from a rise in a core temperature of 0.01 °C (Pickering, 1932). A change in body heat content of as little as 2 kcal can induce an appropriate vasomotor response (Snell, 1954). The responsiveness of the core receptors is dependent on skin temperature. In a bath kept at 28 °C the increase in blood flow per 1 °C rise in core temperature is about one-third that in a bath at 39 °C (Cabanac and Massonnet, 1977). Figure 5 illustrates the interaction between skin and core temperatures in regulating thermal conductance from minimal blood flow levels in resting subjects in cold conditions up to maximal flows at high work rates in hot conditions.

Thermal conductance is, of course, not a purely vasomotor parameter but its usefulness as an indicator of thermoregulatory circulatory response has been stressed by Robinson (in Hardy, 1963). At environmental temperatures ranging from 12 to 48 °C the steady-state thermal conduction of resting subjects increases from a value of 15 to 46 (Hardy and Stolwijk, 1966). These values are increased by moderate exercise (Robinson, 1963) with values of 20 at 9.5 °C (ET) to 130 at 35 °C (ET). The findings shown in Fig. 6 indicate a seven- to nine-fold increase in thermal conductance flow from minimal to maximal vasodilatation.

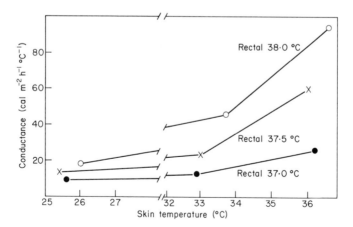

FIG. 5   Curves of conductances plotted against skin temperature for constant rectal temperatures of 38.0, 37.5 and 37.0 °C. (From Wyndham, 1965).

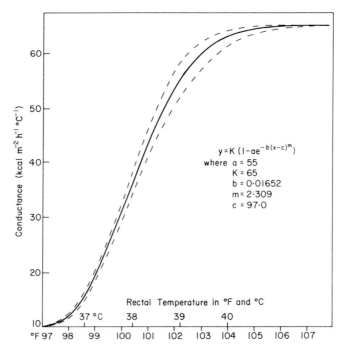

FIG. 6   Conductance versus rectal temperature, average 4 h. (From Wyndham, C.H., Strydom, N.B. *et al.* (1965). *J. Appl. Physiol.* **20**, 37–45.)

Direct plethysmography of the hand reveals the flexibility of its vascular bed. When the subject is cold but not shivering a value of 1.0 ml per 100 ml of tissues is obtained; in the comfortable zone this rises to 4.0–10 ml, and with body heating it may reach 30–40 ml per 100 ml of tissue.

Cold vasoconstriction does not completely cut off the blood supply to the skin. Oxygen is still needed even though at low temperatures the metabolic requirements are small. Even with the minimum flow reaching the skin at deep body temperature the heat loss would still be appreciable, and is reduced still further by constriction of the superficial veins. The venous blood returns mainly along the venae comitans in close apposition with the artery; this facilitates heat exchange between the two. Thus the more distal parts of the limbs cool gradually so that the blood entering the foot and hand does so at a temperature very little above that of the surface. The pre-cooling of the arterial blood markedly reduces the total thermal gradient between the artery and the outside air, cutting down the heat loss accordingly.

Severe cooling of the hands or feet provokes a paradoxical vasodilatation with a large loss of heat which alternates with periods of vasoconstriction—the "hunting" reaction of Lewis (1930).

Full adjustment to cold involves widespread cutaneous vasoconstriction with great diminution in the blood contained in superficial veins. Until the blood volume is redistributed by splanchnic vasodilatation there is commonly an initial elevation of blood pressure. According to Bazett (in Newburgh, 1968) full vasoconstriction requires reduction in blood volume.

The increased skin blood flow consequent on thermally induced vasodilatation involves two vascular adjustments as shown by Kerslake (1972). Firstly there is opening of some arterio–venous anastomoses deep to the capillary bed. The fall in temperature along the length of the artery is diminished because the blood flow is higher and because for a given skin heat loss the temperature difference between the artery and vein is less. The effect is to raise the skin temperature and increase heat loss. Secondly, there is the opening of superficial veins; the skin temperature over these is raised with increased heat loss: the effect of returning more of the capillary blood through a dilated superficial vein, and less through the venae comitans is to reduce pre-cooling by exchange with the adjacent artery, so increasing the temperature in the superficial veins and of the skin.

Marked dilatation of the superficial veins necessitates a large increase of blood to the skin accompanied by splanchnic vasoconstriction. There may be a small fall of blood pressure. An increase in blood volume occurs after some days as part of the general phenomenon of acclimatization.

Shivering

Under conditions of maximum vasoconstriction the maintenance of a steady heat loss and core temperature depends on the provision of appropriate insulation, which will vary to some extent with the amount of subcutaneous fat, but will mainly be dependent on suitable clothing thickness. If the total insulation is inadequate, skin and core temperatures will fall. At the "critical temperature" metabolic rate begins to increase, due first to an increased muscle tone which causes stiffness and even rigidity. For a nude resting man, the critical air temperature is about 27 °C, but this can be lowered considerably by increasing thickness of clothing.

The evidence about the initiation and control of shivering is contradictory; both skin temperature and deep body temperature affect the onset and the extent of the shivering response. Shivering can be demonstrated with surface cooling lowering skin temperature and before deep body temperature falls, but when that occurs shivering is intensified. The rigors which occur in patients suffering from malaria resemble shivering, and these rigors are associated with a feeling of intense cold even when deep body temperature is rising. The sensation of cold is due to an intense cutaneous vasoconstriction and a steep fall of skin temperature.

Thermal receptors in the skin provide the main stimuli for shivering; if these stimuli are absent shivering does not readily occur with moderate rates of fall of deep temperature. Also shivering can be induced by a fall in skin temperature even with some increase in central temperature (Bazett, 1968).

Sweating

The number or eccrine glands which can be activated by a maximal heat load is of the order of 2 million. Like the other thermoregulatory responses sweating can be evoked by both cutaneous and central thermoreceptors. When the skin temperature is maintained steady in the resting subject the threshold for sweat onset is very close to that for vasodilatation. With increasing thermal stimulation sweat gland output increases by the recruitment of additional numbers of glands and by an enhanced output per gland. In hot and humid conditions, sweat rate can be so high that unevaporated sweat drips off without evaporating. Although as far as temperature regulation is concerned such drippage appears to be waste, this is not entirely correct. Brebner and Kerslake (1969) examined the relationship between the percentage wetted area of the skin and the volume of drippage. They showed that drippage begins when 60–70% of the skin is wetted. It had been accepted that drippage only occurred when all the skin surface was wet, but Brebner and Kerslake (1969) demonstrated that 100% wettedness can never be achieved. They concluded that to attain high sweat rates it was necessary to have a degree of drippage. Thus the area of wettedness of the skin surface can be enlarged up to the point when the skin is almost covered by a film of sweat (Fig. 7).

The relative roles of central and peripheral control of sweating is of some complexity. Over a wide range of external temperature—"the prescriptive zone"—the core temperature is stabilized for a particular level of heat production (Nielsen, 1938; Lind, 1963) (Fig. 8). The increase in evaporative requirement over this range is brought about by increase in skin temperature, though at the beginning of any work period the initial rise of core temperature will have contributed to the sudomotor response. If the work rate is altered in a given environment there is little change in skin temperature (Fig. 9) but there is a high correlation between core temperature and sweat output. Beyond the "prescriptive zone" core temperature and skin temperature both rise.

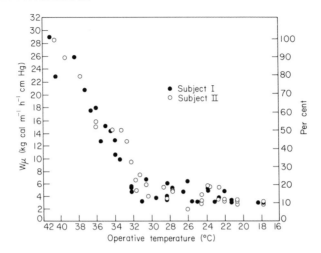

FIG. 7   Relation of the degree of wettedness of skin surface to operative temperature. (From Gagge, A.P. (1937). *Am. J. Physiol.* **120**, 277-285.)

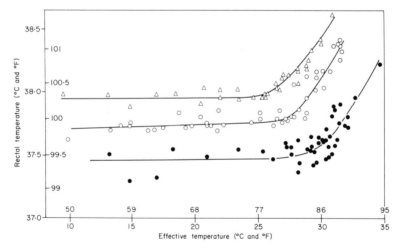

FIG. 8   Equilibrium rectal temperatures of one subject working at energy expenditures of 180 (●), 300 (○) and 420 (▲) kcal h⁻¹ in a range of climatic conditions (CET). (From Lind, 1963.)

If both core and skin temperature are altered by appropriate combination of work load and heat exposure their relative effects on sweat output are as shown in Fig. 10: the greater sensitivity of sweat response to changes in core temperature is evident. The involvement of both stimuli can be easily demonstrated in the sweating subject by sudden cooling of an area of skin

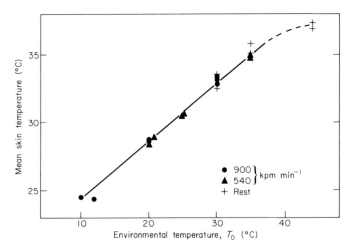

FIG. 9   Mean skin temperature at various levels of environmental (operative) temperature, three metabolic levels; 900 kpm min^{-1}, ●; 540 kpm min^{-1}, ▲; rest, +. (From Nielsen, 1969.)

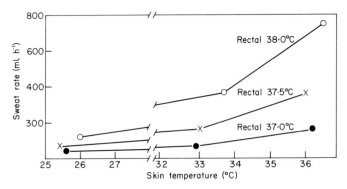

FIG. 10   Curves of sweat rates plotted against skin temperature for constant rectal temperatures of 38.0, 37.5 and 37.0 °C. (From Wyndham, 1965.)

which induces an immediate though transient inhibition (Kuno, 1934). A rapid reflex enhancement of sweating follows local or general application of a warm stimulus (Kuno, 1934) and lasts as long as the additional warmth is applied (Kerslake, 1972). There is some evidence that skin warming has also a direct local effect on the sweat glands.

It is clear from many investigations that increase in sweating is closely paralleled by increase in thermal conductance (see Figs 5 and 10). There is some evidence that vasodilatation at high sweat rates is potentiated by the release from the sweat glands of bradykinin-forming enzyme (Fox and Hilton, 1958).

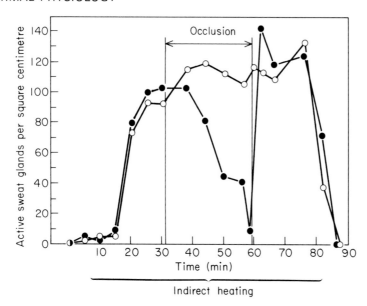

FIG. 11 The effect of arterial occlusion on sweating from the arm. The subject was heated by immersing the legs in warm water. ○, control arm; ●, arm circulation occluded. (From Collins, K.J., Sargent, F. and Weiner, J.S. (1959). *J. Physiol.* **148**, 615–624.)

The dependence of sweat activity on the circulation can be readily demonstrated by a few minutes of arterial occlusion (Fig. 11); the ischaemia particularly affects the fluid output by the gland and not its osmotic work which is an anaerobically based process (Hubbard and Weiner, 1969).

From his observations in the desert Adolph (1947) concluded that a water deficit as high as 11% of body weight did not influence sweat secretion. This contrasts with the earlier findings of Gregory and Lee (1936), or those more recently of Ellis *et al.* (1954) of significantly lower sweat rates associated with progressively greater water deficits. Lowering of the sweat rate by dehydration is offset, however, by the higher temperature which ensues and which will tend to increase the sweat rate.

After a period of secretion the sweating mechanism may show a progressive falling off. This is not so marked in hot dry conditions with adequate water replacement. The decline in sweating, often termed fatigue or hidromeiosis, is commonly seen when the skin remains wet as in hot humid environments or, even more strikingly, during prolonged immersion in warm baths. In hot dry conditions, while sweating over the general body surface remains steady, sweating within an armbag where the skin becomes wet undergoes a marked suppression (Fig. 12). Hidromeiosis can be

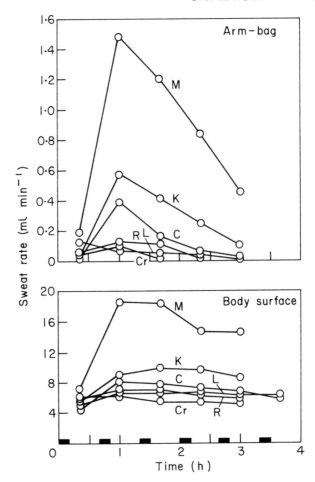

FIG. 12   Sweat rates from the arm and hand, enclosed in an arm bag, and from the general body surface. Six subjects (identified by letters) performed a work and rest routine in a hot environment ($T_0 = 40\ ^\circ C$, $p_a = 4.2$ kPa). Blocks on the abscissa indicate work periods. After the first hour there is a decline in sweat production from the arm, while general sweat production falls only slightly. (From Collins, K.J. and Weiner, J.S. (1962). *J. Physiol.* **161**, 538–556.)

abolished by moving subjects from a hot humid to a hot dry environment. The explanation of the phenomenon is that continually wetted skin reabsorbs water leading to epidermal swelling and poral closure. The subject is thoroughly discussed by Kerslake (1972) who concludes that claims of sweat gland fatigue are difficult to dissociate from the coexistence of hidromeiosis. Ischaemic inhibition of sweating provides an example of synaptic blocking (Collins *et al.*, 1959).

The Central Controller

As already discussed, the homeothermic mechanism comprises a system of thermosensors linked by afferent pathways to a central integrative controller in the CNS; this induces, via effector pathways, appropriate alterations in heat production, conservation and dissipation. The question then arises as to the precise way this system functions, sensitivity and automatically, to cope with changes of heat load imposed on the body. A great deal of investigation and speculation has gone into attempts to answer this question (cf. Fox, 1974). A complementary problem concerns the neuronal organization which gives the central controller its particular integrative and transfer capabilities. No final answer can as yet be given to either question.

The general model widely used to explain the workings of the control system is that of a negative feedback closed loop system. Whatever the nature of the central controller the model must provide the following properties.

(i)   Its response character must be such that swings of core (mean body) temperature are held within narrow limits, with the re-establishment of steady state values.

(ii) It must account for the thermostatic levels characteristic of fever, exercise, the diurnal cycle and acclimatization.

(iii) It must integrate the input information from the separate thermosensors, from periphery or core, whether temperature changes at these sites are in the same or in opposite senses.

(iv) Its functioning must be expressed in quantifiable and predictable relations ("transfer function") between the magnitude of the input stimuli and the magnitude of the effector responses, circulatory, sweating, or shivering, up to the limits of their capacity.

Though some aspects have been greatly clarified, the whole system can be said to be understood at the present time only incompletely.

In models in which the controller incorporates a "set-point" or reference temperature, the deviations between the set-point temperature and the temperature of the body mass (the controlled variable), represents the "load error"; this "load error" furnishes an activating signal which reflects the magnitude of displacement of the controlled variable. The size of the error signal, as registered by the central control elements, bears a direct relation in turn to the magnitude of response of the effectors (vasomotor, calorigenic, sweating). Changes in the controlled variable as a result of internal or external thermal disturbance are detected by the sensors which provide feedback signals to the controlling elements for comparison with the

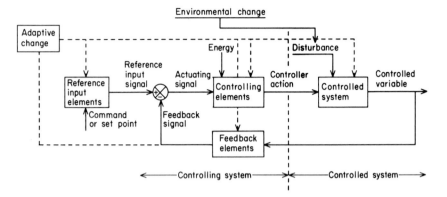

FIG. 13 Block diagram of closed-loop negative-feedback control system with the addition of adaptive change. (From Stolwijk, J.A.J. and Hardy, J.D. (1966). *Pflügers Archiv.* **29**, 129–141.)

reference input from the command or set-point. A block diagram illustrating this closed loop negative feed-back set-point system is shown in Fig. 13.

The functioning of this control loop in physiological terms, based on experimental findings on the dog and monkey, was elaborated by Hammel *et al.* (1963) as embodying proportional control with an adjustable set-point. It has been widely (but not universally) accepted as applicable to man.

In this system the relationship between the load error and the thermal responses (shivering, sweating, skin blood flow) is given by the same general expression:

$$R - R_0 = \alpha R(t_h - t_{set})$$

where $R - R_0$ is the effector response, t_h is the hypothalamic temperature representing the (weighted) controlled variable, t_{set} is the set-point temperature, α_R is the constant of proportionality which differs for each response. It is positive for heat-dissipating responses and negative for heat-conserving and heat-producing responses. The controller which registers the load error is taken to be located in the anterior hypothalamus from where it drives the heat loss and heat production centres located in the anterior and posterior hypothalamus respectively. In man, as in the dog or monkey, the linear relation between the error in the controlled temperature and the size of the effector action is apparent in the findings of a number of workers. Benzinger (1963) reports that at rest and at two rates of steady state work both the rate of sweating and peripheral circulation (as conductance) are linearly related to the value by which deep body temperature exceeds the threshold values for sweating or vasodilatation. These results imply that the controlled variable is the core temperature. Similarly, Cabanac and Massonet

(1977) find for subjects heated up or cooled down in baths that from a threshold core (t_{oe}) evaporative loss increases proportionally to t_{oe} as does the vasomotor (hand blood flow) response; below a threshold, metabolism increases proportionally to the falling t_{oe}. These relations hold also for transient periods and were not influenced by the rate of change of t_{oe}. Transfer functions so far determined are of limited applicability and their linear form would hold only over a restricted range as indicated in Fig. 6.

The virtue of the continuous proportional control system is that it provides a very stable type regulation, as compared to the simple on–off regulator where the temperature oscillates about the set-point. The proportional mode of control can be seen to be modulated or "throttled" and to account very well for the stability of homeothermic "load error" levels seen in steady state exercise and exposure to heat or cold. To account for the temperature levels associated with fever, the diurnal cycle and the Nielsen "plateau" (see Fig. 8), Hammel (1968) postulates that the set temperature is capable of readjustment in response to particular input signals. This would have the effect of minimizing or reducing the change in load-error needed for activation of the proportional controller and would be equivalent to an integral control element in the system. Thus, the action of pyrogen in fever is to raise the set temperature so that the load error due to heating or exercise would function at a higher core temperature. In the case of the Nielsen "plateau" there would be no need for a change in load error if any increase due to rising external temperature were balanced by a change in set-point, induced by the rise in skin temperature. The higher the skin temperature the lower the set-point for sweating. The fall in body temperature during sleep and the rise during the working day are explained by Hammel in terms of an adjustable set-point.

An alternative to the simple set-point is that proposed by Bazett in 1927. This does not required any reference signal but functions as a dynamically balanced negative feedback system. It has been elaborated by Mitchell *et al.* (1972) to account for the same phenomena for which the adjustable set-point was proposed. The basic system is shown in Fig. 14.

Neuronal models

The models described above are essentially physical or engineering analogues. The biological reality of such models must rest ultimately on the actual neurological organisation and properties of the input and output connections and the neural network which make up the anterior and posterior hypothalamic heat regulating centres. Both the advocates of set-point thermostatic systems and of non-reference systems have formulated their models in a neuronal format.

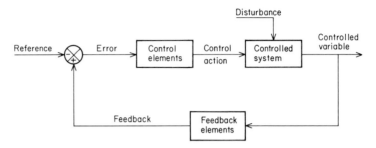

FIG. 14 Block diagram representing a closed-loop negative-feedback control system. (From Mitchell, D., Snellen, J.W. and Atkins, A.R. (1970). *Pflügers Archiv.* **321**, 293–298.)

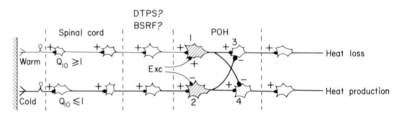

FIG. 15 A neuronal model for the regulation of body temperature in mammals. POH, preoptic hypothalamus; DTPS, diffuse thalamic projection system; BSRF, brain stem reticular; EXC, exercise input. (From Hammel, 1972.)

The model proposed by Hammel (1972) depicts the neuronal connections from the peripheral receptors as well as the network within the controller (Fig. 15). The neurones in the preoptic hypothalamic region are of two kinds in accordance with experimental findings by Hellon and others. Neurones 1 and 2 are thermodependent and are actuated by incoming thermally induced signals. In turn they act on neurones 3 and 4 which are not temperature dependent. These neurones are excited from their own pathway or inhibited from the opposite pathway in accordance with the physiological balance existing between vasomotor heat loss and heat conserving processes. Because neurones 3 and 4 receive more inhibitory than excitatory stimuli from neurones 1 and 2 the temperature thresholds for heat-dissipating and heat-conserving responses are separated. Hammel claims his combination of four neurone types has the basic properties required of a closed-loop, negative-feedback, temperature regulator with a "set-point".

Bligh (1972) postulates additional networks acting as inhibitory influences on the main effector pathways, to account for the separation of the

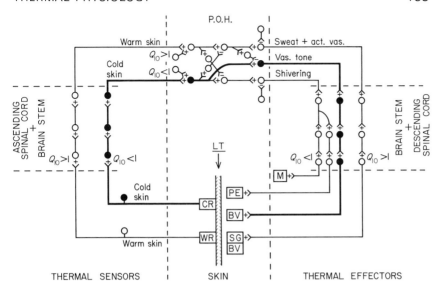

FIG. 16 A schematic and speculative neuronal model for the human temperature controlling system. A heavy line is used to indicate the dominant neural
pathways with an individual in the neutral thermal state. The possible facilitation
of appropriate neural traffic in the cord by local warming and cooling is indicated
by a $Q_{10} < 1$ or $Q_{10} > 1$. WR, warm skin receptor; PE, pilo-erection; SG, sweat
glands; M, muscle; CR, cold skin receptor; BV, skin blood vessels; LT, local
temperature. (From Fox, 1974.)

threshold temperatures for the activation of heat loss by sweating or of heat
production by shivering.

Fox (1974) has combined the features of Hammel's and Bligh's models to
present a comprehensive neural network for the human temperature controlling system (Fig. 16). The cutaneous temperature sensors and effectors
are shown in proximity to emphasize the direct feedback on the thermal
sensors of changes in temperature resulting from effector adjustments.
Changes in skin temperature influence the cutaneous blood flow not only by
a direct action in dilating or constricting the blood vessels, but also by
modulating the effector response to a given level of central neural drive.

Much attention has been paid to elucidating the nature of the neuro–
hormonal transmitters which activate the hypothalamic neural network.
This followed the report by Feldberg and Myers (1963) that injection into
the lateral ventricle or more significantly into the anterior hypothalamus of
noradrenaline acts to reduce body temperature and of 5-hydroxytryptophan
(serotonin) to raise the body temperature of the anaesthetized cat, as well as

in the monkey (Myers and Yaksh, 1969). It had already been known that these monoamines occur in high concentration in the hypothalamic region. Direct demonstration of the release of specific transmitter substances in the hypothalamus during thermoregulation has been reported by Myers and Sharpe (1968).

The earlier simpler inference of the existence of two specific sets of neurones in the primate, one serotoninergic activating heat-conserving pathways and the other adrenergic for heat dissipating pathways has become complicated by the discovery of various species differences and also by the role of acetylcholine (and more recently of prostaglandin E_1).

A neurotransmitter neuronal model (Fig. 17) incorporating a reference set-point and which goes some way to account for normal and pyrogenic temperature regulation has been proposed by Myers (1971).

Heat Balance Equation

The fundamental heat balance equation is

$$M = E \pm C \pm R \pm Cd \pm S$$

Where M = metabolic heat production, E = heat loss by evaporation, C = heat loss or gain by convection, R = heat loss or gain by radiation, Cd = heat loss or gain by conduction, and S = heat loss or gain by storage of heat (in steady state conditions S = zero).

In most environments heat is lost by all channels; the quantity of heat stored in the shell can vary and in this way core body temperature can be maintained at a constant level.

Heat can be gained by solar radiation, and in many industries heat gain by "wild heat", i.e. radiation from hot machines, furnaces or molten materials (glass, ceramics, etc.) can be considerable. Heat gain or loss by conduction is, in general, a small part of the heat balance equation, except during immersion in water.

Heat gain by convection can occur in very hot conditions, e.g. in a desert with high air movement. In such a situation it may only be possible to lose heat by evaporation. The various channels of heat exchange will be considered separately.

Convection

The rate of heat loss by convection is directly proportional to the gradient of temperature from the surface of the skin to the surface of

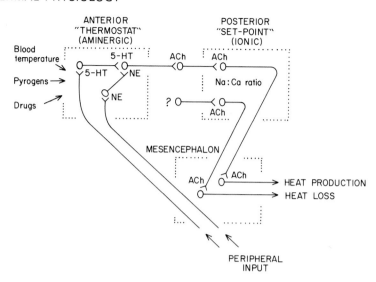

FIG. 17 Diagram of a model to account for temperature regulation under normal conditions as well as during a pyrogen fever. The outflow to the posterior hypothalamic set point is mediated by a cholinergic system which passes through the mesencephalon. 5-HT, 5-hydroxytryptamine; NE, noradrenaline; Ach, acetylcholine. (From Myers, 1971.)

clothing, and thence to the surrounding air, i.e.

$$C \propto t_{skin} - t_{cl} - t_a$$

or
$$C \propto t_{skin} - t_a$$

where t_s = mean skin temperature, t_{cl} = mean surface temperature of clothing, and t_a = temperature of the ambient air.

Heat loss due to convection is described as either free or forced. In still air, at a temperature below that of the skin (or clothed surface) heat is transferred from the body surface to the surrounding air which becomes warmer and rises. This upward-moving envelope of warmed air is the natural convection boundary layer.

In forced convection, air is moving past and round the body and disturbs the natural convection boundary layer. Lewis (1966) used "Schlieren" photography to examine this layer and to determine, amongst other things, the effect of clothing on heat loss by convection. The feasibility of photographing the convective heat flow depends on differences in the refractive index of the air due to the temperature gradients in the moving air stream. In a nude subject, standing in a room kept at about 28 °C, with little air movement, the flow can be seen to be moving upwards from the feet; as it does, the boundary layer becomes thicker and moves faster (Fig. 18). At the shoulders it is several centimetres thick and over the face and head it is approximately 180 mm thick. The flow is fastest about 20 mm above the head, reaching 0.5 m s^{-1}. The convective plume extends some $1\frac{1}{2}$ m above the head, comprising some 10 litres s^{-1} of warmed air.

This pattern is modified by forced convection due either to air movement or movement of the body and limbs. The boundary layer flow will be determined by the speed of the air and the shape of the particular segment of the body.

In free convection, the Grashof number, a non-dimensional parameter, characterizes the boundary layer. It depends upon the viscosity of the air, the vertical height on the body surface, skin and air temperature, and acceleration due to gravity. The nature of the convective air flow is also affected by the dimension and shape of the body, the limbs, torso and head. The velocity of the upward air flow depends on the distance from the skin—there is zero velocity at the skin surface, and then air speed increases

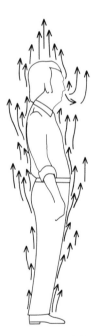

FIG. 18　Convective flow currents in the upright posture, showing how the head is surrounded by warm air. (By courtesy of R. Clark.)

with distance, reaching a maximum about one third of the way across the boundary layer, and then falling to zero velocity at the outer edge.

Heat loss depends upon the horizontal temperature gradient which is largely dependent on the thickness of the boundary layer. Heat loss from a heated horizontal cylinder is greatest at the lowest point where the convective flow is minimal. At the upper surface, convective heat loss is lower. These findings apply to man, whose body can be regarded as made up of cylinders smaller in diameter for the limbs, larger for the torso. The body transfer coefficient (h_c) is an integrated value of the separate segments.

Because of the characteristics of convective heat flow, posture has a significant effect on heat loss. In the supine position the convective heat flow passes upwards and does not form an insulating envelope around the body and head, and heat loss by convection is about 30% higher than in the upright position. Sitting and standing do not differ significantly.

Forced convection

Increased air movement occurs when the body and/or limbs are in motion, or on exposure to wind. In relation to "forced" convection strictly *any* increase in air movement over that of "still" air must be considered as "wind". It is conventional to accept air movements of 0.1 m s^{-1} (or less) as "still" air. As is obvious, there is increased heat loss by convection as air movement increases; what is not so obvious are the relationships concerned. These are affected by the shape of the human body, the movement of limbs in relation to the trunk (as in running) and the laws of aerodynamics. The pattern of air flow (and hence of heat loss) can be illustrated by examining what happens when air flows past a cylinder. At the front leading edge the air flow virtually comes to rest; then the air flows round the cylinder, increasing in speed until the inertia of the air immediately next to the cylinder becomes too great and it breaks away to form a wake at the back of the cylinder. The wake may be turbulent or laminar, depending upon the Reynolds number, defined as $R_e = VD/y$ where V = air velocity in m s^{-1}, D is cylinder diameter in m, and y = viscosity (kinematic) of the air. Heat loss, as in free convection, depends upon the temperature gradient at the surface. Round a cylinder (e.g. a limb or the trunk) the steepest temperature gradients will be at the front leading edge, but at the sides the gradients will be flatter and the heat loss will be lower; in the wake, the heat loss increases again. The average value can be calculated and hence the overall heat transfer convective coefficient (h_c) can be worked out.

The pattern of convective heat loss from the head (Fig. 19) is similar to that from a cylinder. Facing into the wind, the greatest heat loss is from the face and then it falls steeply over the top of the head and increases at the back where the flow breaks away and is either turbulent or laminar, depending upon wind speed. Heat loss from the face varies at different sites: nose, chin or forehead. The results obtained by Clark (1977) are shown in Fig. 19. In part, these findings are due to vasomotor control of the blood flow which varies considerably over face and head. The convective heat coefficient in watts/m^2 °C is related to the square root of the air velocity, i.e.

$$h_c = 8.3\sqrt{V}$$

In a constant air movement, convective heat loss can be described with accuracy, but in real life conditions air flow is rarely uniform, and will usually be turbulent. Furthermore, the individual is often moving about and altering the rate of air flow over the body. An additional complication is the pattern of movement of the limbs in walking and running. This pendulum effect makes it difficult to evaluate the overall heat transfer. Measurements of heat loss from the thighs during running show that this is much greater than would be calculated from determination of the mean velocity of the thigh and using the formula $h_c = 8.3\sqrt{V}$. On athletes running out of doors the overall convective cooling coefficient for the whole body is found to be nearly double that which would be predicted from $h_c = 8.3\sqrt{V}$.

Heat loss in turbulent or gusting air streams can be predicted and have been verified from studies of men exposed to the air flows beneath hovering helicopters. The measured cooling coefficients are more than twice those which would be predicted. At a mean air speed of 13 m s^{-1} the formula $h_c = 8.3\sqrt{V}$ gives a value of 30 W m^{-2} °C, whereas the measured value was 72 W m^{-2} °C. It is not perhaps surprising that men, although clad in full cold-weather clothing, complained of severe chilling when working at low temperatures under helicopters.

Radiation

Heat loss by radiation is not affected by air movement but depends upon the fourth power of the difference of absolute temperatures between skin (or surface of clothing) and surrounding surfaces. The properties of the skin or clothing surfaces have to be considered, that is, the reflectivity and the degree of absorption which, in turn, vary with the wavelength of the incident radiation. Sunlight includes a relatively wide spectrum from ultra-violet to infrared. Human skin has a degree of reflectivity for short-wave radiation but very little in the long wave, including infrared. Skin colour

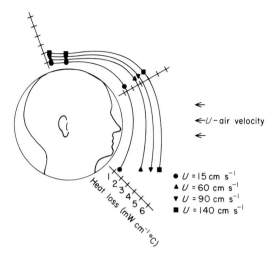

FIG. 19 Heat loss from the head in forced convection facing into the wind, at rates of air flow from 15 to 140 cm s^{-1}. (By courtesy of R. Clark.)

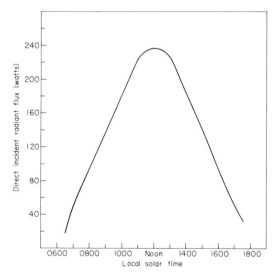

FIG. 20 Direct solar energy, in a tropical region, incident on a prone subject. (From Underwood and Ward, 1966.)

only affects reflectivity of short-wave radiation, being greater in white compared to dark or black skin. As far as longer-wave radiation is concerned, human skin acts as a black body, whatever the skin colour. This means, on the one hand, long-wave radiation is almost completely absorbed and, on the other hand, skin has a high emissivity.

An experiment by Baker (1965) illustrates the effect of a heavily pigmented skin. Two groups of subjects, one white and the other negro, exercised in a dark climatic chamber. There was no difference in body temperature between the two groups. But at the same rate of exercise out-of-doors in the sun, the negro group had a higher body temperature due to the greater absorption of infrared radiation by dark skin.

The amount of solar radiation absorbed depends upon the surface area as "seen" by the sun, which varies with posture and the sun's altitude (Fig. 22). Other factors to be considered will be the reflectivity of clothing and its area, absorption of solar radiation by moisture and dust, diffuse reflection from surrounding surfaces (rock, sand, snow, etc.) and scattered radiation, all of which influence radiation received, as shown in Fig. 20. The largest effective radiation area is that of a standing subject seen by the sun at a low solar altitude, while the smallest area will be presented at noon in the tropics, when only the top of the head and the shoulders can be seen. Measurements of incident radiation made in Aden (latitude 14 °N), with no

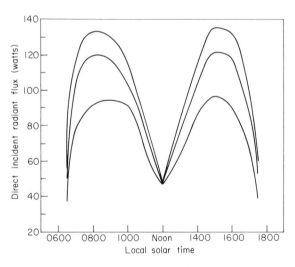

FIG. 21   Direct solar energy on an upright subject (in a tropical region). Upper and lower curves show the maximum and minimum load; the middle curve shows the mean. (From Underwood and Ward, 1966.)

cloud cover, for an upright subject showed the highest level of solar radiation in the morning and afternoon, and the lowest at midday (Fig. 21). In polar regions the altitude of the sun remains low throughout the day and hence the thermal increment due to direct solar radiation can be considerable. In addition, the reflectivity or albedo of snow is almost complete, increasing radiation heat gain. Direct measurements made in the Antarctic

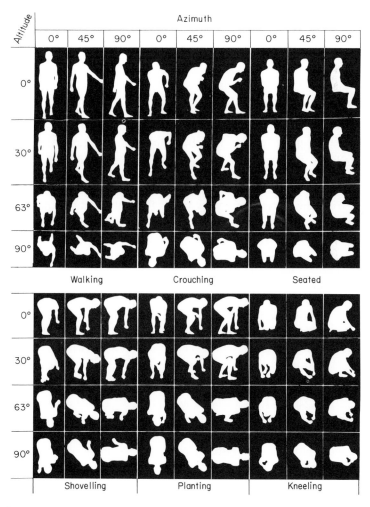

FIG. 22   A selection of silhouettes of a subject in various poses, corresponding to the areas illuminated by the sun's rays at the angles of altitude and azimuth shown. (From Underwood and Ward, 1967.)

(Chrenko and Pugh, 1961) showed that temperature differences of the clothed surface (dark coloured) facing the sun, compared with back, could amount to 30 °C or more.

Heat loss by radiation from the whole body can be larger than heat loss by convection. Indoors, while air temperature may be considered adequate for comfort ($\sim$ 21–24 °C) wall temperatures may be considerably lower, depending upon outside conditions and the construction of the building. Heat loss by radiation to cold surfaces, e.g. windows, may be large enough to make for feelings of discomfort.

Further details concerning heat loss or gain by radiation will be considered in relation to partitional calorimetry.

Evaporation

The energy required for the evaporation of sweat from the surface of the body is derived in the form of heat from the skin. The rate of evaporation depends on the difference between surface vapour pressure and that of the ambient air. Heat is also lost by the evaporation of insensible perspiration derived from the passage of water through the skin, and by the evaporation of water from the upper respiratory tract. The only circumstance in which heat can be gained by this channel is when surface temperature is lower than the dew point, when water condenses on the skin or clothing.

Unlike the regulation exercised over the sweat glands, there is no physiological control of respiratory heat loss, nor of the insensible loss through the skin.

The concept of the "wetted area" was developed to account for variation in the rate of sweating and of sweat evaporation. In very dry conditions, sweat rate and evaporation will be high; the wetted area is small. In humid climates, the wetted area may be large but evaporation will be low. This is turn will be affected by air movement, including the convective heat flow upwards over the body surface.

Calorigenesis

Rhythmical movements of the muscles in the limbs, jaw muscles and eventually trunk muscles increase heat production and diminish cooling. The metabolic rate can be increased for short periods by a factor of five, but this requires violent shivering which cannot be sustained. Over a period of more than 5–10 min metabolic rate may on average be increased three-fold. A nude resting subject immersed in water at 28 °C has a threshold core temperature (T_{oe}) of about 37.3 °C at which value vasoconstriction is already complete. The sensitivity to central stimulation under these conditions is high; the metabolism increases four fold for a 0.3 °C decrease in T_{oe} below threshold.

Although shivering at its maximum may produce heat at a rate comparable to that of strenuous exercise—200 kcal m^{-2} h^{-1}—it is not completely effective in warming the core. Some of the heat produced is transferred to the surface since the arteries travel through or near the contracting muscles *en route* to the skin. Shivering can prevent body cooling if the environment is not too cold, but without completely repaying the heat debt incurred before and during its outset (Spealman 1968).

There has been identified in the posterior hypothalamus a shivering centre which is affected by the deep body temperature as well as by afferent impulses from temperature endings in the skin, via the thermoregulatory centre in the anterior hypothalamus. Impulses from the cortex can affect or inhibit the centre, the output from which reaches the muscles via the tectospinal and ventro–spinal tracts in the spinal cord. Animal experiments show that spinal cooling can initiate shivering, and Bligh has proposed that the irregular pattern of shivering is due to a spinal oscillator.

Another phenomenon which may be associated with shivering is pilo-erection. This is a mechanism of great importance in fur-bearing animals, as the erection of hairs due to the contraction of the pilo-motor muscles traps air. Depending on the length of the hairs in the fur, a greater or lesser layer of still air is held, which provides very efficient insulation. In man, the mechanism persists but is of only trifling physiological significance. Body hair in man is often so slight that the only effect is the appearance of "goose pimples".

Partitional calorimetry

It will be evident that there are considerable difficulties in measuring or calculating the way by which heat flow is partitioned between the three channels. In 1936, Winslow *et al.* developed the technique of partitional calorimetry which made it possible to measure total heat loss and divide this into the quantities lost by evaporation, convection and radiation. The method used depended upon the construction of a nine panelled booth of copper. Air temperature and humidity were accurately controlled and the rate of air movement could be varied from 0.1 to 0.5 m s^{-1}. The chamber was heated by the reflection of infrared radiant heat from the copper walls. Copper reflects about 98% of the infrared radiation, but the copper surfaces remained at approximately air temperature. As Winslow and Herrington wrote: "It was a most unusual experience to sit in the booth and feel the waves of heat from the surrounding surfaces and then find these surfaces cold to the touch". Since wall and air temperatures could be varied over wide ranges, with the one held steady and the other changed, heat loss by radiation could be separated from heat loss by convection. In Fig. 23 is shown the way the proportions of heat loss varies with changes in environmental temperature.

Since then there have been a number of technical improvements which make it easier to determine the partition of heat loss. These include the Schlieren technique for studying convective heat loss, the use of the thermoscan, heat flow gauges and, finally, sensitive and fast responding calorimeters.

Gagge (1972) has summarized the situation as follows. For the successful use of partitional calorimetry the following minimal basic measurements are required: the rate of O_2 consumption, ambient temperature, humidity, mean skin temperature, an internal measure of body temperature and the change in body weight over the exposure period. When radiation is present, a measure made with the globe thermometer is needed. When clothing is used, the clothing surface temperature must be measured and the clothing insulation determined.

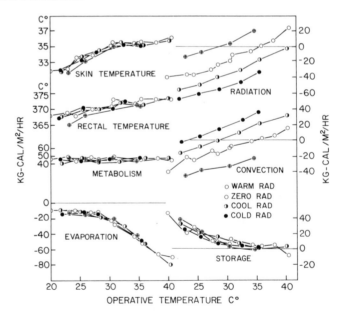

FIG. 23 Temperature values, and values for various factors in thermal balance plotted against operative temperature with mean radiant temperatures above, equal to, and below air temperature. (Winslow, Gagge and Herrington, (1940).) *Am. J. Physiol.* **131**, 79–92.

Thermal Exchange Equations

Detailed mathematical presentation of heat exchange equations for use in partitional calorimetry have been developed by Hardy (1949), Fanger (1970), Gagge (1972) and Kerslake (1972). Burton (in Burton and Edholm, 1955) has provided a simplified presentation which has proved usable and useful.

Burton's formulation:

$$M - S = H + E \qquad (1)$$

where M = metabolic heat production (kcal h^{-1}), S = heat debt (kcal $m^{-2} h^{-1}$), H = non-evaporative heat loss, E = evaporative heat loss.

The total non-evaporative heat loss (H) equals the heat flow through tissues and clothing to ambient air, under the restricted conditions that air temperature and mean radiant temperatures are equal.

$$\frac{H}{A} = \left(\frac{t_c - t_a}{1.21 I_t + I_{cl} + I_a} \right) \qquad (2)$$

where A = DuBois surface area, t_c = core temperature, t_a = air temperature, I_a = insulation of air (in clo units), I_{cl} = insulation of clothing (in clo units), I_t = insulation of tissues (in clo units).

Alternatively

$$\frac{H}{A} = 5.5 \left(\frac{t_s - t_a}{I_{cl} + I_a} \right) \qquad (3)$$

where t_s = mean skin temperature.

The evaporative heat loss is given by

$$\frac{E}{A} = 5.0 \left(\frac{p_s - p_a}{R_{cl} + R_a} \right) \qquad (4)$$

where p_s = average vapour pressure on the skin (mm Hg), p_a = vapour pressure of air (mm Hg), R_{cl} = vapour resistance of clothing (in cm of dead air), and R_a = vapour resistance of ambient air (in cm of dead air).

Fanger's Formulation

This is a comprehensive but simple treatment of heat exchange involved in heat balance at a steady state of the clothed subject. The heat balance is written as a double equation.

$$M_1 - L - E = R + C = K$$

where M_1 = internal heat production, L = dry respiratory heat loss, R = heat loss by radiation, C = heat loss by convection, and K = heat transfer from skin to outer surface of clothed body.

Internal heat production

If M is the energy produced in metabolism, in the working subject a proportion W is attributed to external mechanical work, depending on the external mechanical efficiency,

$$\eta = \frac{W}{M}$$

Thus,

$$M_1 = M(1 - \eta) \text{ kcal h}^{-1} \qquad\qquad\qquad (I)$$

is usually in the range 0.1–0.2 (values of η for different activities are given by Fanger (1970), Table 1).

Dry respiratory heat loss

This is the heat loss due to the difference in temperature between expired (t_{ex}) and inspired air (t_a). Since L is relatively small compared to other losses, a value of 34 °C may be assumed for t_{ex}, although strictly it varies with the temperature and humidity of the inspired air.

$$L = V_{C_p}(t_{ex} - t_a)$$

where C_p = specific heat of dry air at constant pressure = 0.24 kcal/kg °C and V = pulmonary ventilation (kg h^{-1}). It is convenient to relate V to M using a reasonably close approximation (Liddell, 1963).

$$V = 0.0060M \text{ (kg h}^{-1})$$

Thus

$$L = 0.0014M(34 - t_a) \text{ kcal h}^{-1} \qquad\qquad\qquad (II)$$

Evaporative heat loss, E

Total E comprises: E_{re}, the latent respiration heat loss (kcal m^{-2}); E_d, the heat loss by diffusion of water vapour through the skin (kcal m^{-2}), and E_s, the heat loss by evaporation of sweat (kcal m^{-2}).

The latent respiration heat loss (E_{re}) is a function of the pulmonary ventilation V and the difference in water content of expired and inspired air.

$$E_{re} = V(W_{ex} - W_a)\lambda \text{ kcal h}^{-1}$$

where W_{ex} = humidity ratio of expired air (kg water per kg dry air), W_a = humidity ratio of inspired air (kg water per kg dry air), and λ = heat of vaporization of water at 34 °C (575 kcal kg^{-1}). At sea level $W_{ex} - W_a$ can be reformulated simply as $W_{ex} - W_a = 0.029 - 0.00066P_a$ (kg water per kg dry air).
Hence

$$E_{re} = 0.0023M(44 - p_a) \tag{III}$$

Heat loss by diffusion of water through the skin, E_d

$$E_d = \lambda mA(p_s - p_a)$$

where m = permeance coefficient of skin, for which a value of 6.1×10^{-4} kg h^{-1} m^2 mm Hg has been derived. With a small loss of accuracy

$$p_s = 1.92t_s - 25.3 \text{ mm Hg}$$

Hence

$$E_d = 0.35A(1.92t_s - 25.3p_a) \text{ kcal h}^{-1} \tag{IV}$$

Heat loss by evaporation of sweat, E_s

For heat balance under a wide variety of conditions (not only at states of thermal comfort, which is Fanger's principal concern) evaporative heat loss by sweating (E_s) would be obtained from:

$$E_s = E - E_{re} - E_d$$
$$E = \dot{w}$$

where $\dot{w}$ is rate of weight loss (kg h^{-1}).
Hence

$$E = 575\dot{w} \text{ kcal h}^{-1} \tag{V}$$

When the skin is assumed to be maximally wetted

$$E_{max} = Ah_e(p_s - p_a) \text{ kcal h}^{-1}$$

where h_e = evaporative heat transfer coefficient.
According to Kerslake (1972, 1973, 1976) $h_e = 18\sqrt{v}$ kcal h^{-1} m^2 mm Hg.

$$E_{max} = A18\sqrt{v}(p_s - p_a) \tag{Va}$$

Gagge suggests that h_e can be obtained from h_c (the convective heat transfer coefficient) by the Lewis relation (LR).

$$LR = \frac{h_e}{h_c} = 2.2\ ^\circ\text{C per mm Hg (at sea level)}$$

Heat loss by radiation from the outer surface of the clothed body can be expressed by the Stefan–Bolzmann Law.

$$R = A_{\text{eff}}\varepsilon\sigma\,(t_a + 272)^4 - (t_{\text{mrt}} + 273)^4\ \text{kcal h}^{-1}$$

where σ = Stefan–Bolzmann constant = 4.96×10^{-8} kcal m^{-2} h $^\circ$K^4; ε = emittance of the outer surface of the clothed body; t_{mrt} = mean radiant temperature ($^\circ$C); and A_{eff} = effective radiation area of the clothed body (m^2). The emittance for skin is close to 1.0, and most types of clothing have values near to 0.95.

The radiation area is less than the clothed (or nude) surface area:

$$A_{\text{eff}} = f_{\text{eff}} f_{\text{cl}} A$$

where f_{eff} = the effective radiation area factor, i.e. the ratio of the effective radiation area to the surface area of the clothed body. The value 0.71 is an adequate approximation for sitting and standing values; f_{cl} = ratio of the surface area of the clothed body to the nude body; f_{cl} is given by Fanger (1970) in Table 2; substituting for ε, σ, A_{eff} gives

$$R = 3.4 \times 10^{-8} A f_{\text{cl}}\,[(Ct_d + 273)^4 - (t_{\text{mrt}} + 273)^4]\ \text{kcal h}^{-1} \qquad \text{(VI)}$$

Radiation can also be obtained by using the concept of Effective Radiant Field (ERF) introduced by Gagge et. al. (1967). The ERF is the radiant heat gain which the subject would have if his skin temperature equals air temperature. To obtain ERF, however, it is not necessary to know the skin temperature.

$$\text{ERF} = F_{\varepsilon C}\,\sigma\,[t_a^4 - t_s^4]\ \text{W mm}^{-1}$$

where $F_{\varepsilon C}$ = combined emittance and configuration factors (given by Fanger, 1970, pp. 175–195).

A simple direct measure of ERF is given by Gagge (1972):

$$\text{ERF} = A_r/A\,[6.1 + 13.6\sqrt{v}](t_g - t_a)\ \text{W m}^{-2}$$

where v = ambient air movement m s^{-1}.

The first term in brackets (6.1) is the value of h_v for a sphere at average temperature 27 $^\circ$C. The second term ($13.6\sqrt{v}$) is Bedford's formula for the convective heat loss from a six-inch globe; t_g = globe temperature.

Heat loss by convection

Convective heat loss (C) from the surface of the clothed body is:

$$C = A f_{\text{cl}} h_c (t_{\text{cl}} - t_a)\ \text{kcal h}^{-1} \qquad \text{(VII)}$$

For free convection $h_c = 2.05(t_{cl} - t_a)^{0.25}$ kcal h^{-1} m^2 °C; for forced convection $h_c = 10.4\sqrt{v}$.

Heat conduction through clothing, K

Heat transfer between skin and outer surface of clothing involves internal convection, radiation in the air spaces between layers of clothing, and conduction through the cloth itself. The total thermal resistance from skin to outer surface of clothing is termed l_{cl}(clo units). (Clo values are given by Fanger in Table 2.)

$$K = A\left(\frac{t_s - t_{cl}}{0.18l_{cl}}\right)\text{kcal h}^{-1} \tag{VIII}$$

The vapour diffusion resistance of clothing needs to be considered, especially at high rates of sweating.

Heat balance equation

All the eight heat exchange expressions can now be substituted in the full heat loss equation:

$$\begin{array}{cccccccc} (1) & (2) & (3) & (4) & (5) & (6) & (7) & (8) \\ H & - L & - E & - E_{cl} & - E_s & = R & + C & = K \end{array}$$

which becomes

$$\overset{(1)}{M(1-\eta)} - \overset{(2)}{0.0014M(34-t_a)} - \overset{(3)}{0.0023M(44-p_a)}$$

$$\overset{(4)}{-0.35A(1.92t_s - 2.53p_a)} - \overset{(5)}{575W} = \overset{(6)}{3.4 \times 10^{-8}Af_{cl}\,[(t_{cl}+273)^4}$$

$$\overset{(7)}{-(t_{mrt}+273)^4] + Af_{cl}h_c(t_{cl}-t_a)} = \overset{(8)}{A\left(\frac{t_s-t_{cl}}{0.18l_{cl}}\right)\text{kcal h}^{-1}}$$

The values needed are

Environmental	Body	Coefficient
t_a	η	h_c
t_{mrt}	W	h_e
p_a	f_{cl}	h_v
v	t_s	
	t_{cl}	

Acclimatization

Heat

The gradual increase of tolerance for a hot climate of a visitor from temperate regions has been familiar to those living in the tropics for a long time. It has been attributed to habituation—getting used to it—and the development of behavioural adaptations in dress, in activity and in diet. Although all these aspects are relevant, work which originated in the Second World War has shown that substantial physiological changes are mainly responsible for the phenomenon of acclimatization to heat. A majority of the extensive studies carried out in the USA and the UK have involved climatic chamber experiments. The subjects were mostly service men, aged 19–25, who had only experienced the climate of their own country. The results showed that daily exposure, usually for four hours per day, to a hot and humid climate, with a moderate level of physical work, could produce significant changes in body temperature (Fig. 24), heart rate and especially sweat rate, as well as physical performance.

As can be seen on day one, by the end of the four hours' exposure body temperature can have risen to 39.5 °C, heart rate to 180 beats per min, but sweat rate will have doubled. On day one, many subjects experience great

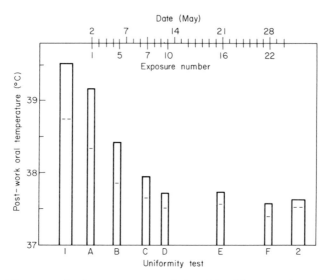

FIG. 24 The mean body temperature of the 15 subjects at the end of four hours' exposure to a hot and humid environment on successive days. The dotted lines on the bars show temperature levels at two hours. As the subjects acclimatize there is a smaller rise of body temperature.

discomfort and distress, and some are unable to complete four hours. On the last day, all subjects remain for the whole time and may be bored, but distress or even discomfort is absent.

The change in sweat rate is particularly striking, and to a large extent is responsible for the smaller change in both core and skin temperatures (Hellon et al., 1956) and hence also of heart rate.

There are two other remarkable features. First, acclimatization develops even if the experiments are carried out in winter when the subjects, on leaving the chamber, will encounter cold outdoor conditions. The second feature is that everyone tested has developed such acclimatization. It may be concluded that even with long-continued life in temperate or cold climates over many generations, without exposure to hot or tropical conditions, the ability to acclimatize is retained.

The effective stimulus to acclimatization was for some time considered to be physical activity in the heat. R.H. Fox disagreed but put forward the hypothesis that it was the rise in body temperature which was responsible for the development of acclimatization, and confirmed this concept by developing and using the technique of controlled hyperthermia. Subjects were exposed to hot and humid conditions such that their body temperature rose rapidly; when it reached the desired level (38 °C usually) the subject put on an impermeable suit and moved to a room kept close to 38 °C. Evaporative heat loss, and hence body temperature, was controlled at a steady level by blowing dry air through the suit. By successive repetition of this procedure, the changes associated with heat acclimatization could be induced. The role of physical work in promoting a degree of heat acclimatization was thus explained by the rise of body temperature.

The increase in sweat rate can be attributed to a training effect on the sweat glands —the more frequent the stimulation the greater the volume of sweat produced (Fig. 25).

The degree of acclimatization produced using controlled hyperthermia was directly proportional to the extent of the body temperature rise and its duration each day.

In recent years, thanks largely to the International Biological Programme (IBP), there have been a number of observations made on indigenous populations, using a test based on the technique of controlled hyperthermia. Some results are shown in Fig. 26. There are considerable differences between peoples living in hot countries; notably the inhabitants of New Guinea have much lower sweat rates than the Nigerians in West Africa. In all places where studies have been made there is a significant difference between men and women, as shown in Fig. 27, with women having lower sweat rates than men. Amongst the reasons for this difference is that men are more physically active, therefore more likely to sweat, hence

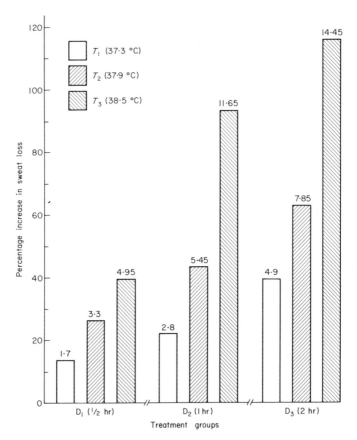

FIG. 25   Percentage increase in sweat loss in subjects acclimatized by different regimes. D1, three groups of subjects with body temperatures raised by half an hour each day; D2, three groups with body temperature raised for one hour each day; D3, three groups with body temperature raised for two hours each day. The degree of acclimatization depends on the intensity and duration of the thermal stimulus. (Figure by R.H. Fox.)

men will have a higher level of acclimatization. A similar difference between men and women who have never been exposed to hot climates has been found in the UK. However, women cyclists who can have high levels of physical activity can have sweat rate responses resembling acclimatized men (Fig. 28).

The process of acclimatization and de-acclimatization has been studied in a group of Indian soldiers, flown to the UK in September after arduous military training in the hot and humid climate of Lucknow. They were tested in climatic chambers on arrival in the UK. They were retested later,

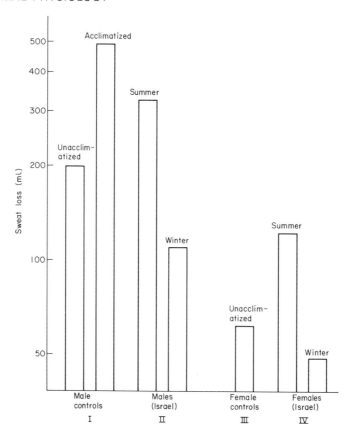

FIG. 26 Sweat loss with controlled hyperthermia. I, in male UK subjects before and after acclimatization; II, in male Israeli villagers in summer and winter; III, in female UK unacclimatized subjects; IV, in female Israeli villagers in summer and winter. (From experiments by R.H. Fox.)

in the winter, and then were re-acclimatized in the chambers. The results are shown in Fig. 29 together with those from British soldiers. On arrival in the UK the Indians could be described as being partly acclimatized on the basis of sweat rates, heart rates and body temperatures. After the winter they had lower sweat rates than their British counterparts and their acclimatization changes, although similar to the British soldiers, ended at a lower sweat rate. From these and the IBP results, as well as the differences between individuals in any one group, it seems probable that genetic factors have some effect on variation in sweat rate (Weiner, 1976).

Other physiological changes which occur during acclimatization include an increase in blood volume; this is partly responsible for the diminished

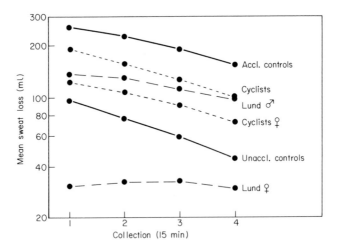

FIG. 27 Sweat loss, in four successive collections at 15-min intervals, during controlled hyperthermia in different groups. "Lund" refers to Swedish subjects studied during the summer. Note, in Figs 26 and 27, the lower sweat rates in women compared with men. (From experiments by R.H. Fox.)

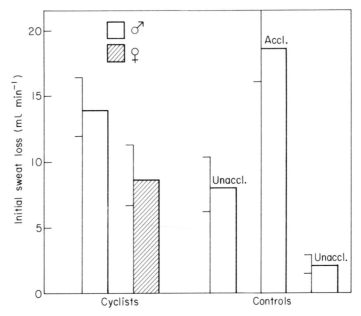

FIG. 28 Sweat loss, during controlled hyperthermia, in male and female cyclists compared with control subjects. (From experiments by R.H. Fox.)

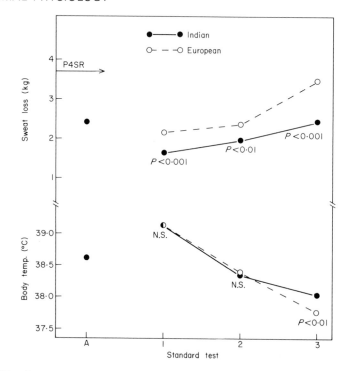

FIG. 29 Comparison of response to heat by Indian and European subjects. Standard test in climatic chamber: A, on Indian subjects immediately after arrival in UK from hot Indian summer. Test 1, sweat rates over four-hour period in chamber, and end rectal temperatures after Indian subjects had spent winter in UK. Tests 2 and 3, carried out about 10 and 20 daily exposures in hot room. The Indian subjects acclimatize more slowly than the UK subjects.

likelihood of circulatory collapse in the heat. There is also a marked fall in the NaCl content of sweat and urine.

Cold

Although unequivocal evidence for physiological acclimatization to heat is easy to obtain, similar evidence in respect of cold is largely lacking. It may be said that man's adaptation to cold is essentially behavioural; first he avoids exposure to cold if at all possible, then he wears suitable clothing if he does have to expose himself and will increase his heat output by muscular work and shivering. By way of contrast, it is easy to demonstrate cold adaptation in many animals, which is achieved by increased insulation provided by a greater growth of fur, particularly by the length of the hairs, and by an increased metabolic rate due to a non-shivering thermogenesis. A number of animals have a store of brown fat, which on exposure to cold is metabolized with a large increase in heat production.

There is some evidence of acclimatization to cold in man, of which the most convincing concerns local changes in the hands. Mackworth (1949) measured the effects of acute cooling of fingers by assessing the loss of sensation and the rate of recovery on stopping cooling. He showed that men working out of doors in severe conditions had less numbness which recovered quicker than the fingers of indoor workers. Such adaptation can be achieved by daily exposure in a cold room with the hand uncovered. The mechanisms may be vascular, as is suggested by observations of the pattern of cold-induced vasodilatation in the fingers. This effect occurs when the fingers are exposed to acute chilling by immersion in ice water, causing immediate intense vasoconstriction; the sensation of cold is rapidly followed by pain which becomes almost intolerable. After approximately five minutes the pain diminishes, often dramatically, due to a sudden vasodilatation with a large blood flow through the fingers. In those whose hands are habituated to cold, the initial vasoconstriction is less and the onset of vasodilatation more rapid, and usually greater than in unadapted subjects. LeBlanc (1957) has also described another cold-induced cardiovascular reflex. Immersion of the face in cold water causes a marked slowing of the heart and vasoconstriction of the hands. In Eskimo subjects this response is greatly reduced and may be absent.

The activity of the thyroid gland as a result of exposure to cold has been shown, in animal work, to increase, with a consequent increased metabolic rate. Studies on man have, in general, been inconclusive. Enhanced sensitivity to the calorigenic effect of noradrenaline has been claimed to occur after prolonged exposure to cold (Budd, 1962; Kay et al., 1970).

It may well be that the lack of evidence of cold-acclimatization in man is due to the reluctance of subjects to endure sufficiently severe cold exposure. Members of an Australian Antarctic expedition were examined by Budd (1962) in a cold climatic chamber in Melbourne, before travelling to the Antarctic where they were studied on three occasions in the same conditions as in Melbourne, where the final cold test took place. On the first and last occasions body temperature dropped gradually in the cold room. In Antarctica this was reversed by the time of the third test when body temperature actually rose. This is at present the most convincing evidence that man, like other animals, can, if exposed to cold for a sufficiently long period, develop at least a degree of general acclimatization.

Exercise

The heat load induced by exercise is proportional to the rate of oxygen usage less that attributable to external work. At his V_{O_2} max an individual could generate as much as 20 kcal min^{-1} (1000 watts) for 3–5 min. This upper limit of aerobic heat production is matched by a correspondingly high potential heat loss from the body utilizing the combined channels of convection, radiation and evaporation. Under favourable conditions of low t_a (say 15 °C), low p_a (say 7 mm Hg, i.e. 50% R.H.), and a moderate air movement, the heat exchange equations indicate heat loss of the same order as the heat generated during work, about 20 kcal min^{-1}. These values are meant merely to show the remarkably high total capacity of the heat dissipation mechanisms, without which body heat gain would be rapid— about 0.3 °C min^{-1}—and heat stroke would be approached in about 15 min. Not only is the heat loss capacity high, it is also rapidly activated.

These features are well exemplified by the performance of marathon runners. At a heat output at about 75% V_{O_2} max (560 kcal m^{-2} h^{-1}), under cool conditions, temperature regulation is not a limiting factor (Buskirk and Beetham, 1960); under warm conditions (22 °C d.b., 13.2 °C w.b.), core temperature can be maintained steadily at 39.5 °C for three hours (Adams et al., 1975). This study (Fig. 30) illustrates the major factors influencing homeothermy at exercise—high environmental heat load, level of work, acclimatization and dehydration.

Total heat load can be altered either by increasing the work load at a fixed ambient temperature or by increasing the heat stress at a fixed work load. How does the control system operate in these two situations? In the first case, T_{core} (Fig. 31), sweat rate and, concomitantly, heat conductance, increase, but T_s remains at a constant level (Fig. 9). In the second case, T_s (Fig. 9) sweat rate and conductance increase but T_{core} remains constant over a wide range (Fig. 31).

The resultant partition of heat flow in the two situations is as shown in Fig. 32. With increasing intensities of work, convection and radiation contribute a constant outflow, the remainder of the heat loss being provided increasingly by evaporation. With T_a increasing and work held constant, convection and radiation heat loss fall progressively until, at room temperature close to skin temperature, evaporation becomes the main channel.

By studying the working subject t_c and t_s can be separated as input stimuli to the regulator system though in fact, as Robinson (1949) first showed, these inputs always interact. In his experiments, lowering the environmental temperature raised the t_c threshold for sweating, but the change in sweat rate per unit change of t_r remained constant.

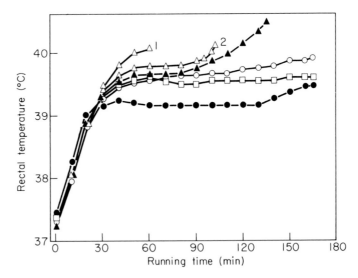

FIG. 30 Changes in rectal temperature (°C) measured using the radio pill during selected runs. △————△¹, one-hour run in hot environment, pre-acclimatization; △————△², maximum-duration run in hot environment, pre-acclimatization; ○————○, marathon distance run in moderate environment, pre-acclimatization; □————□, marathon distance run in cool environment, pre-acclimatization; ●————●, marathon distance run in moderate environment, post-acclimatization; ▲————▲, maximum-duration run in hot environment, post-acclimatization. (From Adams *et al.*, 1975.)

This interaction between t_{core} and t_s is well demonstrated by Nadel *et al.* (1971) in Fig. 33, and has been given expression in mathematical models incorporating proportional control. One such model in general terms is the following (Nadel, 1979):

$$E = [\alpha t_c + 0.1\alpha t_s + y]exp$$

where E = whole body evaporative rate; α = proportional control constants; y = constant term; exp = Q_{10} effect of T_s on sweat gland responsiveness.

On this formulation, there is no difference between exercise and rest on sweat response to drives from t_c or t_s, but this remains a matter of controversy. Nielsen's (1938) finding of a stablilized raised t_c despite rising t_a (Fig. 31) has been taken to mean that exercise elevates the set-point, but most workers regard the raised core temperature as a necessary consequence of a proportional control system—the temperature reaching constancy under the combined action of evaporative, radiative and convective heat loss. At a particular t_a and work load (Fig. 31), the reduction in R and C leads to a shifting upwards of evaporation until maximum thermal

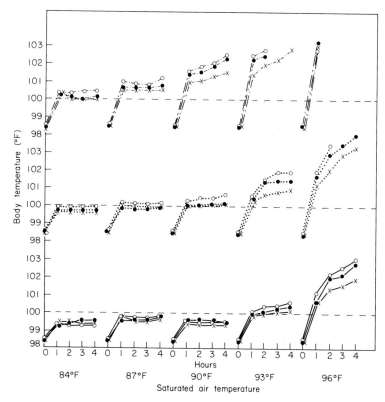

FIG. 31 Body temperatures (mean of 13 men) at each hour of four-hour exposure to above temperatures and *continuous* work: at 100 cal m^{-2} h^{-1} and 400 (X————X); 150 (●————●) and 50 (○————○) ft min^{-1}; at 150 cal m^{-2} h^{-1} and 400 (X · · · · X), 150 (● · · · · ●) and 50 (○ · · · · ●) ft min^{-1}; at 240 cal m^{-2} hr^{-1} and 400 (X · — · — · X), 150 (● · — · — · ●) and 50 (○ · — · — · ○) ft min^{-1}. (From Wyndham *et al.*, 1952.)

conductance (set largely by circulatory limitations) and evaporative capacity are reached.

Despite the high intrinsic ability of the body to cope with heat stress, high external temperatures act to limit the intensity and duration of exercise. Above the effective temperature of about 32 °C, even an acclimatized subject has reached his maximal work output. This limitation is a consequence not only of "saturation" of the heat dissipation mechanisms but also of the concomitant disturbances of the cardiovascular system.

Increase in skin blood flow is a crucial response to both exercise and elevated ambient temperatures. Exercise at a given ambient temperature necessitates an increased blood flow even with a steady or falling t_s (Fig.

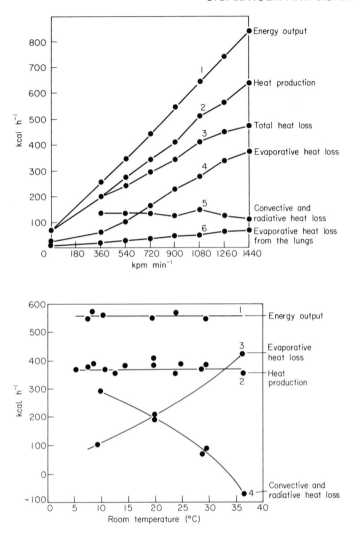

FIG. 32 (a) Heat exchange at rest and during increasing work intensities (expressed in kilopond meters per minute along the abscissa) in a nude subject at a room temperature of 21 °C. See the text for further explanation (b) Heat exchange during work (900 kpm min^{-1}) a different room temperatures in a nude subject. (From Åstrand and Rodahl, 1970.)

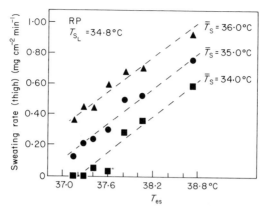

FIG. 33   Effect of T_{es} and $\bar{T}_s$ on local sweating rate with local (under sweat collection capsule) skin temperature $= 34.8\,^{\circ}$C. (From Nadel *et al.*, 1971.)

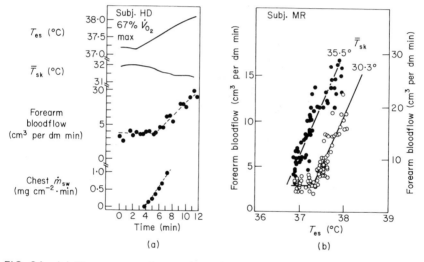

FIG. 34   (a) Time course of oesophageal temperature, mean skin temperature, arm blood flow and chest sweating during a 12-min exercise period at 67% V_{O_2} max. Air temperature 25 °C. (b) Arm blood flow plotted against oesophageal temperature for exercise at two different air temperatures. (From Roberts and Wenger, 1979.)

34). The stimulus to the increased blood flow and the concomitant evaporative heat loss is clearly the increase in t_c. (Fig. 35).

The increased peripheral blood flow during exercise at high temperature carries with it consequences to the cardiovascular system as a whole. The cardiac output in the heat does not increase above that in the cold for the same level of oxygen usage (Williams *et al.*, 1962) but the competition between the increased cutaneous and muscle flows necessitates a number of compensating responses. When skin temperature rises the capacitance of cutaneous venous vessels increases, since sympathetic venomotor tone is reflexly inhibited (Webb–Peploe and Shepherd, 1968). A given distending pressure will produce a larger increase in arm venous volume than at cooler temperatures. The large volume of blood in the dilated cutaneous vessels progressively reduces central venous pressure. The reduced cardiac filling in turn reduces the cardiac stroke volume and the output is maintained by an increased heart rate. But this compensation fails when the combined stress of exercise and heat raises the heart-rate to 180–190 beats a minute, as Rowell *et al.* (1966) found when their subjects worked at 60% $\dot{V}_{O_{2\,max}}$ at 43 °C. With an unsustained cardiac output and in the face of widespread vaso- and venodilatation, blood pressure will fall if reflex regional vasoconstriction mediated by arterial and venous baroreceptors and by reduced central venous pressure proves inadequate. The instability of the circulation due to "peripheral pooling" during or after work at high temperatures has long been known, and can be demonstrated by subjecting the subject to prolonged periods in the upright posture, when vasovagal syncope may ensue (Weiner, 1938).

Exercise itself brings about both splanchnic and cutaneous vasoconstriction. It seems likely that during work in the heat, baroreflexes from low central venous and arterial pressures operate to over-ride in some degree the cutaneous vasodilator reflexes. As Roberts and Wenger (1979) put it: "the disadvantage of decreased skin blood flow is that it forces a higher body temperature by decreasing core-to-skin heat flow, but the advantage of cutaneous vasoconstriction lies in the maintenance of blood pressure and muscle flow". Another manifestation of the competition between muscle flow and cutaneous flow during work under heat stress is that "excess" lactate appears at significantly lower levels of work in heat than in comfortable conditions (Williams *et al.*, 1962).

Another threat to the stability of the circulatory system is the occurrence of progressive dehydration due to sweating without adequate water replacement. It is precisely when the cardiovascular system needs to cope with the combined demands of the working muscles and the skin circulation that a normal blood volume needs to be maintained. Many investigations have revealed the superior work performance and better-controlled body tem-

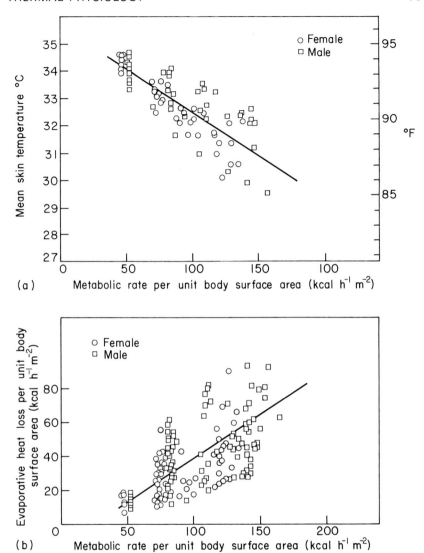

FIG. 35 (a) Mean skin temperature as a function of the activity level for persons in thermal comfort. In order to maintain thermal comfort the ambient temperature was lower the higher the activity level. (b) Evaporative heat loss as a function of the activity level for persons in thermal comfort. There are more points plotted than in (a) since data by McNall *et al.* are included in the analysis (McNall, P.E. Jr, Jaax, J., Rohles, F.H., Nevins, R.G. and Springer, W. (1967). *ASHRAE Trans.* **73**, I.). (From Fanger, 1970.)

peratures of well-hydrated subjects. Unacclimatized marathon runners (Adams *et al.*, 1975), for example, who suffered a fluid deficit of the ordernof 4% in "hot" runs, experienced a fall in heat conductance, a higher rectal temperature (see Fig. 30) and a poorer performance as compared to the hydrated condition. Acclimatization (with water replacement) also improves physiological performance, partly because of the improved responsiveness of the sweating mechanism and partly because of the expansion in blood volume.

Salt and Water Balances

Salt balance

The adjustments in the heat regulatory system necessary for life and work in hot climates involve also the metabolic and endocrine systems.

Sweating is inevitably associated with alteration in the salt balance of the body, simply because of the loss of salt contained in the sweat. There is a large literature on the composition of eccrine sweat (e.g. Robinson and Robinson, 1954; Weiner and Hellman, 1960). The over-riding fact about human thermogenic sweat is that it is hypotonic to blood (when collected by a satisfactory method) and that in its composition as well as in osmotic pressure it differs from blood plasma and extra- or intracellular fluids. Sodium chloride is the main solid constituent and is responsible for about 90% of the osmotic activity.

There would seem to be a simple relationship between intensity of sweating and increased salt loss. But the relationship between sweat rate, salt intake, salt loss via the skin and the salt content of the body, i.e. the factors making up the salt balance, is complex.

In unacclimatized individuals, heat exposure produces sweat of varying salt concentrations. It increases with the rate of sweating and elevation of the skin temperature and also with duration of the sweating.

The level of dietary intake is perhaps the major factor determining the salt content of sweat. On changing to a low salt diet the sweat within a few days becomes diluted as a measure complementary to the action of the kidney for restoring salt equilibrium (Conn, 1949; Weiner and van Heyningen, 1952).

If a negative salt balance develops, the fall in plasma sodium indicates an additional secretion of aldosterone which then acts first on the kidney and then on the sweat gland (Collins, 1966) to increase the reabsorption of sodium and chloride ions. An increased secretion of aldosterone has in fact been detected in men under conditions of heat exposure.

During the acclimatization process a gradual reduction in the salt content of sweat is usual. The mechanism of this change lies in the negative salt balance which can develop as a result of a sudden increase of skin loss of salt due to sweating, when the salt intake remains unchanged and at a moderate level (say 8–10 g per day).

In acclimatization the skin temperature falls progressively (Hellon *et al.*, 1956) and this could be expected to reduce the chloride level. But at the same time the sweat rate is at a higher level and this would tend to bring about a more concentrated sweat. It seems clear that the salt balance is the major factor determining the level in sweat.

After a week or so of heat exposure the compensatory adjustments by the kidney and the sweat glands will usually restore a steady salt equilibrium even on intakes which are often lower than those in temperate climates (Gleibermann, 1973). McCance *et al.* (1971) found that their Cambridge and Sudanese subjects (male and female) tok an average of 13 g of salt in Cambridge, but only about half this amount in Khartoum. This amount provided not only for palatability but also a sufficient safety margin above the daily obligatory losses.

Water balance

Under heat load and particularly during the acclimatization period the water balance is profoundly altered. Whereas in temperate climates under ordinary living conditions the cutaneous (and respiratory) water losses range from 700 to 2000 ml per day, averaging about 1200 ml per day (McCance *et al.*, 1971), it is possible in hot climates for the sweat loss to approach ten times that average figure.

In ordinary life in the tropics the daily water losses are of the order of two or three times those in a temperate climate (Doré *et al.*, 1975; McCance 1971). A useful guide to the range of 24 h sweat losses in the desert and in the tropics is given by Adolph (1947, p. 323).

Where water for drinking is unrestricted there is a clear tendency for daily balance between intake and loss (by sweating and the kidney) to be established. Water balance is regulated to within $\pm 1\%$ of body weight from day to day, on either moderate or heavy sweat outputs. This holds true even when there are marked changes in calorie balance, and therefore body weight. In the experiments of Edholm *et al.* (1964) in Aden, the day-to-day water balance was maintained within $\pm 1\%$ of body weight in 80% of the 60 values recorded. Adolph showed that a 24 h period was necessary for negative water balance to be restored. During active sweating, even when subjects are allowed free access to water, they undergo a period of temporary dehydration or "voluntary" dehydration. The thirst mechanism operates along with urinary restriction to restore the water deficit. Urinary restriction is brought about by the increased secretion of antidiuretic hormone (Hellman and Weiner, 1953) and by the diversion of blood from the kidney (Kenney, 1952). Diuresis is a familiar effect of cold. Frequency of micturition is due to an increased secretion of urine, because of a suppression of the antidiuretic hormone of the posterior pituitary. In some conditions, with subjects wearing a protective immersion suit and floating in water kept at 5 °C, there will be gradual cooling of the posterior surface of the body and some fall of body temperature. Urine secretion can then rise from under 1 ml min^{-1} to 6 ml or even more.

Comfort

The indoor environment can be controlled using air-conditioning systems to maintain almost constant temperature, humidity and air movement. Thermal radiation cannot always be completely controlled if there are large glazed surfaces, but with proper design new buildings can be constructed so that any desired indoor climate is provided. Outdoors, thermal comfort can be achieved in a wide variety of conditions, using suitable clothing matched to activity levels.

The main physiological problem is to assess the factors which make for sensations of comfort or discomfort. The first conclusion is that it is the absence of discomfort which is the most important element in determining thermal comfort. The second conclusion is that there is a significant correlation between mean skin temperature and thermal sensation. There is also a considerable spread between individuals in the same environment in feelings of comfort. However, the preferred mean skin temperature is about 33 °C. There are no sensory endings in the skin subserving the sensation of comfort as such, and thermal sensations from the skin are confined to heat and cold. As with other sensory nerve endings, thermal ones adapt after a period. The sensations are relative, as illustrated by the experiment in which the right hand is placed in hot and the left in cold water. After a few minutes both hands are put into the same bowl of tepid water. The right hand will feel cold and the left hand warm, although both are at the same temperature. The temperature of the face is critical; if it is too hot or too cold, then even if the rest of the body is at a comfortable level the overall thermal judgement will be either too hot or too cold.

Although skin temperature is the main determinant of thermal sensation, core temperature certainly plays a part. In most conditions of ordinary life, core or deep body temperature remains almost constant apart from the circadian variation; hence it is skin temperature which varies and which determines thermal comfort.

The complexity of the relationship between sensation and the level of skin and core temperature can be illustrated by an experiment in which the subject was immersed in a very well-stirred bath. Initially, this was kept at 35.5 °C which felt neutral, even though deep body temperature tended to fall slightly. The bath was next heated rapidly to 41 °C; at first the subject enjoyed the hot water but, as body temperature rose, he gradually became uncomfortable. When the aural thermometer registered 38 °C, bath temperature was lowered to this level, so maintaining a deep body temperature of 38.1–38.3 °C. The subject complained of severe discomfort, although with a normal body temperature such a bath would have felt pleasantly warm. The bath was then cooled to 25 °C, with immediate relief of discomfort even though deep body temperature was still raised.

An experiment by Cabanac illustrates the point. His subject, sitting at rest lightly clad, on a bicycle ergometer, was asked to state whether a bowl of water was unpleasantly hot or cold, or comfortable. When the air temperature was 21 °C, the water at 20 °C felt cold, and the preferred temperature was 41 °C. When the subject was riding hard with a high rate of heat production, water at 41 °C was described as too hot, whereas 20 °C was the favoured temperature.

Preferred conditions have been studied in worldwide surveys using the Bedford questionnaire, in which subjects are asked if thermal conditions are comfortable, whether they would prefer it to be a little warmer or a little cooler, or whether they are too cold or warm, or much too cold or too warm. This provides a seven-point scale. From the results of such surveys, it is possible to construct a "comfort zone" in which the majority of people are thermally satisfied. The general experience has been that within such a comfort zone there are always a few who will feel too hot or too cool. When surveys are carried out in different climatic zones results show that preferred temperatures range from 17 °C up to 31 °C, the higher value being characteristic of the hotter climates. As well as these wide contrasts, the following appear to affect comfort conditions: sex, with women tolerating low temperatures better than men; age, children with a high metabolic rate do not complain of cold when adults do; old people, who may have blunting of temperature sensation, can be at risk in cold or cool conditions. Much bigger differences are due to clothing and physical activity. These findings pose considerable problems for environmental designers and engineers.

However, these are not the only findings. Extensive studies were carried out in the USA by Nevins and Fanger (1968) in climatic chambers and by Fanger (1970) in Denmark. The subjects were asked to state if they were comfortable or wanted to raise or lower the temperature. Similar clothing providing 0.6 clo units of insulation were worn by all subjects and they remained seated throughout the period of observation. Under these precisely controlled conditions similar comfort votes were obtained for all subjects. There were no significant differences between men and women, and body build or the thickness of subcutaneous fat did not affect the results. In a limited number of experiments, no effects of age could be demonstrated. There is evidence that in old people thermal discrimination is reduced, and they take longer to choose their preferred temperature (Collins, 1980). Furthermore, a group of subjects habituated to a hot climate had virtually the same temperature preferences as everyone else. Fanger concludes that normal, healthy human beings sitting at rest and wearing 0.6 clo units are all thermally comfortable at 25 °C, with walls and air at the same temperature, air movement about 0.15–0.2 m s^{-1} and relative humidity, between 30 and 75%.

Amongst the possible reasons for the contrast between Fanger's findings and the results of many worldwide surveys are the strict controls in the climatic chamber. In ordinary life, conditions are seldom so consistent; temperature, air movement are not constant, and there is variation in activity even in office conditions where people will sit or stand or walk about. The most likely explanation is that individual differences are exaggerated when the Bedford scale is used and minimized in the climatic chamber. In general, for sedentary or light work, a temperature of 21.24 °C represents the comfort zone for a large majority. Subjects acclimatized to high temperatures will probably accept warmer conditions. Another reason is that the Bedford scale is relatively crude, with about 3 °C separating each category; only large differences between samples are significant because of marked individual variation. It is possible that the subjects used in the Copenhagen experiments were not sufficiently varied to show up thermal differences.

Over a wide range of activity, clothing and environmental conditions the flexibility of thermoregulatory responses ensures (in the steady state) the maintenance of a near constant internal body temperature. The flexibility of the thermal exchanges that are involved depends essentially on appropriate adjustments of the two variables–mean skin temperature (t_s) and evaporative heat loss (E). But over this wide range of homeothermy there is only a narrow band within which thermal comfort can be experienced; the corresponding values of t_s and E will therefore be within quite narrow limits.

Finger (1970) has determined experimentally the values of t_s and E for comfort of subjects at different activity levels. His regression analysis of the data in Fig. 35(a) shows that for constant comfort, t_s decreases with increasing activity ($H/A = $ kcal m^{-2})

$$t_s = 35.7 - 0.32\frac{H}{A}\ °C$$

and from Fig. 35(b) the sweat secretion needed for comfort increases with activity (for the sedentary subject E is zero) according to the regression

$$E = 0.42 \left(\frac{H}{A} - 50\right)\ \text{kcal h}^{-1}$$

Fanger has introduced these two comfort relations into the general heat balance equation, so converting it into a thermal comfort equation. From this he has specified the combinations of activity level, clothing insulation, air movement, ambient temperature and humidity which create optimal comfort (Fanger, 1970 Figs 3–11).

Fanger has further extended the thermal comfort equation so that it specifies or predicts degrees of discomfort (on a seven-point scale) for groups of subjects, as a function of activity, clothing, air temperature, mean radiant temperature, air velocity and air humidity.

In outline, Fanger's treatment is as follows.

The thermal balance equation can be written as a double transfer equation:

$$H - E = K = R + C$$

where H = metabolic level; E = evaporative and respiratory heat loss; K = heat loss by conduction through clothing; R = radiation heat loss from surface of clothing; and C = convective heat loss from surface of clothing.

This double heat transfer equation is a function of the following form:

$$f\left(\frac{H}{A}, I_{cl}, t_a, t_{mrt}, P_a, v, t_s, \frac{E}{A}\right) = 0$$

where A = duBois surface area; I_{cl} = clo value of clothing; t_a = air temperature; t_{mrt} = mean radiant temperature; P_a = vapour pressure of ambient air; v = air velocity; t_s = skin temperature; and E = evaporative heat loss.

Of these terms, $t_s = f(H/A)$, and $E/A = Af(H/A)$, specify comfort relations as pointed out above, and by substituting these values we obtain the comfort equation of the form

$$f\left(\frac{H}{A}, I_{cl}, t_a, t_{mrt}, P_a, v\right) = 0$$

Thus the comfort equation contains all the variables which need to be specified for thermal comfort (or discomfort), namely: clothing, (I_{cl}); ambient conditions (t_a, t_{mrt}, P_a, v); and H (activity level).

For the detailed and expanded derivation of these comfort equations the reader is referred to Fanger's exposition.

Behaviour

A number of distinct behaviour patterns affecting thermal exchange can be identified: (1) postural changes, i.e. huddling in the cold or extending limbs in the heat as the sunbather does, so altering the surface area exposed; (2) moving from harsh environments (hot or cold) into better situations, i.e. seeking the shade in hot sunny climates or shelter from the wind behind trees, shrubs or buildings; (3) the use of clothing and shelter to protect from heat or cold; (4) changes in purposeful activity, increasing muscular work in the cold by stamping the feet and swinging arms, or by minimal physical effort in the heat. The speed of walking appears to be related to environmental temperature. Work which involves muscular effort in hot climates is often limited to the cooler times of the day, Agricultural workers in the tropics, who work only in the early morning, hav a low level of heat acclimatization. (5) As an extension of the use of shelter, an important behavioural aspect of human culture is the deliberate modification of the thermal environment (Wulsin, 1968). Technology can provide complete indoor air-conditioning with control of temperature, humidity, air movement and shelter from solar radiation.

Clothing

The main function of clothing from a thermal point of view is to provide insulation. Unlike animals, which utilize fur for variable insulation, man has to use materials, fabrics or skins of animals. Clothing can also be used as protection from solar radiation, and as a means of increasing heat loss by convection and evaporation. This is achieved by the use of light weight, loosely woven white fabrics which have a high reflectivity; when moving about with two (or more) layers made from these fabrics there is a bellows action which increases convective heat loss. Birnbaum and Crockford (1979) have described a "ventilation index" which includes a measure of the volume of air within the clothing assembly, and the rate of air exchange. To calculate the index, nitrogen is fed into the clothing assembly until the oxygen level has fallen below 10%. The nitrogen inflow is stopped, the rate of return of oxygen to atmospheric levels is monitored, and from the oxygen trace the exchange rate can be calculated.

The pith or *sola topee* was for long popular with Europeans living in the tropics, as it was then believed that there were dangerous actinic rays in sunlight which had remarkable powers of penetrating the skull and directly affecting the brain. There are, of course, no such rays and heating of the head is due to the direct effect of radiation from the sun and surrounding surfaces.

The insulation of clothing depends upon the air trapped between the fibres of the fabrics used, and between the layers of clothing. Still air is an effective insulator but natural convection currents will carry heat away from the body suface. Convection currents are greatly reduced or abolished when air is confined in narrow spaces: so-called "dead air". The insulation amounts to 1.9 clo units cm^{-1}, or a conductivity value of $k = 0.0083$ kcal cm^{-2} s °C cm. By filling the space between two layers of fabric with material like cotton wool or feathers, insulation of any desired value can be achieved by increasing the thickness. The filler immobilizes the air trapped between the layers so that it becomes "dead air". This has led to the aphorism that "The thermal insulation of clothing is proportional to the thickness of dead air enclosed". It follows that the material of which clothing is made is not critical as far as insulation is concerned: cotton is as effective as linen, silk or man-made fibres. The response to compression or to the relief of pressure is important since it is the thickness of the dead air which matters when as a result of pressure, from sitting or lying, the thickness is reduced; also, the rate of recovery from compression is relevant. Cotton wool is as good as down feathers uncompressed. Under equal load the cotton wool will be compressed rather more than the elastic feathers; with decompression the cotton wool remains compressed whereas the feathers recover their full thickness. Hence the practical superiority of down over cotton wool or kapok.

Air movement between the layers of clothing can greatly increase heat loss (by convection). Effective cold-weather clothing requires a windproof outer garment. A plastic cover provides complete impermeability as well as being waterproof. However, plastic is impermeable to water vapour; evaporated sweat (insensible and sensible) condenses on the inner surface. Except for short periods, a plastic coverall is unsatisfactory.

The movement of moisture through clothing can be affected by the materials used. In any case, in cold or cool weather there can be a considerable temperature gradient from the skin to the clothing surface; water evaporated from the skin surface can condense in the outer layers of clothing. Apart from affecting the insulation, if the atmospheric temperature is below 0 °C there is the likelihood of water freezing in these outer layers. For this reason, one of the hazards of polar clothing is the condensation of sweat produced as a result of hard physical work. Currently, the best solution is provided by very tightly woven cloth which allows water vapour through but a minimum amount of air.

For cold-weather clothing the coverall should be dark since heat gain from solar radiation can be considerable even in polar regions.

When clothing becomes soaked, the dead air is displaced by water which has a much higher ($\sim$ 230 $\times$) thermal conductivity. If the clothing is not wind-proof the wearing of saturated clothing out of doors is virtually useless, as the insulation will be reduced to that provided by the fibres of the fabric which is almost negligible, and additional heat is lost by evaporation. The effect of rain, snow or sleet on temperature balance can be critical for climbers or hill walkers, even in the so-called temperate climate of Britain.

The changes of the moisture content of clothing are related, not to absolute, but to relative humidity. At low or moderate temperatures when the water content of even saturated air is low, the relative humidity may be 100% and clothing will take up water, with a heat regain, and a heat loss when water is given off to the atmosphere. The amount of heat transferred depends on the fabric. Wool has a high heat regain and in rainy weather an initially dry coat may gain up to 15–30 kcal in the course of an hour.

Another characteristic of fibres used in making cloth is the degree of wicking, or the extent of water transport along a fibre. Wool has a high wicking index, whereas man-made fibres do not wick at all.

The design of protective and cold-weather clothing

It is necessary to provide special clothing for severe conditions which may be found in nature or as part of an industrial environment, for example in steel works, with intense radiation from molten metal, or the "wild heat" (radiation) of engine rooms. Extreme conditions may be experienced in polar regions or at very high altitudes. Other examples of special clothing are the anti-G suits provided for air pilots and the suits they may require to protect them against heat or cold (see Chapter 4).

Some features of cold-weather clothing have already been mentioned: others can be illustrated by examining Eskimo dress. This frequently consists of caribou skin worn as a double layer, with the hairs on the outside layer pointing outward and the hairy side on the inner layer directed towards the body. This is more effective than the use of fur with hair twice as long as the single caribou layer, for a reason which can be found in engineering principles. The bending of a beam under load is proportional to the length squared. This can be applied to the designing of clothing using artifical fur.

Caribou hairs are hollow, and this also increases the effective insulation. Eskimo clothing can also be easily adjusted by opening the front of the clothing and, if necessary, letting it fall to the waist. This is an important provision to overcome the problem of a constant clothing insulation with a varied heat production.

Contemporary cold-weather clothing needs to be designed for polar travellers, for the armed forces and for the inhabitants of northern countries such as Canada, Russia, Siberia and northern Scandinavia. Such clothing includes windproof coveralls, light-weight fabrics, with fillers made of down or suitable man-made materials. The layer principle is used, i.e. multiple layers, each relatively thin, rather than one or two very thick layers. In this way insulation can be varied relatively easily.

A particular difficulty has been the protection of the extremities—hands, feet, head/face. Fairly satisfactory boots, with thick insoles and two or more layers of socks are effective, but hand protection is complicated by the shape of the fingers. When insulation is added to thin cylinders, heat loss increases because of the larger surface area. Very thick insulation must be provided before heat loss is diminished. It is easier to provide mitts enclosing all the fingers, but even then the required thickness of the mitts limits the use of the fingers.

For the head, many types of helmet have been designed but these leave the face unprotected. Covering the face, leaving openings for the eyes, is unsatisfactory without provision of low resistance to respiration, and some means of dealing with the moisture contact of expired air. Heat exchangers have been devised, the inspired air being warmed by the expired air, but in practice these have not proved very acceptable. The commonest solution is the parka, covering both head and neck, attached to whatever jacket is worn, and with wide margins, like a monk's cowl, and is frequently lined with fur with a thick wire in the edge which can be moulded as required. Goggles for protecting the eyes from cold and from snow blindness are still unsatisfactory. Many techniques have been tried to prevent misting of goggles; none is ideal.

The limitations of clothing insulation are due to the bulk and overall thickness of clothing which makes movements clumsy and difficult. An average thickness of 25–30 mm, providing about 4.0–4.5 clo units, represents the upper practical limit. Greater insulation, up to 8 clo, can be provided in sleeping bags.

A further limitation is that heat exchange is not exclusively through the clothing. There are parallel channels from the respiratory tract and from exposed areas such as the face. At low temperatures, depending upon the respiratory volume, up to one-third of the total heat production may be lost by these extra-clothing pathways.

The use of electrically heated gloves and socks for the feet have helped to overcome some cold-weather clothing limitations. Relatively low wattages are needed to maintain the warmth of the hands, owing to vasodilatation.

Protection against heat can be achieved by using double layer impermeable suits through which air can be blown at the required temperature. Fairly simple designs have been used in engine rooms, where the radiation heat load is high, to distribute cool air to the limbs and the trunk. Air supply has to be available at an adequate pressure with well-insulated pipes which can be plugged into the suit. Venturi systems are used to assure a supply of cold air.

More sophisticated designs have been developed for suits for air-crew, with a complex tube system distributing the air so that skin temperature in the different body regions is kept at a comfortable level.

Temperature and Performance

In both severe heat and severe cold, the ability to carry out work, both physical and mental, can be impaired. This can be due to changes in body temperature or to the cooling of extremities, with loss of sensation in fingers or fall in muscle temperature altering speed of contraction and relaxation. Impairment of efficiency can also be due to the hampering of movement by the wearing of gloves and clothing in the cold. The discomfort and irritation caused by excessive sweating in the heat can modify performance. The relationship between environmental conditions and the performance of skilled tasks has been studied extensively. The number of errors made in specific tasks is affected by environmental temperature. Experiments show that fewer signals are missed at an effective temperature of 28 °C than either in hotter (33.5 °C) or cooler (20 °C) conditions.

Raised body temperature has a significant effect on mental performance, but this varies according to the task; mental arithmetic scores fall, while auditory vigilance improves according to Wilkinson. One must be cautious in assuming that effects of temperature are always in the same direction.

Extensive studies were carried out by Colquhoun (1970) on the influence of circadian body temperature changes on mental performance, employing both calculation test and a vigilance test. He concluded that in a normal body temperature range (36–37 °C) performance in the calculation test rose and fell with rise and fall of body temperature. In the vigilance task, the detection rate was also related in th same way but the latency of response was less clearly affected.

Since in some industries severe heat can be experienced, investigations

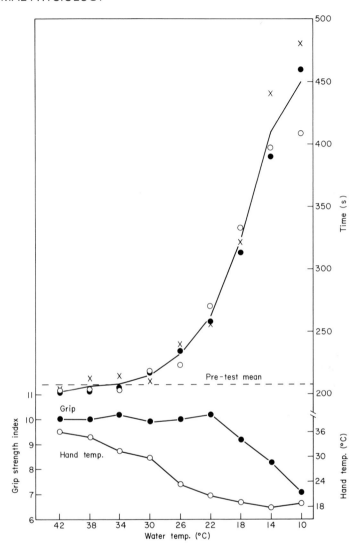

FIG. 36   Effect of immersion of arms and hands in water at different temperatures on typing performance. The time to type a passage increases as the temperature falls. Hand grip strength also declines.

have been carried out to determine safe exposure times, i.e. the period an individual can work in a particular environment before collapsing (Bell *et al.*, 1971). There is considerable individual variation and it is still difficult to predict whether an individual will be tolerant or intolerant. The thermal conditions studied ranged from 37 °C dry bulb and 30 °C wet bulb to 83.4 °C dry bulb, 41.2 °C wet bulb. The wet bulb temperature was critical. Bell *et al.* found that the most useful scale describing the severity of the environment was 0.23 dry bulb + 0.77 wet bulb. Collapse was considered imminent when body temperature exceeded 39 °C and heart rate was either irregular or about 200 beats per min. At 40 °C dry bulb, saturated with water vapour, the tolerance time for 99% of subjects was 16 minutes, but 95% would be safe for 20 minutes, 75% for 26 minutes and 50% would be able to work for 32 minutes. In the studies by Bell *et al.* the subjects were unacclimatized and worked at a moderately demanding physical task (18.6 kJ min^{-1}). Ordinary clothing was worn.

Local temperature of fingers, hands and arms can affect performance. A skilled typist did a set job after 15 minutes' immersion of the arms in water kept at temperatures ranging from 36 °C to 15 °C. Typing speed and errors were constant between 36 and 34 °C; there was a small decline in performance at 33 °C and with greater cooling the errors increased and speed diminished. When fingers are cooled to the point of numbness, fine tasks become virtually impossible to perform. As shown by the typing experiment, work falls off at even mild cooling (Fig. 36).

Hypothermia

It is more difficult to lower body temperature than to raise it, due to the effectiveness of shivering coupled with the reluctance of most to endure the necessary body cooling. This statement needs qualification, in that it is possible to reach 35.5 °C body temperature without great difficulty in people with normal physique by continued swimming in water at temperatures of 16 °C or lower. However, in long-distance swimmers such a fall does not take place as these swimmers have thick layers of subcutaneous fat which act as insulation against the loss of body heat. More extensive falls of body temperature can occur in those who are lost in winter conditions or immersed in cold water. In old people and in babies, hypothermia has similar basic features. The clinical accepted definition of hypothermia is a body temperature of 35 °C or below. When such a body temperature is reached, there will usually have been considerable peripheral cooling, hence muscle temperature can be as low as 30 °C. Muscular weakness is then common; the hill walker or climber may easily stumble or fall; walking becomes slow and unsteady. This is the point of danger for the man or woman out of doors in bad weather with inadequate clothing, because heat production drops but heat loss continues and body temperature falls further.

Temperature regulating mechanisms exist in the newborn child. Exposed to cold, cutaneous blood vessels constrict and heat production increases. This is not due to shivering, which is absent in the baby, but to the metabolism of brown fat which persists in the human infant for the first six months of life. This special fat is found at the back of the neck, and between the shoulder blades; there are often pockets in the abdomen and the upper thorax. Brown fat differs from ordinary fat in that it is richly supplied with blood vessels and autonomic nerves, stimulation of which initiates metabolism of the contained fat, releasing heat. Blood flow increases, conducting heat around the body, but in particular raising the temperature of blood supplying the brain.

If body cooling is sufficiently severe these thermal defences fail and the babies' body temperature falls. Although infants withstand the effects of hypothermia better than adults, the danger is that the condition may not be recognized. Hypothermic babies commonly have red cheeks. At low temperatures the blood flow through facial skin will be low, but the blood contained in the vessels is not reduced and remains red, owing to the effect of cold on the haemoglobin dissociation curve.

In the elderly, temperature regulation may be impaired. Temperature discrimination diminishes with age (Collins et al., 1980), and the sensation of cold is not aroused as easily as in younger adults. Peripheral regulatory mechanisms, including shivering, sweating and vasomotor control are commonly affected with increasing age. There may be a lack of shivering but adequate sweating, or vice versa. Vasoconstriction due to cold may occur but not vasodilatation in the heat, or vice versa. Sometimes there may be a constant blood flow without any response to warming or cooling. These thermal deficiencies suggest that the essential change is a gradual failure of the autonomic system.

Amongst young people, the perils of those lost in the winter has been mentioned. There are also the dangers of immersion in cold water. Because of the heat capacity of water, heat can be lost very rapidly from the body. Survival time, wearing a life jacket, is related to water temperature. When sea temperature is 15 °C, average survival time is five hours, and at 0–5 °C, few last for as long as an hour. However, these times are affected by the clothes worn which usually trap some air and retain a degree of insulation. A waterproof coverall, irrespective of thickness, can prolong survival, probably for several hours. Another factor is the amount of subcutaneous fat. Cross-channel swimmers, who survive for up to 20 h in sea temperatures of 15–17 °C, were studied by Pugh and Edholm (1955) who found that all those who were successful were fat, yet they must be considered great athletes capable of astonishing endurance. The secret of their survival is the combination of high thermal insulation, as fat is a poor conductor, and a high rate of energy expenditure. When swimmers become fatigued, stroke rate falls and so does heat production. At the time of removal from the water body temperatures of 35 °C or less were recorded. Before being taken out, such swimmers often become confused and even hallucinated, going in circles, seeing strange monsters, and refusing to come out. Rather similar confusion has been noted in hill walkers beginning to suffer from hypothermia.

It is not possible to give precise indications of the effects of particular levels of body temperature, as there are individual variations. Below 35 °C, muscle weakness becomes evident; below 34 °C confusion is usual; commonly there are visual errors. Below 33 °C shivering is unusual because of low muscle temperature and then temperature regulation fails, with body temperature falling rapidly. Consciousness becomes clouded at body temperature levels of 32 °C and at 30 °C there is deep unconsciousness. Death may occur at temperatures of 28–25 °C and is due to ventricular fibrillation. Survival, however, has been reported in an individual with a body temperature close to 20 °C. In surgical anaesthesia even lower body temperatures have been achieved; patients were carried through cardiac hazards by the use of a defibrillator.

Recovery from severe hypothermia can be rapid and complete; with acute cases, fast rewarming is effective. If hypothermia comes on gradually, then slow rewarming is more successful. The difference may be due to the changes in blood volume and shifts in body water which occur with continued hypothermia.

Cold injury

Frostbite and immersion foot are the most serious forms of cold injury, but minor damage produces chilblains (pernio) or cracked skin. Frostbite can range from the trivial to the loss of fingers, toes, even hands and feet. It occurs when cooling lowers the temperature to below the freezing point of the tissue concerned. It is the water contained in cells which freezes, at a temperature substantially below 0 °C owing to the high content of salts in cell water. If a frozen part is thawed within minutes, little damage results, apart from some peeling of the skin. Irreversible changes follow thawing in prolonged frostbite. These are due to the changes within the cell resulting from the increased concentration following the effective removal of water by ice formation. There will be a rise in osmotic pressure depending on the extent of freezing; this and other consequential changes may lead to death of the cell. The most evident change is in the circulation; following thawing there is a period of vasodilatation. The premeability of the blood vessels is increased and there is a loss of fluid from the blood into tissue spaces. As a result there may be gross swelling of the affected part, haemoconcentration and a fall in blood flow, which may be completely cut off. If this happens the affected part dies, with the eventual loss of more or less tissue.

Immersion foot is the commonest term to describe the state which has also been named trench foot or, more exactly, peripheral vasoneuropathy after chilling. The condition results from prolonged immersion of feet in cold water or cold mud, at low environmental temperatures. Cases have been reported at water temperatures between 1 and 15 °C. Cold vasodilatation is suppressed by chilling of the body, and the foot blood flow in such conditions would be greatly reduced, possibly with periods of ischaemia. Nerve conduction ceases at a temperature of approximately 9 °C. The majority of observed cases have been in shipwreck survivors who might spend several days with their feet and legs continually cold and wet. After rescue, the feet become swollen and there is difficulty in walking as the feet are heavy and numb "It feels like walking on cotton wool". Subsequently there is pain and hyperaemia with varying degrees of sensory loss. The main damage is to nerves—hence the term peripheral vasoneuropathy. Muscle also may be damaged.

A much milder condition is the chilblain, which may occur in fingers and toes or even ears. It follows prolonged cooling of the extremities, with some body chilling; the affected part becomes red, swollen, painful and tender and frequently there is intense itching.

Cold injury of the respiratory tract has been reported but there is no satisfactory evidence that the trachea or bronchi can be affected by sufficiently low temperatures to cause damage. The upper respiratory tract has a large blood supply and inspired air is rapidly warmed and humidified, and reaches body temperature above the larynx. In dogs, breathing air as cold as −35 °C did not lower the temperature in the lumen of either trachea or larynx.

Heat-Induced Disorders

Heat disorders make their appearance when individuals are exposed suddenly to heat loads at levels to which they are not acclimatized or at which adaptation is not possible. Heat disorders are therefore encountered in rather special and often artificial circumstances, as when temperate dwellers are transported to the tropics or find themselves in badly ventilated compartments of ships or motor cars and particularly in industrial conditions where high levels of air and radiant temperatures are generated. But tropical dwellers can also suffer heat disorders when they encounter abnormal heat waves or need to work in hot industries in their own countries.

The understanding of the aetiology and pathogenesis of heat illness rests to a large extent on an understanding of the normal physiology of heat adaptation. For a full clinical account see Weiner (1973).

Heat stroke: heat hyperpyrexia

Outstanding features are (a) the hyperpyrexia, (b) neurological and mental disturbances and (c) the presence in many but not all cases of a hot dry skin. Insufficiency of sweating due to defective eccrine sweat gland response may well explain the vulnerability of the very young infant and the elderly. Thermal stimuli and intradermal acetylcholine injection have shown that at birth sweat glands respond very poorly (Foster et al,, 1969). Similar tests reveal a marked reduction in responsiveness in those aged 70 and over (Foster et al., 1976)

The uncontrollable rise of body temperature is obviously the consequence of the failure of the heat loss mechanism to dissipate the combined environmental and metabolic heat load. With heat regulation out of physiological control there ensues the cumulative cycle of an increase in temperature stimulating an increase in metabolism and this in turn producing a further rise of body temperature and so on.

What is the nature of this failure in heat regulation? There are two possibilities: (a) a sudden failure of the sweating mechanism, centrally at the hypothalmus; (b) no absolute or total failure of the heat dissipating mechanism, but an incapacity of the sweating mechanism to meet the demands of the heat load.

Primary or total failure of sweating mechanism

In many clinical accounts of heat stroke there are reports of an arrest of sweating in apparently healthy men without any previous history of sweating difficulty. In the nature of the case, however, the sweating performance of these patients cannot be followed quantitatively and it is by no means certain that the appearance of the oft-reported hot dry skin had occurred as a primary event. It is quite likely that the cessation of sweating was at least in some instances a secondary result brought about by damage to the heat centre by the hyperpyrexia itself (Kuno, 1956). This is comparable to the way hypoxia first stimulates, then "wrecks", the respiratory centre.

Nevertheless, acute cessation of sweating is known to occur (Bannister, 1959), but whether this type of anhidrosis is long-lasting enough to cause heat imbalance and heat stroke is not certain. Arrest of sweating through a central action is certainly brought about by pyrogens. The pyrexia has been shown to be accompanied by peripheral vasoconstriction. Thus such conditions as malaria or encephalitis can certainly predispose to heat stroke in hot climates; and in these cases the failure of sweating could be attributed to a central action. However, many heat stroke cases occur outside known malarial areas. Failure of sweating through a pyrogen-like action can be true only of a minority of cases.

Insufficiency of sweating mechanism

Gilat *et al.* (1963) have documented the evidence to show that many cases of heat stroke occur without arrest of sweating. The continuance of sweating while control of body temperature is lost points to the failure of evaporative heat loss to balance the heat input. This failure may arise in two different ways, for both of which evidence is available. These are: (a) factors making for insufficiency of sweating, i.e. a reduction in sweat output; (b) overloading of the normal sweating capacity.

There are a number of factors which may reduce normal sweat output by interference with sweat secretion. The absence or reduction in the number of sweat glands as in ectodermal dysplasia would undoubtedly expose the sufferer to a high risk of heat stroke, but this very rare defect cannot account for the high incidence of heat stroke such as occurs in heatwaves or in particular kinds of hot industry. In any case, heat stroke patients on recovery are known to regain their sweating ability. A second factor known to interfere with sweat secretion is the damage done to the skin by prickly heat. The anhidrosis so produced is part of the syndrome of anhidrotic heat exhaustion; conceivably this could predispose to heat stroke. But the prior occurrence of prickly heat in heat stroke patients has rarely been described. The third factor is the occurrence of fatigue of the sweat glands. During prolonged sweating and particularly in hot humid conditions when the skin remains wet, a falling-off of sweating is noticeable. It is conceivable that in some individuals this falling-off may be so marked that heat stroke is induced.

Disturbance in central nervous function can in favourable cases be reversed by cooling; any recrudescence of the hyperpyrexia will result in the reappearance of the central nervous effects. Post-mortem examination reveals pathological changes in the central nervous system including degeneration of neurones, congestion, oedema and petechial haemorrhages. The degree of damage is related to the duration of hyperpyrexia. Damage by the hyperpyrexia to the heat centres as postulated by Kuno (1965) seems consistent with the pathological findings. Similarly, the respiratory centre, which is first stimulated by heat, would undergo damage by excess of heating, thus accounting for the respiratory failure commonly seen. Heat stroke patients are often in a state of acidosis, as evidenced by increased blood lactate, reduced alkali reserve, and a highly acid urine with increased lactate content. Presumably these are the results of the very severe muscular spasms and convulsion.

The occurrence of blood coagulation defect has been recognized as a serious concomitant of heat stroke and may indeed by the underlying cause of widespread tissue damage through local haemorrhage.

Elevation of serum enzymes—aspartate transaminase (SGOT); alanine transaminase (SPGT) and lactate dehydrogenase (LDH)—resulting from tissue damage is of diagnostic value (Kew et al., 1971).

Heat syncope: circulatory deficiency heat exhaustion

This is a condition characterized by disturbance of the vasomotor system with signs and symptoms of syncope. The distinguishing features are the occurrence of a faint or near-collapse under hot conditions without the signs and symptoms of water or salt deficiency or of anhidrosis; the body temperature is usually not greatly raised. The syncopal signs may appear during or immediately after the heat exposure or after the subject has been standing for a while or working in the heat. The subject is pale, and in working individuals there may be a cyanotic hue; breathing is irregular or shallow with sighing and yawning; complaints of weakness, giddiness, nausea, "colic" spells are common. If the patient does faint, the symptoms may be quite transitory and may pass off in a few minutes when he sits or lies down.

Immediately before and during the faint the blood pressure will be found to have fallen and in many, perhaps most, cases the faint is of the vasovagal type with slowing of the pulse rate; the pulse feels weak, the skin is moist and cold.

Vascular "pooling" in skeletal muscles and skin, especially of the lower limbs, is the causal factor. Important predisposing causes are lack of heat acclimatization and poor physical fitness for work, since these are associated with an insufficiency of adjustment of the circulatory system to activity and postural change in the heat.

In hot conditions the peripheral pooling is aggravated since the working subject has widely dilated skin vessels and arterio–venous anastomoses. This orthostatic syncope can be induced after work in the heat or on prolonged standing in the heat. When this is done deliberately (Weiner, 1938) the effects are very similar to those noted clinically. If fainting does not occur, or for a time before it does occur, there is evidence of compensatory visceral and skin vasoconstriction to maintain the blood pressure and heart output. The diastolic pressure and systolic pressure will usually rise, the pulse accelerate and there will be pallor. But if the heart output remains reduced, the blood pressure rise may not be maintained and the pooling may be so great that the blood pressure falls precipitately. At the same time there is a slowing of the heart, thus presenting the features of a vasovagal faint due to the hypoxia stimulating the vagus and therefore inducing bradycardia.

Water-deficiency heat exhaustion

This is a condition characterized by dehydration and attributable to insufficient water replacement.

This variety of heat exhaustion is encountered amongst workers in hot industry, in open air employment in the tropics (e.g. open-cast mining, road-building, etc.); amongst servicemen during training or on active service in hot climates; in individuals adrift in warm climates or stranded in desert conditions; in long-distance athletes running in the heat; in infants during exceptionally hot weather.

The rate of dehydration will depend on the rate of water loss from lungs, skin and kidney in relation to water intake; the rate of cutaneous loss in turn being dependent on environmental conditions, on activity of the subject and on his body size. In the circumstances of the castaway no water for drinking may be available and dehydration will naturally develop relatively fast.

The rate of dehydration on various levels of water replacement can be calculated from the sweat rate induced by the observed thermal conditions, level of activity, amount of clothing, and body size. The time it would take for a certain degree of water depletion (say 5 or 10% of gross body weight) to develop can be estimated. Such "survival" curves define the likelihood of water-deficiency heat exhaustion under specified conditions. For example (Adolph, 1947) on an intake of one litre daily, at 26 °C mean shade air temperature, four days would elapse, or at 32 °C two days would elapse before a 10% body weight deficit develops—the limit of ability to work; a 20% deficit (maximum endurance limit) would be reached in about eight days at 26 °C, or six days at 32 °C.

The aetiology of this syndrome, i.e. water deprivation, is not in doubt. The lack of water may simply be the outcome of a failure to provide sufficiently for the needs of individuals living or working in hot conditions, but dehydration may occur even when adequate water is available. This "voluntary" dehydration is seen in clinical studies. It is also observed in hot industries when men do not replace by drinking all the water lost by sweating. Only during meals is the water deficit made up. Thus some people consistently fail to correct this deficit and so build up a chronic water debt. Infants who cannot make their needs understood are at risk in hot climates: vomiting will exacerbate the water deficit as will diarrhoea.

Heat exhaustion water deficiency as a clear clinical entity is fully confirmed by experiments such as those of McCance (1936) who reproduced the main clinical and biochemical features of the condition by depriving a group of subjects of water but not of food. The food itself yielded about 600 ml of water a day, sufficient to cover the minimal urinary output, but water loss from skin and lungs reduced the body weight by about 1 kg per day. When the body weight was reduced by about 5% the clinical effects were quite obvious—the mouth and throat dry, the voice husky, swallowing difficult, face pinched, pale and slightly cyanotic, and the subject felt tired and irritable. As in the clinical syndrome, the water loss affected the extracellular volume primarily. There was a gradual decrease in blood volume, and since the excretion of salt was not sufficiently increased, the osmotic pressure of the extracellular fluid and later the intracellular compartment steadily rose. In the first stages of the syndrome the ill effects are attributable to the reduction of circulating blood volume, but in the later and irreversible stages the abnormal rise of osmotic pressure appears to be the responsible factor in bringing about cellular damage.

The effects of the shrinkage of the extracellular compartment and particularly of the blood volume have been followed in detail by Adolph and his colleagues in the desert, i.e. in conditions under which the syndrome actually develops. In the dehydrated subject it was found that the plasma gives up more than its share of water and there is also loss of volume by the red corpuscles. Reduction in the circulating blood volume accounts for the undue acceleration of pulse rate (compared with normal hydrated controls) in the lying or standing position or during work. The pulse rate increment is closely related to the degree of water deficit. Adolph (1947) and others believe that in hot environments death is ultimately due to heat stroke because body temperature control is lost as a result of cellular damage. In cool environments dehydration results in coma and depression of central nervous activity.

Salt-deficiency heat exhaustion

This is a condition in which there has been excessive loss of chloride from the body by sweating without sufficient replacement to maintain normal chloride levels; it is characterized by "asthenic" symptoms—fatigue, dizziness, sleeplessness, vomiting and, in the heat cramps variety, by painful limb and abdominal spasm.

The syndrome is encountered amongst workers in ships' boiler-rooms and iron foundries, and in travellers and service personnel in hot dry deserts and in mining. In the South African mining industry, salt-depletion heat exhaustion occurs with a frequency of about two per 1000 employees per annum. Reduced plasma chloride in 70% of all cases of "exhaustion" in underground workers at the Indian Kolar goldfields has been reported. The salt-deficiency syndrome without cramps is the more common variety.

The syndrome occurs in conditions where the thermal load induces sustained sweating and consequently a sudden and greatly increased salt loss. The usual circumstances comprise a combination of hard physical work and thermal conditions above 26 °C effective temperature. Lack of acclimatization is also a factor; salt deficiency heat casualties were induced in servicemen transferred rapidly from cool conditions to hard training activities in the heat in the USA.

The syndrome can be reproduced in its essentials in laboratory conditions by inducing salt loss through profuse sweating by unacclimatized subjects, while keeping them on a diet adequate in all constituents but low in salt. This makes it plain that the aetiology rests on salt depletion.

There is a demonstrable fall in osmolarity, and this leads to a movement of water into the intracellular compartment, thus affecting the muscle fibres. There is also a continuing loss of water by the kidney. There is a contraction of the extracellular compartment which may affect the plasma component disproportionately. Although the movement of water tends to maintain the plasma colloid osmotic pressure it does not prevent the reduction in blood and plasma chloride. The fall in chloride output from the kidney is a compensatory response. McCance (1936) showed further that in severe salt deficiency the kidney increases its reabsorption of urea leading to a rise in blood urea which is often also observed clinically. All these changes are reversed when salt is given in the diet. Both the experimental and clinical evidence suggest that individual susceptibility is very important. Some individuals lose much more salt in the sweat and the compensating reduction from the kidney may be less in evidence.

Table of symbols used

A,	surface area, m^2	P_a,	vapour pressure of air (mm Hg)
A_{eff},	effective radiation area of clothed body	P_s,	vapour pressure at skin surface (mm Hg)
C,	convection	R,	radiation
C_d,	conduction	R_a,	vapour resistance ambient air (in cm dead air)
c_p,	specific heat dry air at constant pressure $= 0.24$ kcal kg^{-1} °C	R_{cl},	vapour resistance of clothing (in cm dead air)
D,	diameter, m	Re,	Reynold's number
E,	evaporation	S,	heat debt (kcal m^{-2} h^{-1})
E_d,	heat loss by water diffusion through skin	s,	specific heat
E_{re},	latent respiration heat loss	T,	time in hours
E_s,	heat loss by evaporation of sweat	t_a,	temperature of air
E_{max},	maximum heat loss by evaporation of sweat	t_b,	temperature of body
		t_c,	temperature of core
ERF,	effective radiation field	t_{cl},	temperature of clothing
F_{ec},	combined emittance and configuration factors	t_{ex},	temperature of expired air
		t_g,	temperature of globe
F_{cl},	thermal efficiency factor	t_{mrt},	mean radiant temperature
f_{cl},	ratio of surface area of clothed to nude body	t_{oe},	temperature of oesophagus
		t_r,	temperature of rectum
H,	rate of non-evaporative heat loss	t_{ra},	temperature, mean radiant
H_c,	heat flow, core-to-surface, by conduction and convection	t_s,	temperature of skin
		t_y,	temperature of tympanum
h_c,	convective heat transfer coefficient	V,	air velocity
		$\dot{V}$,	pulmonary ventilation (kg h^{-1})
h_r,	radiation heat transfer coefficient	w,	body weight
h_e,	evaporative heat transfer coefficient	$\dot{w}$,	rate of weight loss
		W,	mechanical work
I_a,	insulation of air (in clo units)	W,	watts
I_{cl},	insulation of clothing (in clo units)	ε,	emittance
I_t,	insulation of tissues (in clo units)	η,	mechanical efficiency
K,	degrees Kelvin	λ,	heat of vaporization of water 34 °C (575 kcal kg^{-1})
K,	heat transfer, skin to outer clothing		
M,	metabolic heat production, kcal h^{-1}	W_{ex},	humidity ratio of expired air (kg water per kg dry air)
m,	humidity permeance coefficient	W_a,	humidity ratio of inspired air (kg water per kg dry air)
m_1,	internal heat production		
L,	dry respiratory heat loss	σ,	Stefan–Boltzmann constant (4.96 $\times$ 10^{-8} kcal m^{-2} h K^4)
N_{gr},	Grashof number		

References

Adams, W.C. Fox, R.H., Fry, A.J. and MacDonald, J.C. (1975). *J. Appl. Physiol.* **38**, 1030–1037.

Adolph, E.F. (1947). "Physiology of Maa in the Desert", Interscience, New York.

Astrand, P.-O. and Rodahl, K. (1970). "Textbook of Work Physiology", McGraw Hill, New York.

Baker, P.T. (1958). *Am. J. Phys. Anthrop.* **16**, 287–306.

Bannister, R.G. (1959). *Lancet* **ii**, 313–316.

Bazett, H.C. (1949). *Amer. J. Med. Sci.* **218**, 483–492.

Bazett, H.C. (1968). *In* "Physiology of Heat Regulation and the Science of Clothing", (Ed. L.H. Newburgh), Saunders, Philadelphia.

Bell, C.R. and Watts, A.J. (1971). *Brit. J. Industr. Med.* **28**, 259–270.

Benzinger, T.H. (1959). *Proc. Natl. Acad. Sci., USA* **45**, 645–659.

Birnbaum, R. and Crockford, R.W. (1979). *Appl. Ergonomics* **10**, 88–96.

Bligh, J. (1972). *In* "Essays on Temperature Regulation", (Eds) J. Bligh and R.E. Moore, North Holland Publishing Company, Amsterdam and London.

Bligh, J. (1973). "Temperature Regulation in Man and Other Vertebrates", North Holland Publishing Company, Amsterdam.

Brebner, D.F. and Kerslake, D. McK. (1969). *J. Physiol. (London)* **202**, 719–735.

Budd, G.M. (1962). *Nature, (London)* **193**, 886.

Burton, A.C. (1934). *J. Nutr.* **7**, 481–497.

Burton, A.C. (1935). *J. Nutr.* **9**, 261–270.

Burton, A.C. and Edholm, O.G. (1955). "Man in a Cold Environment", Edward Arnold, London.

Buskirk, E.R. and Beetham, W.P. Jr (1960). *Med. Sport* **14**, 493–506.

Cabanac, M. and Massonnet, B. (1975). *J. Physiol. (London)* **265**, 587–596.

Cabanac, M., Cunningham, D.J. and Stolwijk, J.A.J. (1971). *J. Comp. Physiol. Psychol.* **76**, 94–102.

Carlson, L.D. (1973). *In* "The Pharmacology of Thermoregulation", (Ed. W.H. Weihe), Karger, Basel.

Chrenko, F.A. and Pugh, L.G.C.E. (1961). *Proc. Roy. Soc. B*, **155**, 243–265.

Collins, K.J. (1966). *Clin. Sci.* **30**, 207–221.

Collins, K.J. and Weiner, J.S. (1959). *J. Physiol.* **148**, 592–614.

Conn, J.W. (1949). *Adv. Intern. Med.* **44**, 1–15.

Cooper, K.E. and Kenyon, J.R. (1957). *Br. J. Surgery* **44**, 616–619.

Dawkins, M.J.R. and Scopes, J.W. (1965). *Nature (London)* **206**, 201–202.

Doré, C., Weiner, J.S., Wheeler, E.F: and Hamad El-Neil (1975). *Ann. Human Biol.* **2**, 25–33.

Edholm, O.G. (1972). The effect of acclimatisation to heat on water intake, sweat rate and water balance. *In* "Advances in Climatic Physiology" (Eds S. Ito, K. Ogata and H. Yoshimura), Igaku Shoin, Tokyo.

Edholm, O.G., Fox, R.H., Goldsmith, R., Hampton, I.F.G. and Pillai, K.V. (1965). *J. Physiol. (London)* **177**, 15–16P.

Edholm, O.G., Fox R.H. and Macpherson, R.K. (1957). *J. Physiol. (London)* **139**, 455–465.

Edwards, R.J., Belyavin, A.J. and Harrison, M.H. (1978). *Aviat. Space, Env. Med.* **49**, 1289–1294.

Eichna, L.W., Berger, A.R., Rader, B. and Becker, W.H. (1951). *J. Clin. Invest.* **30**, 353–359.

Ellis, F.P., Ferris, H.M. and Lind, A.R. (1954). *J. Physiol.* (*London*) **125**, 61–62*P*.

Fanger, P.O. (1970). "Thermal Comfort", Danish Technical Press, Copenhagen.

Feldberg, W. and Myers, R.D. (1965). *J. Physiol.* (*London*) **177**, 239–245.

Foster, K.G., Ellis, F.P., Dore, C., Exton-Smith, A.N. and Weiner, J.S. (1976). *Age Ageing* **5**, 91–101.

Foster, K.G., Ginsburg, J. and Weiner, J.S. (1970). *Clin. Sci.* **39**, 823–832.

Foster, K.G., Hey, E.N. and Katz, G. (1969). *J. Physiol.* (*London*) **203**, 13–29.

Fox, R.H. (1974). *In* "Recent Advances in Physiology", (Ed. R.J. Linden), Chap. 8. Churchill Livingstone, Edinburgh and London.

Fox, R.H. and Goldsmith, R. (1962). *Ergonomics* **5**, 415.

Fox, R.H. and Hilton, S.M. (1958). *J. Physiol.* (*London*) **142**, 219–232.

Gagge, P. (1972). *In* "Physiological Adaptations", (Eds M.K. Yousef, S.M. Horvath and R.W. Bullard), Academic Press, London and New York.

Gale, C.C., (1972). *The Physiologist* **15**, 141–145.

Gilat, T., Shibolet, S. and Sohar, E. (1963). *J. Trop. Med. Hyg.* **66**, 204–212.

Gleibermann, L. (1973). *Ecology Food Nutr.* **2**, 143-156.

Gregory, R.A. and Lee, D.H.K. (1936). *J. Physiol.* (*London*) **86**, 204–218.

Hammel, H.T. (1968). *A. Rev. Physiol.* **30**, 641–710.

Hammel, H.T. (1972). *In* "Essays on Temperature Regulation", (Eds J. Bligh and R.E. Moore), North Holland Publishing Company, Amsterdam and London.

Hammel, H.T., Jackson, D.C., Stolwijk, J.A.J., Hardy, J.D. and Stromme, S.B. (1963). *J. Appl. Physiol.* **18**, 1146–1154.

Hardy, J.D. (1970). (Ed.) "Temperature: Its Measurement and Control in Science and Industry", Part 3. Reinhold, New York.

Hardy, J.D. and DuBois, E.F. (1938). *J. Nutr.* **15**, 477–495.

Hardy, J.D. and Stolwijk, J.A.J. (1966). *J. Appl. Physiol.* **21**, 1799–1806.

Hellman, K. and Weiner, J.S. (1953). *J. Appl. Physiol.* **6**, 194–198.

Hellon, R.F. (1970). *In* "Physiological and Behavioural Temperature Regulation" (Eds J.D. Hardy, A.P. Gagge and J.A.J. Stolwijk), Thomas, Springfield, Ill.

Hellon, R.F. (1971). Central thermoreceptors and thermoregulation. *In* "Handbook of Sensory Physiology". Vol. III/I: Enteroceptors, (Ed. E. Neil), Springer Verlag, Berlin.

Hellon, R.F., Jones, R.M., Macpherson, R.K. and Weiner, J.S. (1956). *J. Physiol.* (*London*) **132**, 559–576.

Hensel, H. and Witt, I. (1959). *J. Physiol.* (*London*) **148**, 180–187.

Hensel, H. (1966). "Allgemeine Sinneophysiologie, Hautsinne, Geschmach, Geruch", Springer. Verlags, Berlin.

Hess, W.H. and Stoll, H. (1944). *Helvet. physiol. Pharmacol. Acta* **2**, 461.

Hubbard, J.L. and Weiner, J.S. (1969). *J. Appl. Physiol.* **27**, 715–720.

Iggo, A. (1969). *J. Physiol.* (*London*) **200**, 403–430.

Iggo, A. (1970). *In* "Physiological and Behavioural Temperature Regulation", (Eds J.D. Hardy, A.P. Gagge and J.A.J. Stolwijk), 391. Thomas, Springfield, Ill.

Isenschmidt, R. and Krehl, L. (1912). *Arch. Exp. Path. Pharmak.*, **70**, 109–121.

Joy, R.T.J. (1963). *J. Appl. Physiol.* **18**, 1209–1212.

Kang, B.S., Han, D.S., Paik, K.W., Park, Y.S., Kim, J.M., Kim, C.S., Rennie, D.W. and Hong, S.K. (1970). *J. Appl. Physiol.*, **29**, 6–9.

Kenney, R.A. (1952). *J. Physiol.* (*London*) **118**, 25*P*.

Kerslake, D.McK. (1972). "The Stress of Hot Environments", Cambridge University Press, London.

Kew, M.C., Bersohn, I and Seftel, H.C. (1971). *Trans. Roy. Soc. Med. Hyg.* **65**, 325–330.

Kuno, Y. (1934, 1956). "The Physiology of Human Perspiration" 1st ed Thomas, Springfield, Ill. 2nd edn Churchill, London.

LeBlanc, J. (1975). "Man in the Cold" Thomas, Springfield, Ill.

LeBlanc, J. and Potvin, P. (1966). *Canad. J. Physiol.* **44**, 287–293.

Lewis, H.E. (1966). *J. Physiol. (London)* **188**, 6–7P.

Lewis, T. (1930). *Heart* **13**, 177–206.

Lind, A.R. (1963). *J. Appl. Physiol.* **18**, 51–56.

McCance, R.A. (1936). *Lancet* **i**, 643–704.

McCance, R.A., Hamad El Neil, Nasr El Din, Widdowson, E.M., Southgate, D.A.T., Passmore, R., Shirling, D. and Wilkinson, R.T. (1971). *Phil. Trans. Roy. Soc., Lond. B*, **259**, 533–565.

Mackworth, N.H. (1953). *J. Appl. Physiol.* **5**, 533–537.

Magoun, H.W. Harrison, F., Brobeck, J.R. and Ranson, S.W. (1938). *J. Neurophysiol.* **1**, 101–122.

Mitchell, D., Atkins, A.R. and Wyndham, C.H. (1972). *In* "Essays on Temperature Regulation", (Eds J. Bligh and R.E. Moore), 37–54. North Holland Publishing Company, Amsterdam and London.

Myers, R.D. (1971). *In* "Pyrogens and Fever", (Eds G.E.W. Wolstenholme and J. Birch), 131–145. Ciba Foundation Symposium. Churchill Livingstone, London.

Myers, R.D. and Sharpe, L.G. (1970). *In* "Physiological and Behavioural Temperature Regulation", (Eds J.D. Hardy, A.P. Gagge and J.A.J. Stolwijk), 648–666. Thomas, Springfield, Ill.

Myers, R.D. and Yaksh, T.T. (1969). *J. Physiol. (London)* **202**, 483–500.

Nadel, E.R., Bullard, R.W. and Stolwijk, J.A.J. (1971). *J. Appl. Physiol.* **31**, 80–87.

Nevins, R.G., Rohles, F.H., Springer, W. and Feyerheim, A.M. (1966). *ASHRAE Trans.* **72**, 283–291.

Newburgh, L.H. (1968) (Ed.) "Physiology of Heat Regulation and the Science of Clothing", Saunders, Philadelphia.

Nielsen, B. (1969). *Acta. Physiol. Scand., Suppl.* **323**, 1–74.

Nielsen, M. (1938). *Skand. Arch. Physiol.* **79**, 193–230.

Pappenheimer, J.R., Eversole, S.L. Jr., and Soto-Rivera, A. (1948). *Am. J. Physiol.*, **155**, 458–467.

Pickering, G.W. (1932). *Heart* **16**, 115–135.

Pickering, G.W. and Hess, W. (1933). *Clin. Sci.* **1**, 213–223.

Piironen, P.P. (1970). Sinusoidal signals in the analysis of heat transfer in the body. *In* "Physiological and Behavioural Temperature Regulation", (Eds J.D. Hardy, A.P. Gagge and J.A.J. Stolwijk), Thomas, Springfield, Ill.

Pugh, L.G.C.E. (1964). *Lancet*, **i**, 1210–1212.

Pugh, L.G.C.E. and Edholm, O.G. (1955). *Lancet*, **ii**, 761–768.

Roberts, M.F. and Wenger, C.B. (1979). *Med. Sc, Sports* **11**, 36–41.

Robinson, S. (1949). *In* "Physiology of Heat Regulation and the Science of Clothing", (Ed. L.H. Newburgh), 193–231. Saunders, Philadelphia.

Robinson, S. (1963a). *In* "Temperature: Its Measurement and Control in Science and Industry", (Ed. J.D. Hardy), 287–297. Reinhold, New York.

Robinson, S. (1963b). *Paediatrics* **32**, 691–702.

Robinson, S. and Robinson, A.H. (1954). *Physiol. Rev.* **34**, 202–220.

Rothman, S. (1954). "Physiology and Biochemistry of the Skin", University of

Chicago Press, Chicago.
Rowell, L.B., Marx, J.H. Bruce, R.A., Conn, R.D., and Kusumi, F. (1966). *J. Clin. Invest.* **45**, 1801–1816.
Snell, E.S. (1954). *J. Physiol.* (*London*) **125**, 361–372.
Spealman, C.R. (1968). *In* "Physiology of Heat Regulation and the Science of Clothing", (Ed. L.H. Newburgh), Saunders, Philadelphia.
Underwood, C.R. and Ward, E.J. (1966). *Ergonomics* **9**, 155–168.
Ward, E.J. and Underwood, C.R. (1967). *Ergonomics* **10**, 399–410.
Webb-Peploe, M.M. and Shepherd, J.T. (1968). *Circulation Res.* **23**, 701–708.
Weddell, G. and Miller, S. (1962). *Ann. Rev. Physiol.* **24**, 199–222.
Weiner, J.S. (1938). *J. Industr. Hyg. Toxicol.* **20**, 389–400.
Weiner, J.S. (1973). *In* "Medicine in the Tropics", (Edited A.W. Woodruff), 505–521. Cassells, London.
Weiner, J.S. and Hellman, K. (1960). *Biol. Rev.* **35**, 141–186.
Weiner, J.S. and Van Heyningen, R.E. (1952). *Br. J. Industr. Med.* **9**, 56–64.
Williams, C.G., Bredell, C.A.G., Wyndham, C.H., Strydom, N.B., Morrison, J.F., Peter, J., Fleming, R.W. and Ward, J.S. (1962). *J. Appl. Physiol.* **17**, 625–638.
Winslow, C-E. A., Herrington, L.P. and Gagge, A.P. (1936). *Am. J. Physiol.* **116**, 641.
Winslow, C-E. A. and Herrington, L.P. (1949). "Temperature and Human Life", Princeton University Press, New Jersey.
Wit, A. and Wang, S.C. (1968). *Am. J. Physiol.* **251**, 151–160.
Wulsin, F.R. (1968). *In* "Physiology of Heat Regulation and the Science of Clothing", (Ed. L.H. Newburgh), Saunders, Philadelphia.
Wyndham, C.H. (1951). *J. Appl. Physiol.* **4**, 383–395.
Wyndham, C.H. (1965). *J. Appl. Physiol.* **20**, 31–36.
Wyndham, C.H., Bouwer, W.M., Devine, M.G., Paterson, H.E. and MacDonald, D.K.C. (1952). *J. Appl. Physiol.*, **5**, 299–307.

Acceleration

P. Howard

"... the heavy and the weary weight of all this unintelligible world."

The emergence of life from the oceans dictated a sharp swerve in the trajectory of evolution. The degree of complexity of physiological systems had, until then, been governed almost entirely by "first-order" physical characteristics such as mass, area and volume. High-speed predators needed to be streamlined to be efficient, but to most organisms immersed in and permeated by sea water shape was of small moment.

The complication introduced by a terrestrial existence was gravity which, despite its universal existence, was then encountered for the first time as an unbalanced influence. Itself only an acceleration, it conspired with mass to produce weight, which in turn impeded movement and distorted form. Flat, immobile creatures could function well enough on land once shell or skin had solved the problem of drying out, although total volume then became constrained by the uneasy balance between metabolic need and the laws of gaseous diffusion. Height could only become an exploitable dimension through the development of struts to support an unstable assemblage of fluid-filled bags, and of carapaces or skeletons to give some protection against the structural catastrophe threatened by weight.

But although they ward off the danger of disintegration, such buttresses do nothing to relieve the problem of the forces induced by gravity in tissues, organs and physiological systems—indeed, in some instances they worsen the situation. The evolutionary devices that have been devised to combat the ill-effects of these "local" systematic forces are often ramshackle compromises, which are effective only in a very restricted range of environments. In consequence, most land-based mammals (and especially man) tremble on the brink of gravitational collapse. It is with this physiological precipice that this chapter will be concerned.

Nomenclature and Terminology

"The close affection that grows from common names is ... strong as iron".

Gravity is universal and inescapable, because it results from the existence of masses. Its physiological effects can be almost negated by immersing the body in water (or more strictly, in a fluid of the same mean density as the body) but, as will be shown, the cancellation is not complete. The so-called "gravity-free" flight of orbital satellites is not an escape, but merely a state of balance between the gravitational attraction of the earth (*g*) and the centrifugal acceleration resulting from the circular motion around it.

The oblateness of the earth and the presence of density anomalies (mascons) within its structure cause geographical variations in the magnitude of terrestrial gravity. For all practical purposes these local perturbations can be ignored, and a "standard" value of 9.80665 m s^{-2} is internationally recognised. High accuracy is rarely justified, and common usage accepts 981 cm s^{-2} or the even less exact imperial equivalent of 32 ft s^{-2}. It is convenient to express all other accelerations, including the gravitational attraction of other planets, in terms of the earth's constant.

This convention allows the weight of an object on earth to have the same numerical value as its mass and, by the same token, permits all accelerations to be described in terms of the change of weight by which they are detected, experienced and measured. Although g has the dimensions of acceleration (lt^{-2}), weight is a force (mlt^{-2}). If mass remains constant, as it usually does, it is permissible to speak of gravitational force. There are two major exceptions, the second of which has some practical importance. Particles accelerated to relativistic velocities, close to the speed of light, do not have invariant masses; they violate the tenets of Newtonian physics. Rockets burn fuel as they rise and so lose mass; a constant force (thrust) will accordingly lead to a steadily increasing acceleration.

In aviation physiology, accelerations are classified according to the major anatomical direction in which they act, by means of the signed coordinate system shown in Fig. 1. The inhomogeneities of the body result in considerable differences between what would be called, in an engineering context, the static and the dynamic strengths. For that reason, forces must also be categorized by the time for which they act. The division between "long-duration" and "short-duration" or impact accelerations is set arbitrarily at one second. In fact, most physiological interest attaches to durations one or two orders of magnitude above and below that point of transition. This chapter will deal only with the former category, because the results of short-acting forces are of primary interest to anatomists and pathologists.

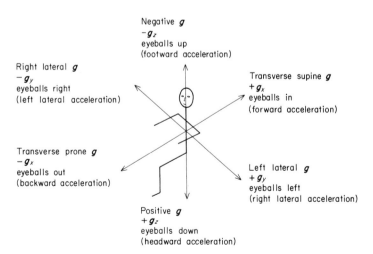

FIG. 1   The "body-orientated orthogonal coordinate" system used to describe the direction of inertial forces. (Terms in brackets refer to the direction of the applied acceleration.)

The word "acceleration" should properly be reserved for situations where the velocity is increasing; a loss of speed or momentum constitutes deceleration. The physiological effects stem from the forces generated by the change, and the body takes little account of speed alone. A forward acceleration ($+ g_x$) is functionally the same as a backward deceleration ($- g_x$), and the shorthand notation permits the use of acceleration as a generic term. However, the familiar assumption that "acceleration" and "force" can be used as interchangeable terms exposes a more significant anomaly. In the physical world, acceleration and the force that produces it act in the same direction—a vehicle struck from behind will be propelled forwards. The occupants of such a vehicle will feel that they have been thrust back into their seats; that is, they will experience a force in a direction opposed to the acceleration. Their sensations and their systems are responding to Newton's third law, which states that to every action there is an equal and opposite reaction or "inertial force". Thus, the preoccupation of the acceleration physiologist (or the gravitational biologist) is not with accelerations, but with the inertial forces induced by changes of the velocity vector, expressed in multiples of the unitary terrestrial gravity.

Accelerations of Long Duration

"... if travail you sustain, into the wind shall lightly pass the pain".

It is usual, in textbooks of aviation medicine, for the various types of sustained acceleration to be discussed separately, as if they were individual syndromes. Although this approach is perfectly valid in relation to the occurrence of accelerative forces in flight, it tends to obscure the fact that the basic physiological responses to gravitational stress are consistent and coherent, with the limited range of exposure to 1 g in different orientations of the body poised in a small central region of a continuum.

In many respects it is more logical to consider in turn the reactions of a given biological system to positive, negative, and transverse accelerations. This stratagem avoids much repetition, and emphasises the generality of the responses. Although it may seem to require that unwonted attention be given to some physiological cloisters, it will be adopted here.

Aetiology

Prolonged and frequent exposures to a positive acceleration of 1 g are a predominant feature of the normal state of wakefulness, but this state is rarely experienced for long in its purest form; when the full length of the body is perfectly aligned with the force vector. Those who are so exposed (for example, soldiers standing rigidly at attention) not infrequently demonstrate the precarious compromise that exists between posture and gravity. Sustained exposures to positive accelerations ($+g_z$) in excess of 1 g are, however, part of all normal flights, for centrifugal forces are generated whenever the direction of travel is changed. The span of such experience is large. A light aircraft flying at 100 miles per hour generates 2.24 g in a turn with a radius of 300 ft; one complete circuit takes 13 s. An airliner should not expose its passengers to a resultant force of more than 1.25 g; at a speed of 400 mph its radius of turn must accordingly be no less than 8600 ft, and one revolution takes at least 92 s. A fighter aircraft which flies at 800 mph and turns in the same radius as the airliner will complete its circle in 46 s, but will expose its pilot to $+5\,g_z$.

The simplest example of negative acceleration ($-g_z$) is encountered during inverted flight or, on earth, when standing on one's head. Significant forces in this direction should not be routinely sustained in aviation, for their effects are both unpleasant and hazardous. They arise during turning manoeuvres in which the underside of the aircraft is directed towards the centre of rotation; that is, when the head of the pilot points away from the axis. Aerobatic displays and competitions usually include items of which "outside" loops, rolls and turns are an integral and spectacular part; accelerations of up to $-6\,g_z$ are sometimes experienced in such demonstrations, but human tolerance for the stress usually limits the duration to a few seconds only.

Transverse (g_x) and lateral (g_y) accelerations are characteristically associated with motion in a straight line and not with curvilinear flight. They are therefore encountered in the more conventional forms of transport, such as trains and motorcars, as well as in the take-off (and landing) of aircraft. Because the time integral of acceleration is velocity, the duration and/or the magnitude of g_x and g_y forces are limited to very low values, save in the special case of the launch of manned space vehicles. For example, a motorcar which attains a speed of 60 mph in eight seconds from a standing start has an average forward acceleration of only one-third g, while a rocket with a constant thrust of 4 g will, after burning for two minutes, reach a velocity of more than 10 000 mph. The limited speeds that can be achieved on earth, and the reciprocal relationship between the size and the duration of the applied force, make sustained acceleration in the g_y and g_x axes of little physiological importance. They do have great practical relevance in crashes. The sportscar used above as an example might expend its hard-won momentum over a stopping distance of four feet, thereby exposing its unfortunate occupants to a mean force of more than 30 g for about one-tenth of a second. The effects of such exposure can hardly be described as physiological.

Physics and Biophysics

The equations that relate velocity, acceleration, time, distance and radius of turn are readily derived from basic physics and simple geometry, but they are quoted here for ease of reference. The symbols used are: $g =$ acceleration expressed in terms of the mean terrestrial gravity (32 ft s^{-2} approximately); $a =$ acceleration (ft s^{-2}; dimensionally lt^{-2}); $t =$ time in seconds; $u =$ initial velocity in ft s^{-1}; $v =$ final velocity in ft s^{-1}; $s =$ distance in feet.
Then, by definition,

$$v = at \tag{1}$$
$$v^2 - u^2 = 2as \tag{2}$$
$$s = \tfrac{1}{2}at^2 \tag{3}$$

For radial or centrifugal forces, ω represents angular velocity in radians per second, and r is the radius of turn in feet. The most commonly used general equations are:

$$g = \frac{v^2}{32r} \tag{4}$$

$$g = \frac{\omega^2 r}{32} \tag{5}$$

It should be noted that these are statements of physical laws, whose truth is independent of the orientation of the body within the aircraft. Positive, negative, and transverse forces can all be calculated by judicious use of one or more of these equations.

The words "acceleration" and "force" are improperly but usefully treated as synonyms. The distinction between them is unimportant when a force is used to impart an acceleration to a single homogeneous mass, or when an acceleration is measured by the apparent weight given to a known mass. The whole physiological significance of sustained accelerations, however, stems from the fact that the body is *not* a uniform system. Instead, a general and evenly distributed increase in the effective gravitational field acts upon the assemblage of masses which constitute the body to generate differential forces. Although this Newtonian rule is true alike for the skeleton and for intracellular structures, its consequences appear most dramatically in the cardiovascular system, in which a relatively incompressible fluid is contained in vessels whose static and dynamic characteristics are greatly influenced by their transmural pressure—it must be remembered that pressure, like weight, is a force.

The absolute hydrostatic pressure exerted by a column of fluid above the point of measurement can be expressed as the product of the vertical height of the column (h), the density of the fluid (q) and the component of the overall acceleration which acts in the direction in which h is measured. Because pressure is measured as force per unit area, the diameter of the vessel does not enter into this equation, although the *weight* of the column will contain a term representing the mean diameter. For an ideal fluid in a rigid, open-ended tube, h and q are constants, and the pressure is then a linear function of the acceleration, a. However, in closed systems, the mean pressure is not necessarily zero, and it is more satisfactory to derive the pressure *difference* between two points in the tube as:

$$\Delta p = aq\Delta h \tag{6}$$

This is a fair representation of the cardiovascular system, provided that pulsatile flow is neglected, and that the vessels are regarded as rigid structures. A simple model of the cardiovascular system might consist of an essentially two-dimensional network of tubes, closed by compliant diaphragms. When this model is placed in the horizontal plane (analogous to an acceleration of 1 g in the x-axis) no hydrostatic differentials are present, and the mean pressure $\bar{p}$ applies throughout the length. When the system is rotated to the vertical position, $(+ 1\ g_z)$, pressure differences appear within it, in accordance with the hydrostatic equation. They are not, however, disposed symmetrically about the midpoint of the cylinder because the "stretchability" of the diaphragms allows the pressure at the foot to deviate more from the distributed average than that at the head. Nevertheless, the simple relationship between the hydrostatic forces at any two points within the circulation remains unaltered with respect to an external frame of reference.

This elementary model yields results of great physiological significance.

(a) For a given orientation of the body, the pressure gradient between two directly connected points within the circulatory system is a linear function of the applied acceleration.

(b) Positive accelerations will cause a net displacement of fluid away from the head towards the feet; for negative forces, the converse is true.

(c) There must be, within the real cardiovascular system, a level at which the absolute pressure is unaffected by sustained acceleration (and also by changes of posture). At all sites above this, pressures will be lower than the average and will fall further as the applied acceleration increases, while more dependent regions will experience rises in pressure proportional to their vertical separation from the so-called "hydrostatic indifference point" (Gauer, 1961). Although it is highly improbable that the hydrostatic indifference point exists as a single invariate site, (because its position must vary with the extent of active changes in the capacitance and resistance of blood vessels), the concept is of some importance because it helps to determine the responses of the circulation to increased gravitational forces. The reference level can have been of little significance to the design of the cow, but some knowledge of its position might have profoundly affected the evolutionary development of cardiovascular compensatory mechanisms in man. Unfortunately, there is very little evidence that the Almighty was a good physiologist.

Tolerance for Sustained Acceleration

"He that endureth to the end shall be saved."

The delineation of tolerance for accelerations is confounded by a multiplicity of possible end-points, and by the fact no single sign or symptom (save the sensation of increased weight) is common to all types of gravitational stress. Indeed, different criteria must be used to define the tolerable limits for forces acting even in one axis and in one sense. Thus, at the lower end of the timescale, intolerance for very high positive accelerations is denoted by the onset of unconsciousness. In the range of 15–20 s duration, the limit is set by loss of vision. When the exposure lasts for many minutes, fainting or fatigue are the limiting factors while, at the extreme, human tolerance of 1 g is authoratively defined as three score years and ten.

These problems are most severe in the case of forces acting in the $+g_z$ axis, but similar difficulties of definition afflict $-g_z$ and, to a smaller degree $\pm g_x$. Despite them, it is possible to draw a reasonably satisfactory smooth curve through data from different sources, and so to obtain useful guidelines for overall tolerance of sustained accelerations in each direction. Representative graphs obtained in this way are shown in Figs 2 and 3. The reliability of the curve is, of course, proportional to the number of observations, and in this respect the plots for $+g_z$ and $\pm g_x$ have the most

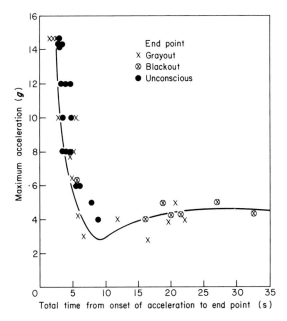

FIG. 2   Pooled data (from centrifuge experiments) on the times of tolerance of relaxed, seated subjects for $+g_z$ accelerations. The increased resistance at times longer than 10 s is due to the recruitment of cardiovascular compensatory reflexes.

credibility. Hardly any measurements have been made of tolerance for sustained $\pm g_y$ forces.

It has to be remembered that the boundaries defined by such curves are both broad and physiological. They cannot be used for predicting the effects of acceleration upon the performance of a task requiring, for example, the precise manipulation of a control or even such coarse activity as raising an arm to the head. These practical limits must be individually determined, although a considerable body of data concerning many of them already exists.

Vision

"Now is the time when all the lights wax dim."

Disturbances of vision provide the earliest objective measure of the physiological effects resulting from long-term positive acceleration, and the

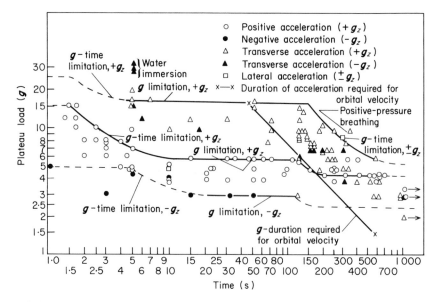

FIG. 3   Strength–time curves for accelerations applied in the x- and y-axes. The data are much less reliable than those of Fig. 2, especially for the longer durations, and for $\pm g_y$ forces.

pattern of response is consistent enough to form the basis for most methods of determining tolerance.

The visual sensations that accompany exposure to gradually increasing gravitational stress are variously described as veiling or fogging, followed by loss of colour and of contrast, by restriction and darkening of the field of view, and finally by a total loss of vision but with retention of hearing and of consciousness. These symptoms are matched by measurable impairment, two phases of which are usually defined as the thresholds of tolerance (Leverett and Zuidema, 1957). The determination of the end-point is made by means of an array of three signal lamps mounted about one metre in front of the subject. He stares at the central light, and must respond to the frequent but random illumination of the peripheral pair by switching them off. Failure to do so provides one index of his tolerance; inability to respond to the central signal gives a second and more reliable one. The first condition is usually called grey-out; a somewhat misleading description of a state in which the peripheral fields are reduced to less than a specified but arbitrary limit of about 90° of arc. The more semantically accurate term PLL (peripheral light loss) is sometimes used, but it is not generally popular. Black-out is the time-honoured name for the condition in which the contraction of the fields extends to embrace the fovea. By analogy with

PLL, this state is sometimes referred to as CLL (central light loss) but the more familiar word dies hard.

The thresholds for grey-out and black-out depend not only upon the orientation of the head and the body with respect to the acceleration, and upon the magnitude and duration of the force, but also upon the rate of its onset. The changes take time to develop, and if a high stress is applied suddenly unconsciousness may supervene before any visual disturbances are noticed. In general, the visual syndrome appears within 8–12 s of the application of a sufficient $+g_z$ acceleration, but the impairment may later recede if the force is sustained.

A study of the thresholds of visual loss resulting from positive acceleration was made by Cochran et al. (1954), in 1000 relaxed subjects sitting upright in a centrifuge. Loss of vision for the "standard" peripheral lights occurred at 4.1 $\pm$ 0.7 g, and black-out at 4.7 $\pm$ 0.8 g; unconsciousness was produced at a mean level of 5.4 $\pm$ 0.9 g. Although the standard deviations are reasonably small, the range of observations was wide. Grey-out was sometimes seen at 2.2 g and consciousness was occasionally lost at 3.0 g; the other extremes for the development of these two states were 7.1 g and 8.4 g respectively. This wide scatter of apparent tolerance can hardly be explained in terms of the physiological mechanisms that lead to grey-out and black-out.

The primary factor is undoubtedly hydrostatic. In erect man, the vertical distance between the heart and the brain (or eye) is about 12 inches and, as eqn (6) shows, the perfusion pressure at the base of the skull is inevitably 25 mm Hg (or 30 cm blood) less than that at the aortic valve. Moreover, the differential is increased by a further 25 mm Hg for every rise of 1 g in the total gravitational field. When the central retinal artery enters the globe of the eye its internal pressure is opposed by the intraocular tension, which normally amounts to about 25 mm Hg. Thus, if the mean aortic pressure is assumed to be 100 mm Hg, the driving force available to the retinal circulation at $+1$ g_z is approximately 50 mm Hg; that is, one-half of the pressure generated by the left ventricle. This offers a substantial reserve for the perfusion of the retina in all normal circumstances, and clinical observations show that vision is maintained when the intra-ocular pressure is raised by a modest amount, as in mild glaucoma. Indeed, a factor of safety must have been incorporated teleologically, so that sight would not be lost during transient physiological episodes of hypotension. The margin is, however, too small to cater for a serious decline in the central or local arterial pressure, and the retinal blood supply is significantly embarrassed by an otherwise innocuous accelerative force of $+2$ g_z. At that level, the peripheral retina is considerably under-perfused and, because the retinal vessels are end-arteries, the visual field is impaired to a degree that becomes progressively more extensive as the acceleration is increased. Even from the first, central visual function is slightly affected, and a total failure of vision (black-out) occurs when, or just after, the pressure in the central retinal artery falls below that in the eye. In a passive hydrostatic man this situation should be reached at a sustained acceleration of about 3 g, a prediction that accords well with the lowest values reported by Cochran et al. In most subjects, however, the threshold of visual loss is set one to two g higher by a number of purposive and involuntary compensations.

The interpretation of grey-out and black-out as peripheral events caused by failure of the retinal circulation has strong experimental support. Lambert and Wood (1946) demonstrated that vision was lost when the mean arterial blood pressure at eye level was reduced to about 20 mm Hg. They also showed that the entire sequence of events from veiling to black-out can be reproduced by external pressure on the eye, which raises the intraocular tension. The black-out caused by $+g_z$ acceleration can also be relieved, in part at least, by applying external suction to the eye, and so restoring the transmural pressure in the retinal vessels. However, the fact that the haemo-dynamic factors associated with black-out also apply to the cerebral circulation has led to many postulates that the failure of vision contains a central nervous component (Greenleaf et al., 1966; Beckman et al., 1962; Poppen and Drinker, 1950). Beckman et al. (1962) claimed to have shown that voluntary movement of the eyes was restricted, both in the extent and the accuracy of alignment, during black-out. The evidence that this impairment (if it is real) implies a critical reduction in the cerebral blood flow is very slender.

Ophthalmoscopy and retinal photography have allowed a correlation to be established between changes in visual function and morphological alterations in the retinal vessels (Duane, 1954; Leverett et al., 1967). When vision was clear, no significant ophthalmoscopic changes were observed during acceleration, but loss of peripheral vision and early grey-out were associated with visible pulsation of the arterioles in the extremities of the field. Black-out was characterized by collapse of the arteries, and, in some instances, bt so-called "arterial spasm". The retinal veins also emptied at this time. When central and peripheral vision were restored by compensatory adjustments (the sustained acceleration remaining constant) these changes were reversed, with restoration of pulsation in the arterioles and temporary over-distension of the veins.

There is strong circumstantial evidence that the loss of vision is a local hypoxic phenomenon. Howard (1964) studied the relationship between the threshold for loss of central vision and the brightness of the light used as a stimulus and a signal. He found that the detection of a small and very dim target at the fixation point was impaired at $+2\ g_z$ or even less, but that an intense flash could be perceived at accelerations considerably higher than the black-out threshold as determined by conventional techniques. When the bright stimulus was presented to an eye that had been temporarily blinded by acceleration or by external pressure the subject did not respond, but an after-image of the source became visible when visual function returned to normal after removal of the accelerative stress. From these and other data, it was concluded that the cause of the failure of vision was an hypoxic disturbance of conduction, probably in the ganglion cell layer of the retina. The exact site of the lesion is difficult to determine, and it is largely of academic interest.

The pressure changes produced at the eye by $-g_z$ acceleration are opposite in sense to those induced by $+g_z$ forces. Because the so-called "hydrostatic indifference point" is sited approximately at the level of the aortic valve, the length of the effective column of blood between the heart and the base of the skull does not exceed 30 cm even in the fully inverted position. This means that, if the mean aortic pressure is 100 mm Hg, the tension in the central retinal artery will be 150 mm Hg at $-2\,g_z$, and 200 mm Hg at $-4\,g_z$ (which is close to the normal level of human tolerance for forces in this direction). Except in very long exposures to acceleration, the intraocular pressure probably remains virtually constant, so that the net arterial pressure at the retina is approximately 25 mm Hg less than these calculations would indicate. Hypertension of this degree is well within the range of clinical experience and, as a transient event, it does not place the eye or the brain at hazard. The visual symptoms resulting from $-g_z$ forces are, accordingly, very minor in comparison with those produced by positive acceleration.

The classical analogue of black-out is "red-out" but the aetiology of this condition (and even its existence) is very much in doubt. It is certain that the descriptions given by pilots of looking through a red mist or of complete obscuration of vision during exposure to negative g are not attributable to haemorrhage or other pathology within the retina. From an analysis of photographic records made during aerobatic manoeuvres, Henry (1950) postulated that gravitation of the lower eyelid over the cornea restricted vision, and that day-light filtering through the lid might give the impression of seeing red. Sieker (1952) correlated reports of red-out with rupture of small blood vessels in the conjunctivae and concluded that blood-stained tears could partly explain the phenomenon. Diminished vision was reported in 40% of exposures to $-3\,g_z$, and conjunctival haemorrhage occurred in the same proportion of subjects. When the conjunctival vessels were protected by pressurized goggles against the high transmural pressures developed during negative acceleration, no haemorrhages occurred, and the reports of red-out also disappeared. Complaints of diminished vision (diplopia and loss of visual acuity) persisted, but they were made by only ten per cent of subjects at 3 g, and by 30% at 4 g. A questionnaire survey of British aerobatic pilots (Howard, 1964) produced no evidence for red-out, although most of those interviewed had, at some time, suffered from varying degrees of conjunctival haemorrhage.

Simple hydrostatic considerations dictate that disturbances of retinal function should not occur at moderate levels of transverse acceleration, and this prediction is borne out in practice. (It is for this reason that the reclined posture is sometimes adopted as a protective measure against $+g_z$ forces.) If the acceleration is sufficiently great, however, and the vertical distance between heart and eye is significant, black-out can result from accelerations in the g_x axis. An ill-documented study by Buhrlen (1937) reported the characteristic visual loss when subjects were exposed to 14 g in the chest-to-back direction, while Kotovskaya (1970) measured black-out thresholds as high as 18 g in fully reclined subjects, without apparent disturbance of consciousness.

Although grey-out and black-out provide the most dramatic evidence of impaired function, other disturbances of vision can be demonstrated during sustained acceleration in any of three axes of the body. White (1958, 1960) made exhaustive measurements of threshold luminances at the fovea and in the peripheral fields, and found that retinal sensitivity was significantly depressed, even at $+2$ g_z. Smedal et al. (1963) and Jaeger et al. (1964) showed that these changes were accompanied by a fall in the sensitivity of contrast discrimination, and that visual acuity was also reduced. There seems to be no doubt that these minor visual effects can be attributed, at least in part, to mechanical distortion of the eye. White and Jorve (1956) found that the relationship between binocular visual acuity and acceleration was almost the same for $+g_x$ and $+g_z$ forces, and Henry (1950) noted a qualitatively similar pattern during exposures to negative acceleration. (He also found that repeated exposure to -3 g_z gave rise to diplopia, which he attributed to a disturbance of balance in the extraocular muscles secondary to oedema in the tissues of the orbit.) On the other hand, Vartbaranov et al. (1969) claimed that significant alterations of visual acuity appeared at $+6$ g_z, and at -6 or -8 g_x, but that no impairment was evident at less than $+10$ g_x. These authors dismissed the suggestion by White and Jorve (1956) that distortion of the globe and, possibly, displacement of the lens of the eye might be responsible for the diminished acuity, and it is certainly true that Smedal et al. (1963) were unable to demonstrate deformation of the lens or the cornea during exposure to 8 g_x. They attributed the inconsistent changes that they found to the accumulation of tears on the cornea; increased lachrymal secretion occurs not uncommonly on the centrifuge, but the reason for it is obscure.

Cardiac Function

"... and make my seated heart knock at my ribs, against the use of Nature."

There are strong haemodynamic reasons for expecting the heart to be adversely affected by gravitational stress, especially when it acts in the g_z axis.

A change in the pulse rate is one of the most consistent effects of $+g_z$ acceleration. For a given individual the tachycardia is almost a linear function of the applied force, up to levels of $+5$ or $+6\,g_z$, but the peak value achieved shows great individual variation. It may lie between 130 and 180 beats per min, and rates in excess of 200 beats per min are occasionally reached (Lambert and Wood, 1946; Tikhomirov, 1969). When reflex cardiovascular readjustment to the stress is complete some 6–8 s after onset, the tachycardia declines, and when the acceleration ceases, a considerable bradycardia usually ensues. This can be blocked by the injection of atropine, and is certainly of vagal origin (Life and Pince, 1969). During very prolonged exposure to $+g_z$ forces the heart rate may slow abruptly. This often betokens incipient circulatory collapse in the form of vaso–vagal syncope, but it occasionally occurs without other signs or symptoms.

The increased heart rate that accompanies positive acceleration is a direct result of the stimulation of the arterial baroreceptors by a falling systemic pressure. It is therefore not surprising that $-g_z$ forces, which raise the pressure in the carotid sinus, cause profound bradycardia. (Armstrong (1943) claimed that negative accelerations of 4 g increased the heart rate to about 155 beats per min, but that author appears to have been unfortunate in his choices of experimental animal and of anaesthetist.) Rapidly applied forces lead to asystole which may persist for 9 s at $-3\,g_z$ (Jasper and Cipriani, 1945; Sieker, 1952) or even until the experimenter loses his nerve and stops the centrifuge. If the stress is gradually applied, the heart slows but does not stop, and the degree of bradycardia is then in rough proportion to the maximum acceleration. Henry (1950) showed that the changes could be greatly reduced, in dogs, by vagotomy or by denervation of the carotid sinuses; these operations also abolished the tachycardia that can normally be recorded upon return to $\pm 1\,g_z$. The small remaining responses could be attributed to activity in other baroreceptor areas, such as the aortic arch.

The response of the pulse rate to transverse accelerations depends to a large extent upon the precise orientation of the trunk and legs in relation to the g-vector. Postures that can be maintained with comfort always involve an element of head-up or head-down tilt with respect to the acceleration, and failure to appreciate this fact has led to conflicting reports of tachycardia and of bradycardia in apparently similar situations. For the practical purposes of aviation, transverse acceleration can be regarded as an expedient alternative to exposure to high $+g_z$ forces, achieved by progressive supination of the body from a normal sitting position. The behaviour of the heart rate in these circumstances *is* consistent; $+g_z$ forces always evoke a tachycardia, but the extent of the change seen at a given acceleration decreases sharply as the angle between the spine and the vertical increases (Glaister and Lisher, 1977; Tikhomirov, 1969).

The pressure generated at the level of the aortic valve remains essentially constant and, although the stroke output falls with positive and transverse accelerations (Howard, 1959; Lindberg *et al.*, 1960; 1962) the weight of each unit volume of blood is raised in proportion to the gravitational field. Thus, the cardiac work is increased, and further embarrassment may be caused by the concomitant tachycardia. Objective evidence of "cardiac strain" has often been sought, and sometimes found, in the electrocardiogram. The interpretation of the changes produced in the electrical activity of the heart by positive acceleration has been a fruitful source of argument for more than 30 years. Ham (1943) was among the first to describe the reduction in amplitude of the R and S waves that commonly occurs, but he placed no sinister interpretation upon this finding. Gauer (1950) claimed that the flattening of the T wave in limb leads, which was sometimes accompanied by pronounced elevation of the S–T segment at high accelerations, might indicate coronary insufficiency, especially as the spatial pattern of the R and S waves at the same time was suggestive of right ventricular preponderance. However, Pryor *et al.* (1952) showed by vector analysis that all the changes noted by Gauer could be attributed to a shift in the electrical axis of the heart. Pryor also pointed out that myocardial embarrassment of the degree postulated by Gauer would almost certainly give rise to severe arrhythmias but that, apart from occasional ventricular or supraventricular extrasystoles, no serious disturbances of cardiac rhythm had ever been observed.

In 1956, Zuidema *et al.* reported the occurrence of potentially serious changes in the ECG of four out of five subjects who were exposed to moderate but prolonged accelerations. The usual crop of sporadic ventricular extrasystoles was accompanied, in this series, by a high incidence of ectopic beats from atrial and nodal foci, and one man experienced severe chest pain that was highly suggestive of ischaemic heart disease. The many hundreds of exposures since made in laboratories throughout the world to equal or more severe stresses have failed to reveal a similar pattern of abnormality although changes in the ECG have been meticulously documented (Cohen and Brown, 1969; Shubrooks, 1972). Zuidema's subjects may have been an abnormal population; his findings have not been satisfactorily explained on other grounds.

Until very recently there was no unequivocal objective evidence of significant impairment of cardiac function during acceleration. However, Burton and MacKenzie (1976) demonstrated that subendocardial haemorrhages were consistently produced in pigs by exposure to $+9\,g_z$ for periods of 15–90 s. The lesions were limited in extent and rarely involved the bulk of the cardiac muscle, but the Purkinje fibres seemed to be preferentially affected. It is of interest that the changes could be completely prevented by the administration of beta blocking drugs before the exposure to acceleration.

Electron microscopic studies by Lindsay et al. (1976) revealed redistribution of mitochondria and nuclei within the cardiac muscle cells of swine subjected to high levels of $+g_z$ acceleration, with myofibrillar disruption in the region of the Z-band. Other changes included rupture of the lyosomal membrane, leading to the release of hydrolytic enzymes into the cell. Later work (Dowell et al., 1976) suggested that "tissue-damage" enzyme activity persisted for up to two weeks after the period of accelerative stress, and that the metabolism of damaged cells was enhanced.

Even more recently, it has been found that the incidence of subendocardial haemorrhage in pigs decreases with repeated exposures to $+g_z$ at intervals of up to 30 days, and that there is a strong correlation between the extent of the lesions and the plasma catecholamine level in the immediate post-acceleration period (Leverett, personal communication, 1978).

The significance of these pathological findings for man is uncertain, but it is known that the concentration of noradrenaline in human plasma is greatly increased after centrifugal acceleration and, in an era when the capability of military aircraft for sustained manoeuvres is rapidly being enhanced, the possibility of cumulative damage to the myocardium must not be ignored (Erickson et al., 1976).

The Cerebral Circulation

" 'In my youth', Father William replied to his son, 'I feared it might injure the brain'."

Loss of consciousness resulting from failure of the cerebral circulation sets the ultimate limit to tolerance of positive acceleration, and forces the body to escape from the stress by (involuntary) adjustment of its posture. The surprising thing is not that collapse occurs, but that it is so long delayed and then so precipitate in its appearance.

The simple hydrostatic model of the cardiovascular system described earlier implies that the blood flow to the brain should fall to zero at about $+4\,g_z$, but it also suggests that cerebral perfusion should be progressively compromised by lesser degrees of stress. When black-out is reached the systolic blood pressure at the eye is no more than 20 mm Hg, and the arterial pressure in the cortex must be even lower than this. Yet, despite the loss of vision, the other senses seem to be but slightly affected, and intellectual function is apparently unimpaired until the final catastrophe supervenes. It must be concluded that the model is inadequate, and that other mechanisms conspire to preserve the cerebral circulation. Anatomical and physiological factors have both been implicated.

The brain is contained in a rigid box, and cushioned by cerebrospinal fluid which is subject to the same hydrostatic laws as the blood. A passive fall in the intracranial blood pressure is accordingly balanced by a similar decline in the CSF pressure—the forces across the walls of the cerebral vessels remain essentially constant, and there is no mechanical incentive for them to collapse. Thus, provided that the systemic circulation has power enough to force blood to the level of the base of the skull, cerebral perfusion can continue. The determining factor in preserving the circulation to the brain then becomes not the arterial pressure alone, but the arterio–venous pressure difference. The same hydrostatic forces that act upon the carotid arteries affect the jugular veins and, for as long as they remain patent, the internal pressure of the latter will be subatmospheric. This will increase the effective cerebral perfusion pressure and blood entering the cranium will be positively sucked into the veins. The end-point will only be reached when the pressure in the vessels on one or other side of the system falls below the critical level for closure, causing them to collapse and bringing the circulation to a halt.

The siphon theory has been used, *faute de mieux*, to explain the maintenance of consciousness during $+g_z$ acceleration, but the direct evidence for it is not strong. Giacobini (1942) claimed that blood flow through the carotid arteries and the jugular veins was significantly reduced at $2\,g$, but the radiographic method that the used probably failed to distinguish between flow and content. Jasper and Cipriani (1945) observed the cortical vasculature through a cranial window, and found that acceleration produced pallor of the surface of the brain, with emptying of the smaller vessels. At the same time, the veins showed some congestion. The authors concluded that the behaviour of the cerebral circulation was largely passive—the blanching was due to a failure of arterial supply and the congestion to obstruction of venous outflow by collapse of the jugular veins. (It is interesting that the sequence of events reported by Jasper and Cipriani was very similar to that seen in the retina by Leverett *et al.* (1967).)

Rushmer *et al.* (1947) demonstrated that the pressure in the cerebrospinal fluid varied as predicted over a wide range of positive and negative accelerations, and that the venous pressure at head level behaved in a similar fashion. Akesson (1948) showed that the sheath of tissue in which the jugular veins are embedded markedly increases their resistance to collapse, and Henry *et al.* (1949) measured pressures as low as -40 mm Hg in the deep veins of the neck in men exposed to $+4.5$ g_z. Although this fall was not sufficient to offset the concomitant arterial hypotension completely, it probably resulted in an effective cerebral perfusion pressure of 60–70 mm Hg. In later experiments (Henry *et al.*, 1951) it was found that the oxygen content of jugular venous blood fell slightly during positive acceleration; some reduction of flow to the brain was, therefore, inferred.

The major physiological mechanism for the preservation of the cerebral circulation is autoregulation. There is ample evidence, from studies both in animals and in humans, that the blood flow to the brain remains within very narrow limits despite wide variations in the systemic arterial pressure (Rapela and Green, 1963; Häggendal *et al.*, 1967). This relative constancy, which is attributable to local vasodilatation, is preserved to hypotensive levels of 65 mm Hg or less; that is, to within the range found on the centrifuge by Henry *et al.* (1949) (Fig. 4).

Objective measurements of the cerebral blood flow during positive acceleration have been disappointingly few. Using a radioisotope clearance method, Howard and Glaister (1964) showed that the total perfusion was reduced at 2, 3 and 4 g, but their analysis suggested that the flow to the grey

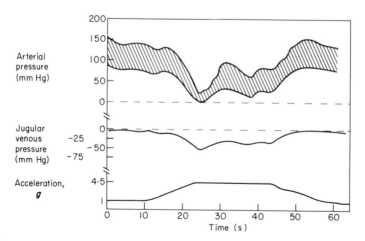

FIG. 4 Pressures in the radial artery (supported at eye-level) and in the jugular bulb during exposure to $+4.5$ g_z. The sub-atmospheric pressure in the vein acts as one limb of a siphon, to sustain cerebral blood flow. (Redrawn from Henry *et al.*, 1949.)

matter was preserved, at the expense of the less functionally vital white matter. The long time required for each determination (25 min) precluded experiments at higher accelerations and cast doubt upon the continued existence of a steady state even at the levels used. Some very recent studies in animals have indicated that cortical blood flow remains adequate up to at least $+7\,g_z$ and that autoregulation plays a large part in maintaining this degree of perfusion (Laughlin, 1978).

Unconsciousness occurring within a few seconds of the onset of $+g_z$ acceleration can fairly be blamed upon the hypotensive disruption of the cerebral circulation, but collapse sometimes results from prolonged exposure to moderate forces that are well within the accepted limits of tolerance. In these instances, the precipitating cause is peripheral circulatory failure. The heart rate, which has usually increased slowly but steadily throughout the period of acceleration, falls abruptly to about 30 beats per minute, and the blood pressure also drops sharply. These events, together with the pallor and sweating that accompany recovery, are typical of the vaso–vagal syndrome (Barcroft and Swan, 1953). They result, at least in part, from the continuing loss of circulating blood volume by pooling and transudation of fluid in the legs.

No measurements have been reported of the cerebral blood flow during exposure to negative accelerations, but there is no reason to suppose that dramatic changes occur. In monkeys, Duane et al. (1952) found that the rise in jugular venous pressure caused by high levels of $-g_z$ forces greatly exceeded the arterial hypertension. This discrepancy can be explained in terms of the different levels of the "hydrostatic indifference points" on the two sides of the circulation. Duane et al claimed that at $1-2\,g$ the effective perfusion pressure of the brain was reduced to about 20 mm Hg, but the relevance of these results to human exposures is small. From the scanty data available, it can be calculated that an applied force of $-4\,g_z$ would lower the "trans-cerebral" vascular pressure differential in man by about 50 mm Hg, so producing a situation very similar to that seen at $+4$ to $+5\,g_z$.

Armstrong and Heim (1938) claimed, from observations made on goats subjected to $-5\,g_z$, that cerebral haemorrhage was a serious hazard, but their findings could not be reproduced by others (Gamble and Shaw, 1947; Lombard et al., 1948). The protective role of the cerebrospinal fluid pressure should be independent of the direction in which the acceleration acts, and the transmural forces in the intracranial vessels should not depart far from normal. A credible explanation for the claims of Armstrong and Heim was supplied by Henry et al. (1949), who pointed to the close similarity between the haemorrhagic patterns attributed to negative acceleration and those certainly produced by asphyxia. Henry showed that if the head and neck were bandaged to prevent gross swelling of the soft tissues and the animals were given a tracheotomy, goats could withstand $-15\,g_z$ for periods of up to 30 s without suffering overt lesions in the brain.

In normal men, sustained negative acceleration has never been known to produce intracranial haemorrhage, although the range of human experience is vastly smaller than that of the goat. Cerebral dysfunction has, however, been deduced from changes in the electroencephalogram (Beckman et al., 1953) and metabolic disturbances suggestive of a depression of oxidation were found in animals exposed to -3 and $-5\,g_z$ by Pogrund et al. (1951). These changes are, of themselves, not sufficient to account for the unconsciousness that sometimes occurs, and it is probable that profound bradycardia with or without asystole plays a major part in that event.

Reflex Responses of the Circulation

"So the heart *be right, it is no matter which way the* head *lies."*

The cardiovascular effects of positive acceleration are in many respects similar to those of the hypotension produced by haemorrhage, and it is not surprising that the reflex responses of the circulation are alike in the two conditions. They consist of tachycardia, selective vasoconstriction (with redistribution of the available cardiac output) and, possibly, the mobilization of sequestered blood from the reservoirs in which it is "stored".

A major difference between the physiological reactions to haemorrhage and to acceleration lies in the speed of their induction. Save in the most catastrophic cases of bleeding, haemorrhagic hypotension develops slowly, with the result that changes in the heart rate and the vasomotor tone keep pace, as far as they are able, with the decline in the systemic pressure. The accelerations encountered in flight are, however, almost invariably of rapid onset, and the primary cardiovascular disturbance appears within a second or two by the mechanical processes described above. Reflex compensation cannot be brought into play immediately and there is, accordingly, a brief period of imbalance before an uneasy equilibrium is restored to the circulation.

This temporary disparity between stimulus and response is demonstrated by the time-course of the visual symptoms caused by rapidly applied acceleration. Exposure to a sustained force that is sufficient to produce grey-out or black-out leads to an almost instantaneous fall of systemic pressure at the eye, followed by an impairment of vision after a few seconds. Within a further five to ten seconds, the symptoms abate—black-out may regress to grey-out or even to an apparently complete restoration of vision. Over the same period of time, the progressive hypotension is first checked and then reversed, reaching a new plateau that may not be far removed from the normal level.

When the rate at which the acceleration is applied is very slow, that is, less than about 0.1 g per second, this sequence of events does not occur. Visual signs and symptoms, when they appear, do not moderate. Indeed, if the exposure is very prolonged, they may become more severe, so that grey-out merges into black-out or black-out into unconsciousness. The threshold of visual decrement is, however, set 1.5–2 g higher when the onset of the force is gradual than when it is rapid. This situation is clearly analogous to the "slow bleed" in which the cardiovascular responses are concurrent with the stress.

Tachycardia is the most consistent result of positive acceleration, but its magnitude can vary widely. Armstrong and Heim (1938) recorded the pulse rates of pilots during aerobatic flight, and reported peak values of between 110 and 180 beats per minute after 5–7 s at about $+g_z$. On the centrifuge, Rose et al. (1942) also found that the rate was markedly increased by moderate accelerations, and one of their subjects exhibited a tachycardia of 196 beats per minute at 4.5g. The authors recognized that apprehension and the Coriolis forces produced by rotation on a short radius might have exaggerated the physiological response. Browne (1959) also measured the maximum "instantaneous" heart rate (derived from R–R intervals in successive complexes of the ECG) achieved in centrifuge experiments, and showed a significant effect of the pattern of application of acceleration. In exposures with a rapid onset, the relationship between the cardiac frequency and the applied force was non-linear; from a (high) control value of 90 beats per minute, it rose to 126 beats per minute at $+g_z$, to 132 at 4g and to 138 at 5g. When the rate of application of acceleration was low (0.1 g s^{-1}) a linear and less steep relationship appeared, with a slope of about 7 beats per minute per g. Thus, in these experiments, the heart rate reached only 105 beats per minute at 3g, and it was slightly less than 120 beats per minute at 5g. This difference represents the change from transient to steady-state conditions, and it is temporally related to the secondary alterations in the level of systemic blood pressure. However, Howard (1964) demonstrated that the equilibrium reached after 15 s or so at a constant acceleration is more apparent than real, and found that the heart rate increases very slowly but very steadily throughout an exposure to $+3g_z$ for one minute.

Howard and Garrow (1958) measured the blood flow through the human forearm at 1 g and at 3 g, and showed that a considerable reduction took place. The arterial blood pressure in the arm was determined at the same time, to permit calculation of the peripheral vascular resistance, and this proved that active vasoconstriction had, in fact, occurred. Later experiments (Howard, 1964) extended these observations and demonstrated that the resistance to flow in the vessels of the forearm is related directly and almost linearly to the applied acceleration, at least up to $+6\,g_z$. Although the method used did not allow vascular resistance to be measured in absolute terms, its value was approximately doubled at 6g.

Evidence of vasomotor changes in other organs is less direct, and is derived mostly from studies in animals. The most comprehensive study was that of Glaister (1969), who measured cardiac output and its subdivisions in dogs exposed to $+2.6\,g_z$ and $+4.2\,g_z$. The total systemic flow fell by 23% at the lower level of acceleration and by more than 50% at the higher one, with increases in the total peripheral resistance of 38% and 67% respectively.

The perfusion of skin and skeletal muscle was markedly reduced at 4.2 g, and there was post-mortem evidence of ischaemia in some muscles of the limbs. In contrast, the blood flow to the diaphragm was significantly increased, presumably as a consequence of the extra effort of breathing against the exaggerated weight of the chest wall and the viscera.

The greater workload required of the heart was also matched by its blood supply, which, when expressed as a fraction of the total cardiac output, was doubled at 2.6 g and more than trebled at 4.2 g. A similarly striking increase also occurred in the adrenal gland, but flow to the kidney, gut, and liver fell, often dramatically. The overall picture was thus of a redistribution of the diminished systemic output to preserve the blood flow to active organs at the expense of less vital tissues.

Recent unpublished work by Laughlin (1978) has demonstrated intense vasoconstriction in the kidney during acceleration, with a complete cessation of renal blood flow at $+5\,g_z$. Laughlin has also found that the distribution of perfusion within the myocardium is patchy, and that some regions of the left ventricle come dangerously close to ischaemia at high accelerations. For the heart as a whole, his results are not in conflict with those of Glaister (1969).

The reflex responses of the circulatory system to acceleration are usually explained as the result of stimulation of the baroreceptors of the carotid sinus by the fall of blood pressure in the upper part of the body. Evidence for this view was provided by Jongbloed and Noyons (1933), who showed that sectioning the carotid sinus nerves abolished the tachycardia induced in normal animals by acceleration, and by Greenfield (1945), who found that the secondary rise of systemic pressure accompanying sustained exposure did not occur after stripping of the carotid sinuses. Denervation of the aortic arch, on the other hand, appeared to have little or no effect.

There is a strong intuitive expectation that the receptors which initiate reflexes to compensate for gravitational stress should be at the upper end of the cardiovascular system, where the hydrostatic pressure change is greatest, and where they are strategically placed to monitor the perfusion pressure of the brain. At the level of the aortic arch, the arterial pressure is relatively unaltered by acceleration, and any stimulation of the baroreceptors in that region will be correspondingly small. However, there are some objections to ascribing the circulatory changes seen during acceleration entirely to the stimulation of the carotid sinus, although this undoubtedly plays a major rôle in their origin. For example, the compensatory responses to orthostasis are virtually unchanged by blocking of the sinus nerves and the vagi (Edholm and McDowall, 1936). In man, (although not in animals), the carotid sinus reflex appears to be concerned almost entirely with regulation of the heart rate, and to have little vasoconstrictor power (Roddie and Shepherd, 1957). There is, fortunately, no dearth of other receptor sites from which the changes induced by acceleration might be triggered. The mesenteric vascular bed was shown by Heymans et al. (1937) to be a potent source of vasoconstrictor stimuli and pressure/volume receptors in the atria and the lesser circulation may also be involved (Barr er al., 1958).

Finally, pressure changes within the peripheral vessels themselves may induce a reflex vasomotor response, affecting both arterial and venous tone (Yamada and Burton, 1954; Haddy and Gilbert, 1956). Venoconstriction was demonstrated to occur during acceleration in the saphenous vein of the dog (Salzman and Leverett, 1956) and in the veins of both the arms and the legs of man (Hiatt et al., 1958).

One puzzling aspect of the reflex responses to force in the $+g_z$ axis is that they do not appear to exert their maximum possible effect. The change in the total peripheral resistance that occurs at, say, 5 g is by no means as large as the control system is capable of generating, nor is the tachycardia close to the sustained capacity of the stressed heart. The efficiency of the circulation obviously cannot be assessed in simplistic terms of the behaviour of two interacting parameters, but it is strange that mechanisms evolved primarily to protect the cerebral circulation from accelerative forces (if only of 1 g) should not be fully exercised when consciousness is threatened.

Almost nothing is known of the cardiovascular adjustments that occur during exposure to negative accelerations. Bradycardia is one obvious reaction to the stress, and it is extremely likely that vasodilatation occurs in peripheral tissues and organs (Howard, 1964). These changes can be attributed to stimulation of the carotid sinus by a raised transmural pressure (Jongbloed and Noyons, 1933; Henry, 1950). Little evidence of other concomitants of vagal activity has been reported, probably because they have not been sought.

Respiratory Effects

"... in this harsh world Draw thy breath in pain."

Difficulty in breathing, caused by inability to expand the thorax fully against its own increased weight, constitutes one limit of tolerance for transverse acceleration. A similar problem does not occur with forces applied in the g_z axis, partly because the mechanics of the chest prevent it, and partly because disturbances of vision and of consciousness pre-empt respiratory distress. The pattern and efficiency of pulmonary ventilation are certainly altered by quite moderate accelerations, irrespective of the direction in which they act, but the effects are remarkably small when the lung is considered as a whole.

The most consistent finding is an increase in the respiratory rate, which may be associated with a reduced tidal volume (Zechman *et al.*, 1959) but which is, at least for accelerations of up to $+5\,g_z$ and to $+8\,g_x$, a component of a true hypoventilation (Rosenhamer, 1967; Cherniak, 1959). The origin of this response is obscure, but the influence of psychological factors is suggested by the finding that, in anaesthetized animals, hypoventilation or apnoea is the rule (Greenfield, 1945).

The total lung capacity and its subdivisions behave in the fashion that would be expected on physical and anatomical grounds. The downward pull of the viscera with $+g_z$ acceleration lowers the diaphragm and increases the functional residual capacity, but the largest effect is seen during the "physiological" range between $-1\,g_z$ and $+1\,g_z$, where it can exceed 700 ml (Colville *et al.*, 1956). Greater degrees of stress produce progressively smaller rises of the FRC, however.

Glaister (1971) has compared the diaphragm to a weight mounted between two unequal springs—a relatively large displacement (and hence a large change in lung volume) can initially be produced by a small force acting in either direction, but thereafter the resistance to stretch or compression of the springs rapidly increases. This model explains most features of the dynamic behaviour of the anatomical compartments of the lung, and it identifies the diaphragm as the primary determinant of lung volume. In support of this contention Glaister (1961) demonstrated a fall of 1 cm (measured at the oesophageal hiatus) in the resting level of the diaphragm at $+3\,g_z$, with an accompanying increase of 300 ml in the lung volume; at $+4\,g_z$ the figures were 2 cm and 500 ml, respectively. However, at the latter value of positive acceleration the total lung capacity was reduced by about 5%, secondary to a diminution of inspiratory capacity. Very little alteration of residual volume or of anatomical dead space could be detected, but the vital capacity was significantly restricted at all accelerations greater than $+3\,g_z$.

Virtually nothing a known of the effects of $-g_z$ accelerations upon the mechanics of respiration, although it is fair to assume that the action of such forces will be the opposite of that seen during positive acceleration. The results reported by Wilson (1927) indicate that relatively large changes are produced by inversion (that is, by the transition from $+1\,g_z$ to $-1\,g_z$). Henry (1950) inferred that higher degrees of negative acceleration caused no major respiratory disturbance, but is very probable that his subjects were preoccupied with more important problems than the minutiae of breathing.

Fewer data exist for forward acceleration, but progressive increases in the minute volume of ventilation have been found at levels of acceleration up to $+8\,g_x$ (Cherniack et al., 1961). At $+12\,g_x$ the pulmonary ventilation may be lower than at $8\,g$, because the fall in tidal volume is not commensurate with the rise in respiratory frequency (Golov, 1966). The restrictions of vital capacity and of expiratory reserve volume are much greater during exposure to accelerations in the g_x direction than at similar levels of $+g_z$, because of the mechanical inefficiency of the chest wall when the forces act transversely. At $+4\,g_x$ the vital capacity has only two-thirds of its resting value, and at $+5\,g_x$ it has fallen by as much as 73% (Cherniack et al., 1959). The expiratory reserve has virtually disappeared at $+g_x$, and the fact that the lung volume at end-expiration is then equal to the residual volume has considerable bearing upon the mechanisms of functional (as opposed to anatomical) respiratory impairment.

Most of the changes seen with forward acceleration can be attributed to the increase in the effective weight of the thoracic cage which must be lifted with each breath. This factor is of far smaller moment in positive acceleration, where its influence is masked to a great extent by the rôle of the diaphragm. Moreover, the diaphragm tends to *rise* in response to $+g_x$ forces, which push the abdominal viscera upwards. It is of some physiological interest that the residual volume remains almost invariant during exposure to relatively high forces in either the x- or the z-axis. Campbell (1958) also drew attention to this constancy in other circumstances, and all the evidence suggests that the residual volume is independent of the properties of the respiratory muscles and rib cage; rather, it appears to be a result of the inherent structure of the lung tissues.

The simple hypothesis that in the erect posture the lung can be regarded as a very compliant spring resting upon a diaphragm which behaves like a piston implies that in dependent regions the pulmonary tissues will be compressed by the weight of those above them, and leads to the conclusion that the alveoli at the bases will be smaller than those at the apices. This has been confirmed by direct measurement in animals (Glazier et al., 1967) and indirectly in man (Sutherland et al., 1968). In the horizontal lung, the alveolar diameters at the "anatomical" base and apex are the same, but the uppermost alveoli are larger than the lower ones; this is true for prone, supine and lateral decubiti (Kaneko et al., 1966). In the head-down ($-g_z$) posture, regional differences are minimal, probably because the vertical gradient of pleural pressure is virtually abolished in this situation.

These findings merely confirm the intuitive belief that the lung is not only supported within the thorax, but also suspended, and hence that the compression of its lower parts is matched by stretching in the upper zones. The resulting distribution of alveolar sizes has a very significant effect upon the dynamics of respiration. The elastic properties of the different regions of the lung are the same, but the relationship between alveolar volume and transpulmonary pressure is strongly sigmoid (Milic–Emili et al., 1967). As a result, the distribution of inspired gas is grossly uneven during normal quiet breathing. (At full inspiration the volume of individual alveoli is nearly equal throughout the lung, and inequalities of ventilation disappear). Milic–Emili and his colleagues (Milic–Emili et al., 1966; Kaneko et al., 1966) have shown that the change in ventilation per alveolus increases by about 2% for each centimetre of vertical distance from the uppermost point, and that the magnitude of this change is largely unrelated to posture.

The effects of increased acceleration upon regional pulmonary ventilation can also be considered in terms of the spring model (Glaister, 1970). The weight of each unit mass of the lung increases in proportion to the applied force, so that the spacing of the coils of the spring decreases still further at the bases and, because the upper zone cannot descend, the apical coils are expanded and separated. In this model, the spacing of the coils represents alveolar size, and at first, this distortion exaggerates the normally existing disparities in the regional distribution of inspired gas. The difference in ventilation between the apex and the base is doubled at $+2\,g_z$ (Bryan et al., 1966) and tripled at $+3\,g_z$ (Glaister, 1965); these figures are in accord with the changed gradient of intrapleural (i.e. trans-pulmonary) pressure.

As the acceleration is increased to even higher levels, however, this linear relationship fails. The lowermost coils of the pulmonary spring close completely and the upper level of the compaction moves progressively up the lung in response to increasing gravitational stress. The analogy is not perfect, for the volume of the alveoli does not fall to zero in these circumstances—instead, they become closed off at a finite volume which is determined by the elasticity of the tissue and by the mechanics of the air sacs and the airways. They then lie on the flat toe of the curve relating volume to transpulmonary pressure. The important physiological consequence is that these lung units are deprived of ventilation, and can play no part in metabolic gas exchange. At the same time, and for the same reason, the volume of the apical alveoli rises towards the elastic limit as they approach the flat shoulder of the expansion curve. Extrapolation from the data of Jones *et al.* (1969) suggests that when the acceleration exceeds $+9$ g_z, alveoli at the top of the lung may attain their maximum capacity during quiet breathing, while those at the base (and for a substantial distance above it) are at their minimum possible size.

Similar considerations govern the behaviour of regional ventilation in response to $\pm g_x$ and $\pm g_y$ forces, when due allowance is made for the vertical extent of the lung in the plane of the acceleration. Attempts to generalize the concept of the simple spring, to make it independent of posture and of lung shape, have been made by Proctor *et al.* (1968) and by Krueger *et al.* (1961), who proposed, respectively, models based upon compound springs and upon a fluid-filled bag. The advantages of the more complex analysis that these systems permit are generally small.

The efficient distribution of ventilation is not the primary purpose of respiration, and the disturbances associated with acceleration would be of small concern if metabolic gas exchange was not impaired. That process requires the regional ventilation to be finely matched to regional perfusion, but blood pressure and blood flow in the lung are subject to the same influences as those in other parts of the cardiovascular system.

The perfusion of the erect lung at $+1$ g_z has been intensively studied since Orth (1887) deduced that hydrostatic effects must lead to an uneven distribution of blood flow. It has since been shown that the isolated lung can reasonably be described as a three-decker sandwich, the layers of which are set by the pressures in the pulmonary arteries, the pulmonary veins, and the alveoli (Permutt *et al.*, 1962; Bannister and Torrance, 1960). In the upper zone the intra-alveolar pressure is greater than that in the regional arteries which, because of hydrostatic differences, are at a lower pressure than the central pulmonary artery. In this region, then, there is no sustained flow, although at its lower margin the respiratory variations in alveolar pressure and the pulsatile arterial pressure permit a fluctuating perfusion.

In the middle zone the alveolar pressure is less than the mean arterial pressure, but greater than that in the pulmonary veins. The blood flow in this zone is directly determined by the arterio–alveolar pressure gradient, with the veins playing no active part. Because the arterial pressure increases steadily with vertical distance from the apex, so is the descent through zone 2 characterized by progressively increasing perfusion of the alveoli.

In the lower zone, venous and arterial pressures are both greater than intra-alveolar pressure and, as in the systemic circulation, blood flow is determined by the arterio-venous pressure difference. This should lead to even perfusion of the entire zone, but the high compliance of the relatively unsupported pulmonary vessels allows them to distend in response to the transmural pressure, which increases as the base of the lung is approached. The change of perfusion with height is, however, much less steep in this region than in the middle zone.

The effects of acceleration upon regional perfusion within the lung should be readily predictable. The "hydrostatic indifference point" lies about 5 cm caudad of the hilum, and it can be assumed that pulmonary arterial and venous pressures at that level are unaltered by accelerative forces. At $+1\,g_z$ the unperfused zone at the apex should have a vertical extent of about 5 cm; at $+3\,g_z$ it should increase to approximately 15 cm. Glaister (1967) provided experimental proof of the theory, and showed that the upper limits of detectable perfusion in human subjects at $+1$, $+2$, and $+3\,g_z$ were situated $3\frac{1}{2}$, 9 and 14 cm from the apex, respectively.

Because of the complex shape and architecture of the lung it is more physiologically significant to express these results in terms of ventilated volume, and Glaister (1964) demonstrated that at $+3\,g_z$ the alveolar dead-space had risen to occupy about one-third of the total lung volume. He also found a good linear relationship between the measured regional blood flow and the calculated mean pulmonary arterial pressure at all levels. An unexpected, but not invariable, finding was the existence of a region of decreased blood flow at the lower edge of the lung. This appeared to coincide with the zone of "gas-trapping" in which ventilation of the alveoli was prevented by mechanical distortion of the airways, and it is possible that kinking of blood vessels by the compressed parenchyma was also a factor.

The overall picture of the lung in high gravitational fields is, then, of a system in which the functional size of the alveoli decreases with vertical distance from the upper surface, while relative ventilation (which is inversely related to alveolar diameter) increases to a limit in the same direction. Perfusion is always deficient in the uppermost zone and the level to which blood can flow declines in proportion to the acceleration. At the extreme base of the lung both ventilation and perfusion may behave anomalously. However, separate consideration of these two parameters gives little indication of the efficiency of gas exchange. A much better yardstick is offered by the regional ventilation/perfusion ($\dot{V}_A/\dot{Q}$) ratio, and when this is used, the effects of accelerative forces upon lung function appear to be rather less dramatic. At $+1\,g_z$ the gradient of perfusion between base and apex is balanced by a similar pattern of change in ventilation and, although the matching is far from perfect, it is generally the case that the regions with the highest blood flow also receive a greater-than-average fraction of the inspirate. At higher levels of acceleration the mismatching becomes much more pronounced, and at $+3\,g_z$ the $\dot{V}_A/\dot{Q}$ ratio reaches infinity in the upper third of the lung (ventilation without perfusion) and may be zero at the base (perfusion without ventilation). Similar changes occur with accelerations in other axes; at $+5\,g_x$ ventilation/perfusion ratios range from 10 or more in the region immediately beneath the sternum to zero in the paravertebral zones.

Acceleration atelectasis first came to attention in 1958, when the pilots of some high-performance aircraft began to report respiratory symptoms during and soon after flight. The commonest complaint was of dull pain in the lower part of the chest, with difficulty in drawing a deep breath. Attempts to do so provoked a dry cough which sometimes relieved the pain. Recovery was usually very rapid, but some discomfort might persist for up to 24 hours. This syndrome was at first attributed to the very low humidity of the gas in the oxygen system—the aircrew were, for the first time, required to breathe undiluted oxygen for the whole duration of each sortie. However, the association with positive acceleration soon became apparent, and it was shown that both factors must be present for the condition to develop.

Chest radiographs of affected pilots, taken immediately after flight, showed obliteration of the costophrenic and cardiophrenic angles and diffuse shadowing of the bases. These changes were considered by some to indicate regional pulmonary oedema (Wood et al., 1963) and by others to denote absorptional atelectasis (Green, 1963). Pulmonary function tests showed a gross limitation of the total lung volume and the inspiratory capacity, but the residual volume and the functional residual capacity were not significantly altered. It was found that some reduction in the vital capacity could be produced by exposure to positive acceleration alone, but the change was considerably greater when 100% oxygen was inspired, and it was augmented still further when an inflated anti-g suit was worn. Green and Burgess (1962) found that these three factors acting in concert could reduce the vital capacity by as much as 60%.

Although the high capillary blood pressures produced at the base of the lung by high accelerations (up to 100 mm Hg at $8\,g$) clearly predispose to the transudation of fluid into the alveoli, and although the permeability of small vessels is known to be increased by high local concentrations of oxygen (Ohlsson, 1947), pulmonary oedema is now largely discredited as the direct cause of the development of acceleration atelectasis. The most compelling argument against it was provided by Glaister (1966) who showed that a syndrome identical to that reported by aviators could be induced at 1 *g* in normal subjects breathing pure oxygen at very low lung volumes. Absorption must, therefore, be the over-riding factor, but the possibility that when the applied acceleration is very high the collapse may be facilitated by fluid exudation cannot be entirely excluded.

The sequence of events leading to the atelectasis is reasonably clear (Fig. 5). Positive acceleration exaggerates the regional differences in ventilation throughout the vertical extent of the lung and may, of itself, reduce the size of some alveoli at the base to a critical degree. The inflation of the anti-*g* suit raises the diaphragm, further compressing the lung tissue, and the smaller airways are closed off. Metabolic exchange with the pulmonary capillary blood continues, but if the gas trapped in the occluded, non-ventilated, regions contains a significant fraction of nitrogen (more than about 20%) full absorption does not take place. If, however, the alveoli contain no inert gas when closure occurs, the exchange can continue to completion; there is a further shrinkage of their volume and the unit soon becomes atelectatic. It remains so, even after the accelerative force is removed, until a deep breath or a cough raises the pressure in the airway supplying the collapsed segment above the critical opening pressure of the alveoli.

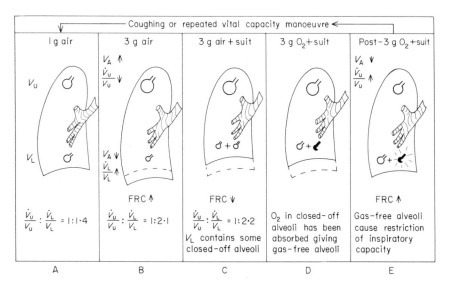

FIG. 5   The mechanics of basal atelectasis resulting from $+g_z$ accelerations, showing the critical dependence of the syndrome upon the composition of the alveolar gas, and the effect of the anti-*g* suit. The process is reversible at any stage.

The total volume of the lung involved in the collapse is very small in comparison with the measurable reduction in vital capacity, indicating that the latter must have a strong reflex component (Glaister, 1966).

It is apparent that the mechanism by which atelectasis is produced will also operate when accelerations are applied in other axes of the body, despite the fact that an inflated anti-g suit is not usually worn for such exposures. Data on the effects of $-g_z$ accelerations are understandably scanty, but Green and Burgess (1962) were able to provide some evidence of impaired pulmonary function in men suspended upside down while breathing oxygen at low lung volumes. Transverse accelerations, also, are rarely sustained for long enough to pose a hazard from atelectasis, but space-flights involve prolonged exposures to high levels of force in the g_x axis, and it is not uncommon for the astronauts to breathe pure oxygen throughout the launch and re-entry phases. Forward acceleration reduces the functional residual capacity (Cherniack *et al.*, 1961) and respiration is maintained at a low lung volume. The alveoli in dependent units (that is, in the para-vertebral regions) are reduced in size to an extent similar to that produced by comparable positive accelerations, and the blood flow to these segments is also relatively increased. Gas-trapping and gas absorption can certainly occur, and signs and symptoms of acceleration atelectasis have been reported in subjects exposed on the centrifuge to simulated rocket launches (Nolan *et al.*, 1963). No definite identification of the condition has yet been made during actual spaceflight, probably because evidence for it has not been sought.

Irrespective of the direction in which they act, accelerations lead to an exaggeration of the normal ventilation/perfusion inequalities in the lung, and effects upon the oxygenation of the blood might therefore be predicted. The implications of these gross inequalities are lessened by the sigmoid shape of the oxyhaemoglobin dissociation curve, which allows a remarkable degree of mismatching to be accommodated. At $+3\,g_z$ the arterial oxygen tension is 94 mm Hg, compared to the accepted value of 98 mm Hg at $+1\,g_z$, and this slight change produces no significant reduction in the oxygen saturation of the blood leaving the lungs. However, the gas tensions in unventilated alveoli must, once equilibrium has been reached, equal those in venous blood and perfusion of these regions represents physiological waste. Some venous admixture, or shunting, is present even at $+1\,g_z$, and the effect is exaggerated at greater accelerations (Wood *et al.*, 1963; Barr *et al.*, 1959). Henry (1950) demonstrated a fall of the arterial saturation to about 65% in an anaesthetized dog exposed to $-7g_z$ for 2 min, but it is probable that asphyxia contributed to this profound change.

In men subjected to positive accelerations of 4–5 g_z for 15 s Gauer *et al.*

(1949) measured arterial oxygen saturations of 90%, while Wood *et al.*
(1963) showed that the decline was progressive, leading to values of less
than 80% after 4 min at $+5$ g_z. Barr (1962) found that the desaturation was
reduced, but not abolished, by the breathing of oxygen, that it was more
pronounced when an anti-g suit was worn, and that repeated runs had a
progressively greater effect. The majority of observations in man has been
made with forces applied in the g_x axis, and Steiner *et al.* (1961) measured
arterial oxygen saturations of 84% at $+6$ g_x and 75% at $+8$ g_x (the value at
$+1g_x$ was 96%). By plotting data from a variety of sources Glaister (1970)
found that the relationship between arterial oxygen saturation and acceleration was approximately linear with a gradient of approximately 2.5% per
g. Neither the direction nor the time of application of the force seemed to
exert a clear influence.

The desaturation can be attributed to the intrapulmonary shunting, with
"venous admixture" of the arterial blood, and considerable support is given
to this hypothesis by the finding that breathing 100% oxygen does not
entirely prevent the condition. Oxygen does, however, delay its development; Hoppin *et al.* (1965) found that in exposures lasting for less than two
minutes the decrease in arterial saturation was 20% greater when air was
breathed, but that with prolonged acceleration this difference was reduced
to 10%. Oxygen breathing thus appears to be a possible method of
sustaining the saturation of the arterial blood in aviators who must be
exposed to high accelerations, but the physiological advantage is more
theoretical than real. Moreover, any potential benefit is completely overshadowed by the rôle of oxygen in the development of acceleration
atelectasis.

These observations can all be explained in terms of the distribution of
$\dot{V}/\dot{Q}$ ratios throughout the lung and of the mechanism of atelectasis.
Increasing the alveolar oxygen tension (by enrichment of the inspirate) will
obviously enhance the respiratory efficiency of the critical regions in which
ventilation is an inadequate match for the prevailing perfusion, and it is to
this that the beneficial effect of oxygen-breathing can be ascribed. The
greater part of the venous admixture probably arises from basal alveoli
which have been occluded and are therefore unventilated but, in the short-
term, their gas content at the moment of closure has a significant effect
upon oxygen exchange. Local arterial desaturation will not occur until the
alveolar gas has attained equilibrium with the venous blood, and the time
required for this to occur will be protracted if pure oxygen is breathed. The
initial alveolar oxygen tension before gas trapping is about 110 mm Hg
breathing air, and about 670 mm Hg breathing oxygen; other factors being
equal, the times taken to reach the point of significant venous admixture

would differ by a factor of approximately ten. This reasoning is in good agreement with the observed "protective" effect of high inspired oxygen tensions upon arterial oxygen saturation.

Attempts to estimate the magnitude of the arterio–venous shunting during acceleration are complicated by a simultaneous decrease in pulmonary diffusing capacity which is sufficient of itself to lower arterial saturation to 90% (Power et al., 1965) and by the lack of reliable information on the oxygen content of the mixed venous blood. Some measurements have been made in dogs exposed to $+6\,g_x$ (Banchero et al., 1965), and these suggest shunts of up to 40% of the total systemic flow. From experiments in man, Glaister and Mellemgaard (unpublished observations) concluded that the true right-to-left shunt was small, amounting at $+4\,g_z$ to little more than 5% of the cardiac output. The "physiological" shunt calculated from measurements of the arterial oxygen tension was, however, considerably greater, and in one subject it reached 50%. The difference reflects the relative proportions of completely unventilated (closed or collapsed) alveoli and of those with significant disturbances of the $\dot{V}/\dot{Q}$ ratio.

Protection against Sustained Acceleration

"Protection is not a principle, but an expedient."

Practical methods for raising human tolerance of the accelerations encountered in flight can be classified broadly into two groups; anatomical and physiological. A third possible class—pharmacological—is excluded because, despite intensive studies over more than 40 years, no drug has been found that is of practical use. All the successful techniques act by preventing or counteracting the cardiovascular effects of positive acceleration; that is, by raising the effective blood pressure at head level, or by hindering the redistribution of blood to the periphery, or by both.

The greatest increase in tolerance can be obtained by converting $+g_z$ forces into transverse accelerations, by altering the posture of the body with respect to the applied stress. The theoretical limit is attained in the fully prone or fully supine position, when the head, the heart and the legs all lie in the same horizontal plane, but a useful benefit is conferred by less extreme postures. It was an early observation of fighter pilots that crouching forwards delayed the onset of visual symptoms, and Stewart (1940) found that the majority of subjects could raise the threshold for black-out by at least 1 g simply by flexing their bodies at the hips. The mechanism of this effect is obvious enough—by reducing the vertical distance between the base of the brain and the heart it mitigates the hydrostatic fall of blood pressure at head level. The geometry of the cardiovascular system indicates that a gain of about 0.5 g can be obtained by inclining the trunk forwards to 30° from the vertical. The values obtained by Stewart were significantly greater than this, but the discrepancy can be explained by the increased intrathoracic and intra-abdominal pressures generated by the change of posture (see below).

A fully prone position is far from ideal for the pilotage of an aircraft (although several aeroplanes exploiting the principle were built for experimental flying) and a forward crouch with the head on the knees is hardly practicable. A degree of supination might, however, be acceptable and it should, on theoretical grounds, carry the same advantage as an equivalent amount of forward angulation. Raising the legs so that they are level with the hips was shown to afford protection of about 0.5 g by reducing the pooling of blood in the lower extremities, with a secondary elevation of the arterial pressure and, when this technique was combined with backward tilting of the seat to 65°, accelerations of $+7 g_z$ could be tolerated without loss of vision (Dorman and Lawton, 1956). The effects of small angles of reclination were found to be smaller than had been predicted. The reason is that in the normal sitting position the eye lies 10–15 cm anterior to the plane of the heart. Thus, when supination begins, the vertical distance between the aortic valve and the brain actually *increases* slightly, and it is not until the angle of the spine exceeds about 30° from the vertical that the expected protection becomes manifest. In a recent study Glaister (1977) has shown that the added tolerance for positive acceleration is inversely proportional to the cosine of the back angle. Reclination to 58° raises the grey-out threshold by 1 g, and increments of 2 g and 3 g can be obtained at angles of 69° and 74° respectively. Very much greater protection has been claimed to result from some extreme anatomical adjustments, such as the adoption of a fully-flexed foetal position (von Beckh *et al.*, 1976), but these have no obvious practical application (Fig. 6).

Of the various physiological techniques used for increasing the tolerance of sustained accelerations, voluntary action takes pride of place in terms both of simplicity and of history. The fact that shouting and forcible contraction of the muscles could prevent black-out or delay its onset was well known before 1940 and, although objective measurements were not made, such actions were claimed to give an effective protection of at least 2 g (von Diringshofen, 1942). Muscle tensing reduces the venous capacitance of the limbs and so prevents the pooling of blood; indeed, it can augment the central blood volume by mechanically squeezing the peripheral vessels. This

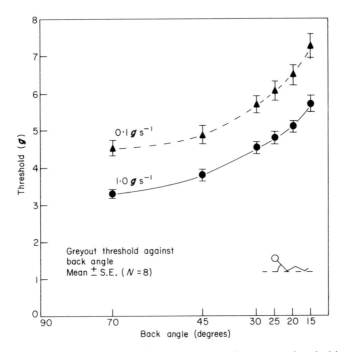

FIG. 6   The relationship between sitting posture and grey-out threshold for $+g_z$ accelerations. The slower rate of onset allows the compensatory reflexes to keep pace with the stress, but does not affect the shape of the tolerance curve.

can hardly explain the immediate action of the voluntary measures, because pooling does not become a significant factor in the production of symptoms for about 20 seconds after the onset of the acceleration. Straining and screaming do, however, markedly raise the intrathoracic pressure and thus have a direct effect upon the blood pressure. If all the force exerted were transmitted unchanged to the arterial tree and sustained there, raising the pressure in the chest by 25 mm Hg would increase the threshold for visual symptoms by 1 g, but the nett effect is rather less than this. Moreover, the performance of a Valsalva manoeuvre (a forced expiration with the glottis closed) has an *adverse* effect upon tolerance, because the initial rise in the blood pressure is followed within a few seconds by a decline to reach a systolic level that is often below normal. If the expiration is prolonged, the high venous pressure that accompanies the raised arterial pressure results in the more rapid and extensive pooling of blood in the peripheral reservoirs. Wood and Lambert (1952) showed that a protracted Valsalva manoeuvre could lead to loss of consciousness at accelerations as low at 3 g, but that if the intrathoracic pressure was raised in a series of short bursts interrupted by rapid inspirations, tolerance might be raised by as much as 2 g. This technique has achieved formal recognition as the M-1 manoeuvre, and is widely used in the USA for the training of pilots of high performance aircraft. It has the great disadvantage that it is tiring, and if it is carried out conscientiously it diverts attention from the primary task of flying. These objections can be overcome by the use of positive pressure breathing, which allows the intrathoracic pressure to be raised throughout each respiratory cycle to a level appropriate to the applied acceleration. In practical systems, the sensitivity of the gas regulator is usually set between 5 and 10 mm Hg g^{-1}, with a "cut-in" level of 2 g, so that at 5 g the mean pressure in the chest rises to about 15–30 mm Hg. Pressure breathing is a more efficient procedure than the M-1 manoeuvre because its action is continuous, but during long exposures it does increase the rate and magnitude of blood pooling. Accordingly, it is best considered as an adjunct to the protection given by posture or by anti-g garments.

Anti-g suits work on the same principle as elastic stockings; that is, they prevent distension of the veins and capacity vessels of the legs by applying external counter-pressure. (Tight bandaging of lower limbs was, in fact, employed by Japanese pilots in World War II as a method of raising tolerance for acceleration.) Franks (1940) realized that the distribution of hydrostatic pressures in the vessels could only be properly balanced by a similar gradient at the surface of the limbs, and he devised a closed garment made of inelastic fabric and filled with water. This suit, which was reputed by its inventor to raise the black-out threshold by 3 g or more, was self-balancing, in that it provided the theoretically ideal disposition of pressures to all parts of the legs at all levels of acceleration. It was, however, cumbersome and inconvenient, and its true efficiency was doubtful. Wood *et al.* (1943) demonstrated a gain of less than 1.5 g when the Franks suit was used on the centrifuge and this finding has recently been reconfirmed by Glaister (unpublished observations, 1977). An interesting physiological disadvantage of the water-filled suit was the induction of a diuresis in the wearer. This can, with hindsight, be attributed to the fact that the garment exerts counter-pressure even at 1 g. By so doing, it displaces some of the blood normally contained in the legs, increasing the central volume and initiating the "Gauer–Henry reflex". (A similar syndrome affects weightless astronauts—see p. 233.)

The use of a water-bath to provide hydrostatic counterpressure during exposure to accelerations was suggested in Germany in 1934 (Gauer, 1950). Experiments on a primitive centrifuge showed that the black-out threshold was raised by 1.7 g when the body was immersed to the level of the lower ribs. This is a surprisingly small degree of protection for such a theoretically efficacious technique, and Greenfield (1945) showed, in cats, that the acceleration required to reduce the systemic pressure at the head to zero was increased from 4 g to 10 g when the animals were supported in water to heart level. However, other reports tended to confirm the German results, although clear vision was maintained at 16 g by one individual almost totally submerged in a water bath (Gray and Webb, 1960). The only consistent finding from the conflicting accounts appears to be that protection is very much higher when the counterpressure includes the thorax than when it extends only to the xiphisternum.

Most of the problems associated with the Franks suit were overcome by substituting an air-filled garment, attached through a g-sensitive valve to a source of compressed gas. The first suits of this type consisted of a series of independent bladders which, in order to provide the "ideal" distribution of pressures to the legs, were connected through a manifold of individual reducing valves to the supply. The mechanical complexity of this arrangement was finally abandoned when Lambert et al. (1944) demonstrated that the protection given by a single-pressure garment was as great as that of the pneumatic gradient suit. The same authors found that the best results were obtained when the suit was inflated to between 1 and 2 lb in$^{-2}g^{-1}$ (52–100 mm Hg). As a logical extension of the belief that an ideal anti-g suit should entirely prevent the pooling of more blood in the limbs and at the same time raise the systemic blood pressure, Wood et al. (1943) proposed that pneumatic cuffs should be applied to the roots of the arms and legs and be inflated, during acceleration, to a pressure sufficient to cause arterial occlusion. They reported that the black-out threshold was raised by about 3 g when this arrangement was used, but "customer resistance" was overwhelmingly great.

In its current form, the anti-g suit consists of five interconnected bladders; one over the abdomen and one over each thigh and each calf. They are contained in an inextensible fabric which encircles the lower half of the body with the exception of the ankles and feet, the knees and the buttocks and perineum. The suit is inflated with air or with oxygen whenever the resultant acceleration exceeds 2 , to a pressure of 1.25 lb in^{-2} per g (65 mm Hg). When properly fitted and used, the garment confers a protection of between 1.2 g and 1.5 g. There is some evidence that better results can be obtained by using a non-linear relationship between suit pressure and acceleration and also by "pre-inflation" (that is, filling the dead space of the garment with gas) but the influence of these factors is small and non-proven.

The mode of action of the anti-g suit is complex. The pressure applied to the legs undoubtedly reduces the sequestration of blood but, as noted above, that process is of little importance during brief exposures to moderate accelerations. Moreover, a garment without an abdominal bladder gives little benefit. Counter-pressure to the abdomen alone is effective, but its value is significantly less than that of the whole suit. It is possible, therefore, that pressurisation of the legs is required to prevent the influx of blood from the splanchnic vessels when the abdominal bladder is inflated.

The protective effect of the anti-*g* suit can be partially explained by a reduction in the hydrostatic column between the heart and the head. The diaphragm descends almost passively during acceleration, the displacement being caused, in the main, by the inertia of the liver (Hyde, 1962). The downward movement averages 1 cm at 2 *g* and 2 cm at 4 *g* (Glaister, 1961), and the diaphragmatic surface of the heart is similarly displaced. Inflation of the anti-*g* suit prevents the descent (Rushmer, 1947) and, if the pressure in the abdominal bladder is high, may even reverse it (Sieker *et al.*, 1953).

Calculation shows that the elevation of the diaphragm *per se* can only account for about 0.3 *g* of the protection given by the suit, and even when this is added to the estimated 0.3 *g* gained from the prevention of venous pooling, it still falls far short of the 1.2 *g* advantage that the garment confers. The difference can be ascribed to the elevation of systemic blood pressure that results from an increased intra-abdominal pressure (Rushmer, 1947) and to the raised resistance to arterial inflow caused by the external pressure on the lower extremities. All these factors sustain the venous return and the cardiac output, and so maintain the circulation to the eye and the brain. The physiological penalty imposed by the anti-*g* suit is paid in the respiratory system, where its mechanical interference with lung function increases the incidence and severity of acceleration atelectasis.

Methods of protection against forces acting in the $-g_z$ direction have not been vigorously sought, because the need for them is small and rare. Postural adjustment is known to be effective (Henry, 1950) but its practical value is not great. The ill-effects of negative accelerations are largely the result of the high intravascular pressures generated in the head and neck, and they can be mitigated by applying counterpressure to these regions. Pressure breathing by means of a mask alone has little effect, but a pressurized helmet enclosing the entire head prevents the facial suffusion and the conjunctival haemorrhages that are characteristic of exposures to $-g_z$ accelerations. If the relief is to be more than symptomatic, the helmet must extend down over the neck far enough to cover the carotid sinuses.

Weightlessness

"What should such fellows as I do, crawling between heaven and earth?"

Sustained positive acceleration can be cheaply experienced by anyone who cares to visit a fairground, but prolonged exposure to weightlessness is the prerogative of the few who are chosen to represent nation or science in the cosmos.

Formally defined as the condition in which the external forces acting on a body sum to zero, weightlessness is popularly referred to as the gravity-free state; a term born of the false belief that an orbiting spacecraft has escaped from the influence of the earth's gravitational field. In fact, the fundamental principle of Newtonian physics decrees that such release is impossible, and the inverse square law allows the magnitude of the force at any given point above the surface of the earth to be calculated. Thus, at a height of 200 km (about the minimum safe altitude for spaceflight) the value of g is 94% of the standard, while at the so-called geostationary distance of 37000 km it has fallen to approximately 2.2% of the terrestrial normal. The remaining gravitational attraction is, however, precisely balanced by the centrifugal force generated as the spacecraft circles the earth, and if the velocity of the vehicle is increased, so its orbital altitude will expand to preserve the equilibrium. The simple equations derived in the early part of this chapter are as valid for a satellite as for a whirligig. Distance and speed (and hence period of rotation) determine the local value of "g", with the added requirement that the resultant must be numerically equal to the attractive force exerted not only by the earth but, theoretically, by all other bodies in the universe.

Information about the physiological effects of weightlessness comes from three sources; from measurements made on animals and astronauts in orbital flight, from extrapolation to zero of results obtained during exposures to *increased* accelerations, and from experiments designed to simulate the "gravity-free" state on earth. The reliance that has been placed upon the latter in attempts to explain the long-term consequences of spaceflight has been so great (and, often, so ill-founded) that some description of the techniques is desirable.

Elementary physics predicts that the primary effect of weightlessness upon the cardiovascular system will be an abolition of all hydrostatic pressure gradients, for if A is zero in eqn 6, Δp must also be zero. Basic physiology suggests that the equilibrium of pressures in the circulation will result in the redistribution of blood, with a reduction in the volume of gravitationally dependent regions and an augmentation of the "central blood volume". Essentially similar shifts accompany the change from the upright stance to recumbency; that is, in the substitution of transverse g for positive g. The crudest simulation is, therefore, the supine posture, but its veracity is clearly limited and it has had little application to the study of "acute" weightlessness. (As will appear, it has been exploited extensively as an analogue of long-term exposures.)

Many of the objections to decumbence can be overcome by immersion in water. This method allows the hydrostatic columns of the cardiovascular system to be balanced by precisely equal external pressures, because the relationship between the depth of water and the pressure that it exerts obeys the same law as the blood in the vessels. (This is, of course, the rationale of the water-filled anti-g suit described on p. 227). Moreover, the component of transverse g inseparable from the supine posture is eliminated, making total immersion an apparently ideal physiological parallel to weightlessness. It has been extensively employed as such, with results that have been well summarized by McCally (1968). The technique has some serious defects, however. It cannot mimic the behaviour of the weightless vestibular system and, more important, it does not provide hydrostatic indifference in the pulmonary vessels. Indeed, the need to provide breathing equipment for totally immersed subjects inevitably introduces disturbances of intrapulmonary pressure which can attenuate (or, more usually, exaggerate) the effects of the simulated weightlessness (Howard et al., 1967).

The opportunity to carry out formal evaluations of cardiovascular function in astronauts has not yet arisen, because of the "operational" nature of most space-flights and of the limitations of non-invasive methods of measurement. In early missions the electrocardiogram was recorded as a rudimentary form of medical monitoring, and reports were confined to analysis of the pulse frequency during launch, weightlessness and re-entry (Kas'yan and Chekhonadski, 1967; Berry et al., 1966). Not unexpectedly, it was found that the heart rate, which had increased during the accelerative phase, fell to pre-flight levels and below after orbital flight had been established, and then rose markedly again when re-entry began. Secondary measures of cardiac performance, such as the phonocardiogram and the relative time relationships within the ECG complex showed little or no deviation from normal, although Parin et al. (1967) claimed to have found evidence of reduced myocardial contractility. Measurements of arterial blood pressure by "classical" methods showed a moderate decrease both in systolic and in diastolic levels, with some narrowing of the pulse pressure (Berry, 1969; Volynkin et al., 166) and a fall in the cardiac output was, reasonably enough, postulated by these investigators. All the results could have been predicted from existing knowlege of cardiovascular physiology reinforced by extrapolation from the known effects of positive acceleration. (A pretty demonstration of this type of analysis was provided by Glaister (1977) who showed that the linear responses of the heart rate to acceleration applied in various body axes could be extended back to meet at a single point at "zero-g") (see Fig. 7).

During re-entry after prolonged periods of weightlessness the tachycardia is commonly greater than the magnitude of the applied acceleration would suggest. This is one sign of a state that has become known as "cardiovas-

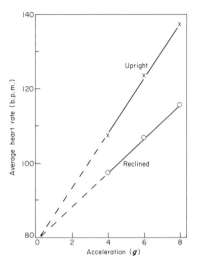

FIG. 7   Changes in heart rates as a function of applied acceleration, for two sitting postures (X, upright; 0 reclined). Extrapolation indicates that at zero-*g* the heart rate is slowed, and unaffected by the position of the body—a result confirmed during the true weightlessness of spaceflight.

cular de-conditioning", later manifestations of which include postural hypotension of varying degree (Dietlein and Judy, 1966; Gazenko and Gyurdzhian, 1968). The syndrome does not appear until after the return to normal gravitational conditions, and its severity and duration bear a fairly close relationship to the length of the preceding period of weightlessness (Kalinichenko *et al.*, 1970). When it was first noted, the de-conditioning was thought to be secondary to the prolonged immobility and lack of exercise (the so-called "hypokinesia") associated with a small space cabin. The importance of these factors was demonstrated by laboratory experiments in which weightlessness was simulated by weeks of bed-rest, sometimes with the limbs confined in casts to prevent voluntary muscular activity (Lawton, 1962; McCally, 1968). Cardiovascular deconditioning could indeed be induced by such heroic procedures, but the introduction of a vigorous regimen of in-flight exercise did not significantly mitigate its severity in actual spaceflights (Hoffler and Johnson, 1975). Moreover, the postural intolerance that results from several days of true weightlessness or several weeks of recumbency could be produced by a few hours of immersion in water, proving that mechanisms other than loss of muscle tone must be implicated.

One crucial result of the abolition of hydrostatic pressure gradients in the limbs is an increase in the volume of blood contained in the central "pool" of the thorax, and the change in the size of that compartment was estimated by Gauer *et al.* (1967) to amount to 400 ml or more. Gauer *et al.* (1961) had shown that an augmentation of the central volume by infusion rapidly led to a diuresis which, by lowering the total circulating blood volume, reduced the degree of filling of the central veins towards their homeostatic "set-point". Measurements from astronauts after flight have revealed reductions of up to 13% in the plasma volume (Berry, 1970), and there is good evidence that the disturbance is mediated by inhibition of the secretion of anti-diuretic hormone, which is involved in the "Gauer–Henry reflex". Similar changes of fluid balance are brought about by immersion, especially if it entails negative pressure breathing. The loss of fluid accounts for the greater part of the reduction in body mass experienced by most spacemen, and an increased consumption of water with a relative oliguria are characteristic findings immediately after flight (see Fig. 8).

The orthostatic intolerance does not disappear completely when the metabolic equilibrium is restored, and this indicates that the diuretic reflex does not act alone. Other possible factors that have been suggested include a paradoxical rise in the capacity of the blood vessels of the legs, increased vascular permeability with loss of plasma into the intercellular space, and a form of "disuse atrophy" of the normal vasoconstrictor responses to postural change. It is claimed that worthwhile protection against cardiovascular deconditioning can be gained from exposing the weightless limbs to periodic venous distension by applying subatmospheric pressures to the

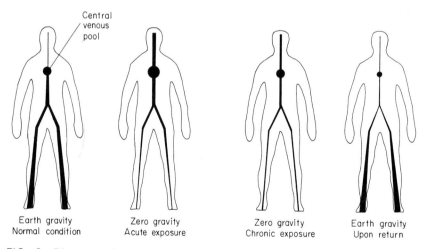

FIG. 8   Diagram of the postulated mechanism of "cardiovascular deconditioning". Explanation in the text.

lower half of the body (Cramer, 1971), and this technique was used with some success in flights of the Apollo series (Hoffler and Johnson, 1975).

A wide variety of metabolic and biochemical changes has been reported to result from weightlessness (Pestov and Gerathewohl, 1975), but they have generally been minor. Many of them can probably be attributed to dietary factors or to disturbances of fluid balance, but derangements of calcium balance have caused some alarm. Early studies by Mack *et al.* (1967) revealed gross reductions of radiographic density in the os calcis of astronauts even after short periods of weightlessness, and these were thought to result from the mobilisation of minerals from bones that would normally be weight-bearing. (Similar changes are seen in the legs of bed-ridden patients.) Significant rarefaction was, however, also found in the radius and in the phalanges of astronauts but not in those of subjects confined to bed for up to 30 weeks (Vogel *et al.*, 1974). More closely controlled studies made before and after later spaceflights confirmed that 10–12 days of true weightlessness produced losses of density in the os calcis greater than those seen after 10 weeks of bed rest; the radius and ulna were depleted in the astronauts but not in the earth-bound controls. Reliable estimations of calcium balance were not obtained, but deficits of 0.5–1.0 g per day were suggested by the concomitant metabolic data. There was also evidence of active resorption of bone, indicated by a rise in the level of plasma hydroxyproline (Alexander *et al.*, 1975). The mechanism of the mineral loss remains obscure, but the implications of a progressive drain of calcium, which is almost certainly not confined to the bones so far examined, are obviously serious. The successful extension of the duration of spaceflights to six months or more indicates, however, that the loss does not continue unabated, or reach catastrophic proportions.

Vestibular disturbances occur in spaceflight, as they do in almost every other mode of transport. In the absence of a true (gravitational) vertical, occasioned by the inability of the weightless otolith organ to signal direction, the astronaut relies upon his vision and his body image to distinguish "up" and "down". The convention is only valid for the frame of reference set by his immediate surroundings, and his free movement within the spacecraft or the unexpected discovery that the earth is above his head can give rise to severe but temporary spatial disorientation.

Motion sickness is a more serious problem. Despite careful selection and screening, and the habituation to vestibular stimuli produced by extensive previous experience of conventional flight, the majority of cosmonauts and astronauts have reported symptoms of motion sickness, and several have vomited. In most cases, the symptoms have occurred soon after the start of the weightless phase of flight and have been related to sudden movements of the head or body. Adaptation has usually been rapid (Hawkins and

FIG. 9   The human centrifuge at the RAF Institute of Aviation Medicine. It has a radius of 30 ft and can impose accelerations of up to 30 *g* upon subjects in the cabins at the ends of the arms.

Zieglschmid, 1975). The occurrence of this condition in individuals who have a comparatively high resistance to other types of motion can be explained in terms of the interaction between the semicircular canals and the otoliths (Yuganov, 1965). A change of posture or a rotation of the head normally stimulates both sets of receptors but, in the weightless state, only the canals respond fully to such movements. It is postulated that at 1 *g* impulses from the gravity-sensitive otolith impose a bias on those from the semicircular canals, and so moderate their impact upon the central nervous system. By this hypothesis, it is the action of the unmodulated impulses upon the vestibular nuclei that results in nausea and vomiting. Some support for this view was provided by the experiments of Benson and Bodin (1965), which showed that responses to angular accelerations (a function of the canals) were attenuated by centrifugal accelerations greater than 1 *g* (which affect the otoliths alone). Extrapolation of Benson's data indicates that rotational forces would, indeed, have a disproportionately great effect during weightlessness. The exact mechanism by which "space sickness" is produced remains to be discovered, but all the findings are in accord with the Sensory Conflict Theory (Reason and Brand, 1975).

REFERENCES

Akesson, S. (1948). *Acta Physiol. Scand.* **15**, 237–244.
Alexander, W.C., Leach, C.S. and Fischer, C.L. (1975). *In* "Biomedical Results of Apollo". (Eds R.S. Johnson, L.F. Dietlein and Berry C.A.). National Aeronautics and Space Administration, Washington, D.C.
Armstrong, H.G. (1943). "Principles and Practice of Aviation Medicine" 2nd edn. Williams and Wilkins, Baltimore.
Armstrong, H.G. and Heim, J.W. (1938). *J. Aviat. Med.* **9**, 199–215.
Banchero, N. Cronin, L. Nolan, A.C. and Wood, E.H. (1965). *Aerospace Med.* **36**, 608–617.
Banister, J. and Torrance, R.W. (1960). *Q.J. Exp. Physiol.* **45**, 352–367.
Barcroft, H. and Swan, H.J.C. (1953). (Eds). "Sympathetic Control of Human Blood Vessels", Edward Arnold, London.
Barr, P.-O. (1962). *Acta Physiol. Scand.* **54**, 128–137.
Barr, P.-O., Bjurstedt, H. and Coleridge, J.C.G. (1959). *Acta Physiol. Scand.* **47**, 16–27.
Beckh, H.J., von, Voge, V.M. and Bowman, J.S. (1976). Paper to 47th Scientific Meeting, Aerospace Medical Association, Miami, FA.
Beckman, E.L., Duane, T.D. and Coburn, K.R. (1962). *In* "Visual Problems in Aviation Medicine", (Ed. A. Mercier), AGARDograph No 61. Pergamon, London.
Beckman, E.L., Ziegler, J.E., Duane, T.D. and Hunter, H.N. (1953). *J. Aviat. Med.* **24**, 377–392.
Benson, A.J. and Bodin, M.A. (1965). IAM Report No 323. RAF Institute of Aviation Medicine, Farnborough.
Berry, C.A. (1969). *Aerospace Med.* **40**, 245–254.
Berry, C.A. (1970). *Aerospace Med.* **41**, 500–519.
Berry, C.A., Coons, D.O., Catterson, A.D. and Kelly, G.E. (1966). *In* "Gemini Midprogram Conference, including Experimental Results". National Aeronautics and Space Administration, Washington, D.C.
Browne, M.K. (1959). Report No 1046. Flying Personnel Research Committee, London, Air Ministry.
Bryan, A.C., Milic–Emili, J. and Pengelly, D. (1966). *J. Appl. Physiol.* **21**, 778–784.
Buhrlen, L. (1957). *Luftfahrtmedizin* **1**, 307–325.
Burton, R.R. and Mackenzie, W.F. (1976). *Aviat Space Environ. Med.* **47**, 711–717.
Campbell, E.J.M. (1958). "The Respiratory Muscles and the Mechanics of Breathing", Lloyd–Luke, London.
Cherniak, N.S., Hyde, A.S. Watson, J.F. and Zechman, F.W. (1961). *Aerospace Med.* **32**, 113–120.
Cherniak, N.S., Hyde, A.S. and Zechman, F.W. (1959). *J. Appl. Physiol.* **14**, 914–916.
Cochran, L.B., Gard, P.W. and Norsworthy, M.E. (1954). Report No 001.059.02.10 US Naval School of Aviation Medicine, Pensacola, FA.
Cohen, G.H. and Brown, W.K. (1969). *J. Appl. Physiol.* **27**, 858–862.
Colville, P., Shugg, C. and Ferris, B.G. (1956). *J. Appl. Physiol.* **9**, 19–24.
Cramer, D.B. (1971). Paper to AIAA/ASMA Symposium on Weightlessness and Artificial Gravity, Williamsburg, VA.
Dietlein, L.F. and Judy, W.V. (1966). *In* "Gemini Midprogram Conference, includ-

ing Experimental Results", National Aeronautics and Space Administration, Washington, D.C.

Diringshofen, H. von, (1938). *Luftfahrtmedizin* **2**, 281.

Dorman, P.J. and Lawton, R.W. (1956). *J. Aviat. Med.* **27**, 490–496.

Dowell, R.T., Sordahl. L.A., Lindsey, J.N. and Stone, H.L. (1976). *Aviat. Space Environ. Med.* **47**, 1171–1173.

Duane, T.D. (1954). *Arch Ophthalmol.* **51**, 343–355.

Duane, T.D., Weschler, R.L., Ziegler, J.E. and Beckman, E.L. (1952). *J. Aviat. Med.* **23**, 479–489.

Edholm, O.G. and McDowall, R.J.S. (1936). *J. Physiol.* **86**, 8*P*.

Erickson, H., Sandler, H. and Stone, H.L. (1976). *Aviat. Space Environ. Med.* **47**, 750–758.

Franks, W.R. (1940). CAM Report No C.2829. National Research Council, Canada.

Gamble, J.L. and Shaw, R.S. (1947). *Fed. Proc.* **6**, 109.

Gauer, O.H. (1950). In "German Aviation Medicine in World War II". **1**, 554–583. Department of the Air Force, Washington, D.C.

Gauer, O.H. (1961). In "Gravitational Stress in Aerospace Medicine". Gauer, O.H. and Zuidema, G.D. eds. Boston; Little-Brown.

Gauer, O.H., Eckert, P., Kaiser, D. and Lincenbach, H.J. (1967). In "Proceedings of Second International Symposium on Basic Environmental Problems of Man in Space", (Ed. H. Bjurstedt), Springer, New York.

Gauer, O.H. Henry, J.P. Martin, E.E. and Maher, P.J. (1949). *Fed. Proc.* **8**, 54.

Gauer, O.H. Henry, J.P. and Seiker, H.O. (1961). *Proc. Cardiovasc. Dis.* **4**, 1–26.

Gazenko, O.G. and Gyurdzhian, A.A. (1968). *Proc. Am. Astronaut. Soc.* **20**, 1091–1108.

Giacobini, E. (1942). *Scritta Ital. di Radiobiol.* **8**, 297–314.

Glaister, D.H. (1961). *Rev. Med. Aeronaut.* **21**, 28–29.

Glaister, D.H. (1964). Flying Personnel Research Committee Report No 1231. Ministry of Defence (Air Force Dept), London.

Glaister, D.H. (1965). *J. Physiol.* **177**, 73*P*–74*P*.

Glaister, D.H. (1966). Ph.D. Thesis, University of London.

Glaister, D.H. (1967). *Proc. R. Soc. B.* **168**, 311–334.

Glaister, D.H. (1969). *Proc. XVII Internat. Congr. Aviat. Space Med.* (Oslo 1968).

Glaister, D.H. (1970). "The Effects of Gravity and Acceleration on the Lung", AGARDograph No 133. Technovision Services, Slough.

Glaister, D.H. (1977). IAM Scientific Memorandum No 129. RAF Institute of Aviation Medicine, Farnborough.

Glaister, D.H. and Lisher. (1977). Report No. 1362 F.P.R.C. Ministry of Defence (Air).

Glazier, J.B., Hughes, J.M.B., Maloney, J.E. and West, J.B. (1967). *J. Appl. Physiol.* **23**, 694–705.

Golov, G.A. (1966). *Bull Expl. Biol. Med. USSR.* **62**, 1240–1243.

Gray, R.F. and Webb, M.G. (1960). Report No NADC-MA-5910. US Naval Air Development Center, Johnsville, PA.

Green, I.D. (1963). IAM Report No 230. RAF Institute of Aviation Medicine, Farnborough.

Green, I.D. and Burgess, B.F. (1962). Report No 1182. Flying Personnel Research Committee. Ministry of Defence (Air Force Dept), London.

Greenfield, A.D.M. (1945). *J. Physiol.* **104**, 5*P*–6*P*.

Greenleaf, J.F., Matter, M. and Bosco, J.S. (1966). *Aerospace Med.* **37**, 34–39.
Haddy, F.J. and Gilbert, R.P. (1956). *Circ. Res.* **4**, 25–32.
Häggendal, E., Löfgren, J. Nilsson, N.J. and Zwetnow, N. (1967). *Acta Neurochir.*
16, 163.
Ham, G.C. and Patterson, J.L. (1943). CAM Report No. 139. National Research
Council, USA.
Hawkins, W.R. and Zieglschmid, J.F. (1975). *In* "Biomedical Results of Apollo",
(Eds R.S. Johnson, L.F. Dietlein and C.A. Berry), National Aeronautics and
Space Administration. Washington, D.C.
Henry, J.P. (1950). Technical Report No. 59–53. USAF Wright Air Development
Centre, Dayton, Ohio.
Henry, J.P., Gauer, O, Kety, S.S. and Kramer, K. (1951). *J. Clin. Invest.* **30**,
292–300.
Henry, J.P., Gauer, O, Martin, E.E., Kety, S.S. and Kramer, K. (1949). *Fed. Proc.*
8, 73.
Henry, J.P., Gauer, O.H., Martin, E.E., Simons, D.G. and Maher, P.J. (1949).
Memorandum Report MCREXD-695-74N. Air Materiel Command, US Air
Force.
Heymans, C., Bouckaert, J.J. and Wierzuchowski, M. (1937). *Arch. Int.
Pharmacodyn.* **55**, 233.
Hiatt, E.P., Leverett, S.D. and Bondurant, S. (1958). *Fed. Proc.* **17**, 70.
Hoffler, G.W. and Johnson, R.L. (1975). *In* "Biomedical Results of Apollo", (Eds
R.S. Johnson, L.F. Dietlein, C.A. Berry), National Aeronautics and Space
Administration. Washington, D.C.
Hoppin, F.G., Sever, R.J. and Hitchcock, L. (1965). Report No NADC-MR-6519.
US Naval Air Development Centre, Johnsville, PA.
Howard, P. (1959). *J. Physiol.* **147**, 49*P*.
Howard, P. (1964). Ph.D. Thesis, University of London.
Howard, P., Ernsting, J., Denison, D.M., Fryer, D.I., Glaister, D.H. and Byford,
G.H. (1967). *Aerospace Med.* **38**, 551–563.
Howard, P. and Garrow, J.S. (1958). *J. Physiol.* **143**, 83*P*.
Howard, P. and Glaister, D.H. (1964). *J. Physiol.* **171**, 39*P*.
Hyde, A.S. (1962). Technical Report No 62–106. USAF Aerospace Medical
Research Laboratory Wright Patterson AFB, Dayton, Ohio.
Jaeger, E.A., Severs, R.J., Weeks, S.D. and Duane, T.D. (1964). *Aerospace Med.* **35**,
969–972.
Jasper, H.H. and Cipriani, A (1945). *J. Physiol.* **104**, 6*P*–7*P*.
Jones, J.G. Clarke, S.W. and Glaister, D.H. (1969). *J. Appl. Physiol.* **26**, 827–832.
Jongbloed, J. and Noyons, A.K. (1933). *Plüg. Arch. Ges. Physiol.* **233**, 67–97.
Kalinichenko, V.V., Gornago, V.A., Machinskiy, G.V., Zhelburova, M.P.,
Pometov, Yu D. and Katkovskiy, B.S. (1970). *Kosm. Biol. Med.* **4**, 68–77.
Kaneko, K, Milic–Emili, J., Dolovich, M.B., Dawson, A. and Bates, D.V. (1966). *J.
Appl. Physiol.* **21**, 767–777.
Kas'yan, I.I., Vasil'yev, D.V., Maksimov, D.G., Akulinichev, I.T., Uglov, A.Ye,
Baykov, A.Ye and Chekhonadskiy, N.A. (1967). *Izv. Akad. Nauk. SSSR* **1**,
104–118.
Kotovskaya, A.R. (1970). Cited by Vasil'yev and Kotovskaya (1975).
Krueger, J.J., Bain, T. and Patterson, J.L. (1961). *J. Appl. Physiol.* **16**, 465–468.
Lambert, E.H., Code, C.F., Baldes, E.J. and Wood, E.H. (1944). CAM Report No
248 National Research Council, USA.

Lambert, E.H. and Wood, E.H. (1946). *Med. Clin. N. Am.* **30**, 833–844.
Laughlin, M.H. (1978). Personal communication.
Lawton, R.W. (1962). *Astronaut. Sci. Rev.* **4**, 11–18.
Leverett, S.D. Jnr, Kirkland, V.E., Schermerhorn, T.J. and Newson, W.A. (1967). Paper to Annual Scientific Meeting, Aerospace Med Assocn, Washington, D.C.
Leverett, S.D. Jr and Zuidema, G.D. (1957). *Meddel. Fran. Flyg-och Navalmed. Namden* **6**, 33–39.
Life, J.S. and Pince, B.N.N. (1969). *Aerospace Med.* **40**, 44–49.
Lindberg, E.F., Marshall, H.W., Sutterer, W.F., McGuire, T.F. and Wood, E.H. (1962). *Aerospace Med.* **33**, 81–91.
Lindberg, E.F., Sutterer, W.F., Marshall, H.W., Headley, R.N. and Wood, E.H. (1960). *Aerospace Med.* **31**, 817–834.
Lindsey, J.N., Dowell, R.T., Sordahl, L.A., Erickson, H.H. and Stone, H.L. (1976). *Aviat. Space Environ. Med.* **47**, 505–511.
Lombard, C.F., Roth, H.P. and Drury, D.R. (1948). *J. Aviat. Med.* **19**, 355–364.
Mack, Pauline B., Lachance, P.A., Vose, G.P. and Vogt, F.B. (1967). *Am. J. Roentgenol. Rad. Ther. Nucl. Med.* **100**, 503–511.
McCally, M.S. (Ed.) (1968). "Hypodynamics and Hypogravics—the Definitive Work on the Biomedical Aspects of Weightlessness", Academic Press, New York and London.
Milic–Emili, J., Henderson, J.A.M., Dolovich, M.B., Trop, D. and Kaneko, K. (1966). *J. Appl. Physiol.* **21**, 749–759.
Milic–Emili, J., Henderson, J.A.M. and Kaneko, K. (1967). *J. Nucl. Biol. Med.* **11**, 63–68.
Nolan, A.C., Marshall, H.W., Cronin, L., Sutterer, W.F. and Wood, E.H. (1963). *Aerospace Med.* **34**, 797–813.
Orth, J. (1887). "Atilogisches und Anatomisches uber Lungenschwindsucht", Hirschwald, Berlin.
Parin, V.V., Bayevskiy, R.M., Volkov, Yu N. and Gazenko, O.G. (1967). *In* "Kosmischeskoy Kardiologii", Meditsina Izdatel' stvo, Leningrad.
Permutt, S., Bromberger-Barnea, B. and Bane, H.N. (1962). *Med. Thorac.* **19**, 239–260.
Pestov, I.D. and Gerathewohl, S.U. (1975). *In* "Foundations of Space Medicine and Biology", Vol. II, Book 1, Chapter 8. (Eds M. Calvin and O.G. Gazenko), National Aeronautics and Space Administration. Washington, D.C.
Pogrund, R.S., Ames, S.W. and Lombard, C.F. (1951). *J. Aviat. Med.* **22**, 50–59.
Poppen, J.R. and Drinker, C.K. (1950). *J. Appl. Physiol.* **3**, 204–215.
Power, G.G., Hyde, R.W., Sever, R.J., Hoppin, F.G. and Nairn, J.R. (1965). *J. Appl. Physiol.* **20**, 1199–1204.
Proctor, D.F., Caldini, P. and Permutt, S. (1968). *Resp. Physiol.* **5**, 130–144.
Pryor, W.W., Sieker, H.O. and McWhorter, R.L. (1952). *J. Aviat. Med.* **23**, 550–562.
Rapela, C.E. and Green, H.D. (1963). *Circ. Res.* *14–15 Suppl I*, 205–211.
Reason, J.T. and Brand, J.J. (1975). "Motion Sickness", Academic Press, London and New York.
Roddie, I.C. and Shepherd, J.T. (1957). *J. Physiol.* **139**, 377–384.
Rose, B.W., Kerr, W.K. and Kennedy, W.A. (1942). CAM Report C-2390. National Research Council, Canada.
Rosenhamer, G. (1967). *Acta Physiol. Scand.* **68** Suppl 276.
Rushmer, R.F. (1947). *Am. J. Physiol.* **151**, 459–468.

Salzman, E.W. and Leverett, S.D. (1956). *Arc. Res.* **4**, 540–545.

Shubrooks, S.J. Jr (1972). *Aerospace Med.* **43**, 1200–1206.

Sieker, H.O. (1952). Technical Report No 52–87. USAF Wright Air Development Center, Dayton, Ohio.

Sieker, H.O., Martin, E.E., Gauer, O.H. and Henry, J.P. (1953). Technical Report No 52–317. USAF Wright Air Development Center, Dayton, Ohio.

Smedal, H.A., Rogers, T.A., Duane, T.D., Holden, G.R. and Smith, J.R. (1963). *Aerospace Med.* **34**, 48–55.

Steiner, S.H. and Mueller, G.C.E. (1961). *J. Appl. Physiol.* **16**, 1081–1086.

Steiner, S.H., Mueller, G.C.E. and Cherniack, N.S. (1961). *J. Appl. Physiol.* **16**, 641–643.

Stewart, W.K. (1940). Report No 177. Flying Personnel Research Committee. London; Air Ministry.

Sutherland, P.W., Katsura, T. and Milic–Emili, J. (1968). *J. Appl. Physiol.* **25**, 566–574.

Tikhomirov, Ye P. (1969). *Kosm. Biol. Med.* **3**, 71–75.

Vartbaronov, R.A., Eshanov, N.Kh., Kotovskaya, A.R. and Suvorov, P.M. (1969). *In* "Aviakosmicheskaya Meditsina", (Ed. V.V. Parin), Akad. Nauk. SSR Moscow.

Vasil'yev, P.V. and Kotovskaya, A.R. (1975). *In* "Foundations of Space Biology and Medicine", Vol. II, Book 1, Chapter 5. (Eds M. Calvin, M and O.G. Gazenko), National Aeronautics and Space Administration, Washington, D.C.

Vogel, J.M., Rambaut, P.C. and Smith, M.C. (1974). Report No TMX-58110. National Aeronautics and Space Administration, Washington, D.C.

Volynkin, Yu M., Akulinichev, I.T., Vasil'yev, P.V., Voskresenskiy, A.D., Kas'yan, I.I. and Maksimov, D.G. (1966). Kosm. Issled. **4**, 755–768.

White, W.J. (1958). Technical Report No 58–333 USAF Wright-Patterson AFB, Dayton, Ohio.

White, W.J. (1960). Technical Report No 60–34 USAF Wright Air Development Center, Dayton, Ohio.

White, W.J. and Jorve, W.R. (1956). Technical Report 56–247 USAF Wright Air Development Center, Dayton, Ohio.

Wilson, W.H. (1927). *J. Physiol.* **64**, 54–64.

Wood, E.H., Code, C.F. and Baldes, E.J. (1943). CAM Report No 207. National Research Council, USA.

Wood, E.H. and Lambert, E.H. (1952). *J. Aviat. Med.* **23**, 218–224.

Wood, E.H., Nolan, A.C., Donald, D.E. and Cronin, L. (1963). *Fed. Proc.* **22**, 1024–1034.

Yamada, S. and Burton, A.C. (1954). *J. Appl. Physiol.* **6**, 501–505.

Yuganov, Ye M. (1965). *In* "Problemy Kosmicheskoy Biologii", Vol. 4, (Ed. N.M. Sisakyan, Akad. Nauk. SSSR, Moscow.

Zechman, F.W., Cherniack, N.S. and Hyde, A.S. (1959). *J. Appl. Physiol.* **15**, 907–910.

Zuidema, G.D., Cohen, S.I., Silverman, A.J. and Riley, M.B. (1956). *J. Aviat. Med.* **27**, 469.

CHAPTER 5
High Altitudes and Hypoxia

D. Denison

Introduction

Oxygen lack is one of the most important aspects of human physiology. It is experienced naturally by people living at high altitudes, it is a potential or actual hazard for everyone who flies, and it is the primary consequence of anaemia and many pulmonary and cardiovascular disorders which are common causes of morbidity and mortality in man. The effects of oxygen lack are bewildering in number and depend on its severity and rate of onset, and on its duration. They range from subtle learning difficulties to loss of consciousness within a few seconds. However, although the subject is vast it is also well understood; few facts are required to make sense of the topic and predict its features with some precision. To do this one needs to know something about the biochemical properties of the oxygen molecule and how it is distributed to the tissues. These points will be discussed before the effects of hypoxia in man are described.

Biochemical Properties of the Oxygen Molecule

The oxygen molecule has a paradoxical nature since we bathe in it without apparent harm although it is a powerful poison. With the exception of fluorine, it is the most corrosive of the elements, because it has a very high avidity for electrons and will attract them from any adjacent molecule. This is the essential nature of the process of oxidation. In this respect, all substances can be ranked in order of their tendency to give or receive electrons. The property is expressed as the voltage needed to prevent electrons flowing between hydrogen in simple solution and the substance in question. The voltage required is known as the *oxidation–reduction* or *redox* potential of the system. On this scale (Fig. 1) the strongest reductant exerts an electron pressure of -3.1 V and the most powerful oxidant exerts an electron attraction of $+3.1$ V. Ordinary diatomic oxygen has a *low* redox potential, (-1.3 V), and so actually repels electrons and therefore is difficult

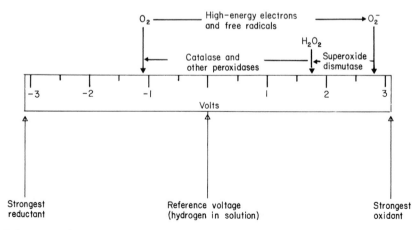

FIG. 1   The redox potentials of the oxygen molecule and its immediate derivatives. Enzymes quenching the oxidative capacity of the latter are also shown.

to ionize. It can be ionized by electrons travelling at high speeds. The superoxide ion (O_2^-) that is formed has a high potential $(+2.8$ V) and rapidly attracts a further three electrons before it is fully reduced. This inevitably disrupts other molecules lying close by.

The removal of electrons from molecules is almost always exothermic and in some circumstances the energy released by the ionization of a single oxygen molecule is sufficient to ionize another and chain-reaction formation of superoxide ions, i.e. combustion, occurs. In oxygen-rich environments this process allows fires triggered by trivial ignition sources to consume the body of a man in a few seconds (Denison *et al.*, 1968).

The presence of oxygen in the atmosphere constitutes a serious threat to living things because it forms superoxide ions in cell fluids. As a result, a series of protective enzymes has evolved. The most important of these is *superoxide dismutase* which is found in almost every mammalian cell. It converts the superoxide ion to hydrogen peroxide which is somewhat less corrosive (redox potential, 1.77 V). *Catalase* and other *peroxidases*, that are also present in most cells, break down the hydrogen peroxide to molecular oxygen and water. Naturally formed superoxide ions that are not quenched in this way oxidize neighbouring molecules spontaneously. This process of random destruction and loss of molecular information is an important part of biological ageing and of oxygen toxicity.

In the course of evolution other enzymes have developed which are capable of using oxygen in a controlled way. They can be divided into three groups; *oxidases* which combine electrons and molecular oxygen directly, with the formation of water; *oxygenases* which insert respired oxygen into biochemical substances, forming new compounds; and *mixed-function oxidases* or *hydroxylases* which behave as an oxidase to one atom of the oxygen molecule and as an oxygenase to the other. In aerobic organisms oxygen is mainly used as the ultimate electron sink, permitting an enormous number of biochemical reactions to proceed. However, it is also used synthetically, which is an aspect of respiration that has been largely overlooked until recently, although emphasised by Mason in 1957 and Cohen in 1972.

Energy is released whenever electrons are removed from molecules. In most oxidase reactions this all appears as heat, but in one it is coupled to another process that traps part of the energy—in the conversion of ADP to ATP. The coupled reaction, *oxidative phosphorylation*, is governed by cytochrome a_3 oxidase, which occurs in mitochondria. It is the principal mechanism of controlled energy release in almost all animals and varies rapidly over a many-fold range, continuously matching energy supply to energy needs. Oxidative phosphorylation is responsible for some 80% of oxygen consumption in man at rest. Almost all of the remainder is consumed outside mitochondria by other oxidases. They degrade a wide variety of unwanted metabolites, particularly amines. A small but critical fraction of basal oxygen uptake is consumed by oxygenases in the synthesis of several compounds, notably the amine transmitters of the nervous system and some immediate precursors of steroid hormones.

Spontaneous oxidation of cellular material by superoxide ions and peroxides, (which is also known as non-enzymic or auto-oxidation), has great biological significance but normally accounts for an extremely small part of the oxygen consumed. It follows that almost all of the oxygen used by cells is handled, and therefore controlled, by enzymes. Some of these are found free in the cytosol. Most are bound to the inner surface of membrane-covered organelles, namely mitochondria, perioxosomes, microperioxosomes and the endoplasmic reticulum. These structures are significant because they present barriers to the passive movement of oxygen within cells and also because the tubular organelles may provide "fast lanes" for its travel by facilitated diffusion.

The velocities of oxygen-consuming reactions, like those of other enzymatic processes, can be described by Michaelis–Menten kinetics. The simplest form of the Michaelis–Menten Law, shown in Fig. 2, predicts a hyperbolic relationship between reaction velocity and substrate concentration. The relationship can be defined by a single characteristic, the *Michaelis constant*,which is the substrate concentration at which the reaction proceeds at half-maximum velocity when all other factors are held constant. In the case of oxygen dissolved in the cell fluids surrounding

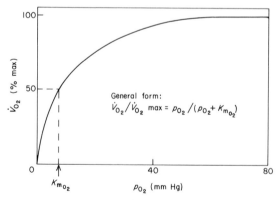

General form:

$$\dot{V}_{O_2}/\dot{V}_{O_2}\,\text{max} = p_{O_2}/(p_{O_2} + K_{m_{O_2}})$$

FIG. 2 The simplest form of the Michaelis–Menten equation, applied to oxygen-consuming processes. Oxygen consumption, $(\dot{V}_{O_2})$, is on the ordinate and substrate concentration, (p_{O_2}), is on the abscissa. The substrate concentration at which the reaction proceeds at half its maximum velocity is known as the reaction's Michaelis constant for oxygen, $(K_{m_{O_2}})$.

organelle membranes, there is a linear relation between concentration and partial pressure, so the Michaelis constant for oxygen, (K_{mO_2}), of any intracellular oxygen-consuming process can be expressed as that partial pressure of oxygen in solution at which the reaction in question proceeds at half its maximal velocity. This constant defines the *oxygen affinity* of the reaction system. High-affinity systems will have a low K_{mO_2} and vice versa. In practice oxygen affinities are difficult to determine because Michaelis constants measured *in vitro* differ from those existing *in vivo* and there is no simple law to relate the two. Oxidative phosphorylation appears to have a very high affinity for oxygen, $(K_{mO_2}$ of 0.1–1.0 mm Hg) and the reactions controlled by the oxygenases and the other oxidases seem to have much lower affinities, $(K_{mO_2}$ of 5–250 mm Hg).

This observation has important consequences that are illustrated in Fig. 3. It shows Michaelis curves for a series of hypothetical oxygen-consuming reactions with K_{mO_2} of 0.1, 1, 5, 10, 25, 50, 100 and 200 mm Hg. These span the values believed to exist in man. As substrate concentrations fall the reaction velocity of the major consumer (oxidative phosphorylation, K_{mO_2} 0.1–1.0 mm Hg), proceeds unabated to very low oxygen tensions indeed. By contrast, the reaction velocities of the low-affinity systems are already

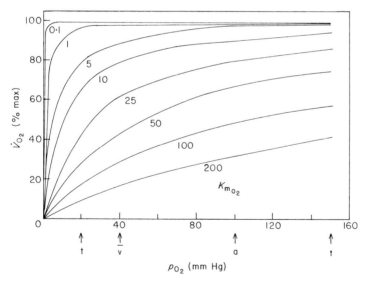

FIG. 3 The reaction velocities of eight hypothetical reactions with Michaelis constants for oxygen, (K_{mO_2}), of 0.1, 1, 5, 10, 25, 50, 100 and 200 mm Hg. Oxygen consumption, $(\dot{V}_{O_2})$ is on the ordinate, and substrate concentration, (p_{O_2}), is on the abscissa. Also shown are usual values for moist-inspired (I), arterial (a), mixed venous (v) and cerebral tissue (t), oxygen tensions.

limited by oxygen supply at arterial oxygen tensions and drop further as the p_{O_2} is reduced. Thus it is possible to have marked falls in the velocities of some critical reactions (e.g. oxygenations), although the consumption of the whole body is largely independent of oxygen supply. Another consequence is that the velocity of a particular reaction may vary significantly between cells close to the arterial ends of capillaries and those close to their venous ends. Several oxygen-consuming reactions of different affinities can occur in the same cell, so there will also be competition for oxygen within the cell, which will be won by the system with highest affinity. The points that have been mentioned so far are discussed in more detail by Denison (1981).

The Distribution of Oxygen in Tissues

It is intuitively obvious that a fall in ambient oxygen pressure must lower oxygen tensions all the way from the lungs to the intracellular organelles. It is also evident that the extent of the internal changes will depend upon the amount of oxygen stored in the body and on the ease with which oxygen moves in and out of the system. At first sight easy movement and a large storage capacity seem to be entirely desirable attributes, however the penalty of easy movement is vulnerabilty to external change, and the drawback of a high storage capacity is a long refilling time, which is important in recovery from hypoxic episodes, when the ambient supply is restored. The object of the present section is to describe a quantitative model of oxygen distribution which can be used to predict and analyse these changes with some precision. This cannot be done unless the flow of CO_2 is considered as well.

Since the major oxygen-consuming reaction in man has a very low K_{mO_2}, oxygen uptake of the body as a whole is almost independent of the oxygen pressure in the surrounding air. Equally, in steady-state conditions, metabolic CO_2 production is determined by the nature of the fuel which is being burnt and is influenced very little by ambient levels of CO_2. Because this is so it is possible to construct a model of tissue respiration quite simply. The model is illustrated in Fig. 4. It represents the systems that transport oxygen and CO_2 between ambient air and the loci of cellular respiration, as a series of compartments which are linked by tubes. The two gases behave as fluids flowing from one end of the system to the other. The mass flows are imposed by the tissues and are assumed to be independent of the pressure and composition of ambient air. The flows ($\dot{M}_{O_2}$ and $\dot{M}_{CO_2}$) are shown entering and leaving the tissue compartment, at the right of the figure. Partial pressures of oxygen and CO_2 are represented by the fluid levels in each compartment.

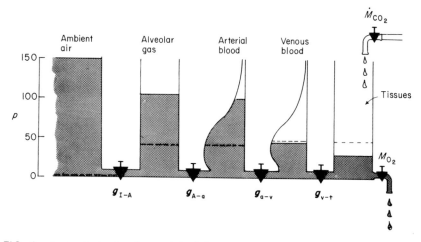

FIG. 4   A simple hydraulic model of the transfer of oxygen and CO_2 between the environment and the tissues. The system is represented by a series of compartments with capacitances equivalent to their widths, linked by conductances (G_{a-v}, etc.), mass flows, ($\dot{M}_{O_2}$ and $\dot{M}_{CO_2}$) are imposed by the tissues on the right of the figure, and the partial pressures generated in each compartment by these flows are shown as fluid levels, to be read against the pressure scale in the left of the figure (mm Hg).

In real life the gases are transferred to and from the tissues by a combination of convection, diffusion and facilitated diffusion, and by a series of "pumping" processes which raise the partial pressures of O_2 and CO_2 at some points along the chain. *Diffusion* is entirely passive but it is only effective over very short distances indeed. *Facilitated diffusion*, which depends upon the thermal agitation of carrier molecules, is also passive and only effective at short range. *Convection* consumes energy which has to be supplied to the chain. It includes two long distance carriers, ventilation and perfusion, which are easily perceived, but also many less obvious intermediate and short-range transport processes, such as the mixing of haemoglobin molecules as the red cell is deformed on its passage through capillaries, the flows of interstitial fluids which are caused by filtration and vascular pulsations, and the stirring of intracellular fluid due to the rapid changes of mitochondrial size with respiration. *Partial pressure pumps* also consume energy, although it is not easy to determine how much. Quantitatively, the most important of these are the Bohr, Haldane and thermal effects that normally promote respiration, especially during exercise. Some vertebrates have spectacular oxygen pumps, as in the rete mirabile which inflate the swim bladders and others which oxygenate the avascular retinae, in bony fish. It is possible that subdued forms of these devices remain to be found in man.

Although the model shown in Fig. 4 is elementary, a great deal can be learnt by studying and re-studying the properties of its elements in turn. Clearly, in steady states, that is when the oxygen and CO_2 flows are constant, the falls in pressure along the system will be decided by the properties of the tubes, and will be entirely independent of any properties of the compartments. However the latter will determine the time-courses of any changing states.

Each tube represents all those factors that determine the ease with which gas is transferred from one compartment to another. This can be expressed as mass flow per unit of partial pressure difference, $(d\dot{M}/dp)$, between one compartment and the next. This characteristic of the tube is its *conductance* for oxygen or CO_2. Conductance is the reciprocal of resistance, $(dp/d\dot{M})$, which is a measure of the opposition to transfer. Physiologically it is more fruitful to think in terms of conductances than resistances, because the factors which facilitate gas transfer are then identified overtly.

In contrast to these tubes, the model's compartments represent all those factors which buffer the partial pressure disturbances caused by changes in ambient conditions or respiratory demands. This functional property, *capacitance*, can be expressed as the mass of gas needed to change the P_{O_2} or P_{CO_2} of the compartment by one unit of partial pressure, (dM/dp). (Note the difference between a mass of gas, M, and a mass flow, $\dot{M}$.) This characteristic, which is represented by the width of the compartments in the model, can vary with partial pressure, as the odd shapes of the containers in the diagram shows. Capacitance is the reciprocal of *elastance*, (dp/dM) which is a measure of the system's physical responsiveness to change. Here also it is more helpful to think of capacitances than elastances, because positive features of the transport system are identified explicitly by doing so.

This simple model was described by Farhi in 1965 and, in one form or another, has been developed and used by many investigators since. With it one can obtain a surprisingly detailed insight into the physiology of respiration, quite quickly. We can begin by considering gas transfer under normal conditions at ground level, taking each element of the model in turn, and then use the model to predict the effects of hypoxia.

Ambient air

This tank sets the levels about which the O_2 and CO_2 pressures in the rest of the system are poised. At sea level, in middle latitudes, barometric pressure normally varies between 730 and 780 mm Hg, and the water vapour in the air exerts a pressure of about 10 mm Hg. The latter may rise to 30–35 mm Hg in very humid climates or fall almost to zero in extremely dry ones. Dry air always has a composition very close to 20.93% O_2, 0.9% argon, and 0.03% CO_2 in nitrogen. As the air is inspired it is warmed and moistened by the pharynx, and even the coldest and dryest inspirate is usually at body temperature and saturated with water vapour by the time it reaches the medium-sized bronchi. Thus the oxygen pressure, $p_{I,O_2}*$, in moist inspired air entering the alveoli is

$$p_{I,O_2}* = 0.2093 \times (p_B - 47) \text{ mm Hg} \tag{1}$$

or, more generally

$$p_{I,O_2}* = F_{I,O_2} \times (p_B - p_{H_2O}*) \tag{2}$$

where F_{I,O_2} is the fractional concentration of oxygen in dry air, and $p_{H_2O}*$ is the vapour pressure of air saturated with water at body temperature. (Respiratory physiologists use a set of symbols that have been agreed internationally and are very convenient. They are summarized in Table 1

TABLE 1   A list of the principal symbols used by respiratory physiologists to describe various features of gas exchange. These are discussed in detail in an important paper by Piiper *et al.* (1971).

Principal symbols	Examples
M = mass of substance (in moles).	$p_{v_{O_2}}$ = tension of oxygen in mixed venous blood
$\dot{M}$ = mass flow of substance (mol min^{-1}).	
V = volume of substance (conditions of measurement must be stated).	V_I = volume of inspirate
	$\dot{V}_{E_{N_2}}$ = volume of nitrogen expired per minute
$\dot{V}$ = volume flow of substance (dV/dt).	$\beta_{t_{CO_2}}$ = capacitance of tissue for CO_2
p = partial pressure of substance	$C_{A-a_{CO}}$ = alveolar–arterial conductance for carbon monoxide.
c = mass concentration of substance (M/V).	
β = capacitance of compartment for a substance (dM/dp).	
G = conductance of a route for a substance $(d\dot{M}/dp)$.	

Subscripts
A = alveolar
I = Inspired
E = Expired
Ē = mixed expired
a = arterial
v = venous
c = capillary
t = tissue

and described in an important paper by Piiper *et al.*, 1971.) The standard value of $p_{I,O_2}*$, (when $p_B = 760$ mm Hg and body temperature is 37 °C) is 150 mm Hg. Changes in barometric pressure at sea level actually cause moist-inspired p_{O_2} to vary between limits which are approximately 145 and 155 mm Hg. Usually the CO_2 pressure in inspired air is taken as zero. Under normal circumstances these are the "fixed" pressures about which the levels of O_2 and CO_2 in the rest of the system are set.

The atmosphere is held in place by gravitational attraction. Since air is compressible its density, and therefore pressure, increase as the Earth's surface is approached. The mean pressure air exerts at sea level varies because the atmosphere is more extensive equatorially than over the poles. It also fluctuates with temperature which affects both the pressure at sea level and the diminution of pressure with height. Temperature variations with altitude are more marked over the equator than the poles.

Atmospheric pressure decays almost exponentially with height, roughly halving with every 18 000 ft of ascent. Since the oxygen concentration of air remains constant to very high altitudes indeed, its partial pressure also diminishes with height at almost the same exponential rate. Because of the rapid fall of water vapour pressure with altitude the decays of total pressure and oxygen pressure are not matched precisely but the differences are very small. For many purposes it is convenient to use the properties of some "standard" atmosphere, e.g. the one defined by the International Civil Aviation Organization. Its pressure–altitude characteristics are illustrated and compared with the 18 000 ft "rule-of-thumb" in Fig. 5. Although the standard atmosphere is very useful it is important to remember that significant departures from it occur in air over high mountains and generally over polar and equatorial zones. In this chapter all altitudes will be expressed in feet because many references will be made to studies in aviation, where this is a safety convention.

The alveolar air compartment

The concept of steady-state respiration, i.e. that the alveolar compartment is ventilated by a constant stream of CO_2-free gas, is very convenient but mythical, because ventilation is tidal. Expiration can be regarded as a breath-hold at diminishing lung volume, and inspiration as a breath-hold at increasing lung volume which is accompanied by dilution. "Steady-state" respiration is the balanced alternation of these two unsteady states. The oscillations in p_{a,O_2} and p_{a,CO_2} that they produce are proportional to $\dot{M}_{O_2}$ and $\dot{M}_{CO_2}$ and to the breath duration, and they are inversely proportional to the effective volumes in which alveolar oxygen and CO_2 are distributed.

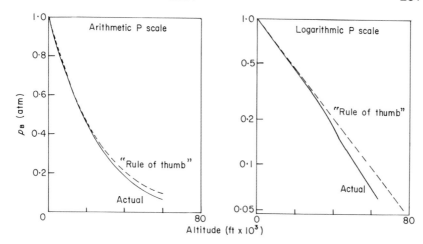

FIG. 5   A comparison of the pressure–altitude characteristics of the ICAO model atmosphere with the simplest "rule of thumb" model, in which barometric pressure decays exponentially, halving every 18 000 ft.

These are not the same for the two gases since, as for any gas in the lung, one part is dispersed in alveolar air and another part is dissolved in alveolar tissue fluids. The capacitance of the alveolar compartment is directly proportional to this effective distribution volume, V_{dist}, where:

$$V_{\text{dist}} = V_{\text{A}} + \alpha V_{\text{t}} \qquad (3)$$

V_{A} is the alveolar gas volume, α is the solubility of the particular gas in tissue at 37 °C and V_{t} is the volume of tissue in the lung. In normal lungs V_{t} is about 10% of alveolar gas volume, and the solubilities of oxygen and CO_2 are 0.03 and 0.60 vol/vol atm^{-1} respectively, so the term V_{t} is insignificant for oxygen but raises the effective lung volume by some 6% for CO_2.

The amount of oxygen stored in the lungs of a normal subject at ground level varies between about 200 ml at residual volume and 800 ml at total lung capacity. These volumes should be compared with a basal oxygen uptake of about 250 ml min^{-1} and an aerobic capacity of 3–4 litres min^{-1} on severe exertion.

The alveolar compartment has an important role in "steady-state" respiration, buffering the respiratory oscillations in alveolar p_{O_2} and p_{CO_2}. In otherwise constant circumstances, the larger the lungs, the less these oscillations will be. At sea level they are of little importance because they occur on the plateau of the oxyhaemoglobin dissociation curve, but at high altitudes swings of the same magnitude can be critical since they occur on steep parts of the curve. This is especially true of the large oscillations of slow deep breathing. Conversely, at any given level of alveolar ventilation, the partial pressure swings can be minimized by rapid shallow breathing at a high lung volume.

The alveolar air–arterial blood conductance

Five factors are known to facilitate gas transfer between alveolar air and systemic arterial blood. These are: complete gas mixing within individual alveoli; complete equilibration across their membranes; rapid rates of chemical combination reactions within red cells and plasma; perfect matching of ventilation and perfusion in all alveoli; and no flow of blood that bypasses ventilated alveoli entirely. These topics, which have been the subject of many researches, are dealt with in detail by West (1977), Piiper and Scheid (1971), Forster and Crandall (1976) and Wagner (1977).

Normal people at sea level show an alveolar–arterial oxygen pressure gradient of some 11 mm Hg, of which 5 mm Hg are usually attributed to ventilation–perfusion inequality, another 5 mm Hg to anatomical shunt flow, and a nominal 1 mm Hg to the other factors combined. Alveolar–arterial pressure gradients for CO_2 are of great theoretical interest but usually are very small and can be ignored. The small pressure gradients for both gases imply high conductances ($Mx/(p_{A_x} + p_{a_x})$). These are complex functions of alveolar gas mixing, pulmonary diffusion capacity, ventilation–perfusion equality, red cell equilibration velocities and total blood flow/anatomical shunt flow ratios.

The arterial and venous blood compartments

The capacitance of whole blood, for oxygen and CO_2, is proportional to its volume and haemoglobin content, and to the shape of the blood–gas dissociation curves. The latter, which differ for each gas, vary with the partial pressure, hence the curved shapes of these compartments in Fig. 4. It follows that they cannot be expressed by a single value but, over the normal physiological range, they average about 2 ml per mm Hg p_{O_2} and 10 ml per mm Hg p_{CO_2} for each litre of blood. Roughly one-sixth of the blood lies in the arterial tree and the remainder is held on the venous side of the circulation. The volume of blood lying in the capillaries at any one time is negligible.

A healthy man resting at ground level, with a normal circulating blood volume of 5–6 litres, has about 200 ml of oxygen stored in arterial blood and 600 ml stored on the venous side. This total volume of 800 ml is equivalent to 3–4 min of basal oxygen consumption. On heavy exertion the arterial store will stay constant but the venous reserve will fall to 300 ml.

This total, of 500 ml, is equivalent to less than 10 s of maximal oxygen consumption.

The arterial–venous conductance

Circulating blood is the other long-distance convective carrier. It transports the gases between the tissues and the lungs and distributes them over the exchange surfaces at each site. In steady states the mass flow of transported gas will equal the product of blood flow $(\dot{Q})$ and the arterio–venous gas content difference, $(c_{a_x} \pm c_{v_x})$:

$$\dot{M}x = \dot{Q}(c_{a_x} \pm c_{v_x}) \quad \text{(Fick's Law)} \tag{4}$$

Since the a – v partial pressure difference $(p_{a_x} \pm p_{v_x})$ is related to the content difference by the slope (s) of the blood gas dissociation curve over the content range concerned, this equation can be rewritten to obtain the conductances for oxygen and CO_2:

$$\dot{M}x/(p_{a_x} \pm p_{v_x}) = \dot{Q}s \tag{5}$$

It is often helpful to bring out the role of haemoglobin in the transport of both gases, by using s^*, the slope of the dissociation curve per gramme of haemoglobin. The conductances are then seen to be:

$$\dot{M}x/(p_{a_x} \pm p_{v_x}) = \dot{Q}s^*\text{Hb} \tag{6}$$

Since $\dot{Q}$ and Hb are common to the transport of both gases but the effective solubility (s^*) is some five times greater for CO_2 than O_2 the arterio–venous conductance is correspondingly higher. This has important consequences which are discussed later.

The venous–tissue conductance

The transfer of oxygen and carbon dioxide between blood and tissue has exact parallels to gas exchange in the lung, i.e. its conductances are determined by the diffusion characteristics of the capillary bed, tissue fluids and microsomal membranes, by convective flows of interstitial and intracellular fluids, and by the precise matching of blood flow to point-to-point variations in metabolism. This $\dot{M}/\dot{Q}$ matching is entirely analogous to $\dot{V}/\dot{Q}$ matching in the lung.

At this stage, when the simple model shown in Fig. 4 has been described almost completely, it is neccessary to justify one of the model's principal features, namely that the gas tensions within the tissues are primarily determined by, and correspond closely to, the p_{O_2} and p_{CO_2} in the immediately adjacent veins. To do this the distribution of oxygen between the capillary and individual cells has to be examined in some detail. This is difficult to study directly because capillary architecture is complex, the regions served by individual vessels are very small and, after passing through several membranes, the oxygen is consumed in microscopic organelles. As a result many workers have tackled the problem mathematically, using arguments which are summarized in Fig. 6 and go as follows:

1. Actual beds are too complex to model directly, so consider an orderly array of parallel vessels, itself difficult to study.
2. Allot each vessel a cylinder of tissue, recognizing that this leaves some gaps between cylinders, that cannot be dealt with.
3. Take any one cylinder and its central vessel, slice it into a series of imaginary discs and consider the diffusion of oxygen along one radius, allowing for the greater amount of tissue at the periphery than at the centre of the disc, and obtain a p_{O_2}–distance curve for this radius, using Fick's diffusion equation.
4. Rotate the curve to construct a frequency histogram for the p_{O_2} distribution of the whole disc; repeat the procedure for each disc in

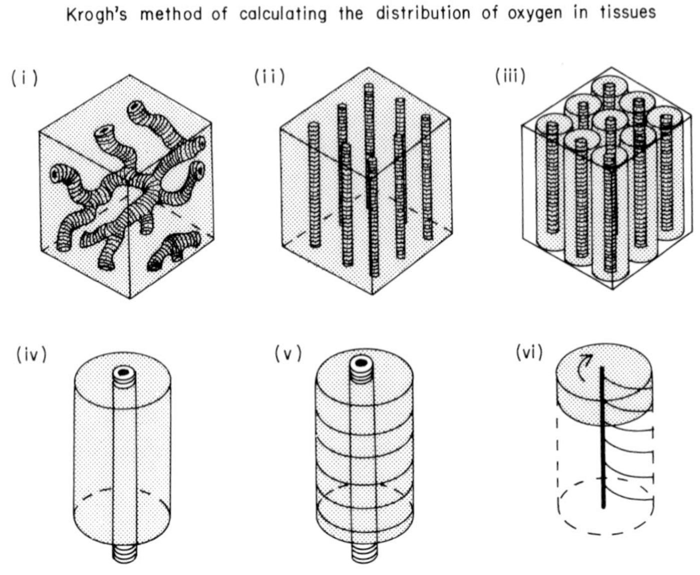

Krogh's method of calculating the distribution of oxygen in tissues

FIG. 6   A pictorial summary of the features and failings of Krogh's approach to calculating the distribution of oxygen in tissues (see text for further details).

turn, and so build up an O_2 distribution map and p_{O_2}—frequency histogram for the cylinder as a whole!

Laborious as this sounds, these sums were done most brilliantly in 1919 without the aid of computers, by their first author, Auguste Krogh. However, they have an important weakness, which he recognized. It is that they only allow oxygen to diffuse radially from the vessel, whereas in fact it is free to diffuse through the tissue in any direction. That is a problem which cannot be solved analytically but can be dealt with nowadays by analogue models or by numerical computers. On the whole the analogue simulations of tissue gas exchange have been less satisfactory because of the difficulties of modelling alinearities of the oxygen and CO_2 dissociation curves and because of the Bohr and Haldane effects and the problems of finding suitable physical counterparts for point-to-point variations in tissue oxygen consumption. Much of our present understanding has been obtained with digital computers, where these features can be modelled with ease.

Some numerical models directly imitate the haphazard nature of diffusion, setting up a two- or three-dimensional matrix of oxygen-consuming points, allowing a large number of oxygen molecules arriving continuously in a vessel to take "random walks" through the matrix, and observing the steady-state distribution that results. Other models take a similar array of points, allot oxygen pressures from Krogh cylinder calculations as a starting condition and then calculate the oxygen flux at each point over a small interval of time from the p_{O_2} gradients, and then repeat the process many times until equilibrium is achieved. These are the "finite difference" or "progressive relaxation" techniques.

Studies of this sort emphasize the importance of *axial*, or more strictly para-axial, diffusion, as is shown in Fig. 7. The left-hand graph illustrates a plane of tissue radiating from the central capillary of a Krogh cylinder. The rate of perfusion, dimensions and other features of the blood vessel and tissue cylinder are characteristic of well-perfused parts of the cerebral cortex in man. The incoming blood has a p_{O_2} of 96 mm Hg and a p_{CO_2} of 40 mm Hg, which are normal for sea level. Blood leaving the venous end of the capillary has a p_{O_2} of 40 mm Hg and a p_{CO_2} of 46 mm Hg, which are also typical of normal mixed venous blood (in rest). The p_{O_2} isopleths on the tissue plane, map out the distribution of oxygen pressure that would exist if oxygen were distributed through the cylinder by radial diffusion alone. The graph on the right of the figure shows the radically different distribution that is obtained when oxygen is permitted to diffuse through the tissue freely, under otherwise identical conditions.

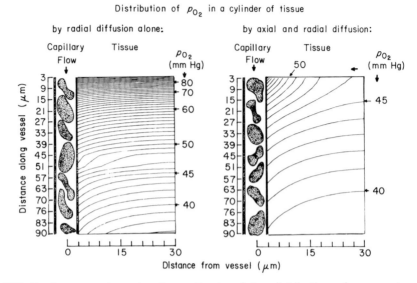

FIG. 7 A progressive relaxation estimate of the distribution of oxygen in a "Krogh cylinder" of brain tissue, before and after allowing for para-axial diffusion in the tissue. Each graph shows p_{O_2} isopleths inscribed on a tissue plane radiating from the central capillary to the edge of the cylinder (model of Bransby and Denison, 1976).

Study of Fig. 7 shows a very useful conclusion can be drawn from this distribution, particularly when it is weighted by rotation about the capillary axis to obtain the whole-cylinder distribution. It shows that the great bulk of the tissue served by a single capillary will have oxygen tensions that lie very close to the P_{O_2} in blood at the venous end of the capillary. This unexpectedly simple relationship arises because almost all of the oxygen used by the cylinder leaves the blood immediately it arrives at the entrance to the capillary, and floods outwards and downwards parallel to the capillary. As a result the highest flows, and thus the steepest falls in p_{O_2}, occur in a small cone of tissue at the arterial end of the capillary.

We are now in a position to return to the model of gas exchange shown in Fig. 4 and summarize the determinants of tissue oxygen and CO_2 pressures under steady-state and dynamic conditions. These are shown in Fig. 8(a). Although many factors contribute to the overall inspired air-to-microsomal *conductances* for oxygen and CO_2, under normal circumstances only two of these are critical, both at rest and during moderately severe exertion. They are *alveolar ventilation* for the transfer of CO_2 and *blood flow* for the transport of oxygen. In healthy people at sea level, alveolar ventilation is the prime determinant of tissue p_{CO_2} and local blood flow is

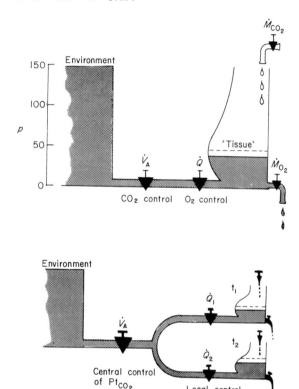

FIG. 8 (a) The upper diagram shows a simplified form of the gas transport model of Figure 4, which emphasises the homeostatic roles of ventilation and blood flow. (b) The lower diagram indicates a consequence, i.e. that there is regional control of tissue p_{O_2} but competition between tissues for use of the single central control of tisue p_{CO_2}. In both graphs $\dot{Q}$ represents blood flow and $\dot{V}_A$ indicates alveolar ventilation.

the main determinant of tissue p_{O_2}. This rule holds over a wide variety of circumstances and is largely due to the difference between the effective solubilities of the two gases in blood. This concept is illustrated in Fig. 9 which shows how venous, and therefore tissue, O_2 and CO_2 tensions are affected by changes in perfusion and ventilation. The graph demonstrates that alterations in perfusion produce wide swings in venous p_{O_2} but have little affect on venous p_{CO_2} and that changes in ventilation cause marked fluctuations in venous p_{CO_2} but only alter its p_{O_2} slightly. One consequence of this phenomenon is that there is regional control of tissue p_{O_2}, by local blood flow, but competition between tissues for use of the lung, to regulate local p_{CO_2}, (as indicated by Fig. 8(b)).

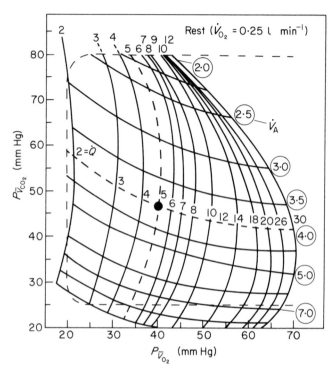

FIG. 9 A graph showing how variations in blood flow, $(\dot{Q})$ and alveolar ventilation, $(\dot{V}A)$, affect venous p_{O_2} and p_{CO_2} (from Denison, 1968).

The Predicted Effects of Hypoxia

One feature of the model is that it allows us to explore situations that connot be observed in real life, e.g. those effects that would be seen if the body made no response to hypoxia at all. In this way we can obtain a scale against which to judge the effectiveness of the responses that actually occur.

The alveolar air equation predicts that if ventilation is held constant a given fall in moist-inspired oxygen pressure, p_{I,O_2}*, wil produce an equal drop in alveolar oxygen pressure, p_{A,O_2}. This will have a complex but small effect on the alveolar–arterial oxygen gradient because the conductance of this barrier is high. Consequently the arterial oxygen tension will fall by much the same amount as the inspired and alveolar oxygen pressures. The venous, and *mean* tissue oxygen tensions which are closely tied, will show smaller reductions because of the shape of the oxyhaemoglobin dissociation curve.

These points are brought out in Fig. 10(a) which plots various body oxygen tensions against altitude. The tensions have been calculated using the barometric pressure versus altitude curve shown in Fig. 5 and the "physiological" oxyhaemoglobin dissociation curve given in Fig. 10(b), assuming that p_{A,CO_2}, R and the alveolar–arterial oxygen tension gradient were constant at 40 mm Hg, 0.85 and 10 mm Hg p_{O_2} respectively. The venous oxygen pressures are shown for three metabolism/blood flow ratios, producing p_{v,O_2} levels of 20, 30 and 40 mm Hg at sea level.

Several conclusions can be drawn from the figure, particularly with the help of the Michaelis curves presented in Fig. 3. Firstly, ascents which cause immediate falls in arterial oxygen pressures have little effect upon venous, and therefore mean tissue, oxygen tensions until altitudes above about 10 000 ft have been reached. This is due to the flat-topped shape of the oxyhaemoglobin dissociation curve. Below this ceiling the reductions in arterial p_{O_2} will mainly affect cells close to the entrances of systemic capillaries. Here the falls are sufficient to affect biochemical reactions of low affinity and there is some evidence that certain cells are preferentially located at these sites.

Above 10 000 ft the venous, and so mean tissue, oxygen tensions fall at much the same rate in all beds, and almost as rapidly as the moist inspired values. If the minimum acceptable p_{O_2} in any vein is taken to be 10 mm Hg then, in the absence of any response, tissues like the heart, brain and working muscle with venous oxygen tensions of 20–25 mm Hg at sea-level, will fail at about 12 500 ft. Similarly, if the minimum acceptable p_{O_2} in mixed venous blood is assumed to be 17.5 mm Hg then a resting man, with a normal sea-level p_{v,O_2} of 40 mm Hg could not survive above 15 000 ft.

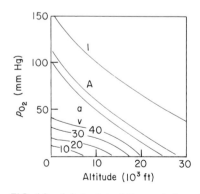

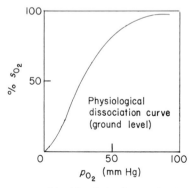

FIG. 10   (a) A plot of the variations in venous p_{O_2} with altitude in tissues having various oxygen-consumption/blood-flow ratios,[2] ($\dot{M}/\dot{Q}$), assuming the blood had; (b), the physiological HbO_2 dissociation characteristics shown in the right-hand graph.

Although sustained muscular work depends upon oxidative phosphory-lation, which has a high affinity for oxygen, its upper limit is determined by the maximum oxygen-carrying capacity of the transport system. In normal people this is set by the ceiling to cardiac output. On exposure to hypoxia, the aerobic capacity declines progressively with arterial desaturation, for two reasons. Firstly because of the fall in oxygen delivery per ml of arterial blood, and secondly because maximum cardiac output itself falls as oxygen delivery to the myocardium fails. These effects can be calculated. Figure 11 illustrates the results that would be expected if arterial oxygen tensions and saturations followed the curves shown in Fig. 10, minimum venous oxygen tensions for cardiac and striped muscle were 10 mm Hg, and the maximum cardiac output was proportional to the maximum a − v content difference for oxygen. It indicates that if ventilation is set to hold alveolar p_{CO_2} at 40 mm Hg aerobic capacity will fall gradually with ascent to 10 000 ft and will drop much more steeply thereafter.

These calculations give no guide to the time-course of events. If the air pressure surrounding an unresponsive individual is rapidly reduced over a period of a second or two, and then held constant at the lower level, the partial pressures of all the alveolar gases but water vapour will fall in strict proportions with equal rapidity, and the arterial tensions will follow just as speedily. The pressure of water vapour remains constant because it evolves from the moist surface of the lung very quickly. Within the tissues arrival of the altered blood is delayed by the individual lung-to-organ transit times. Of these the most important is the lung-to-brain delay which is normally

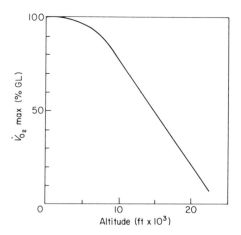

FIG. 11   A graph showing the fall in aerobic capacity with altitude that would be predicted from the fall in arterial p_{O_2} and a proportionate reduction in cardiac output due to the hypoxaemia.

about seven seconds. The tissues equilibrate to the now conditions with time-constants determined by the characteristics of each bed, but essentially by the solubilities of these gases and the blood flow/tissue mass ratios. Of these the most critical is again the brain, which has a time-constant for oxygen of 3–5 s and therefore is more or less fully equilibrated some 6–10 s after the arrival of the front of desaturated blood, i.e. about 15 s after the decompression.

If the decompression is sufficiently extreme the alveolar oxygen pressure will then fall below that in mixed venous blood. Oxygen will pass from the blood to the lungs, which will counteract the fall in alveolar p_{O_2} until recirculation occurs. Figure 12 illustrates the time courses the change in alveolar p_{O_2} would follow in a model resting man. Obviously, many factors influence the changes in partial pressure for each gas, nevertheless it is generally true that rapid decompressions impose a two-stage hypoxic insult that will affect the brain within 15 s of onset and be fully developed in a matter of minutes.

If the p_{O_2} of ambient air falls, while metabolic demand remains constant and the body makes no response at all, there will be equal falls of p_{O_2} all

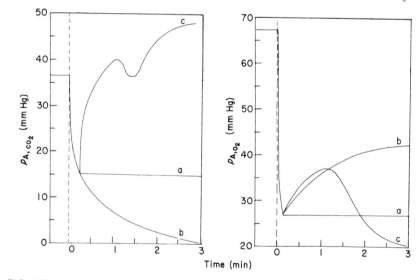

FIG. 12 The changes in alveolar gas tensions that would be expected to occur in a resting man abruptly decompressed from 8000 ft to 27 000 ft, assuming normal initial values, ($p_{a_{O_2}} = 65$ mm Hg, $p_{A_{CO_2}} = 40$ mm Hg); (a) if his lungs were bloodless and obstructed, (b) if his lungs were ventilated but not perfused, and (c) if he had a normal alveolar ventilation and blood flow but made no protective responses at all. Note the peaks in $p_{A_{CO_2}}$ and $p_{A_{O_2}}$ which are reversed after a minute when effective recirculation occurs. Later $p_{A_{CO_2}}$ rises as ventilation fails.

along the transport chain. There are three possible mechanisms for mini-mizing the effects of these falls, namely; increasing the conductances, augmenting partial pressure pumps or raising the oxygen-affinities of the oxygen-consuming processes.

Particularly in the Western world, very little is known about the extent of the latter two mechanisms in man, but some research on these topics and on the recruitment of alternative anaerobic processes has been reported from Eastern Europe by Barbashova (1967 and 1969) and from South America by Reynafarje (1971). The biochemistry of hypoxia is a subject that badly needs further study. The impetus and insight for this may well come from investigations on marine mammals. These creatures can swim actively throughout a breathold dive of 45 minutes or more. While it is clear that conventional aerobic metabolism of their heart and brain persists due to a remarkable withdrawal of blood flow from the rest of the body it is not known how the latter, particularly the working muscles, endure what must be very severe tissue asphyxia.

In the West, physiological researches on hypoxia have concentrated almost entirely on the mechanisms which ease the transfer of oxygen from the environment to the cell. The conductances responsible for this transfer, which are shown in Fig. 8, can be grouped in three categories:

(i)   those responses that can be rapidly mobilized but are energetically expensive or carry severe physiological penalties ($\dot{Q}$, $\dot{V}_A$);

(ii) those responses that take some hours or days to deploy, but consume little energy once they are developed (Hb, s^*, a com-pensated rise in $\dot{V}_A$, ? K_{mO_2});

(iii) those responses that take many days to develop or can only occur *in utero* or in childhood (diffusion characteristics of the systemic and pulmonary beds, possibly some of the short range convective pro-cesses and perhaps the oxygen affinities of some processes).

On this basis one would expect these three groups of responses to pre-dominate in acute, chronic, and life-long hypoxia respectively.

The Observed Effects of Hypoxia

Introduction

The preceding sections have outlined what is known of the biochemical properties of the oxygen molecule and of the manner in which it is transported to the tissues and these facts have been put together to provide a conceptual framework on which actual observations can be hung. Such observations are obtained from people who experience oxygen lack in four ways; on mountains, in flight, during some diseases and experimentally. These conditions are neither comparable nor are they as simple as they may seem.

Life at high altitudes differs from that at sea level in many ways. In general the winds are stronger, air temperatures are lower, sunlight is fiercer, diurnal changes of climate are greater and the diet, work habits, cultures, genetic backgrounds and diseases of people indigenous to mountains are quite distinct from those of folk living in low-lying plains. This chapter concerns only one of the environmental differences, that of oxygen supply. It is an aspect that has been studied in great detail, particularly as a model for pathological conditions in which oxygen lack is a critical feature. Many excellent accounts of the topic are available, for example by Bert (1877), Ravenhill (1913), Barcroft (1925), Houston (1947), van Liere and Stickney (1963), Weihe *et al.* (1963), Margaria (1967), Wulff *et al.* (1968), Dill (1968), Rennie (1976), Clegg *et al.* (1970), Lenfant and Sullivan (1971), Cohen (1972), Mazess (1975), Ward (1975), Heath and Williams (1977), Kellogg (1978) and Baker (1978).

At first sight it would seem that relatively few people encounter altitude hypoxia naturally since the fraction of land above 3500 ft is quite small and that part above 10 000 ft is much smaller still (Fig. 13). However, large areas of lowland, such as Central Australia, Northern Russia and the Sahara, are deserted, and several high-altitude zones, notably the Tibetan

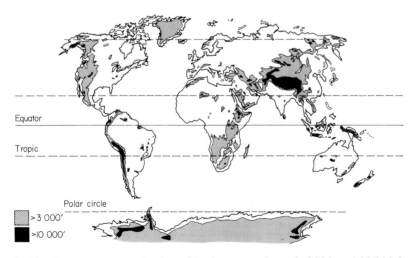

FIG. 13 An equal area projection of land masses above 0, 3000 and 10 000 ft.

and Ethiopian plateaux, the Andean Altiplano and the East African uplands are quite flat and well populated. Conservative estimates suggest one in seven people live in mountainous regions and one in twenty (i.e. more than 20 million people) live at altitudes greater than 10 000 ft. Many communities live at up to 16 000 ft and work regularly as high as 18 000 ft. Numerous mountaineers have climbed above 20 000 ft without oxygen and quite recently two men (Peter Habeler and Reinhold Messner) reached the summit of Everest, which is at 29 028 ft, without its aid (Habeler, 1979).

Nowadays, of course, resident highlanders and mountaineers are by no means the only people at risk from altitude hypoxia. During the course of a year many more people fly above the height of Everest than live on the high-altitude surfaces below it. Modern transport planes cruise at 30 000–65 000 ft and can reach these heights in half an hour or so, while militiary aircraft can do so within a few minutes. All commercial and militiary vehicles flying above 5000–10 000 ft are fitted with equipment that maintains the occupant's alveolar oxygen pressures above 65 mm Hg, but these devices can fail, suddenly exposing the occupants to the real environment around them. The abruptness of this change, sometimes taking as little as 1–2 s to complete, is quite crucial because it overtakes any physiological, as opposed to behavioural, capacity of the body to respond defensively at all. Aviation hypoxia has also been the subject of many excellent reviews, notably by McFarland (1946, 1971), Armstrong (1949), and Ernsting (1965, 1978).

Some years ago the last-named author (John Ernsting) asked me to strangle him repeatedly firstly in an air-filled room at ground level and then when he breathed pure oxygen at a pressure of four atmospheres. We set about it by wrapping a sphygmomanometer cuff around his neck and connected the cuff, by wide-bore tubing and a Douglas tap, to an oil-drum filled with compressed air. (The arrangement provided a satisfyingly abrupt onset to events.) In the air-filled room at ground level his head and neck became ischaemic and the rest of his body asphyxiated, whereas in the compression chamber his head became ischaemic though hyperoxic and the rest of his body hypercapnic (CO_2 excess) but not asphyxiated. The distinctions were not entirely semantic since he regularly lost consciousness after seven or eight seconds in the first situation but was still fully conscious after 20 seconds of strangulation in the pressure chamber. However, at no stage was any part of him anoxic, or even simply hypoxic.

If this seems surprising it is because, although the subject of oxygen lack is very well understood, some of the terms associated with it are occasionally used imprecisely, which can be confusing. *Dysoxia* is the general label coined by Eugene Robin for any derangement of oxygen supply, whether too little or too much and whether simple or complicated by a host of factors. *Hypoxia* describes oxygen lack relative to normal conditions at ground level. It cannot be defined simply as "inadequate oxygenation" since there are some processes that would proceed faster and others that would be affected adversely if more oxygen were made available to healthy people at sea level. Paradoxically, even in normal circumstances cells are oxygen deficient and intoxicated by oxygen at the same time.

The most commonly abused term is *anoxia*. Strictly it means a *complete* absence of oxygen. This is not compatible with life in vertebrates and even experimentally is almost impossible to achieve. *Asphyxia* is another specific form of oxygen lack, due to inadequate ventilation which causes blood and tissue CO_2 tensions to rise as the oxygen supply fails. The hypercapnia greatly modifies the effects of the hypoxia. *Ischaemia* is another cause of oxygen lack, due to impaired blood flow. This limits the supply of many substances besides oxygen and the accumulation of many others, besides CO_2, and thus is quite distinct from simple hypoxia.

Altitude exposure causes *hypobaric hypoxia*, while breathing oxygen-deficient mixtures of air at ground level leads to *hypoxic hypoxia*. A fall in the effective haemoglobin content of blood, whether due to anaemia, methaemoglobinaemia, carbon monoxide poisoning or other haemoglobinopathies, produces *anaemic hypoxia*. Sometimes ischaemic hypoxia is called *stagnant hypoxia*. Many poisons, such as cyanide, block cytochrome oxidase, causing *histotoxic hypoxia*. Strictly, oxygenase and mixed function oxidase inhibitors should be regarded in the same light. Their dependence on oxygen has been overlooked until recently.

Clearly, many diseases and drugs are capable of causing hypoxia, but of them anaemia and the common heart and lung disorders are much the most important. In anaemia blood leaves the lung with a normal oxygen tension but a reduced oxygen content. In the capillary there is a greater fall in p_{O_2} for a given O_2 uptake and so the venous and mean tissue oxygen tensions are abnormally low. Carbon monoxide poisoning has very similar effects but also alters the shape of the dissociation curve and inhibits some uptake mechanisms.

Surprisingly, the common obstructive airways diseases (asthma, bronchitis and emphysema), cause arterial hypoxaemia long before they cause asphyxia. This occurs because the obstruction is *unevenly* distributed, which permits hyperventilation of less obstructed alveoli to compensate for the hypercapnia that exists in the more obstructed spaces. Because of the shape of the oxyhaemoglobin dissociation curve the hyperventilation cannot compensate for their hypoxia. Systemic arterial hypercapnia is a late and grave feature of these diseases, indicating that most airways are severely involved and the remainder are too few to permit its correction. The important concepts of ventilation/perfusion inequality which have illuminated our understanding of these diseases, and of normal gas exchange in the lungs, are outside the scope of this chapter but have been very clearly described by West (1977a, b, and 1979).

Asphyxia results whenever hypoventilation is generalized, whether it is due to tracheal obstruction, to neuromuscular disorders such as polio or phrenic palsy, or to very severe diffuse airways obstruction. Alveolar membrane disorders (e.g. sarcoidosis, fibrosing alveolitis and hyaline membrane disease) are much less common. Because of its high solubility, carbon dioxide passes through the membrane very readily, and so these disorders, also, cause arterial hypoxaemia rather than asphyxia.

Amongst the vascular disorders, ischaemia can be caused by passive compression, active constriction, mural thickening or lumenal obstruction of blood vessels or by hypotension due to heart failure, blood loss or generalized vasodilatation. Cardiovascular disorders can cause hypoxia by two other mechanisms. When the left heart fails to maintain a low pulmonary venous pressure the lungs become oedematous. The heart failure is usually due to systemic hypertension, coronary occlusion or rheumatic valvular disease. At first the oedema fluid fills the interstices at alveolar junctions and does not interfere with gas exchange. As individual alveoli fill with fluid, reflex vasoconstriction shunts the incoming blood elsewhere and gas exchange remains well preserved until a majority of the alveoli are involved, when hypoxia and later asphyxia develop. (The topic of pulmonary oedema is the subject of an excellent monograph by Staub, 1978.) Hypoxaemia also occurs when systemic venous blood is shunted through abnormal routes, by-passing the lung. This is usually due to congenital malformation. Although such conditions are very uncommon they are of great physiological interest because they provide models of life-long hypoxia. A good introduction to this difficult subject is that by Shinebourne and Anderson (1978).

Two methods are commonly employed to make people hypoxic experimentally. One is the breathing of oxygen-poor mixtures under laboratory conditions at ground level, and the other is decompression in an altitude chamber. The first of these, which is inexpensive and much the simplest, differs from breathing air at low pressure in several ways that are often unimportant but sometimes significant:

1. It usually requires the subject to breath from a mouthpiece or oronasal mask, that may affect the pattern of ventilation.
2. The change in inspired oxygen concentration (F_{I,O_2}) disturbs the relation of p_{A,O_2} to p_{A,CO_2} predicted by the alveolar air equation.
3. It requires the body to accumulate rather than evolve tissue nitrogen, which perturbs the composition of alveolar air further, until nitrogen equilibrium is re-established.

It is also difficult to achieve a truly abrupt onset of a controlled degree of hypoxia because that depends upon the rate and depth of the subject's breathing relative to his end-expiratory lung volume. With these reservations noted, a great deal has been learnt from this simple laboratory procedure.

The alternative method of suddenly making people hypoxic is by rapid decompresion in an altitude chamber. These vessels exist in many forms but share the common principles illustrated by Fig. 14. For most decompressions, which take several seconds or minutes to complete, the subject is studied in a pressure vessel which is large enough to contain medical attendants, technicians and much equipment. The vessel is decompressed by one or more large vacuum pumps, connected to a reservoir. The decompression is regulated by an "ascent" control valve between the test chamber and the reservoir. "Descent" is regulated by another control valve which admits ambient air. The test chamber is always fitted with a second inlet valve which can be controlled from inside and outside the chamber. The bore of this emergency descent valve is sufficient to return the chamber interior to a sea-level pressure even if the pumps are working and the ascent valve is jammed open. The chamber also has a mandatory oxygen supply with sufficient access points and suitable regulators for all of the occupants.

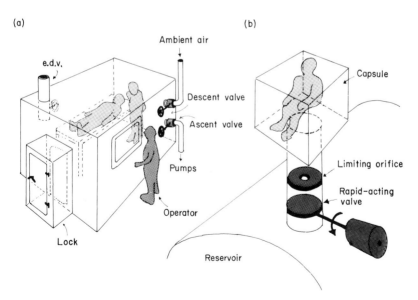

FIG. 14 A sketch of the essential features of (a) a routine decompression chamber, and (b) a vessel for extremely rapid ascents. The abbreviation e.d.v. represents the emergency descent valve.

Sometimes it is neccesary to decompress people faster than the pumps can evacuate a large chamber. In these circumstances the subject sits in a much smaller test vessel, connected via an additional rapidly acting wide-bore valve to a larger reservoir, typically something like 50 times its volume, as shown in Fig. 14(b). At the start of the study the reservoir is pumped down to a pressure which is slightly lower than the intended test pressure. The small chamber is then taken to base altitude using the normal controls. Once this is reached the normal ascent and descent valves are closed, the subject is warned that decompression is about to take place, and following a few seconds "count-down" which he can hear, the rapid-decompression valve is opened. At this stage air floods from the test chamber to the reservoir and the chamber pressure falls exponentially. The rate of this fall is determined by the size of a pre-set limiting orifice. In this manner it is possible to decompress subjects to very high altitudes with great accuracy in 1–2 s or less.

This form of decompression, like that of submarine escape, or free ascent in diving, carries a risk of lung rupture, which can be considered in the following way. The lungs can be thought of as a series of balloons attached to a branching system of tubes, with the whole arrangement enclosed in a compliant bag representing the chest wall, as indicated in Fig. 15(a). When this arrangement is freely ventilated at base altitude the absolute pressures in all its parts will be very nearly the same and almost equal to barometric pressure. On decompression the pressures in the balloons will fall more slowly than that outside, at rates determined by the pressure–volume characteristics of the balloons and by the ease with which they can vent through the airways that serve them. In any balloon, with a capacity C, an initial volume V, and the pressure–volume characteristic shown in Fig.

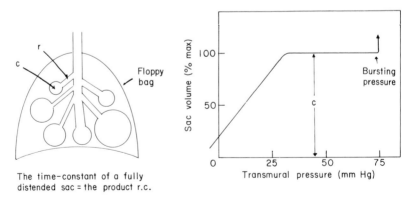

The time-constant of a fully distended sac = the product r.c.

FIG. 15 A simple model of the lung, suitable for considering the risk of lung rupture on rapid decompression (see text for further details).

15(b), served by an airway of resistance R, the pressure in the balloon will fall almost as rapidly as the external pressure until the expanding gas has filled it completely. This will occur when the external pressure has fallen at least in the ratio of V to C. (Some gas escapes through the airway during this time.) Thereafter the internal pressure will fall more slowly and exponentially, with a time-constant that is proportional to the product RC, i.e. to the resistance of the pathway and the capacity of the balloon. If the time-constant is short relative to the decompression time the pressure within the balloon will closely follow that outside. However if it is long, (i.e. if R or C is high), a large differential pressure will build up until the balloon bursts.

As a guide, healthy people can empty their lungs from total lung capacity to residual volume in 2–3 s, implying an overall time-constant for emptying of about one second. Since the bursting pressure of the normal lung is 75–100 mm Hg, subjects that are fit and trained, who have breathed out to a low lung volume, can tolerate a decompression over a three- to four-fold pressure range in 1–2 s and, providing they breath out continuously, a further fall in pressure of 25–50 mm Hg per s thereafter. By contrast, someone breath-holding at total lung capacity, when the elastic recoil pressure of the lung is 25 mm Hg, could not tolerate a decompression exceeding 50 mm Hg without risk of lung rupture. In elective decompressions, subjects are taught to breath out to low lung volumes immediately before decompression and ensure that their glottis is open throughout the procedure.

From this brief summary it is obvious that there are many patterns of hypoxia that may be mild and prolonged as in residence a few thousand feet above sea level, or brief and profound, as in the interval between loss of a canopy and automatic correction of hypoxia in a fighter aircraft. The effects that are produced depend upon the pattern of the hypoxic insult and upon the responses that are made to it.

The duration and rate of onset of oxygen lack are most important factors, both for effects and for responses. In very sudden hypoxia the body can fail without mobilizing any responses at all. During acute exposures of a few minutes or hours there is only enough time for rather expensive and unsatisfactory increases in ventilation and blood flow to be deployed. Chronic exposures permit slower effects to be unmasked but there is also time for more efficient and less costly remedies such as increased red cell mass and compensated hyperventilation to develop. The most efficient responses are seen *in utero* where the alveolar and systemic capillary beds can be remodelled to facilitate diffusion. This plasticity, which accounts for several features of people exposed to life-long hypoxia, is lost quite soon after birth.

Although it is helpful to distinguish *effects* from *responses* this is not always easy to do. For example people who have been at high altitude for some time develop blood with more cells that are rich in haemoglobin. The increased oxygen-carrying capacity of the blood is a beneficial response but the tendency of the blood to sludge and obstruct vessels is an obvious disadvantage which is better thought of as an effect, or at least as a price that has to be paid.

The remainder of this chapter describes the physiological changes that are seen in man during hypoxia. It is in three parts: the first concerns acute exposures of a few minutes or hours; the second describes exposures over periods from several hours to many years; and the third discusses the changes characteristic of life-long hypoxia, which distinguish immigrant from native highlanders.

Acute Hypoxia

The acute changes brought about by hypoxic episodes lasting a few minutes to a few hours, are dominated by cardiorespiratory responses and neurological effects. Mild and moderate degrees of acute hypoxia cause an increase in cardiac output that appears within 15–30 s (see Fig. 21(c)), and persists at a constant level, or regresses slightly thereafter (van Liere and Stickney 1963; Green 1965). The initial response is mediated by arterial chemoreceptors and later this is sustained by products of hypoxic metabolism such as lactate and adenosine. At altitudes below 10 000–12 000 ft, when arterial p_{O_2} is still on the plateau of the oxyhaemoglobin dissociation curve, there is little immediate change in ventilation, but a slight increase develops later. When the hypoxia is more severe some hyperventilation appears immediately. Despite these responses rapid impairments of cerebral function occur, even at low altitudes. When arterial p_{O_2} falls below 65 mm Hg these are accompanied by reductions in aerobic capacity. Hermansen and Saltin (1967) found that the changes in ventilation, heart rate, blood lactate and blood pH with work load were markedly affected by acute exposure to altitude but bore a constant relation to the exertion when it was expressed as a fraction of the maximal possible at each altitude. Their findings are summarized in Fig. 16. Other examples of the neurological and cardiorespiratory effects of acute hypoxia are shown in Figs 17–27 and discussed in some detail below.

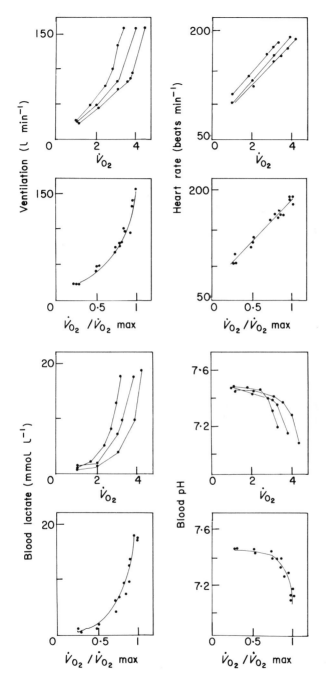

FIG. 16   A graphical summary of the observations about exercise-related effects of acute hypoxia in normal subjects, that were made by Hermansen and Saltin (1967). The upper graphs plot changes seen during progressive exertion in one normoxic and two hypoxic states against actual oxygen consumptions. The lower graphs show the same data plotted against oxygen consumption expressed as a fraction of the maximum achieved in each circumstance.

Cardiovascular and respiratory changes

As predicted from the model discussed in previous sections, there is little immediate change in ventilation until altitude exposures exceed 10 000 ft. At this altitude moist-inspired p_{O_2} is about 105 mm Hg and alveolar p_{O_2} is around 60 mm Hg. However, heart rate and cardiac output rise continuously directly the inspired p_{O_2} falls below sea-level values, but the increased flow does not prevent a fall in venous, and so tissue, p_{O_2}. These points, which have been demonstrated by many investigators, are evident in the results of a study summarized in Fig. 17. It shows respiratory and cardiovascular data obtained from four normal subjects exercising at a constant light work load, equivalent to driving a bus or piloting a plane, while they breathed gas mixtures corresponding to altitude exposures of 0, 6000, 10 000 and 15 000 ft. Measurements were made between the tenth and thirtieth minute of each exposure.

When the hypoxia is more prolonged a slowly developing hyperventi-

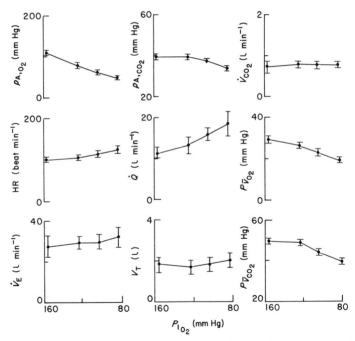

FIG. 17 Observations on the actual responses of normal men lying supine and exercising against a constant work load, while breathing hypoxic gas mixtures equivalent to exposures to 0, 6000, 10 000 and 15 000 ft. Symbols are defined in Table 1. (Details of the experiment are given in Denison, 1968 and 1970.)

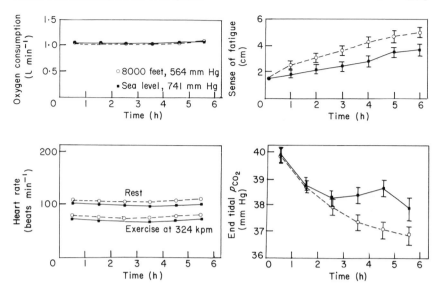

FIG. 18 Observations of oxygen consumption, heart rate, alveolar p_{CO_2} and subjective sense of fatigue in four trained subjects exercising at against a fixed light load in a decompression chamber close to ground level (741 mm Hg), and at an effective altitude of 8000 ft. Note the identical V_{O_2}, slightly raised heart rate, slowly developing hyperventilation and definitely increased sense of fatigue that appear at 8000 ft. Symbols as in Table 1 (Denison, unpublished data).

lation becomes apparent (Fig. 18(d)). It is particularly curious that these changes can occur at low altitudes where there is very little fall in arterial oxygen content. Although many investigators have found that people who are at rest or are exercising lightly do not show a ventilatory response to hypoxia until p_{A, O_2} falls below 60 mm Hg, this may not be true at heavier work loads. Hornbein and Roos (1962) studied two subjects exercising at work levels equivalent to an oxygen consumption of 1.5–2.0 litre min^{-1}, and found that a decrease in p_{A, O_2} of 6 mm Hg, from 100 to 94 mm Hg, was sufficient to cause an increase in steady-state exercise ventilation. They found, in one subject, that this response could be attenuated by pharmacological blockade of the sympathetic nerves to the carotid and aortic bodies, and have suggested that exercise may cause the hyperventilation that appears as exposure to hypoxia is prolonged. It is interesting that, at ground level, oxygen administration causes a 15–20% reduction of normal resting ventilation in man (Stockley, 1977).

The ventilatory responses are certainly mediated by the arterial chemoreceptors since they appear within four seconds of exposure, i.e. before hypoxic blood can have reached the brain stem (Downes and Lambertsen,

1966), and are not seen in patients whose carotid bodies are unperfused (Wade et al., 1970), or have been resected, (Lugliani et al., 1971). The responses to moderate and severe hypoxia develop with a half-time of 30–120 s, are quicker for depth than for rate of breathing and, when the oxygen supply is restored, they disappear about twice as quickly (Reynolds and Millhorn, 1973). The hypocapnia that results from hyperventilation limits the response to hypoxia, but also induces a lactic and pyruvic acidaemia, which partly corrects the respiratory alkalosis in the first few hours, before renal compansation becomes effective (Edwards and Clode, 1970; Garcia et al., 1971).

The vascular responses to hypoxia are equally rapid. Ernsting, working at the RAF Institute of Aviation Medicine in Farnborough, has made several studies of brief profound hypoxia in man. In one, three men vigorously over-ventilated for brief periods while breathing nitrogen. Voluntary over-ventilation with nitrogen for 16 s lowered alveolar p_{O_2} below 10 mm Hg for eight seconds. During and immediately following these periods, in which he made some notable measurements of the oxygen content and saturation of jugular venous, pulmonary arterial and femoral arterial blood, Ernsting observed rapid rises in pulse rate, arterial blood pressure and calf blood flow. The heart rate increased abruptly and reached a maximum in about 30 seconds. It then returned to basal levels almost as quickly. Arterial blood pressure and calf blood flow showed equally rapid changes (Ernsting, 1963). Figure 20(c) plots the heart rates observed in a very similar study, which employed rapid decompression and so did not require any voluntary over-ventilation.

In the systemic circulation generally, hypoxia is a major vasodilator and hypercapnia has relatively slight effects, however in the lung and brain these stimuli behave differently. In the former hypoxia is a potent *vasoconstrictor* and in the latter blood flow is perdominantly controlled by CO_2. These atypical vascular responses, which are critical features of bodily reactions to hypoxia, are discussed below.

Pulmonary circulation

Hypoxia causes vasodilatation in most vascular beds. The mechanism is still disputed but is probably mediated by adenosine. However, in the lung, oxygen lack produces a rapid and rapidly reversible vasoconstriction. The phenomenon was observed first in the cat (von Euler and Liljestrand, 1946), and has since been elegantly demonstrated as a local reflex in many animals whose lungs particularly lend themselves to regional studies, e.g. sheep (Edwards and Learoyd, 1967) and the coatimundi (Grant et al., 1976). It has also been convincingly shown in man (Arborelius, 1966). Before birth when it is most powerful, it is responsible for diverting blood away from the unaerated lung, After birth it persists as a mechanism that constantly matches local blood flow to local ventilation. In healthy people oxygen-breathing at sea-level causes a detectable fall in pulmonary vascular resistance, suggesting there is some hypoxia-induced tone in the vessels under normal conditions (Harris and Heath, 1977). When alveolar hypoxia affects the whole lung, as on ascent to high altitudes, the entire bed constricts and pulmonary hypertension results (see later).

The effective stimulus for constriction is alveolar rather than pulmonary–arterial hypoxia but the vessels which respond are the small arteries a short distance upstream of the alveolar bed. This has led many people to suspect a local neurohumoral mechanism, however the work of Staub and Schulz (1968), Duling and Berne (1970) and Ernest (1977) has shown that oxygen readily diffuses through the walls of small arteries and arterioles, so the response may well be due to a direct action of hypoxia on the vessel walls. It is remarkable that there is a significant degree of tone under normal conditions at sea level. This provides further evidence of the body's ability to detect variations in p_{O_2} at high tensions, presumeably via reactions of low oxygen affinity, and has led to a search for likely oxygen sensors. It is possible that these will include the neuroepithelial bodies, or Feyrter cells, that are distributed through the lung in man, are more abundant in the foetus than the adult, are histologically similar to the carotid body, and hypertrophy in hypoxia, but have no known function as yet, (Lauweryns and Goddeeris, 1975; Moosavi et al., 1973).

Many transmitters, such as histamine, prostaglandins, bradykinin, sero-tonin and angiotensin II, have been suggested and then abandoned as possible mediators of the response. At present we are ignorant of its mechanism. The vessels have adrenergic and cholinergic innervations but so far there is little information on the activity of these nerves in man. It is not certain whether the response arises because oxygen causes active vasodila-tation or its lack cause active constriction. There is some evidence that calcium is involved since "Verapamil", which blocks its entry into muscle cells, selectively inhibits the response (Reeves et al., 1979). It is also blocked by volatile anaesthetics but not by intravenous ones, is not seen in liver failure (where abnormal a–v fistulae develop in the lung) and is absent in people with some inherited anomalies of the autonomic nervous system. The topic, which is the subject of much research at present, has been reviewed recently by Fishman (1976) and Hughes (1977). Its role in the hypertension and oedema that are seen at high altitudes is discussed later.

Cerebral circulation

Most vascular beds are more sensitive to hypoxia than to CO_2 excess but in the brain the carbon dioxide pressure of incoming blood is a major determinant of vessel calibre. The effects of alterations of p_{O_2} and p_{CO_2} on cerebral blood flow and tissue metabolism have recently been the subjects of a comprehensive review by Lassen *et al.* (1978), and of two most scholarly monographs (Purves, 1972 and Siesjo, 1978) which summarize the vast amount of work that has been done in these fields by many investigators. Although several aspects are still controversial the main features are undisputed. In man, and other mammals, CO_2 is a potent cerebral vasodilator. Over the range of arterial CO_2 pressures which can be tolerated by normal subjects (20–80 mm Hg) there is a roughly linear relation between cerebral blood flow and p_{a,CO_2}, i.e. halving p_{a,CO_2} from 40 to 20 mm Hg halves blood flow and raising p_{a,CO_2} from 40 to 80 mm Hg doubles it. Beyond these limits responses to changes in CO_2 pressures are less marked.

Hyperoxia constricts the cerebral vessels and hypoxia dilates them. When arterial p_{O_2} falls below 50 mm Hg there is a sharp increase in cerebral blood flow. Responses to oxygen and CO_2 pressures appear to be simply additive. Although hyperventilation raises arterial p_{O_2}, the hypocapnia and thus cerebral vasoconstriction it causes leads to cerebral hypoxia. Many, perhaps all, of the functional effects of hyperventilation at ground level can be attributed to hypoxia and are reversed by oxygen administration. Conversely, the cerebral effects of hypoxia can be mitigated by the administration of CO_2. The responses have short time constants of about 20 s and are mediated in part via the arterial chemoreceptors.

It is believed that the vascular responses to CO_2 reflect cerebral sensitivity to changes of pH. The most elementary view supposes that the blood–brain barrier is permeable to CO_2 but not to hydrogen ions and that the freely diffusing CO_2 regulates cerebral tissue pH via the Henderson–Hasselbalch relationship, i.e:

$$pH = pK + \log \frac{[HCO_3^-]}{0.03 \times p_{CO_2}} \tag{7}$$

This immediate influence is gradually modified by changes in CSF and tissue bicarbonate concentrations. Normally the concentration ratio ($[HCO_3^-]/(0.03 \times p_{CO_2})$) is kept very close to 20 : 1, holding arterial p_{CO_2} near to 40 mm Hg. As a rough guide, an abrupt halving of p_{a,CO_2} will raise pH by 0.3 units, and doubling it, from 40–80 mm Hg, will reduce pH by the same amount. This view of cerebral pH control is certainly much too simple. Although it is definite that hydrogen and bicarbonate ions are not in electrochemical equilibrium across the blood–brain and blood–csf barriers, the nature of these dysequilibria are not clear and details of the intracellular buffer systems that exist under physiological conditions are difficult to obtain, especially in man. While it is established that arterial p_{CO_2} is an important regulator of cerebral tissue pH, and vice versa, our understanding of this system is incomplete. Readers wishing to study this important aspect of respiratory physiology further should read the recent work of the research teams that are led, respectively, by Cunningham, Severinghaus, Semple, Dempsey, Lassen, Siesjo and Purves.

The arterial chemoreceptors

The systemic circulation serves some highly specialized chemoreceptors that respond to the p_{O_2}, p_{CO_2} and pH of arterial blood, and are so placed that they can sense its composition almost as soon as the blood leaves the left ventricle, appreciably before it arrives at the systemic periphery. The *aortic bodies*, which are found at the roots of the main thoracic arteries, obtain blood from small arteries nearby and are innervated by vagal fibres that ascend to the nodose ganglion in the medulla. The *carotid bodies*, which are larger, lie in the bifurcation of each common carotid artery, obtain blood from the occipital and ascending pharyngeal arteries and are innervated by fibres which travel to the petrous ganglia in the medulla via the glossopharyngeal nerve (Fig. 19(a)).

The functions of the arterial chemoreceptors were discovered separately by de Castro (1928) and Heymans and Heymans (1929). They act mainly as

(a)

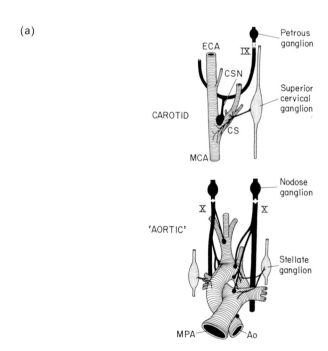

FIG. 19(a)   The anatomy of the arterial chemoreceptors.

(b)

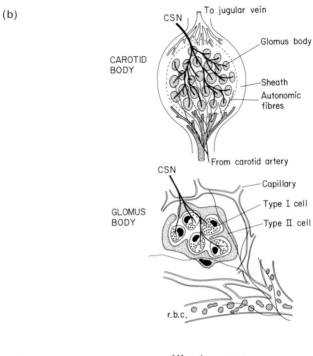

(c)

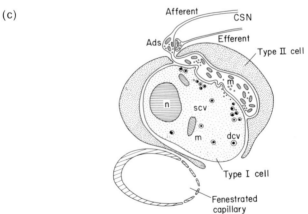

FIG. 19   Diagrams indicating (b) the histology, and (c) the probable innervation of the arterial chemoreceptors. Figs. (a) and (c) are adapted from the data and drawings of Howe and Neil (1971) and Verna (1975).

a protection against acute hypoxia which directly depresses the activities of the respiratory and circulatory driving centres in the brain stem. Although the existence of this function is very well established, the mechanisms by which it is achieved are still the subject of vigorous controversies that are clearly evident in the review by Howe and Neil (1971) and in the international proceedings edited by Torrance (1968), Purves (1975), Paintal (1976) and Acker et al. (1977).

Each organ consists of a highly vascular medulla enclosed by a fibrous sheath. The medulla contains cellular islands known as glomus bodies (Fig. 19(b)). Histologicaly, the glomera consist of two main cell types. Type I glomus cells are large and rounded, have an oval nucleus and contain many electron-dense vesicles. These "chief" cells lie in a basket of flattened "sustentacular" or Type II cells that have no vesicles and which imperfectly separate the chief cells from the interstitial spaces and vascular lacunae beyond. The Type I cells are innervated by nerve cells with at least two types of endings (Fig. 19(c)), but whether these are afferent or efferent is still disputed (Verna, 1975). The dense vesicles are rich in catecholamines, particularly dopamine. In rats, acute hypoxia depletes the carotid body of about 60% of its dopamine within 30 minutes, whether the carotid sinus nerves are intact or not. Return of the oxygen supply restores the amine to its original levels after some two hours. The dopamine is probably discharged from the vesicles by exocytosis and almost certainly behaves as an inhibitory transmitter (Hellstrom, 1977). More prolonged hypoxia causes hypertrophy of the Type II cells with vacuolation of the vesicles. This has also been observed in man (Arias–Stella, 1969; Arias–Stella and Valcorcel, 1973). Pathological enlargement (*chemodectomas*) with hyperplasia of Type II cells and dense-cored vesicles is rare but much commoner at high altitudes than at sea level (Saldana et al., 1973). These points are discussed in some detail by Heath and Williams, (1977).

General

Providing hypoxic episodes are separated by normoxic intervals of comparable duration there is no carry-over or evidence of acclimatization. An example of the cardiorespiratory effects of repeated exposures to mild hypoxia is given in Fig. 20. It shows further data from the experiment that provided the data for Fig. 18. Four normal subjects were exposed to an effective altitude of 8000 ft for six hours on five consecutive days. During exposures they exercised on a cycle ergometer at a constant light work load for 30 minutes in each hour. The barometric pressure of the experiment (565 mm Hg) was chosen because it is the minimum pressure permitted in

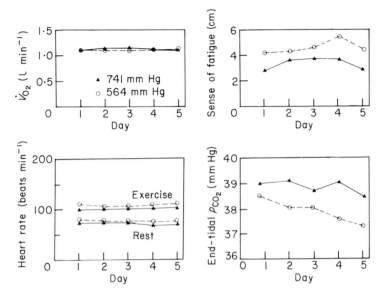

FIG. 20 The oxygen consumption, heart rate, alveolar ventilation and subjective sense of fatigue observed in four trained subjects, exercising, on five
succesive days, against a constant work load, in a decompression chamber at
pressures close to ground level (741 mm Hg) and at an effective altitude of 8000
ft (565 mm Hg). Note there is no evidence of any adaptation to hypoxia from
one day to the next (Denison, unpublished data).

passenger aircraft cabins and also because it is equivalent to the altitude
(8000 ft) of Mexico City which was the site of the 1968 Olympic Games,
and so has been employed by many other investigators. The stable increase
in heart rate and the slowly developing hyperventilation that did not vary
from day to day are typical of such studies (e.g. Turner *et al.*, 1969).

If the onset of hypoxia is rapid and severe, the alveolar–arterial oxygen
gradient is reversed and oxygen flows into the lungs from pulmonary
arterial blood until recirculation occurs, as predicted in the model discussed
previously (Luft *et al.*, 1949; Ernsting *et al.*, 1960; Rahn, 1963; Ernsting,
1963). A hypothetical example of this phenomenon was shown in Fig. 12.
Observations from an actual study are given in Fig. 21. When oxygen lack
is severe the reflex cardiorespiratory responses are swamped by the direct
effects of hypoxia on the heart and brain stem, various arrythmias and
conduction defects appear, cardiac output falls, a profound compensatory
vasoconstriction occurs and eventually the heart arrests. Quite often this
pattern of events is overtaken early in its course by vasovagal bradycardia
and vasodilatation. When this defensive response occurs in aircrew or
parachutists, who are strapped in an upright posture and so cannot faint,
the effects of the hypoxia are exaggerated.

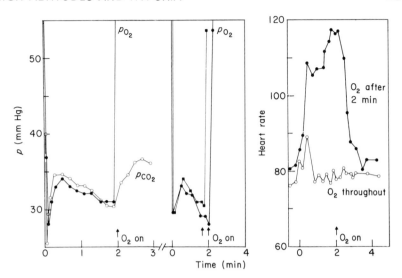

FIG. 21 (a) The alveolar O_2 and CO_2 tensions observed in one normal subject, following rapid decompression from 8000 ft to 27000 ft. Note the reflux of oxygen and CO_2 which reverses some 30 s after decompression, and that the hyperventilation ceases almost immediately after the oxygen supply is restored. Some of the rise in pA_{CO_2} is due to hypoxic depression. (b) The central graph shows the mean curve of the alveolar p_{O_2} observed in seven subjects decompressed over the same range on two occasions when the oxygen supply was restored after 90 and 120 s respectively. (c) The graph on the right shows the mean heart rates observed in the seven subjects on the second occasion, (O_2 after 120 s), and on decompression to 27000 ft while breathing oxygen throughout. Note that hypoxia causes a rapid rise in heart rate, which explains why the alveolar p_{O_2} peak reverses after 30 s, and the equally rapid slowing of the pulse when the oxygen supply is restored (Denison, unpublished data).

Neurological Effects of Acute Hypoxia

The early balloonists and aviators, who made rapid ascents to high altitudes, had noted that many of their sensory and mental functions dimmed on ascents above 12000 and were quickly restored by the administration of oxygen. These findings were studied experimentally by several investigators in the 1930s. Prominent amongst these were Ernst Gellhorn from the University of Illinois and Ross McFarland from the Harvard Fatigue Laboratory. In a review of his studies Gellhorn observed that "divers functions of the central nervous system, as visual intensity, discrimination, the latent period and intensity of negative after images, auditory acuity, muscular coordination, associations and simple mental tasks (addition of two digits, etc.) are impaired under approximately the same degree of oxygen deficiency" (Gellhorn, 1937). In some most important experiments, he observed that the visual and mental changes of hypoxia when breathing 8–9% O_2 could be almost completely reversed by simultaneously breathing 3% CO_2. He rightly supposed that this was due to the raising of cerebral venous p_{O_2} by an increase in blood flow. At the same time he drew attention to the advantage of using the eye as an accessible and readily tested extension of the brain.

Visual effects of hypoxia

At Harvard, McFarland made a particular study of dark adaptation during hypoxia. He showed that impairments of adaptation could be detected at altitudes as low as 7400 ft, and that they were very rapidly reversed by oxygen administration (Fig. 22, McFarland and Evans, 1939). At the same time, McDonald and Adler (1939), demonstrated that the effects of hypoxia and of Vitamin A deficiency upon night vision were independent though additive. Both groups concluded that oxygen lack could not be affecting the photochemical reaction but must be acting elsewhere, probably on retinal or central neurones. Subsequently Goldie (1942) found that hypoxic deterioration of night vision could be detected at 4000 ft. This has recently been confirmed by Pretorius (1970).

The effects at these low altitudes are too slight to be of much operational importance in aviation (Pierson, 1967), but are of great physiological interest because they occur at such high arterial oxygen tensions. The mechanisms are not yet known, but it may be significant that retinal

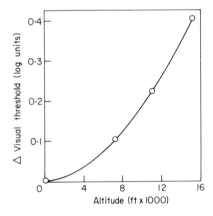

FIG. 22 A plot of the impairment of night vision seen in normal subjects exposed to various altitudes. The ordinate shows the mean elevation of the threshold for night vision in log units (adapted from McFarland and Evans, 1939).

interneurones are peculiarly rich in the neurotransmitter *dopamine* which is both made and destroyed by oxygen-dependent enzymes and is only found in comparable concentrations in the systemic arterial chemoreceptors (Krnjevic 1974). The effects of hypoxia on dark adaptation are more marked on cone than on rod vision and amongst cones is more evident at the periphery than the centre of the retina (Ernest and Krill, 1971). This could be due to oxygen losses by diffusion through arterial walls as the blood travels away from the central retinal artery (Ernest, 1977). Low concentrations of carbon monoxide affect the eye in a similar way to hypobaric hypoxia. According to McFarland (1946) and Magdelano (1968), smoking three cigarettes influences night vision to the same extent as does ascent to 8000 ft. These effects are also more marked at the retinal periphery (Johnston, 1965; Johansson and Jansson, 1965).

Cerebral effects of hypoxia

Increasing hypoxia impairs psychomotor performance progressively. At low altitudes (10 000 ft), the only *objective* functions which are known to be affected concern the learning and performance of novel tasks. Above 12 000 ft there are clear deteriorations in sensory acuity, vigilance, judgement, speed of response and manual dexterity. These are particularly dangerous because their onset is insidious. Subjects who are hypoxic have little or no insight into their condition, often believe they are performing normally or even better than usual, and rarely take any corrective action unless they have been especially well trained to do so. It is perhaps the greatest single hazard in aviation. At altitudes above 20 000 ft consciousness is lost with increasing rapidity; within a few minutes at 25 000 ft, a minute or so at 30 000 ft, and 15 seconds at 40 000 ft and above (Fig. 23).

The altitude threshold for cerebral deterioration may be as low as 5000 ft (Denison *et al.*, 1966). At 8000–10 000 ft impairments in the performance of novel tasks have been demonstrated repeatedly (McFarland 1946; Dugal and Fiset 1950; Ernsting *et al.*, 1962; Denison *et al.*, 1966; Kelman and Crow, 1969; Ledwith 1970; Frisby *et al.*, 1973; Billings, 1974). An example of the cerebral effects of an acute exposure to a low altitude (8000 ft) is given in Fig. 24, which records the choice-reaction times of a group of subjects who were confronted with an initially unfamiliar but repetitive visual reorientation task while breathing air, or oxygen-enriched air, in a decompression chamber at a pressure of 565 mm Hg. Mild hypoxia caused them to learn the task more slowly but did not impair their ability to perform the task once it had been learnt. The effects were more marked on the complex than on the simple elements of the task. Studies of this sort can

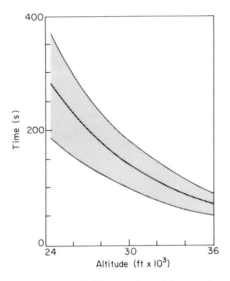

FIG. 23 The mean and ranges of "times of useful consciousness" observed in normal subjects suddenly exposed to various altitudes. (based on the data of McFarland).

be interpreted in at least three ways. Unlearnt and complex task elements may be more affected simply because they involve more neurones, or because they utilize more hypoxic areas of the brain, or because they rely on biochemical reactions of lower oxygen affinity.

Another subtle effect of mild hypoxia is an increased sense of fatigue. Most passenger aircraft are designed to have maximum "cabin altitudes" of about 8000 ft. Crew members, flying regularly, quite often complain that they feel more tired at maximum than at lesser cabin altitudes. An experimental demonstration of this effect has been included in Fig. 18. It shows that, under strictly controlled conditions, well-trained subjects felt as tired after some three hours at 8000 ft as after six hours under identical circumstances at ground level.

The mechanisms underlying the cerebral effects of mild hypoxia are completely unknown. Animal studies have shown that hypoxia alters catecholamine content and turnover in the brain (Davis and Carlsson, 1973), and interferes with synaptic transmission long before axonal conduction is affected. It is possible that the psychomotor effects of hypoxia arise in the same way and are due to less obvious disruptions of amine transmitter formation and decay (Siesjo, 1978).

The operational consequences of exposures to altitudes below 10000 ft are debateable; perhaps the most reasonable view to take at present is that

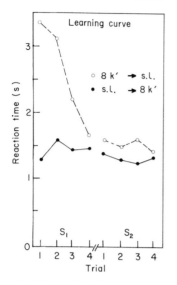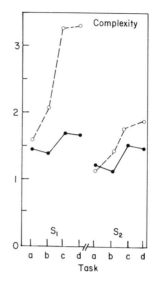

FIG. 24   The left-hand graph compares the learning curves of two groups of subjects who learnt the same, novel visual-reorientation task while breathing gas mixtures equivalent to air at sea level or air at 8000 ft. Choice-reaction times are on the ordinate and succesive trials are along the abscissa. Note that hypoxia interfered with their ability to learn but not with performance once the task was learnt. The graph on the right shows the same data arranged in order of task complexity. Note that hypoxia interfered more with the learning of difficult than of simple elements of the task (from the data of Denison, *et al.*, 1966).

this degree of hypoxia will not affect aircrew in normal flight when they are practising over-learnt skills, but many prejudice the crew's ability to cope with novel or unexpected problems, as in an emergency. There is no doubt that exposures to altitudes above 12 000 ft directly interfere with flying skills (Gold and Kulak, 1972). These matters are well reviewed by Ernsting (1978).

In brief profound hypoxia the deterioration and recovery of consciousness is very rapid and closely mirrored by changes in the electroencephalogram. Many years ago Davis observed that when men breathed hypoxic mixtures (7.8–11.4% O_2) at ground level, there was an initial increase in overall EEG activity, then short bursts of alpha activity appeared, which were replaced by irregular delta waves as cyanosis developed. These waves became much larger as consciousness was lost. An example of these changes, obtained in a recent study of rapid decompression, is shown in Fig. 25. Gellhorn found that these electrical changes, like those in visual and psychomotor performance, could be prevented or diminished by the simultaneous administration of 3% CO_2 (Gellhorn and Heymans, 1948;

Effects of acute hypoxia on the E.E.G. in man

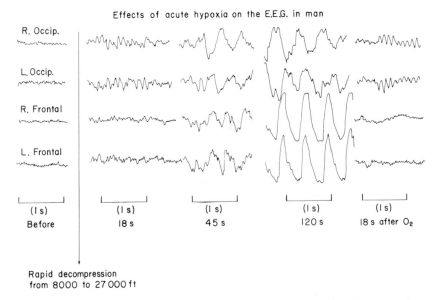

R. Occip.

L. Occip.

R. Frontal

L. Frontal

(I s)	(I s)	(I s)	(I s)	(I s)
Before	18 s	45 s	120 s	18 s after O₂

Rapid decompression
from 8000 to 27 000 ft

FIG. 25   Tracings of a four-lead EEG recorded from a normal subject at various times after a single rapid decompression from 8000 ft to 27 000 ft. The oxygen supply was restored immediately after the trace at 120 s (Denison, unpublished data).

Gellhorn, 1950, 1953). These and many similar studies up to 1960 are summarized in a major monograph edited by Gastaut and Meyer (1961).

The earliest EEG change that occurs in hypoxia is the development throughout the cortex, of alpha-like activity (8–16 cps) which is very similar to that seen over the occipital cortex when the subject's eyes are closed. The spindles of alpha-like activity due to hypoxia are best seen in the frontal cortex where there is little natural activity. Normocapnic subjects do not show them on acute exposures unless alveolar p_{O_2} falls below 30–38 mm Hg, but they are rapidly produced by hyperventilation, even at sea level. Meyer *et al.* (1965) found these appeared whenever cerebral *venous* p_{O_2} fell below 19 ± 2.5 mm Hg.

The EEG is a particularly complex physiological signal that is not easily analysed by eye, but it can be simplified electronically in various fashions. One widely-used principle is frequency power spectrum analysis (Byford, 1965). This technique determines the power dissipated in various frequency bands of the EEG (e.g. 1–2, 2–4, 4–8, 8–16, and 16–32 cps). Figure 26, which is a typical record, shows how the electroencephalogram of a normal subject was affected by various pulses of brief profound hypoxia. There are

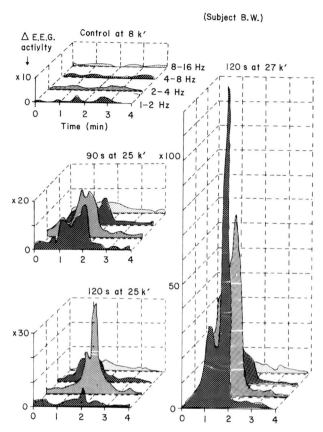

FIG. 26 Frequency power spectral analyses of the EEG recorded from a single normal subject exposed, by rapid decompression, to various pulses of brief profound hypoxia. Time from decompression (in minutes), is on the abscissa, and power dissipated in various frequency bands, expressed as multiples of the control outputs, is on the ordinate (Denison, unpublished data, method of Byford, 1965).

four particular features to note in the figure. First, due to the lung-to-brain transit delay of 7–8 s and the similar delay while oxygen dissolved in brain tissue is metabolized, there was no increase in activity until 15 s after the onset of hypoxia. Secondly, 15–30 s later again there was a transient fall-off in increased activity, which is due to the buoying-up of alveolar p_{O_2} by oxygen from systemic venous blood, as in Fig. 21. Once recirculation occurs, and alveolar p_{O_2} begins to fall further, the EEG activity rises steeply. Thirdly, the total power output rises with the degree of hypoxia. Lastly, there is a progressive shift to slower wave activity as hypoxia intensifies.

Using this technique Ernsting *et al.* (1973) found that, during brief profound hypoxia, there was a linear relation between alpha-band activity and
the extent of the fall in alveolar p_{O_2} below 30 mm Hg (Fig 27(a)).

Various attempts have been made to relate fluctuations of consciousness
to these changes in the EEG but all share a common difficulty, that
meaningful tests of consciousness take time and are not easily adapted to
the study of rapid events. In an ingenious experiment Walsh (1952), made
normoxic subjects respond to an irregularly flashing light, in an attempt to
relate variations in visual reaction time to those of occipital alpha activity,
but he found no connection. Ikegami (1967) presented 11 healthy subjects
with a series of problems in mental arithmetic at ground level, and gave
them 7.2% oxygen to breathe part of the way through the study. Their EEG
records were submitted to power spectrum analysis. He found that in the
second minute of hypoxia 3–4 cps activity appeared in the frontal leads
without any slowing of problem response times. At the third minute
slowing of the EEG was more marked and response times began to
lengthen. By the fourth minute groups of 2–3 cps activity appeared
and responses were much slower. At the fifth minute many questions were
missed entirely although subjects could respond when their names were
shouted.

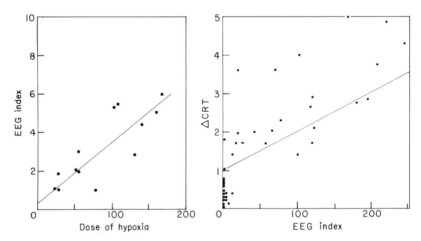

FIG. 27   (a), A plot of 8–16 Hz frontal EEG activity against intensity of hypoxia
in three subjects rapidly decompressed from 8000 ft to 40 000 ft. The abscissa is
scaled in units of the area of the pA_{O_2}–time plot below 30 mm Hg (adapted from
Ernsting *et al.*, 1973). (b) A plot of EEG activity, on the abscissa, against mean
choice-reaction time on the ordinate, in seven subjects rapidly decompressed to
25 000 ft or 27 000 ft. Reaction times are expressed as multiples of control levels.
Although the altitude was lower, the periods of hypoxia were much longer than
in (a), hence the apparent discrepancy between the two EEG scales (Denison,
unpublished data).

Over several years, Ernsting, Gedye, Byford and Denison, working at Farnborough, developed various automated tests that could follow more rapid and severe changes in useful consciousness. The most recent of these are a visual re-orientation task and a target step-tracking test, which are capable of sampling consciousness at three-second intervals (Fig. 28), and distinguishing between the effects of abrupt exposures to 25 000 and 27 000 ft, within 20 s of decompression, i.e. about 5 s after the hypoxic front develops in the brain (Fig. 29). Their studies show a close temporal and quantitative correlation between EEG activity and psychomotor ability (Fig. 27(b)), and show that consciousness is lost whenever alveolar p_{O_2} falls below 30 mm Hg for more than 140 mm Hg s, during abrupt exposures in normal unacclimatised subjects (Fig. 27(a)). This threshold, which is very sensitive to alveolar p_{CO_2}, may rise to 40 mm Hg during moderate hyperventilation.

Chronic Hypoxia

Several other effects of oxygen lack appear if men remain exposed to altitudes of 11 000 ft or above are held there for more than an hour or two.

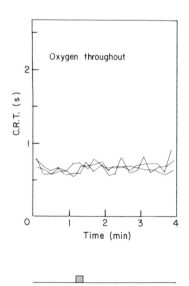 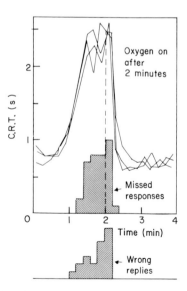

FIG. 28   The effects of rapid decompression from 8000 ft to 27 000 ft on the complex choice-reaction times of a fully trained normal subject, (a) on three occasions when breathing oxygen throughout, and (b) on three occasions when oxygen was withheld until two minutes after decompression (Denison, unpublished data).

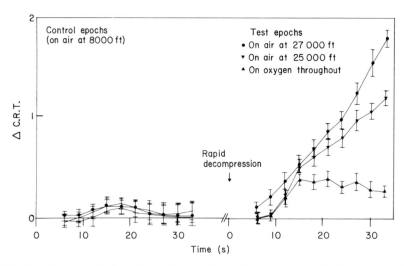

FIG. 29   The cumulative effect of acute hypoxia on complex choice-reaction in seven normal and fully trained subjects, (a) when breathing air at 8000 ft, and (b) immediately following rapid decompression to 25 000 ft or 27 000 ft. The ordinate is scaled in multiples of mean response times under control conditions. The transient slowing 12–15 s after decompression on oxygen is probably due to hypocapnic constriction of cerebral vessels. Note how significant departures from the oxygen-breathing curve appear after 15 s (Denison, unpublished data).

Three derangements are particularly important: namely, pulmonary vaso-constriction, pulmonary oedema and cerebral oedema. However, if the same degree of hypoxia arises gradually, a number of protective responses which can only be mobilized slowly have time to appear, and these disturbances are limited or suppressed altogether. The two principal responses to chronic hypoxia are a compensated form of hyperventilation and an increase in total circulating haemoglobin.

A striking example of their effectiveness can be obtained by reconsidering Fig. 29 and the text related to it. The figure shows how unacclimatised *normocapnic* men, *at rest*, who are suddenly decompressed to a pressure of 256 mm Hg, exhibit a deterioration of consciousness which appears within a few seconds and progresses to loss of consciousness within a minute or so. If the men had been hyperventilating or exerting themselves, even lightly, at the time of decompression they would have lost consciousness earlier still. By chance this pressure is almost exactly equal to that at the summit of Everest, which is 29 028 ft above sea level but has an effective altitude of about 27 000 ft. On May 8th 1978 it was scaled successfully without oxygen by two Austrian climbers, Peter Habeler and Reinhold Messner, six weeks after leaving Katmandu which is at 3600 ft (Fig. 30). It is certain that as

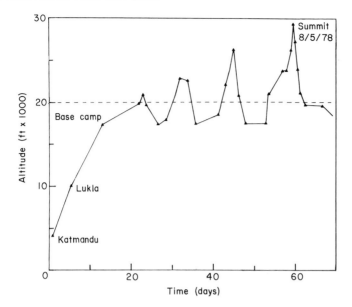

FIG. 30   A plot of the altitude–time exposure of Peter Habeler and Reinhold Messner who succeeded in climbing Everest without oxygen on May 5th 1978 (from the data of Habeler, 1979).

they approached the summit climbing in arduous terrain, as quickly as they did, these men had alveolar CO_2 tensions below 20 mm Hg, and probably as low as 10 mm Hg. During the period between leaving Katmandu and arriving at the summit substantial changes must have occured in Habeler's and Messner's capacity to resist hypoxia which distinguish them from the men described by Fig. 27. These changes are discussed below.

Hyperventilation

In the winter of 1921–22 a group of British and American physiologists made an expedition to Cerro de Pasco, the third largest city in Peru, to investigate "the physiological conditions which made muscular and mental effort possible at high altitudes", (Barcroft et al., 1922). The town, which at that time had 12 000 inhabitants, lies on an incline rising from 14 000 to 16 000 ft. Barcroft and his colleagues calculated that since the barometer at Cerro was usually at 458 mm Hg the moist-inspired oxygen pressure would be 88 mm Hg and, if alveolar p_{CO_2} was held at 40 mm Hg, their alveolar oxygen pressures would be about 38 mm Hg. On arrival at Cerro they studied their responses and observed that alveolar p_{CO_2} actually ranged from 25–31 mm Hg which raised alveolar p_{O_2} from 38 to 50–55 mm Hg, "a gain of some 15 mm oxygen pressure". This finding, similar to those seen by many high-altitude physiologists before and since, is particularly interesting because it illustrates a response which is potentially available immediately but is actually developed gradually over several days. Clearly there must be some drawback to using it directly but the precise nature of that disadvantage is not yet clear.

Hypoxia stimulates respiration via the periphereal chemoreceptors but depresses its central drive. When the hypoxia is mild or moderate the periphereal stimulus outweighs the central depression and hyperventilation develops. When hypoxia is severe the central depression predominates and breathing becomes periodic and then fails entirely. In man, most of the periphereal drive comes from the carotid bodies and is largely abolished by their resection. After this procedure (chemodectomy) people make a slight ventilatory response to hypoxia if they are also hypercapnic. It is thought that this residual stimulus comes from the aortic bodies. In mild and moderate hypoxia the hyperventilation appears quite gradually and takes some days to develop. At first the drive to increased ventilation is purely hypoxic and is completely reversed by oxygen administration, as in Fig. 21, but later it is sustained by some other mechanism and the provision of oxygen has little effect.

In 1968 the Olympic Games were held in Mexico City at an altitude of 8000 ft, and this stimulated much work at the time on the physiological effects of mild prolonged hypoxia. One such study was initiated by the Italian National Olympic Committee and conducted by Scano and Venerando (1968). They took 19 champion athletes in training, made a series of respiratory measurements on them in Rome, first at ground level, and again during a brief exposure at 8000 ft in a decompression chamber. A few days later they took the athletes to Mexico City and made the same "ground-level" measurements repeatedly over the subsequent 40 days at that altitude. The findings are summarized in Figs. 31 and 32 which have been constructed from their data. The acute exposure in the chamber dropped alveolar p_{O_2} from a ground level value of 107 mm Hg to an average at 8000 ft of 69 mm Hg, but this had a trivial effect on alveolar ventilation and p_{CO_2}. Over the first few days of exposure to the same altitude, in Mexico City, alveolar ventilation almost doubled and alveolar p_{CO_2} fell to 31 mm Hg holding alveolar p_{O_2} at raised values close to 79 mm Hg. This result is typical of many studies of ventilation at altitude that are summarized in the important review by Lenfant and Sullivan (1971) from which the next set of graphs (Fig. 33) have been adapted. In essence Lenfant and Sullivan concluded that the ventilatory response to altitude exposure generally

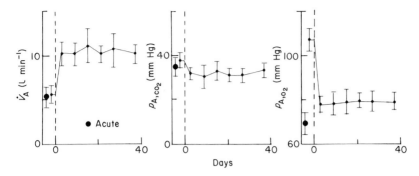

FIG. 31   The alveolar ventilation ($\dot{V}$A), and O_2 and CO_2 tensions seen in 19 Olympic athletes from Italy in Rome (sea level), and then during five weeks in Mexico City. Also shown are findings during an acute (decompression chamber) exposure to 8000 ft before leaving for Mexico (from the data of Scano and Venerando, 1968).

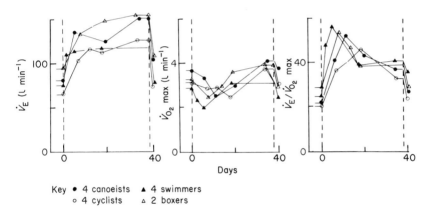

Key • 4 canoeists   ▲ 4 swimmers
     o 4 cyclists   △ 2 boxers

FIG. 32   Further measurements on 14 Olympic athletes from Italy during five weeks at Mexico City (8000 ft). Measurements made at in Rome at sea level before and after the visit are also shown. Symbols as in Table 1 (from the data of Scano and Venerando, 1968).

reached a maximum in four days and soon after very slowly decreased, reaching native resident levels after 40 years.

Extrapolating from his own and other observations West, who was one of the physiologists on Hillary's Everest expedition (West *et al.*, 1962) has calculated that Habeler and Messner's alveolar CO_2 pressures at the summit of Everest were likely to have fallen to 15 mm Hg and possibly to 10 mm Hg, i.e. to values which, even at ground level, would provoke profound cerebral vasoconstriction and loss of consciousness from cerebral

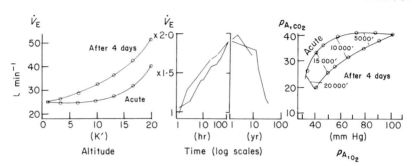

FIG. 33 Plots of the changes in ventilation seen in acute and prolonged exposures at various altitudes (adapted from Lenfant and Sullivan, 1971).

hypoxia in unadapted people. Although it is known that the renal compensation for respiratory alkalosis is the major mechanism which permits and sustains such a severe degree of hypocapnia without loss of consciousness the means by which it does so are still unresolved. In the past few years it has become clear that respiration is controlled in a complex manner as if there were independent regulators during sleep and wakefulness. This subject has been reviewed by Phillipson (1978). As sleep approaches, breathing becomes periodic but during "non-rapid eye movement" sleep it is regular and depressed causing alveolar p_{CO_2} to rise by a few mm Hg and alveolar p_{O_2} to fall by a rather greater amount. When "rapid eye movement" sleep appears breathing becomes irregular and ventilation rises. In healthy subjects at ground level the fall in alveolar and arterial p_{O_2} during sleep is of little account. However, at altitude the falls are more marked and cause disproportionately greater falls in arterial oxygen concentrations because of the shape of the oxyhaemoglobin dissociation curve. This leads to a much commoner incidence of periodic breathing, hypercapnia and other disturbances of sleep (Reite et al., 1975). These are major features of unadapted or partly adapted life at high altitudes.

Thus it seems that mild and moderate hypoxia leads to a slowly developing hyperventilation which can become extreme without causing loss of consciousness, despite the hypoxia, but this ventilatory drive fails periodically during sleep, leading to frequent bouts of transient asphyxia. These in turn cause cerebral vasodilatation and oedema which lead to headaches that aggravate the insomnia.

The Role of Haemoglobin

The model of respiratory gas exchange illustrated in Fig. 8 identifies the conductances which facilitate the flow of oxygen from air to tissue. One of these that can be "opened up" almost immediately, is blood flow which is very effective but consumes energy continuously. Over a period of a few days part of its task is taken over by another mechanism, hyperventilation, that can only be deployed gradually, presumably because of its cerebral effects. It is cheaper to run but it too uses energy continuously. During the subsequent weeks of altitude exposure, the burden of improving oxygen supply is transferred from the heart and lungs to a mechanism which requires much less energy, namely an increased concentration of haemoglobin in blood. Until recently changes in the affinity of haemoglobin for oxygen were also believed to be valuable; however this is no longer thought to be so. The role of haemoglobin in altitude adaptation, which is described below, has been discussed in detail by Lenfant and Sullivan (1971), Bartels and Baumann (1977), Garby and Meldon (1977) and Heath and Williams (1977).

The haemoglobin molecule is a globular structure which consists of four polypeptide strands, so coiled that their hydrophilic groups lie externally, giving a high aqueous solubility which is very necessary to its task. The hydrophobic groups lie internally, creating protective nests that prevent the four iron atoms within from being truly oxidized (methaemoglobin formation). Since each iron atom can bind an oxygen molecule, every haemoglobin tetramer can accept four oxygen molecules. As each one enters it changes the conformation of the haemoglobin favouring entry of the next oxygen molecule (cooperative ligand binding). The explains the sigmoid shape of the oxyhaemoglobin dissociation curve. Like Michaelis–Menten curves it can be described by a single characteristic, the p_{50} or t_{50}, which is the pressure or tension at which a population of haemoglobin molecules is half-saturated with oxygen.

In simple aqueous solution, human haemoglobin has a p_{50} of 11 mm Hg. However, the globular molecule behaves rather like a biscuit-tin lid since it can exist in two stable forms and is readily flipped from one to the other. Protons, CO_2, organic phosphates, and to a lesser extent some other ions, pull it into the taut configuration which has a lower oxygen affinity, and so higher p_{50}, than the relaxed form. Thus in whole blood at 37 °C, with a plasma pH of 7.4 and a p_{CO_2} of 40 mm Hg, human haemoglobin has a p_{50} of 26–28 mm Hg (Fig. 34).

At the onset of hyperventilation the p_{CO_2} and so the p_{50} of blood falls, (i.e. the oxyhaemoglobin dissociation curve "shifts to the left"), but quite

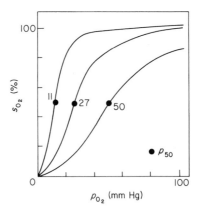

FIG. 34 Theoretical oxygen dissociation curves for haemoglobins with p_{50} values of 11, 27 and 50 mm Hg. These represent haemoglobin in simple solution, in normal blood, and in blood with a gross "shift to the right", respectively.

soon the lactacidaemia of hyperventilation partially reverses this and later the renal compensation of the respiratory alkalosis corrects it almost entirely. Hypoxia has another, independent, effect on the oxygen-affinity of haemoglobin, since it promotes glycolysis in the red cell and so increases the production of a particular organic phosphate, 2, 3-diphosphoglycerate, (2, 3-DPG), wich raises the p_{50} and thus shifts the oxyhaemoglobin curve to the right. When people are exposed to moderately high altitudes (16 000–18 000 ft), the p_{50} of their haemoglobin rises to about 30 mm Hg. This is due partly to the uncompensated hyperventilation and partly to the increase in red cell 2, 3-DPG. The increase in 2, 3-DPG reaches a peak in a couple of days and then declines somewhat as compensated hyperventilation sets in and the p_{O_2} of the circulating blood rises.

More interestingly, Hebbell et al. (1978) have argued that it has not originated as an adaptation to hypobaric hypoxia at all but as a defence against the much more commonly encountered stress of anaemic hypoxia, for which it is entirely appropriate. They believe that a leftward shift is desirable at altitude but that man, unlike the other high-altitude mammals, has not learnt to make it. They have presented some compelling evidence that this is so, taken from two patients with a rare type of haemoglobin (Hb Andrew–Minneapolis) which has a p_{50} of 17 mm Hg. Hebbell and his companions compared these patients with normal siblings and found that when they were exposed to a moderately high altitude the patients showed smaller increases in heart rate and in plasma and urinary erythropoietin levels, no decrement in aerobic capacity and no thrombocytopenia. Not surprisingly, they did hyperventilate, and did raise their red cell 2, 3-DPG levels to the same degree as the controls. West has extended this argument further and proposed that Habeler and Messner were able to reach the summit of Everest because they superimposed a marked leftward shift of the curve due to uncompensated hyperventilation upon a trivial rightward shift of "acclimatized hypoxia" in their swift ascent from a relatively low base-camp altitude. It is interesting that life-long hypoxia has very little effect on the position of the curve. Cerretelli and his colleagues have found that highland Sherpas have a p_{50} of 27.3 mm Hg, which is indistinguishable from that in sea-level man (Samaja et al., 1979). It is also worth noting that life in the womb, which has been described as "Everest in utero" depends upon foetal haemoglobin which has a p_{50} of about 15 mm Hg, suggesting that man has evolved a leftward shift to cope with intra-uterine hypoxia but can neither retain nor recall it when presented with the lack of oxygen at high altitudes.

In contrast to the speculative role of changes in p_{50}, the increase in blood haemoglobin concentration that occurs at altitude is undisputedly of the greatest adaptive value. Its existence was predicted by Paul Bert in 1878 and demonstrated first by Viault (1891). The extent of the increase depends upon the severity and duration of the exposure. Typically, at altitudes of 16 000–18 000 ft, the haemoglobin concentration rises from its sea-level value of 150 gm litre^{-1} to a concentration close to 200 g litre^{-1}, the haematocrit rises, from 45%, to 60%, and the red cell count increases, from 5 million per mm^3, to 7 million per mm^3. These changes occur because hypoxia stimulates the production of erythropoietin which induces bone marrow hyperplasia. The mechanism by which it does so is not known. Although the increases in erythropoietin formation and in iron turnover occur within a few fours of the onset of hypoxia, the new red cells do not appear in the peripheral blood until several days later. Their arrival is heralded by a rise in the reticulocyte count which eventually trebles its sea-level value. Before this true polycythaemia occurs, some increases in haemoglobin content and in haematocrit and red cell count are seen. These are due to a haemoconcentration that is associated with altitude dehydration and expulsion of haemoglobin-rich blood from the spleen. The genuine polycythaemia, which is rather more marked in immigrant than resident highlanders, persists indefinitely at altitude but regresses over a period of a few weeks on descent to sea level.

The 30% increase in oxygen-carrying capacity of blood, which is seen at 18 000 ft, ought to extend aerobic capacity by much the same amount and permit a roughly equivalent fall in cardiac output at any given load. However during recent Italian expeditions to the Himalayas, Cerretelli and his colleagues (1976) have found the observed advantages are less than those predicted. They believe that the polycythaemia, which raises the viscosity of blood and provokes rouleaux formation and sludging, causes ventilation/perfusion inequality in the lung (as reported by Haab *et al.*, 1971) and metabolism/blood flow inequality in the systemic periphery. It is easy to imagine how this could occur by plasma-skimming alone. One of the crucial observations for their argument is the finding that when men at altitude are given oxygen or brought rapidly to sea level they do not achieve the aerobic capacity expected from their well-oxygenated condition and may not even reach their previous sea-level performance. Despite this reservation there is no doubt that polycythaemia is the principal long-term adaptation to sustained hypoxia in lowland men whether they visit high altitudes or are subject to arterial hypoxaemia from heart or lung disease. It is not seen in people with tissue hypoxia arising from restricted blood flow or exaggerated metabolic demands and is not a complication of chemodec- tomy. The hypoxia must therefore be mediated via some other tissue with a high blood flow, probably in the kidneys which are the primary source of erythropoietin. The topic of oxygen-sensing in the body is a fascinating subject that has recently been discussed by Jobsis (1977).

Failures of Adaptation

When people are exposed to gradual hypoxia they undergo the sequence of adaptations of increased blood flow and uncompensated hyperventi- lation, leading to compensated hyperventilation and soon after to an increased concentration of haemoglobin in blood, and they remain "healthy" until altitudes above 18 000 ft are reached. That height probably represents the altitude limit for reasonably complete adaptation to hypoxia. Above that ceiling substantial deteriorations of performance and other pathological consequences of hypoxia are inevitable; below it these pheno- mena can largely be avoided providing the hypoxia develops slowly. However, for any ascent above 10 000 ft if the fall in pressure is too rapid or if some incidental disorder rapidly aggravates the hypoxaemia in any ascent above 10 000 ft, people will show the consequences of incomplete adaptation or of exaggerated responses to the stress. These can take many forms and can vary in severity from a temporary reduction in aerobic capa- city to acute pulmonary oedema. Although the manifestations are protean they can usually be attributed to some mixture of three particular complica- tions: excessive polycythaemia; pulmonary vasoconstriction; and cerebral oedema.

Mild forms of these conditions lasting a few days and disappearing after a few days of rest improved oxygenation are known as Acute Mountain

Sickness. They are characterized by light-headedness, headaches, paraesthesia, lassitude dyspnoea, somnolence and insomnia, and usually resolve on rest alone. More severe expressions merge imperceptibly with the major purely polycythaemic, pulmonary or cerebral forms. The former, which are seen whenever the haemoglobin concentration exceeds 230 g litre^{-1}, can present in many ways that simply depend upon the nature of the particular vessels that are obstructed, and upon any general strain that the increase in blood viscosity puts on the hypoxic heart. The pulmonary and cerebral complications have interesting physiological origins which have been touched on before.

Pulmonary vasoconstriction

In utero, the foetal lung is collapsed, its highly muscular bed of blood vessels is constricted and the right ventricle, which is as powerful as the left, drives blood at systemic pressures. At birth, the lung inflates with air, which raises the oxygen tension of the blood and so promotes relaxation of the pulmonary arterial tree and constriction of the ductus arteriosus. As a consequence the mean pressures in the right heart and pulmonary tree fall over a few days to the level that is seen in the normal adult population (15 mm Hg). Similarly, the muscularization of the ventricle and arteries regresses markedly compared to that of the corresponding systemic structures. The capacity of pulmonary arteries and arterioles to constrict in response to hypoxia, which was employed *in utero* to divert blood away from the unoxygenated lung, persists as an important means of matching regional blood flow to regional ventilation at subsegmental or even perhaps at acinar level. It normally functions at this microscopic local level, continuously optimizing blood flows through the normally unevenly ventilated lung. Because the vasoconstriction only involves a few vessels at a time this "autoregulation" has little effect on the work of the heart. However, if there is a generalized alveolar hypoxia, as on ascent to altitude, then the entire bed of small arteries constricts throwing a severe mechanical load on the heart. Initially it responds by pumping harder and so pulmonary hypertension and right ventricular hypertrophy occur. There is also a marked hypertrophy of arterial muscle due to the hypoxic and mechanical stimuli.

Hypoxia which is prolonged or more severe may lead to two serious consequences—pulmonary oedema and/or right heart failure. The pulmonary oedema is believed to arise because the vasoconstriction, which protects the lung capillaries from the hypertension and so keeps the alveoli dry, is unevenly distributed, exposing some capillaries to abnormally high perfusion pressures. Right heart failure occurs when the hypoxic right ventricle is unable to pump the flows required of it without the assistance of abnormally high right atrial filling pressures. These conditions which are of the utmost interest, both as practical problems on the mountain-side and as thoroughly documented physiological models of very common clinical conditions, are described with great clarity by Harris and Heath (1977) and by Staub (1978). Many recent studies are summarized in the proceedings of an international symposium on Hypoxia, which was held in Banff early in 1979, (Houston, 1979). It includes an annotated bibliography of high-altitude medicine which is a valuable starting-point to further reading.

Cerebral odema

The brain lies in, and occupies 90% of, the cranial cavity, blood occupies 5% and cerebrospinal fluid the remainder. If the brain swells, blood and/or cerebrospinal fluid have to be displaced, and vice versa. Hypoxia causes the brain to swell in two ways; first by depriving the sodium pumps of energy and causing cellular expansion directly (cytotoxic oedema), and secondly by hypoxic vasodilatation (vasogenic oedema). As the oedema develops it causes further vascular obstruction and hypoxia and a vicious circle develops which is aggravated by hypercapnia as the hypoventilation develops. Signs and symptoms depend upon severity and range from asymptomatic retinal haemorrhages (the eye is once more a model for the brain), to severe headaches, papilloedema, loss of consciousness and death.

Life-long Hypoxia

People whose entire lives are spent at high altitudes appear to have two developmental advantages that can only be acquired *in utero*. These are a modest increase in the diffusing capacity of the lung and a corresponding increase in the diffusing capacity of the systemic capillaries. Evidence for these and other is not especially satisfactory because in these respects it is extremely difficult to unscramble the true effects of hypoxia from those of all the other physical, behavioural and genetic forces which operate on folk in mountains. These difficulties, which seem almost insurmountable in human studies, must be appreciated if correct conclusions are to be drawn. They are well described in two essays, by Clegg *et al.* (1970) and Mazess (1975), and in the monographs prepared for the International Biological Programme by Baker (1978) and Shephard (1978). These careful studies, which are of the greatest value, suggest that life-long existence in high mountains exposes people to quite different patterns of physical activity, environmental stress and contact with pathogens, and is associated with considerable genetic and cultural isolation such that it is almost impossible to find any group of sea-level people that can act as satisfactory controls. With that reservation in mind, it seems that native highlanders are somewhat, but not markedly, beter adapted to hypoxia than are long-stay residents who have come up from below, but with the exception of the increases in pulmonary and systemic diffusing capacities, and a possibly greater increase in lung volume, their advantage stems from better developed responses that are available to everybody than to other factors which are peculiar to them. Equally, when judged by their ability to recolonize their environment after depopulation, they are as successful as lowland folk are below. Animal studies of the responses of individual organs to developmental hypoxia, e.g. the increase in alveolar surface area or the decrease in fertility, provide very useful pointers, since they can be conducted with adequate controls, but have to be interpreted with caution since they ignore all the other factors which distinguish life on high mountains from that on lowland plains.

Summary

The oxygen molecule has the curious property of being quite inert until it is ionized. By contrast, the superoxide ion so formed is a very powerful oxidant which inevitably demolishes other molecules close by. When oxygen appeared in the atmosphere, primitive life forms which had developed in an anaerobic environment evolved a series of protective enzymes to quench superoxide ions as they formed in tissues. Later other enzymes were developed which could use the destructure power of the superoxide ion in a controlled manner. In man, and in other vertebrates generally, oxygen is handled by many enzymes and used in many ways. One of these, oxidative phosphorylation, is quantitatively the most important but is also the least sensitive to oxygen lack. Others, which include several critical processes such as the manufacture and destruction of synaptic and blood-borne hormones, consume very little oxygen but are much more affected by its lack.

Hypoxia can be overcome in three ways: by turning to anaerobic processes; by increasing the oxygen affinity of existing aerobic processes; and by easing the access of oxygen to the tissues. Little is known about the first two options at present but there is some evidence that they do occur.

Most research in the Western world has concentrated on the adaptations which reduce the fall in oxygen pressure between inspired air and the sites of intracellular utilization.

These adaptations occur in three stages: first there is an immediate increase in blood flow which is accompanied by a modest hyperventilation. The latter is almost certainly restrained to limit disturbances of cerebral pH. Over the subsequent few days of exposure the hyperventilation can develop further as renal compensation for the respiratory alkalosis takes place. Later still the task of easing the carriage of oxygen is transferred from these energy-consuming processes to a much cheaper one, of increased blood haemoglobin.

These responses, if allowed time to develop, permit people to live successfully at altitudes up to 18 000 ft where the barometric pressure and thus the inspired oxygen pressure are half as great as at sea level. However, if exposures are rapid, various failures of performance or complications of exaggerated adaptation occur. Hypoxic deteriorations of cerebral function are seen at altitudes as low as 4000–5000 ft. At higher altitudes these are much more marked and consciousness is lost after a few minutes of abrupt ascent above 20 000–25 000 ft. By contrast, after some weeks of partial adaptation, people are able to survive several days climbing at 25 000–29 000 ft.

If lowlanders or returning highlanders arrive too rapidly at altitudes above 10 000 ft they can show three exaggerated adaptations: excessive polycythaemia, pulmonary hypertension leading to pulmonary oedema or right heart failure, and cerebral oedema.

People who spend their entire lives at high altitude are slightly better adapted than long-stay residents who were born at sea level. Some of this advantage is due to increased capillary densities in the heart and lungs, which can only be acquired *in utero*. There is no reliable evidence of any genetic adaptation.

Life at high altitudes exposes people to many other stresses than hypoxia, of which cold is the most obvious and limiting, and also to some advantages such as the absence of particular pathogens. Because there are so many unavoidable distinctions between life at low and at high altitudes, studies of the effects of life-long hypoxia must be interpreted with great caution. One of the most telling arguments against highlanders being at a disadvantage is the observation that they appear to be able to recolonize their environment, after depopulation, as rapidly as can lowlanders below.

References

Acker, H., Fidone, S., Pallot, D. Eyzaguirre, C. Lubbers, D.W. and Torrance, R.W. (1977). (Ed.) "Chemoreception in the Carotid Body", 1–296. Springer–Verlag, Berlin.

Arborelius M. Jr (1966). *Scand. J. Resp. Dis.* **62**, 105.

Arias–Stella, J. (1969). *In* "Proceedings of the American Association of Pathologists and Bacteriologists", San Francisco, California.

Arias–Stella, J. and Valcorcel, J. (1973). *Pathologia et Microbiologia* **39**, 292.

Armstrong, H.G. (1949). "Principles and Practice of Aviation Medicine", Williams and Wilkins, Baltimore.

Baker, P.T. (1978). (Ed.) "The Biology of High-altitude Peoples", Cambridge University Press, London.

Barbashova, Z.I. (1967). *Int. J. Biometeorol.* **11**, 243.

Barbashova, Z.I. (1969). *Int. J. Biometeorol.* **13**, 211.

Barcroft, J. (1925). Cambridge University Press, London.

Barcroft, J., Binger, C.A., Bock, A.V., Doggart, J.H., Forbes, H.S., Harrop, G., Meakins, J.C. and Redfield, A.C. (1922). *Philosoph. Transact. (Roy. Soc.).* **13**, 351.

Bartels, H. and Baumann, R. (1977). *In* "International Review of Physiology, Respiratory Physiology", II, 14–107. University Park Press, Baltimore.

Bert, P. (1877). "La Pression Barométrique", (Hithcock, M.A. and Hitchcock, F.A., translators). College Book Company, Columbus, Ohio.

Billings, C.E. (1974). *Aerospace Med.* **45**, 128.

Bransby, M. and Denison, D.M. (1976). *J. Physiol.* **256**, 101*P*.

Byford, G.H. (1965). *Proc. Roy. Soc. B* **161**, 421.

Cerretelli, P. (1976). *J. Appl. Physiol.* **40**, 658.

Clegg, J.E., Harrison, G.A. and Baker, P.T. (1970). *Human Biol.* **42**, 486.

Cohen, P.J. (1972). *Anesthesiology* **37**, 148.

Davis, J.N. and Carlsson, A. (1973). *J. Neurochem.* **20**, 913.

De Castro, F. (1928). *Trab. Lab. Invest. Biol. Univ. Madr.* **25**, 331.

Denison, D.M. (1968). Ph.D. Thesis, University of London.

Denison, D.M. (1970), "A non-invasive technique of cardiopulmonary assessment in man", Agardograph 143. Paris. North Atlantic Treaty Organisation.

Denison, D.M. (1981). *In* "The Scientific Basis of Respiratory Medicine", (Eds G. Cumming and G. Scadding), Heinemann, London.

Denison, D.M., Ledwith, F. and Poulton, E.C. (1966). *Aerospace Med.* **37**, 1010.

Denison, D.M., Ernsting, J., Tonkins, W.J. and Cresswell, A.W. (1968). *Nature (London)* **218**, 1110.

Dill, D.B. (1968). *J. Am. Med. Assoc.* **205**, 123.

Downes, J. and Lambertsen, C.J. (1966). *J. Appl. Physiol.* **21**, 447.

Dugall, P. and Fiset, P.E. (1950). *J. Aviat. Med.* 362.

Duling, B.R. and Berne, R.M. (1970). *Circ. Res.* **27**, 669.

Edwards, A.W.T. and Learoyd, B.M. (1967). *Resp. Physiol.* **2**, 36.

Edwards, R.H.T. and Clode, M. (1970). *Clin. Sci.* **38**, 269.

Ernest, J.T. (1977). *Exp. Eye Res.* **24**, 271.

Ernest, J.T. and Krill, A.E. (1971). *Invest. Opthalmol.* **10**, 323.

Ernsting, J., Gedye, J.L. and McHardy, G.J.R. (1960). *J. Physiol (Lond.)* **155**, 3*P*.

Ernsting, J., Gedye, J.L. and McHardy, G.J.R. (1962). *In* "Human Problems of Supersonic and Hypersonic Flight", (Eds A. Buchanan–Barbour and H.E. Whittingham), 359. Pergamon, Oxford.

Ernsting, J. (1962). "Some effects of brief profound hypoxia upon the central nervous system" *In* "Selective Vulnerability of the brain in Hypoxaemia", (Eds W.H. McMenemy and U.P. Schade), Blackwell, Oxford.

Ernsting, J. (1963). *J. Physiol. (Lond.)* **169**, 292

Ernsting, J. (1965). *In* "A Textbook of Aviation Physiology", (Ed. J.A. Gillies), 270. Pergamon, Oxford.

Ernsting, J., Byford, G.H., Denison, D.M. and Fryer, D.I. (1973). Flying Personnel Research Committee Report 1324, Ministry of Defence (Air), London.

Ernsting, J. (1975). *In* "Oxygen Measurements in Biology and Medicine", (Eds J.P. Payne and D.W. Hill, 231. Butterworths, London.

Ernsting, J. (1978). *Aviat. Space Environ. Med.* **49**, 495.

Von Euler, U.S. and Liljestrand, G. (1946). *Acta Physiol. Scand.* **12**, 301

Farhi, L.E. (1965). *In* "Handbook of Physiology", Section 3: Respiration. Vol. 1, 873. American Physiological Society, Washington, D.C.

Fishman, R.A. (1975). *New Eng. J. Med.* **293**, 706.

Fishman, A.P. (1976). *Circ. Res.* **38**, 221.

Fishman, A.P. (1977). *Adv. Exp. Med. Biol.* **78**, 143.

Forster, R.E. and Crandall, E.D. (1976). *Ann. Rev. Physiol.* **38**, 69.

Frisby, J.P., Barrett, R.F., and Thornton, J.A. (1973). *Aerospace Med.* **44**, 523.

Garby, L. and Meldon, J. (1977). "The Respiratory Functions of the Blood", 1–282. Plenum Press, New York and London.

Garcia, A.C., Lai, Y.L., Attebery, B.A. and Brown, E.B. (1971). *Respir. Physiol.* **12**, 371.

Gastaut, H. and Meyer, J.S. (1961). "Cerebral Anoxia and the Electroencephalogram", Charles C. Thomas, Springfield, Illinois.

Gellhorn, E. (1937). *Am. J. Psychiatry.* **93**, 1413.

Gellhorn, E. (1951). *Am. J. Physiol.* **164**, 748.

Gellhorn, E. (1953). *E. E. G. Clin. Neurophysiol.* **5**, 401.

Gellhorn, E. and Heymans, C. (1948). *J. Neurophysiol.* **11**, 261.

Gold, R.E. and Kulak, L.L. (1972). *Aerospace Med.* **43**, 180.

Goldie, E.A.G. (1942). Unpublished data cited in: Whiteside, T.C.D. (1957), "The Problems of Vision in Flight at High Altitude", Pergamon, Oxford.

Grant, B.J.B., Jones, H.A., Davies, E.E. and Hughes, J.M.B. (1976). *J. Appl. Physiol.* **40**, 216.

Green, I.D. (1965). *In* "A Textbook of Aviation Physiology", (Ed. J.A. Gillies), 264. Pergamon, Oxford.

Haab, P., Held, D.R., Ernst, H. and Farhi, L.E. (1969). *J. Appl. Physiol.* **26**, 77.

Habeler, P. (1979). "Everest—Impossible Victory", Arlington Books, London.

Harris, P. and Heath, D. (1977). "The Human Pulmonary Circulation", Churchill Livingstone, London.

Heath, D. and Williams, D.R. (1977). "Man at High Altitude: the Pathophysiology of Acclimatisation and Adaptation", 1–292. Churchill-Livingstone, Edinburgh.

Hebbel, R.P., Eaton, J.W., Kronenberg, S., Zanjani, E.D., Moore, L.A. and Berger, E.M. (1978). *J. Clin. Invest.* **62**, 593.

Hellstrom, S. (1977). *In* "Chemoreception in the Carotid Body", (Eds H. Acker, S. Fidore, D. Pallot, C. Eyzaguirre, D.W. Lubbers and R.W. Torrance), 122. Springer–Verlag, Berlin.

Hermansen, L. and Saltin, B. (1967). *In* "Exercise at Altitude", (Ed. R. Margaria) 48. Excerpta Medica Foundation, Milan.

Heymans, C. and Bouckaert, J.J. (1930). *J. Physiol. (Lond.)* **69**, 254.

Heymans, J.F. and Heymans, C. (1927). *Arch. Int. Pharmacodyn.* **33**, 273.

Hornbein, T.F. and Roos, A. (1962). *J. Appl. Physiol.* **17**, 239.

Houston, C.S., (1947). *J. Aviation Med.* **19**, 237.

Houston, C.S. (Ed.) (1979). "Hypoxia Symposium 79". The Arctic Institute of North America, (University of Calgary), Calgary.

Howe, A. and Neil, E. (1971). *In* "Handbook of Sensory Physiology", Vol. III. Enteroreceptors. (Ed. E. Neil), 47. Springer–Verlag, Berlin.

Hughes, J.M.B. (1977). *In* "International Review of Physyiology", II, Vol. 14, (Ed. J.G. Widdicombe), 135.

Ikegami, H. (1967). *Int. J. Biometeor.* **11**, 139.

Jobsis, F.F. (1977). *Adv. Exp. Med. Biol.* **78**, 3.

Johansson, G. and Jansson, G. (1965). *Scand. J. Psychol.* **142**, 295.

Johnston, M. (1965). *Life Sci.* **4**, 2215.

Kellog, R.H. (1978). *Resp. Physiol.* **34**, 1.

Kelman, G.R. and Crow, T.J. (1969). *Aerospace. Med.* **40**, 981.

Krnjevic, K. (1974). *Physiol. Rev.* **54**, 418.

Krogh, A. (1919). *J. Physiol. (Lond.)* **52**, 209.

Lassen, N.A., Ingvar, D.H. and Skinhoj, E., (1978). *Sci. Am.* **239**, 62.

Lauweryns, J.M. and Goddeeris, P. (1975). *Am. Rev. Resp. Dis.* **111**, 469.

Ledwith, F. (1970). *Ergonomics* **13**, 465.

Lenfant, C. and Sullivan, K. (1971). *New Eng. J. Med.* **284**, 1298.

Liere, E.J. van and Stickney, J.C. (1963). "Hypoxia", University of Chicago Press, Chicago.

Luft, U.C., Clamann, H.G. and Adler, H.F. (1949). *J. Appl. Physiol.* **2**, 37.

Lugliani, R. Whipp, B.J., Winter, B., Tanaka, K.R. and Wasserman, K. (1971). *New Eng. J. Med.* **285**, 1112.

Magdelano, F. (1968). *Revistade Aeronautico Y Astronautica* **333**, 581.

Margaria, R. (Ed.), (1967), "Exercise at Altitude", Excerpta Medica, Milan.

Mason, H.S. (1957). *Adv. Enzymol.* **19**, 79.

Mazess, R.B. (1975). *In* "Physiological Anthropology", (Ed. E. Damon), 167. Oxford university Press.

McDonald, R. and Adler, F.H. (1939). *Archiv. Opthalmol. (Chicago).* **22**, 980.

McFarland, R.A. (1941). *Am. J. Psychiat.* **97**, 858.

McFarland, R.A. (1946). "Human Factors in Air Transport Design", 69. McGraw-Hill, New York.

McFarland, R.A. (1963). *Ergonomics* **6**, 338.

McFarland, R.A., (1971). *Aerospace Med.* **12**, 1303.

McFarland, R.A. and Evans, J.N. (1939). *Am. J. Physiol.* **127**, 37.

Meyer, J.S., Gotoh, F., Ebihara, S. and Tomita, M. (1965). *Neurol. (Minneap).* **15**, 892.

Moosavi, H. Smith, P. and Heath, D. (1973). *Thorax* **28**, 279.

Paintal, A.S., (Ed.), (1976). "Morphology and Mechanisms of Chemoreceptors", V. Patel Chest Institute, Delhi.

Phillipson, E.A. (1978). *Ann. Rev. Physiol.* **40**, 133.

Pierson, W.R. (1967). *Aerospace Med.* **38**, 993.

Piiper, J. Dejours, P., Haab, P. and Rahn, H. (1971). *Respir. Physiol.* **13**, 292.

Piiper, J. and Scheid, P. (1971). *Ann. Rev. Physiol.* **33**, 131.

Pretorius, H.A. (1970). *Aerosp. Med.* **41**, 560.
Purves, M.J. (1972). "The Physiology of the Cerebral Circulation", 1–420. Cambridge University Press, London.
Purves, M.J. (Ed.), (1975). "The Periphereal Arterial Chemoreceptors", 1–492. Cambridge University Press, London.
Rahn, H. (1963). *In* "The Regulation of Human Respiration", (Eds D.C. Cunningham and B.B. Lloyd, Blackwell, Oxford.
Ravenhill, T.H. (1913). *J. Trop. Med. Hyg.* **20**, 313.
Reeves, J.T., Jokl, P. and Cohn, J.E. (1965). *Am. J. Physiol.* **92**, 813.
Reeves, J.T., Grover, R.F. and Cohn, J.E. (1967). *J. Appl. Physiol.* **22**, 546.
Reeves, J.T., McMurtry, I.F., Wagner, W.W. and Grover, R.F. (1979). *In* "Hypoxia 79", (Ed. C.S. Houston), Arctic Institute of North Amercia, (University of Calgary), Calgary.
Rennie, D. (1976). *Lancet* **ii**, 1177.
Reite, M., Jackson, D., Cahoon, R.L. and Weil, J.V. (1975). *Electroencephalog. Clin. Neurophysiol.* **38**, 463.
Reynafarje, D.B. (1971). *Arch. Inst. Biologia Andina* **4**, 1.
Reynolds, W.J. and Milhorn, H.T. (1973). *J. Appl. Physiol.* **35**, 187.
Saldana, M.J., Salem, L.E. and Travezan, R. (1973). *Human Pathol.* **4**, 251.
Samaja, M., Veicsteinas, A. and Cerretelli, P. (1979). *J. Appl. Physiol.* **37**, 337.
Scano, A. and Venerando, A. (1968). "Studi sulla acclimatazione degli atleti italiani a Citta del Messico", Comitato Olimpico Nazionale Italiano, Rome.
Shinebourne, E.A. and Anderson, R. (1978). *In* "Price's Textbook of the Practice of Medicine", (Ed. R. Bodley-Scott), 711. Oxford University Press, Oxford.
Shephard, R.J. (1978). "Human Physiological Work Capacity", International Biological Programme, Vol. 15. Cambridge University Press, London.
Siesjo, B.K. (1978). "Brain Energy Metabolism", Wiley, Chichester.
Staub, N.C. (Ed.) (1978). "Lung Biology in Health and Disease", Vol. 7. Marcel Dekker, New York.
Staub, N.C. and Schulz, E.L. (1968). *Respir. Physiol.* **5**, 371.
Stockley, R.A. and Bishop, J.M. (1977). *Clin. Sci. Molec. Med.* **53**, 93.
Torrance, R.W. (Ed.) (1968). "Arterial Chemoreceptors", Blackwell, Oxford.
Turner, H.S., Hoffler, G.W., Billings, C.E. and Bason, R. (1969). *Aerospace. Med.* **40**, 971.
Verna, A. (1975). *In* "The Periphereal Arterial Chemoreceptors", (Ed. M.J. Purves), 75. Cambridge University Press, London.
Viault, F. (1891). *Séances Acad. Sci. Paris* **112**, 295.
Wade, J.B., Larson, C.P. Jr., Hickey, R.F., Ehrenfield, W.K. and Severinghaus, J.W. (1970). *New Eng. J. Med.* **282**, 823.
Wagner, P.D. (1977). *Physiol. Rev.* **57**, 257.
Walsh, E.G. (1952). *J. Physiol. (Lond.)* **118**, 500.
Ward, M. (1975). "Mountain Medicine", Crosby Lockwood Staples, London.
Weihe, W. (Ed.) (1973). "The Physiological Effects of High Altitudes", Pergamon, Oxford.
West, J.B. (1977a). "Pulmonary Pathophysiology—the Essentials", Williams and Wilkins, Baltimore.
West, J.B. (1977b). "Ventilation/blood flow and gas exchange", Blackwell, London.
West, J.B. (1979). "Pulmonary Physiology—the Essentials", Blackwell, London.
West, J.B., Lahiri, S., Gill, M.B., Milledge, J.S., Pugh, L.G.C.E. and Ward, M.P. (1962). *J. Appl. Physiol.* **17**, 617.

Wulff, L.Y., Braden, I.A., Shillito, F.H. and Tomashefski, J.F. (1968). Physiological relating to high terrestial altitudes: A Bibliography", Publ 3. Ohio State University Libraries, Columbus, Ohio.

CHAPTER 6
Underwater Physiology

D.H. Elliott

Introduction

A brief history of the development of diving provides a practical basis for the discussion of the different approaches that man has used in his attempts to live and work in an unnatural environment. References to accounts of breath-hold diving occur as early as the fifth century B.C., and archaeological evidence of artefacts that could only be the product of a diving industry of some 3000–5000 years B.C. have been reviewed elsewhere (Davis, 1962; Dugan, 1965). Breath-hold diving continues to be used commercially in Japan and Korea where shell-fishing is done by diving women who make regular breath-hold dives as deep as 40 m (Rahn and Yokoyama, 1965). Such dives are limited in duration and many writers in the sixteenth and seventeenth centuries proposed methods to extend man's capabilities under water.

The majority of these early ideas for breathing under water were physiologically unsound. Some attempted to supply air by a tube from the surface but such attempts to provide air at no more than atmospheric pressure to a submerged man whose chest is compressed by hydrostatic pressure, were doomed to failure. The development of the diving bell during these early years provided the diver with an entrapped volume of air at ambient pressure which he could breathe while under the surface. Edmund Halley, the Astronomer Royal in 1690 spent some 90 min at 60 ft (18 m) in the river Thames in a diving bell, the atmosphere of which was refreshed by weighted barrels of air that were sent down from the surface.

Although preceded by prototypes designed by others, it was Augustus Siebe who is generally acknowledged as inventing the first breathing apparatus to be produced commercially that would give the diver a potentially unlimited duration of endurance. A helmet, not much more than an inverted bucket with windows, was sealed to a leather diving dress. Air was supplied under pressure from the surface to the helmet by a pump and the exhaust gas escaped under the hem of the waist-length jacket. He later modified this equipment so that the air supplied to the helmet was vented through a valve and the diver was enclosed in an airtight suit. With such technology man was brought for the first time into confrontation with many physiological and medical problems not then fully appreciated. In the meanwhile, the development of diving bells and caissons was acknowledged as being the cause of respiratory difficulties, rheumatic pains, paralysis and an alarming number of fatalities.

The apparent inability of a diver to perform his tasks well while breathing compressed air at depths around 200 ft (70 m) was first thought to be due to psychological inadequacies. It was then realized that there was a build-up of carbon dioxide in the helmet of the standard diver and that this required a very large minute volume of gas from the surface which, when compressed to ambient pressure, would be sufficient to flush the helmet and reduce the carbon dioxide to tolerable limits (Haldane and Priestley, 1935). Since the design of pumps to deliver the necessary volumes of compressed air was not adequate for the greater depths, the Royal Navy used a system of partial recirculation through a soda-lime canister. Thus a smaller constant flow of compressed-air would supplement itself, by venturi action, with some supplementary breathing gas drawn out of the helmet, scrubbed of carbon dioxide and returned. Many years ago the Royal Navy's daily issue of rum was credited by some for the seemingly better performance of British divers at depth on compressed air in comparison with the divers of the US Navy who were not provided with a rum ration. A hypothesis of adaptation to narcosis in the British divers could be based upon the similarities of the mechanism of the narcotic actions of alchohol, the volatile anaesthetics and the so-called inert gases. However, such a hypothesis does not acknowledge the probable differences of carbon dioxide levels in the different compressed-air equipments then used by the two navies (Zinkowski, 1971).

The potential value of helium as an alternative to nitrogen as the oxygen diluent for use in deep diving was based upon its lesser solubility than nitrogen in the tissues (Sayers et al., 1925). That the behavioural abnormalities of divers during deep compressed-air diving was due to the narcotic effects of the raised partial pressures of nitrogen (Behnke et al., 1935) led Behnke and Yarbrough (1938) to infer that the operational importance of helium in diving was the improved mental condition of the divers rather than any potential saving of decompression time. Increasingly greater depths were achieved over the next decade using oxy–helium mixtures. The concept of saturation diving, although suggested by Behnke (1942), was not developed until experimental studies were initiated by Bond in 1957 (Penzias and Goodman, 1973). This led to the US Navy series of Sealab dives and so to modern commercial saturation diving in which men may spend 30 days at depth before beginning decompression.

In the meanwhile there had been considerable progress in other techniques used at relatively shallow depths. Different designs of closed-circuit oxygen rebreathing apparatus were produced by various navies but it was not until the sinking of the battleships H.M.S. Valiant and H.M.S. Queen Elizabeth in Alexandria harbour by underwater swimmers of the Italian Navy, that the merits of such diving techniques were realized. The inherent dangers due to oxygen toxicity emphasize the importance of physiological principles in the design of closed and semi-closed underwater breathing apparatus.

However, the development by Cousteau and Gagnan of the demand-valve pressure regulator enabled a diver to breathe at ambient pressure from compressed-air cylinders carried by him, freeing him from dependency upon a surface supply and providing him with relative mobility (Cousteau and Dumas, 1953; Dugan, 1956). This development led to the world-wide use of self-contained diving apparatus for both commercial and recreational purposes.

Some Laws of Physics

Depth and pressure

A general review of the physical laws of nature as they effect man underwater is a necessary basis for the understanding of underwater physiology. By definition man at sea level is exposed to one atmosphere of pressure (1 bar) and there is a relationship between the increasing pressure to which he is exposed while descending through the water to the depth of that water, a relationship which is conventionally simplified so that each additional atmosphere is represented by 10 m (33 ft) of sea water. This pressure relationship is affected by the specific gravity of the water: the accepted value is 1.025 but a value of 1.0197 gives the convenient equivalent of 10 m = 1 bar. For most purposes this is sufficiently accurate. The correct unit of pressure is Newton per square metre ($N\,m^{-2}$) but, since this unit is rather small for convenience, the bar has been adapted for general use and represents $10^5\,N\,m^{-2}$ (Table 1).

Pressure gauges usually indicate zero at sea level and a reading from such a gauge is referred to as "gauge pressure". More correctly, quoted pressure should include the atmospheric pressure at sea level; this is then termed "absolute". Thus, at a depth of 10 m depending on water salinity and temperature, sea water exerts a pressure of 2 bar but this may be recorded as, for instance, 1 kgf cm^{-2} gauge.

Water or gauge pressure is measured with a Bourdon tube. This type of gauge measures the pressure inside a coiled tube relative to the outside which is generally maintained at atmospheric. A pneumofathometer is a gauge at the surface which registers the pressure at depth through a hollow tube. The weight of the gas column in the tube subtracts some pressure

TABLE 1   Equivalent pressures

1	bar	b
1000	millibars	mb
1×10^5	Newtons per square metre	$N\,m^{-2}$
100	kiloPascals	kPa
0.9869	standard physical atmospheres	atm
1.0197	technical atmospheres	at
750.06	Torricelli	torr
10	metres sea water (sp. gr. 1.0197)	m
9.95	metres sea water (sp. gr. 1.025)	m
32.25	feet sea water (sp. gr. 1.025)	fsw
14.51	pounds per square inch	psi

from the gauge reading at the surface and thus, for instance, if the reading at the surface is 83 m of sea water (fsw) the actual water depth with an air-filled pneumofathometer would be approx 85 m, whereas if filled with 10% oxygen/90% helium it would be approx 84 m.

Gas pressure and volume

The ideal gas law is

$$PV = nRT$$

Where P is the absolute pressure, V is the volume of gas, n is the number of moles of gas, R is the universal gas constant and T is the absolute temperature. From this law are derived some simpler laws which are more directly applicable to underwater physiology.

Boyle's Law states that "at a constant temperature, the volume of a given mass of gas is inversely proportional to the pressure." In other words, 1 litre of gas at sea level decreases to a volume of 0.25 litre at 30 m (4 bar). Another special application of the gas laws explains the phenomenon that when the internal pressure of a compression chamber is being raised, there is an associated increase of temperature. Conversely during decompression the interior of the chamber becomes cool and misting of the chamber atmosphere may occur. The pressure reading from a cylinder of compressed air left in the sun will be significantly greater than when the same cylinder is immersed in cold water.

Partial pressure

Dalton's Law states "in a mixture of gases the pressure exerted by each gas is the same as it would exert if it alone occupied the same volume; the total pressure of a gas is thus the sum of the partial pressures of the component gases." In other words, the partial pressure of gas (p_{gas}) is equal to that fraction of the total pressure (p_B) which can be ascribed to that gas.

Example: What is the partial pressure of oxygen in the air at 50 m?

$$p_{O_2} = 0.21 \times \frac{50 + 10}{10}$$
$$= 1.26 \text{ bar}$$

Example: Calculate the percentage of oxygen required to produce a partial pressure of oxygen of 0.21 bar at a depth of 500 m.

$$O_2 = \frac{p_{O_2}}{p_B} \times 100$$

$$= \frac{0.21 \times 100}{(\frac{500}{10} + 10)}$$

$$= 0.41\%$$

(In practice a slightly greater oxygen percentage would be used.)

Buoyancy

Archimedes' Principle states that an object is buoyed up in water by a force equal to the weight of water that it displaces. The buoyancy of a diver is affected by such factors as body composition, whether or not a wet or dry suit is worn, the salinity of the water, the continuously variable lung volume associated with respiration from underwater breathing apparatus and the slight yet significant lightening of the weight of compressed gas bottles during a dive. The problem of varying buoyancy is discussed further in relation to the breath-hold diver for whom it is particularly important.

The Effects of Hydrostatic Pressure on Cells and Organisms

Pressure effects are likely to be of special importance as a limiting factor in deep dives (Fenn, 1969) and are likely to confine man to exposures of less than 100 bar. The physiological basis of such effects is best understood by a study of the biological effects of much greater pressures. All aspects of physiological and biochemical processes to a greater or lesser extent are affected by pressure which acts by changes in molecular volume (Macdonald, 1975). A gas such as argon increases in density in proportion to pressure up to some 3000 bar, but its viscosity only increases six-fold. In contrast, water molecules are already closely packed and while density increases by only 4% with an increase of pressure to 1000 bar, the viscosity of water increases but shows some seemingly anomalous changes compatible with its other molecular properties. The thermodynamic basis of the changes which hydrostatic pressure induces at the molecular level and which lead to acceleration and retardation of chemical reactions, is reviewed by Macdonald (1975). Among the equilibria which may be disturbed by an increase of ambient pressure is ionization. Substances which are only weakly ionized under normal conditions show an increased ionization. Thus the disassociation of magnesium salts in the 200 bar range has been invoked to account for an inhibition of the synthesis of poly-phenylalanine. Those salts which are already almost completely dis-associated will clearly not increase the extent of their disassociation at high pressures, yet the zone of water around individual ions may show changes in properties, as for instance in the case of chloride ions and the red cell membrane. The formation of ionic bonds involves a volume increase and is thus opposed by raised environmental pressure. Other examples at the molecular level include the liquefaction of gels. Cytoplasmic streaming and cleavage are diminished to the extent that, at 400 bar, tadpoles develop abnormally. Examples of reversible depolymerization of proteins include sickle-cell haemoglobin and a number of enzymes. The effect of high pressure on enzymic reactions is complex and is a subject of great academic importance. There may seem to be no direct application of the conclusions of such studies to effects upon man but it should be inferred that there are many important but subtle effects which may be difficult to identify. The responses of various tissues that have been studied show that tissue behav-iour is significantly altered at pressures greater than those to which man is likely to be exposed. The responses of nerves to pressure are many and varied but may be more relevant to human diving. The flux of sodium and potassium in unstimulated nerve is increased. Pressure reverses the action of some anaesthetic agents (Roth *et al.*, 1976), a phenomenon which has been applied in reverse to diving research by using narcotic agents to counteract the effects of pressure. The normal functioning of inexcitable membranes is also probably affected by numerous indirect routes (Macdonald, 1975). The bulk compression of a cell membrane bilayer can indirectly affect the functioning of ion channels and enzymes but Macdonald (1977) suggests that very high pressures would be required to "freeze" the bulk of the bilayer in such cell membranes even though specific regions of the mem-branes able to influence some functional component may themselves be affected. Pressure has been shown also to increase the affinity of haemo-globin for oxygen up to 100 bar. Hydrostatic pressure decreases the solubility of gases, causing their partial pressures to be increased by approximately 14% at 100 bar.

The behaviour of whole animals was studied first by Regnard (1891) who showed how pressure immobilized small crustacea. This work and later studies by others are also reviewed by Macdonald (1975) but of more interest to diving physiology are the comparative studies of mammalian physiology. The Weddell seal would appear to be able to descend to 500 m in a matter of minutes, presumably without serious tremor or convulsions, and it may have some instructive adaptation to its environment. By using hydraulically compressed fluorocarbon-breathing mice Kylstra *et al.* (1967) were able to avoid any effects due to respiratory gases, and rapid compression elicited tremors and spasms in the limbs from 50 bar and convulsions beyond 80 bar. Lundgren and Örnhagen (1976) demonstrated that pressure would paralyse the respiratory muscles of liquid-breathing mice at 125 bar. Such studies indicate a boundary unlikely to be crossed by human experimentation.

Pressures in the range of 50–350 bar can produce profound excitatory and inhibitory changes in the electrical characteristics of neuronal membranes and it must be anticipated that the net effect which pressure exerts on the function of an integrated nervous system will be found to be even more complex. Excitatory or inhibitory effects on single cell function can lead to the opposite effects on the output of the integrated neural network.

The Effects of Pressure upon the Gas-Containing Spaces

Notwithstanding the foregoing account of extreme pressures, the tissues of man under relatively moderate pressures behave like a fluid and can be considered as virtually incompressible. Very small differences in the compressibility of watery and fatty tissues do occur and hypothetically may account for the high pressure neurological syndrome to be discussed later. For practical purposes in diving the relative incompressibility of the tissues needs to be contrasted with the volume changes of any gas in accordance with Boyle's Law. If contained in a space which is surrounded by flexible tissues, such as the alimentary canal, the volume changes of the contained gas on compression and decompression are of no great significance. Other spaces such as the oro-pharynx are readily supplied with additional gas during compression thus avoiding any reduction of volume. If no additional compressed gas can be admitted to a contained space there must be a diminution in volume of that space during descent. This isolation of an air cavity may occur if the natural opening of, for instance, a sinus into the upper respiratory tract becomes blocked. If such a space has relatively rigid walls, as have the paranasal sinuses, it may fill with transudate or possibly blood.

Different effects may result from reduction of environmental pressure. If the contained gas is in a cavity communicating with the respiratory tract it should be free to escape during expansion. If venting is blocked and expanding gas is retained within relatively rigid walls such as the paranasal sinuses, a carious tooth or within a dental filling, the relative over-pressure will cause local pain until the excess gas is absorbed by the blood stream. However, if gas is trapped in a relatively unsupported space, such as a portion of a lung, the expansion may rupture alveoli with some potentially very serious consequences.

Barotrauma

Barotrauma may be defined as tissue damage which occurs as a direct result of a change of barometric pressure. There are a number of anatomical sites where barotrauma can occur but to illustrate the underlying principles of applied physiology it is sufficient to select only two examples, one resulting from compression and one from decompression.

Barotrauma of the middle ear

One of the commonest problems in diving is an inability during descent to equalise the pressure within the middle ear through the Eustachian tube with the increasing pressure outside. The naso-pharangeal end of the Eustachian tube does not open spontaneously during pressurisation and a diver must perform this consciously by some manoeuvre such as swallowing. The Valsalva manoeuvre is commonly used by beginners but, with experience, most divers are able to use the superior pharangeal constrictor muscle and the muscles of the floor of the mouth with the glottis closed in order to "clear the ears" (the Frenzel manoeuvre). The closed tube can be opened against a pharangeal over-pressure of some 10 mm Hg (1.3 kPa), but at some 90 mm Hg (12 kPa) not only is it no longer possible to open the pharyngeal end of the tube but the ear will have become painful. Associated with the pain is some conductive hearing loss. If compression continues without middle ear pressure being equalized, the pain will become severe. Tinnitus and vertigo may occur. Transient vertigo may also result from an unequal pressurization of right and left middle ears, perhaps if just one ear is "sticky". This "alternobaric vertigo" (Lundgren, 1965, 1973) is potentially hazardous for a diver, who has no proprioceptive input while buoyant, and who may have no visual points of reference in turbid or dark water. At a differential pressure of somewhere between 100 and 500 mm Hg (12 and 66 kPa) the tympanic membrane will rupture (Coles, 1964). Thus a diver whose normal ability to "clear his ears" may be temporarily impaired by some local inflammation may be able to descend only a few metres before damage occurs. Forced Valsalva manoeuvre when the tube is blocked can raise cerebrospinal fluid pressure and lead to rupture of the round window with a leakage of perilymph and damage to the inner ear. The pathological and medical aspects of the consequences of such damage are described elsewhere (for instance: Edmonds *et al.* 1973).

Pulmonary barotrauma

In contrast to the middle ear, which is affected by pressure and volume changes during compression, the lungs are more commonly damaged by the pressure–volume changes of decompression. An inability of the lungs to vent the expanding intrapulmonary gases during ascent leads to pulmonary barotrauma. However, expansion to a potentially dangerous volume implies that the respiratory gas must exceed a volume greater than is normal. Thus a breath-hold diver during his ascent is merely re-expanding his lungs to their original volume and he is not exposed to this hazard. To achieve an expanded volume greater than that held previously, the diver must have inhaled gas at a greater environmental pressure. While the diver is approaching the surface at a constant velocity, the rate of expansion of gas is increasing exponentially. A differential pressure of some 80 mm Hg (11 kPa) on arrival at the surface may be more than can be retained by some alveoli and thus "burst lung" has occurred following exposure to less than 2 m depth. Certainly the risk of lung damage is greatest if the diver holds his breath during the decompression. Yet lung rupture can occur even when the diver has been careful to exhale continuously during his ascent and when it is known, by previous X-ray and medical examination, that he has no gross pathological lesion that might impede the venting of alveolar gas. It has been suggested that, in these circumstances, dynamic airway collapse may occur (Schaefer *et al.*, 1958), resulting in closure of the lumen of some airways at the moment that gas flow should be greatest. Studies of air-trapping during head-out immersion suggest that collapse is less likely to occur when the chest is held at a relatively large volume (Dahlback and Lundgren, 1973). Another possible explanation is a diminished pulmonary compliance in susceptible individuals (Colebatch *et al.*, 1977).

The escape of gas along the vascular sheaths to the hilum of the lung, into the pleural sacs or into the pulmonary circulation can lead to very serious medical consequences, the features of which are described elsewhere (For instance: Kidd and Elliott, 1975).

Gas Uptake and Elimination

Until the early nineteenth century, when development of tunnel and bridge building techniques enabled man to work below water-level at a raised environmental pressure, there was no such clinical condition as acute decompression sickness. Decompression sickness is one of the dysbaric illnesses which follow a change of environmental pressure but, unlike pulmonary barotrauma and air embolism in which, as mentioned already, the origin of intravascular gas is alveolar, decompression sickness is the illness which follows a reduction of environmental pressure sufficient to cause the formation of bubbles from the gases which have been dissolved in the tissues. Long before this illness became a hazard for man, bubbles had been seen in animals on decompression to the low pressures of simulated altitude (Boyle, 1670). Among the principal factors determining whether a particular decompression will cause bubbles in an individual are those that determine the total amount of gas dissolved in the tissues before the start of decompression and those that determine the rate of elimination of these gases. For many years the presence of intravascular bubbles was thought to be synonymous with acute decompression sickness until they were detected by the use of ultrasonic techniques in divers without any subsequent manifestations (Evans *et al.*, 1972). Thus, another aspect of the aetiology of decompression sickness relates to the response of the individual to the presence of bubbles and, in particular, the pathophysiological events that occur at the blood–gas interface.

Gas uptake

In any prediction of safe methods of returning to normal atmospheric pressure without incurring significant bubble formation, the first concept to be considered is the rate of gas uptake by the various tissues of the body at raised environmental pressure. The type of gas mixture breathed is a significant factor in such calculations but, for illustrative purposes, exposure to compressed air will be considered rather than to oxy–helium and other gases which provide a more challenging problem.

The inspired partial pressure of nitrogen is a function of depth and at 50 m (165 ft), for instance, the p_{N_2} is 4.74 bar. However, the alveolar p_{N_2} equals that of the inspired air only when the respiratory quotient is 1.0. Henry's Law states that "when a gas and a liquid, or when two liquids, are exposed to each other, gas will diffuse from one to the other until their partial pressures become equal". The volume of gas dissolved in a liquid is directly proportional to the partial pressure of the dissolved gas, the constant of portionality being the solubility coefficient. Thus the volume of nitrogen dissolved in any fluid will increase linearly with increasing p_{N_2} and at any given p_{N_2}, tissues with a greater solubility coefficient, such as fat, will contain a greater quantity of dissolved nitrogen.

During compression, alveolar and thus arterial p_{N_2} increase and the tissues begin to take up nitrogen. This process will continue until all the tissues have achieved equilibrium at the new environmental pressure. The rate of uptake by the tissues varies widely and depends upon such factors as the dynamics of tissue perfusion. The time-constants of uptake are, of course, independent of the pressure and thus the time to equilibrium (which is erroneously called "saturation" by divers) is theoretically a constant and is considered to be about six to eight hours. In practice, the duration of a dive is often well within the time required for complete equilibrium and thus there is a need to predict the volume of gas dissolved in each tissue, be its uptake "fast" or "slow". Many dives do not remain at a constant depth and thus for each change of depth the driving pressure of nitrogen will alter the tendency of gas uptake towards some new state of equilibrium.

Provided that it is remembered that the different "tissues" referred to are not defined upon a regional or even histological basis but represent convenient mathematical concepts, the uptake of gases in diving can be predicted and is analogous to the uptake of gaseous anaesthetic agents. Gas uptake in diving has been well discussed in the literature (for instance Behnke, 1969; Workman and Bornmann, 1975; Hempleman, 1975; Buhlmann, 1975; Hills, 1975). However, when the uptake is considered in more detail at the level of a single hypothetical tissue, the mathematical discussions, such as of radial diffusion during perfusion through a cylindrical unit of tissue, can be placed one step further away from the empirical realities of decompression tables.

Decompression and gas elimination

Not only did Haldane (Boycott et al., 1908) propose the concept of fast and slow tissues for gas uptake during a dive, but he also defined the principles of safe decompression and supported these by animal experimentation. Haldane and his co-workers estimated that a 70 kg man would take up about one litre of nitrogen for every atmosphere of increased air pressure, which is about 70% more nitrogen than an equal amount of blood would dissolve. Thus at any depth, with the weight of blood in man approximately 6.5% of his body weight, the amount of nitrogen held in solution at equilibrium in the tissues would be about 26 times as great as the amount held in the blood alone. On decompression, to avoid the evolution of bubbles in the tissues from the dissolved nitrogen, the reduction of hydrostatic pressure was never permitted by Haldane to exceed a computed nitrogen pressure in the tissues of more than twice the ambient pressure. This 2 : 1 ratio assumed equilibration of the tissues to ambient pressure of the depth, rather than the actual nitrogen partial pressure. Thus the absolute pressure of the maximum depth was halved in order to determine the depth of the first stop during decompression, a stop being defined as the pause by a diver at constant depth for a given period of time in accordance with the predetermined programme of gas elimination. Decompression schedules based upon Haldane's 2 : 1 ratio have not provided adequate decompression in practice but did form the basis for the majority of decompression calculations made during the subsequent decades.

When calculating gas elimination, the analogy between inert gas uptake and gaseous anaesthetic uptake must be abandoned. The elimination of anaesthetic gases may be considered as the inverse of uptake but, even at one atmosphere, nitrogen elimination during oxygen breathing is complex and not multi-exponential (Barnard et al., 1973). The probability of a phase change from dissolved to gaseous inert gas during the early part of decompression from a dive adds a new complexity to the mathematical model of elimination. The concepts of gas micronuclei, bubble generation, growth and decay in relation to the calculation of decompression tables are detailed in the literature already cited.

Intravascular bubbles

The mechanical effects of the bubble are sufficient to explain many of the manifestations of acute decompression sickness, but not all. It is now recognized that bubbles can produce a number of indirect effects in the body as a result of surface activity at the blood–gas interface. The consequences, which have been reviewed elsewhere (Philp, 1974; Elliott and Hallenbeck, 1975), are due to a 40–100 Å (4–10 nm) layer of electrokinetic forces at the blood–gas interface that tends to orientate the exposed globular proteins in such a way that their hydrophilic groups are in the blood while their non-polar groups protrude into the gaseous phase. The resultant disruption of the native secondary and tertiary configuration of the proteins promotes clumping of cells, denaturation of lipoproteins and other events leading to activation of the complement system, kinins and other smooth muscle activating factors (Fig. 1). Of particular interest is the adherence of numerous platelets to the blood–bubble interface (Fig. 2), with the formation of pseudopodia (Philp et al., 1972). A detailed account of the pathophysiology of decompression sickness may be found elsewhere (Elliott and Hallenbeck, 1975). For practical purposes, a case of decompression sickness may be due to failure of the individual to adhere to established decompression tables, a failure of less reliable tables to provide a safe elimination of gases, even when the table has been followed, or an unusual response of the decompressed diver to a "safe" decompression in accordance with normal biological variation.

Counter-diffusion

It has recently been recognized that, if more than one inert gas is used during a dive, a phenomenon called counter-diffusion may occur. When the partial pressures of inert gases are altered at a constant environmental pressure their independent behaviours depend upon their individual solubi-

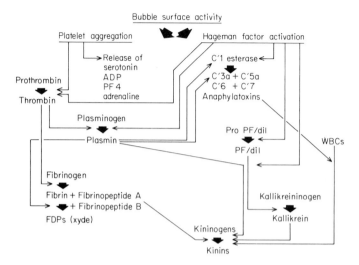

FIG. 1 Some consequences of platelet aggregation and Hageman factor activation initiated by activity at the blood–bubble interface (after Hallenbeck and Elliott, 1975).

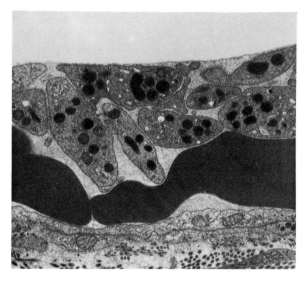

FIG. 2 Electron micrograph (× 15 000) of platelets adhering to a blood–bubble interface and to each other. The bubble is the clear area at the top of the photo. Deformed red cells can be seen at the bottom of the photo (from Philp *et al.*, 1972, with permission).

lities and diffusivities. Superficial counter-diffusion represents diffusion in opposite directions across the skin of two inert gases, one dissolved in the tissues and the other being in the ambient gaseous environment. Deep counter-diffusion represents the diffusion of two inert gases dissolved in the tissues of the body when, for instance, oxy–helium as a breathing gas is replaced by compressed air. In certain circumstances the superficial form is known to produce venous gas emboli without a change of environmental pressure. It follows that any unsuitable change of respiratory gas or the gas passing over the skin may become especially hazardous during a decompression (Lambertsen and Idicula, 1975). In practice, however, such situations can be avoided.

Pulmonary Function

The inadequacies of pulmonary function at depth not only provide a limiting factor for the rate at which physical work can be performed under water, but also may be one of the principal factors in deciding the maximum depth at which man can work in relative safety.

Ventilation may be inadequate at depth as a result of breathing dense gas, the increased work of breathing leading to some degree of hypoventilation, even without the added external resistances of breathing apparatus. There is also an effect upon ventilation due to the vertical gradient of hydrostatic pressure to which the chest is necessarily exposed in the water. This effect is commonly excluded in the laboratory since many studies of human respiration at depth are performed in the dry chamber. A full discussion of pulmonary dysfunction at depth is well presented by Lanphier (1975).

Dead space

In addition to physiological dead space of some 0.3 litres, depending upon body build of the individual and varying with his tidal volume, there is the added dead space of any equipment used. In a helmet with a high rate of free-flow across the diver's face, the amount of exhaled carbon dioxide re-inspired could be negligible but often is not. A mouthpiece or oro-nasal mask is commonly used to supply inspiratory gas only when "demanded" by a relative pressure drop caused by initiating an inspiration. Compared with "free-flow" such "demand breathing" is economical of total gas but adds 0.1 litres or more to the effective dead space. In some military forms of closed and semi-closed breathing apparatus, to-and-fro breathing through a wide-bore corrugated hose, 20–30 cm long, increases functional dead space excessively. Lanphier (1963) estimates that the p_{A,CO_2} is increased by a further 6 mm Hg (0.7 kPa) for each 0.5 litre of added dead space. Any attempt to reduce the lumen of the airway and thus the volume of the dead space in breathing apparatus, be it a snorkel for the breath-hold diver swimming at the surface or a mouthpiece for the conventional diver, must not do so at the expense of increasing by too much the flow resistance as a result of narrowing the diameter.

Ventilatory capacity

The limitation of maximum ventilatory capacity at depth may be related to the increase of gas density. Whether the change of density is a result of a change of pressure or a change of gas mixture, such as by inclusion of sulphur hexafluoride, makes no significant difference to the resulting reduction of maximum voluntary ventilation (MVV; Maio and Farhi, 1967). The MVV is a useful index of the effects of pressure and is roughly proportional to the reciprocal of the square root of density (Fig. 3). Slightly more optimistic predictions have been based upon values for MVV with the use of selected gases at 38.5 bar chosen to provide a density equivalent to that of helium at 150 bar (Wright *et al.*, 1972) but such extrapolation may not be valid (Spaur *et al.*, 1977). While the MVV may be used to predict the maximum *sustainable* pulmonary ventilation, this can represent a work-

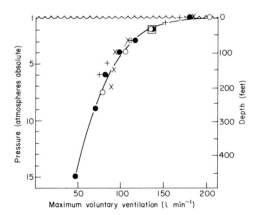

FIG. 3   Maximum voluntary ventilation breathing compressed air at various pressures. The solid circles represent mean values in litres per minute derived from various other studies (after Lanphier, 1975).

load requiring as little as 60% of the MVV at 6 bar (Fagraeus and Linnarrson, 1973). If effort is required to overcome the effects of external resistance due to underwater breathing apparatus, greater values of MVV can be achieved but only for a short while before respiratory fatigue sets in (Vorosmarti and Lanphier, 1975). Thus the response of an individual to an emergency may well be limited by his ability to maintain adequate ventilation. The measurement of maximal work at great depths has demonstrated a wide range of individual variability but in only a few such studies have the subjects been exercised to a level they could not maintain.

Work of breathing

The limitation of ventilatory capacity during exercise at depth is, to a large extent, due to the work of breathing against airway resistance and added external resistance. Little gas flow, if any, in these circumstances is linear in the larger airways and Rohrer's equation, (1915) though useful, has limitations (Lanphier, 1975). Turbulent flow will occur in the airways when the Reynolds number exceeds about 2000, the Reynolds number being the product of average gas velocity, diameter of the airway, gas density and reciprocal of viscosity. In expiration when external breathing resistance is high, and in inspiration the limitation of respiratory flow may be considered as imposed by the power of the respiratory musculature to overcome respiratory resistance. However, a concept of dynamic airway compression was proposed to account for the limitations in the respiratory tract to expiratory flow. In certain circumstances the greater the increase in voluntary expiratory effort the greater the reduction of the lumen of an airway (Mead et al., 1967).

Thus, the one way in which an individual could increase his $\dot{V}_E$ max when limited by maximum expiratory flow would be to respire at higher lung volume. Greater inspiratory work would be required against elastic forces, but the airway diameter would be increased and also, if the inspiratory phase were shortened, expiration could be slower.

Hydrostatic gradient

At the normal end-expiratory position of rest in air the pressure in all the alveoli will be equal to that of the environmental atmosphere. Totally submerged in water, however, the pressure in all the alveoli will again be equal but will correspond to that at some particular level in the pressure gradient of the water surrounding the thorax. This balance point will not be the same in different attitudes of the body. The factors are many (Lanphier,

1975) and the effects complex. For instance, there is a tendency for blood to pool in the great veins and pulmonary vessels rather than the lower portions of the body. The most comfortable relative pressure at which gas is delivered to a diver may be also measured in terms of a horizontal level in the water surrounding the chest. This eupnoeic pressure was found to correspond to the suprasternal notch in all positions except vertical when it was higher, though lowered by hypopnoea (Paton and Sand, 1977). In practice the demand-valve supplies gas at a pressure equivalent to its own depth in the water and its position alters with the varying attitudes of the diver thus supplying gas at greater or lesser pressures than are comfortable.

Lanphier (1975) draws attention to the need, in any explanation of the effects of hydrostatic imbalance on the work and energy cost of breathing, also to take into account the unusual respiratory patterns adopted by divers in such circumstances. The role of the larynx in the regulation of gas flow may be particularly relevant to breathing with a "negative-pressure balance".

The unique circumstances of head-out immersion do not simulate a situation in which the diver commonly finds himself and are discussed further in relation to breath-hold diving.

Alveolar gas exchange

The hydrostatic gradient which tends to elevate the diaphragm and increase the volume of blood in the great veins and pulmonary vessels can alter the ventilation–perfusion ratio, $\dot{V}_A/\dot{Q}$ relationship, of the lungs with a consequential alveolar–arterial difference in oxygen tension, and possibly some carbon dioxide retention. It has been suggested by Miller and Winsborough (1973) that the alterations of the ventilation perfusion ratio at depth are due mostly to an impairment of total lung ventilation. It seems less probable that regional alterations in the distribution of inspiratory gas flow might be attributable to a greater quantity of the more dense gas flowing to those protions of the lungs with a lower airway resistance.

The increased density of respiratory gas could affect the exchange of oxygen and carbon dioxide in the alveoli. The tidal flushing of gas may be relatively unaffected but the alveolar mixing of tidal gas with residual gas is dependent upon diffusion (Cumming et al., 1966) which may therefore be slowed at depth. An extreme form of diffusion limitation is found in liquid breathing, a concept discussed later. A consequence of breathing very dense gas is the limitation to oxygen and carbon dioxide exchange and thus a possible hypoxia with normal p_{I,O_2} (Chouteau, 1972). However this effect appears to be related not to the density of the mixed gas but to the binary diffusion coefficient of oxygen in helium (Lanphier, 1972). It has also been shown that a pause on inspiration before exhaling increases the efficiency of alveolar gas exchange. The role of convection in the alveoli, the effect of the heart beat in inducing gas mixing and the importance of longitudinal dispersion have been reviewed by Johnson and Van Liew (1974).

Oxygen consumption

Swimming and other forms of underwater activity require the expenditure of much energy, and measurements of sustained levels of oxygen consumption (Fig. 4) have maxima of 2.5 litre min^{-1} (Lanphier, 1954), 3 litres min^{-1} (Donald and Davidson, 1954) and 3.4 litres min^{-1} (Morrison et al., 1973). While the sustained $\dot{V}_{O_2}$ max is normally limited by the cardiovascular transport of oxygen, the limitations under water are ventilatory (Miller et al., 1973). Any associated increase of p_{I, O_2} causes only a small increase of the $\dot{V}_{O_2}$ max (Fagreus et al., 1973; Deroanne et al., 1973).

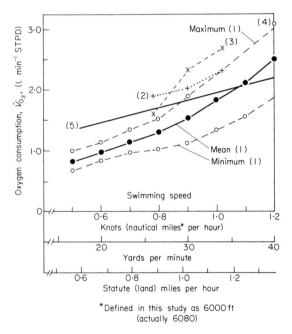

FIG. 4 Oxygen consumption in underwater swimming at various speeds. Several studies are summarised in this graph, the minimum (1), mean (1) and maximum (1) representing the 1954 study by Lanphier (after Lanphier, 1975).

Any study of the transients of respiratory function and control in exercise is not easy even at atmospheric pressure. The effects of the additional variables imposed at raised environmental pressure make the study more complex. There are also the practical problems of studying the actual work done by a submerged man, not only the consequences of the hydrostatic gradient already discussed, but also the influence of water viscosity upon limb and chest movement. Thus in a study of exercise physiology there can be no simple comparison between pulmonary function at the surface and at depth.

Carbon dioxide

At depth, the alveolar and arterial tensions of carbon dioxide should remain around normal values, independent of ambient pressure. A tendency to an increased p_{A,CO_2} in diving may result from both extrinsic factors, such as an elevated p_{I,CO_2}, and intrinsic factors causing inadequate alveolar ventilation.

Inspired partial pressure and carbon dioxide

The normal percentage of carbon dioxide in normal air (0.04%) is of no great significance even at the maximum pressure at which compressed air is normally breathed (6 bar; p_{I,CO_2} 0.0024 bar). At greater depths artificial gas mixtures are prepared, usually using helium as the diluent for oxygen, and these should contain no carbon dioxide. Thus an elevated p_{I,CO_2} is likely to be a result of an inadequate elimination of carbon dioxide from the expired gas which is to be rebreathed.

In some forms of underwater breathing apparatus there is, as has been mentioned, an unavoidable increase in functional dead space and some increase in tidal volume is required in compensation. Most commonly, the exhaled gas is discharged to the water ("open-circuit") and fresh gas is inspired as required ("demand-breathing"). In traditional ("helmet") diving dress there is a constant flow of fresh gas into the helmet, the diver breathes from the gas within the helmet and a constant venting of gas from the helmet takes with it the exhaled carbon dioxide. In such a helmet there must always be some elevation of p_{I,CO_2} and, depending upon the diver's ventilatory response, some carbon dioxide retention. In order to keep this within tolerable limits the mass of gas supplied to the diver must be increased in proportion to his increased ambient pressure thus maintaining the effective volumes needed for flushing. Similarly, divers without breathing apparatus but at depth in a chamber of compressed air require, in the absence of a scrubber, to have their chamber flushed through periodically to get rid of accumulating carbon dioxide. As represented by equivalent volumes measured at the surface, the quantities of gas required for flushing the chamber are increased proportional to the pressure within the chamber. This subject is dealt with in detail by Lanphier (1975) and practical guidance is given in appropriate diving manuals (e.g. US Navy, 1973).

In order to conserve valuable helium many systems incorporate rebreathing circuits with an alkaline neutralizer as an absorbent to remove the carbon dioxide. For instance, the US Navy uses "Baralyme", a mixture of barium, calcium and potassium hydroxides and the Royal Navy uses "soda lime" a mixture of sodium, calcium and potassium hydroxides. Not only do these chemical agents have a limited absorbing capacity but, used as granules packed into canisters, their efficiency depends on the flow patterns through them. They are exothermic and their effectiveness increases with humidity. In closed-circuit and semi-closed-circuit underwater breathing apparatus their use must be associated with the relatively easy task of planning the dive within the effective duration of the carbon dioxide scrubber, usually a matter of some hours. In chamber life-support systems a build-up of p_{I, CO_2} is avoided by an internal or external scrubbing system with provision for changing the absorbent from time to time, thus permitting extended habitation at raised environmental pressure.

Control of ventilation

Inadequate pulmonary ventilation during normal diving may lead to some degree of hypoxia and hypercapnia. In that these divers should be capable of better ventilation under these circumstances, assuming p_{I, CO_2} to be negligible, this condition has been called "carbon dioxide retention" (Lanphier, 1975). In the presence of an increased p_{I, O_2} the response to p_{A, CO_2} is decreased, but this accounts for not more than 25% of the p_{A, CO_2} elevation. Also, the ventilatory response to an increase in p_{A, CO_2} is diminished in the presence of increased breathing resistance.

Some divers demonstrate an inadequate respiratory response to exertion and inspiratory carbon dioxide even at the surface, and have marked elevation of p_{A, CO_2} and p_{A, CO_2} with only moderate work. That this may be an adaptive response is discussed by Schaefer (1975). Other divers, trying to conserve precious gas when scuba diving on demand intentionally hypoventilate at depth by prolonging the inspiratory and expiratory pauses ("skip-breathing").

The tendency to retain carbon dioxide during exercise is found in a proportion of healthy men but is especially associated with extensive diving experience. There have been many studies of these so-called "carbon dioxide retainers" and these have been discussed by Bradley *et al.* (1971), Schaefer (1975) and Florio, Butt and Morrison (1979). Although a diver is motivated to continue working even after his p_{A,CO_2} exceeds 40 mm Hg (5 kPa) he is not only more likely to suffer from carbon dioxide intoxication with dyspnoea, tachycardia, headache, dizziness and light headedness, confusion and possibly unconsciousness, but is also more susceptible to oxygen poisoning and nitrogen narcosis, both of which are discussed later. When investigating an underwater accident, the possibility of a carbon dioxide seizure should be considered and may be distinguished from an oxygen seizure by its preponderance of tonic activity (Stein, 1955). Besides a detailed review of these effects of carbon dioxide, Lanphier (1975) also discusses the chemical control of respiration at pressure and the relationship between the respiratory stimulus and the response in terms of the output of respiratory work. It is also necessary to consider a hypothesis that hydrostatic pressure *per se* may affect the threshold of carbon dioxide sensitivity in the chemical control of respiration (Saltzman *et al.*, 1971) based on their observation that there was high correlation between p_{A,CO_2} and ambient pressure while no correlation was found between p_{A,CO_2} and gas density. However there appears to be no elevation of p_{A,CO_2} at rest even at simulated depths around 50 bar and it was concluded by Spaur *et al.* (1977) that dyspnoea at great depths is of mechanical, not chemical, origin.

Breath-Hold Diving

Perhaps the commonest form of exposure to raised environmental pressure is the technically simple technique of breath-hold diving. Certainly it was the earliest form of diving and it continues to be used by amateur divers for sport, by professional divers for selected tasks of short duration relatively close to the surface and by both amateurs and professionals during their basic training. Although apparently simple, for neither the snorkel nor the facemask are themselves essential, the physiological consequences of breath-holding while swimming underwater are complex. The study of breath-hold diving is of considerable interest to respiratory physiologists as it demonstrates the affect of additional variables upon otherwise normal physiological mechanisms (Rahn, 1965).

The natural buoyancy of the human body is to a large extent dependent upon the volume of air in the thorax. With a deep breath of air the vast majority of individuals are positively buoyant at the surface. To hold the lungs at their vital capacity and then to descend to some 5 m (15 ft) is to reduce their volume by as much as one-third. At some relatively shallow depth every breath-hold diver becomes negatively buoyant and becomes increasingly so the deeper he goes. Thus the first lesson that the breath-hold diver must learn is that he must not rely upon natural flotation because he would not return to the surface. To a certain extent the diver's buoyancy will also be affected by the suit that he may be wearing for thermal protection but, for practical purposes, each ascent will require a conscious effort such as swimming, pushing off from the sea-bed or ascending an anchor chain.

Effects of immersion

As has already been mentioned, the vertical pressure gradient of the water in which the diver is partially or completely immersed has some profound effects upon his pulmonary and cardiovascular state. Before a breath-hold dive the diver is probably floating prone on the surface with his face immersed, breathing through a J-shaped snorkel tube. However, the effects of immersion are more easily studied when they are considered in relation to the vertical attitude. In these circumstances the alveolar gas pressure must be atmospheric yet the hydrostatic pressure at each level of the chest is increasingly greater than atmospheric in proportion to its depth. The net effect of this pressure gradient on the chest in the vertical position is approximately equivalent to some 20 cm of water, an effect referred to as "negative pressure breathing". The pressure upon the chest, together with its effect on the abdomen, reduces the expiratory reserve volume by some 70% and this is associated with a reduction in residual volume, probably due to an increased pooling of some 0.5 litres of blood in the great veins and pulmonary vessels and increased cardiac output. At the same time there is an increase of right atrial pressure as great as 16 mm Hg (2.1 kPa). As a result of the hydrostatic pressure gradient on the chest there is a 60% increase of respiratory work.

Diving bradycardia

A bradycardia occurs as a cutaneous reflex to immersion of the face and appears to be independent of both physical exercise and ambient pressure, though greater in water colder than 15 °C. The heart rate begins to diminish immediately and achieves some 70% of the pre-immersion rate in half a minute. The role of the cold-receptors of the face, the hypothesis that bradycardia is an indirect response of the baroreceptors to a decrease of intrathoracic pressure and increased venous return and the role of the partial pressure of oxygen are reviewed by Hong (1976).

Alveolar gas exchange

The changes of intrapulmonary pressures during the descent and subsequent ascent of a breath-hold dive imply some differences of alveolar gas exchange from those that occur during breath-holding at the surface. An additional factor is the hyperventilation of many divers before they take a final deep breath and descend. The alveolar concentration of nitrogen at depth is slightly greater than at the surface, even though, at its elevated partial pressure, some nitrogen must have been absorbed by the blood. This is attributed to a much faster rate of diffusion of oxygen and carbon dioxide from the alveoli into the blood plasma. The raised partial pressure of oxygen in alveolar gas at depth is sufficient to maintain a continuous diffusion of oxygen into the blood for as long as the diver is on the bottom. In the meanwhile carbon dioxide is not being transferred to the alveoli but is being retained in the blood to the extent that the p_{A, CO_2} will exceed that of mixed venous blood. The increasing p_{A, CO_2} will be interpreted by the diver as the signal to return to surface. As he ascends the p_{A, CO_2} decreases progressively, reducing the oxygen diffusion gradient. Indeed a reversal of the normal direction of oxygen transfer following a prolonged stay at depth has been described (Lanphier and Rahn, 1963). The elimination of the retained carbon dioxide continues after the return to the surface.

If hyperventilation before a breath-hold dive is excessive the changes of gas tensions are aggravated to the extent that loss of consciousness under water may be followed by drowning (Craig, 1961). The elimination of carbon dioxide from the body by pre-dive hyperventilation reduces p_{A, CO_2} to a level at which breath-holding time is greatly prolonged before the "breaking point" is reached.

Because hyperventilation cannot increase the oxygen content of blood significantly, the rate of p_{a, O_2} reduction is unchanged. The breath-hold breaking point can be extended by both physical activity and will-power. Thus oxygen consumption possibly increased by exercise, continues to deplete the oxygen content and, when the diver returns to the surface the p_{a, O_2} can fall to hypoxic levels and he may lose consciousness.

Liquid Breathing

Provided that gas exchange can be maintained within relatively normal limits, the potential advantages of breathing an incompressible liquid are sufficiently great to have justified considerable research. Oxygenated physiological saline solution has been used as a breathing medium at depth, and other animal studies have used fluorocarbon fluids which are more easily oxygenated. Carbon dioxide retention was a limiting factor but this was ameliorated if tris-hydroxymethyl.aminomethane ("Tham") was added as a buffer.

Saline lavage of the lung of a patient with an alveolar proteinosis and of a healthy volunteer have demonstrated the feasibility of this approach and though it may be some time before such techniques are applied, there is no reason to doubt the speculation that liquid breathing could, for instance, provide the basis for human escape through the water from submarines or seabed habitats at very great depths. A detailed review of liquid breathing and of respiration by the use of artificial gills is available elsewhere (Kylstra, 1975).

Nitrogen Narcosis

Although the intoxicating effects of breathing compressed air had been noted by Junot as early as 1835 it was not until 100 years later that this effect was attributed to the specific narcotic action of a raised partial pressure of nitrogen (Behnke et al., 1935). The manifestations of nitrogen narcosis are proportionate to the partial pressure of the inspired nitrogen and, subject to individual variability, begin to be noticed during descent from around 30 m (4 bar). Nevertheless, mental impairment due to nitrogen certainly occurs at less than 3 bar and effects have been reported at 2 bar (Poulton et al., 1964). Narcosis increases to the extent that early reports described a "semi-loss of consciousness" at depths as great as 350 ft (11.6 bar). The signs and symptoms are similar to those of alcoholic intoxication and there may be an impairment of a diver's ability to recognize and cope with a diving emergency when it occurs. Down to some 50 m (6 bar) the individual may find only that concentration is difficult and that there may be a slightly impaired degree of neuromuscular coordination. Before codes of safe diving practice advised against compressed-air diving deeper than some 50 m, experience demonstrated that only very few divers could accomplish useful work at depths greater than 90 m (10 bar). Although compressed air has been breathed at depths as great as 600 ft (180 m; 19 bar) in submarine escape procedures, the duration of this exposure was deliberately kept very short and thus was within the latent period of onset of narcosis. Perhaps the deepest recorded experience is that of Goodman (1963) who describes the glassy appearance of the diver's eyes at 462 ft

(144 m; 15 bar) as suggesting those of the "firmly plastered drinker" and adds that after some 45 s the simple task of assembling pegs had deteriorated to mere fumbling. "Bending forward ever more closely over his 'precious' pegboard, with intermittent bursts of inappropriate laughter and hearty, self-satisfied chuckling, the subject has, after 90 s of air breathing, effectively retreated into a private world."

Quantification of the effects of compressed-air intoxication upon diver performance are numerous. The psychometric tests used differ considerably as do the interpretations of such tests and conclusions reached.

Besides the variability due to differing experimental conditions, other factors also make a contribution to the impaired performance including carbon dioxide synergism (Hesser *et al.*, 1971), possibly associated with hyperoxic potentiation, the psychological effects of immersion in a unique environment and the physical effects of water viscosity and cold. The consequences of nitrogen narcosis and these other factors upon underwater performance are discussed later.

Mechanism

Narcosis leading to anaesthesia is associated with many of the so-called inert gases. The narcotic potency of gaseous and liquid anaesthetics have been related to their lipid solubility, partition coefficients and molecular weight, adsorption coefficients, thermodynamic activity and the formation of clathrates. Thus, xenon is an anaesthetic agent at normal atmospheric pressure, krypton causes dizziness and argon is about twice as narcotic as nitrogen, whereas helium is not significantly narcotic and neon is intermediate between helium and nitrogen (Fig. 5). Hydrogen is not a member of the inert gas series but, in accordance with its lipid solubility, falls between neon and nitrogen as a narcotic agent.

The extensive history of the unravelling of the mechanisms of narcosis and anaesthesia has been reviewed by Bennett (1975). The concept of a blockage of ion exchange across axon cell membrane due to the absorption of inert gas molecules is generally accepted. Pharmacological studies of the prevention of narcosis with agents such as Frenquel (alpha-4-piperidyl benzhydrol hydrochloride) and lithium carbonate tend to support the importance of the water–lipid interface of the cell membrane and cell membrane permeability to cations. Of particular interest is the reversal of anaesthesia by high hydrostatic pressures (Lever *et al.*, 1971) possibly related to the concept of the "free volume" of the lipid cell membrane. Inert gas narcosis will thus occur when the concentration of dissolved inert gas causes the "free volume" in the lipid phase to exceed some critical value (Bennett, 1975).

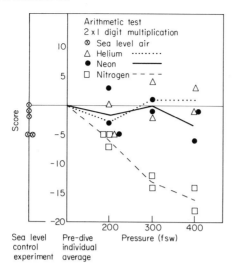

FIG. 5   Decrement in score from arithmetic test at various depths breathing oxy–helium, oxy–neon or oxy–nitrogen (after Bennett, 1975).

Oxygen Toxicity

Oxygen is a toxic gas and, for practical purposes, two types of toxicity need to be considered: pulmonary (Smith, 1899) and neurological (Bert, 1878). The first clinical effect to be described was that of retrolental fibroplasia in neonates. However, there are many other target areas affected by increased partial pressures of oxygen.

At partial pressures greater than normal, the effects of oxygen include a decrease in blood flow to the heart (Lambertsen, 1965) and the brain (Kety and Schmidt, 1948). The evidence for oxygen-induced haemolysis is not so strong and is discussed by Wood (1975).

In continuous exposures to oxygen partial pressures of less than 2 bar the lungs are the first to be affected, though no detectable change occurs at tensions less than 0.5 bar (Clark and Lambertsen, 1971 a), but at partial pressures greater than 3 bar the neurological complications dominate.

Pulmonary toxicity

The first subjective sensation of the onset of pulmonary oxygen toxicity is a dry retrosternal irritation usually associated with coughing. If permitted to progress the tracheal symptoms become worse, inspiration is painful and the frequent coughing becomes associated with dyspnoea (Clark and Lambertsen, 1971 b). Progressive pulmonary oxygen toxicity is correlated with a diminution of vital capacity. Associated symptoms include nausea, vomiting, dizziness, tunnel vision and hearing difficulties. In due course pulmonary oedema may lead to death. On reduction of the oxygen partial pressure some manifestations diminish rapidly though vital capacity may not return to normal for some days.

The damage is proportional both to the partial pressure of oxygen and the duration of exposure, there being a latent period before changes become manifest (Fig. 6) Besides the partial pressure effect, breathing 100% oxygen can itself, by absorption from the terminal airways, lead to atelectasis. The lung changes of pulmonary toxicity may be described as occurring in two phases. During the first phase there is an acute exudation leading to alveolar oedema, intra-alveolar haemorrhage and destruction of endothelial and type I alveolar epithelial cells. This phase is reversible. Subsequently, there is proliferation, involving interstitial fibrosis and hyperplasia of type II alveolar epithelial cells, which is more slowly reversible and may result in some permanent scarring.

Though high-pressure oxygen leads to adrenal hypertrophy and hypophysectomy is protective (Bean and Smith, 1953) the role of the endocrine system in oxygen poisoning is not fully understood.

Oedema produced by exposure to oxygen is not wholly due to haemodynamic factors nor to structural damage of the alveolar septum. It is thought to be due also to an upset in the balance of hydrostatic forces across the septum caused by the direct inactivation of pre-existing surfactant and by impaired or absent surfactant production (Shields, 1977).

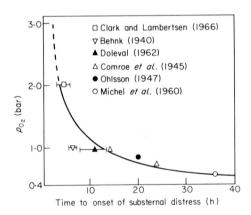

FIG. 6   Time to onset of pulmonary symptoms in man breathing various partial pressures of oxygen (after Wood, 1975).

Prevention of pulmonary toxicity

In an attempt to quantify the cumulative effects of oxygen, the partial pressure of which often varies with time, and thus to be able to predict a safe limit of oxygen exposure for a given dive, the concept of "Unit Pulmonary Toxicity Dose" (UPTD) was introduced (Wright, 1972). The UPTD is defined as being equal to one minute of breathing a gas mixture containing a partial pressure of 1.0 bar oxygen. The UPTD can be calculated for an exposure to any given p_{I,O_2} and the UPTDs for successive p_{I,O_2} levels summed. For greater safety no additional allowance is made for the beneficial effects of intermittently reducing oxygen levels towards normal as, for instance, with periods of air breathing during exposure to pure oxygen. The calculation of total UPTD from published tables is relatively simple and some upper limits of exposure, which are necessarily relatively arbitary, have been recommended.

Oxygen neurotoxicity

At tensions of more than 1.8 bar the consequences of oxygen toxicity are neurological. Though oxygen convulsions are rarely preceded by aura, it is recognized that dizziness, restlessness, disturbances of vision, incoordination, parasthesiae of the arms or legs and twitching of the muscles around the mouth or eyes represent early warnings of oxygen toxicity. The convulsive episode begins with a tonic convulsion and apnoea during which consciousness is lost, followed by a clonic phase. Within a minute or two consciousness is regained but confusion and headache may persist. Apart from the absence of cyanosis due to the high preconvulsive oxygen tension, the fit is essentially identical to grand mal epilepsy as are the electroenephalographic traces.

The threshold tension of oxygen is very variable. Not only are there considerable differences between individuals but the susceptibility of any one person may change from one day to the next (Donald, 1947 a, b). For this reason the practice in some countries of submitting diver candidates to an oxygen tolerance test has no logical basis. Tolerance to oxygen is increased at rest and in a dry chamber, particularly if there are interludes of air breathing, and partial pressures as great as 3 bar may be used therapeutically for durations of several hours.

With exercise and when immersed, the threshold is significantly diminished. Divers swimming while breathing 100% oxygen are, for safety, limited to 8 m depth (p_{I,O_2}, 1.8 bar) and breathing oxygen-rich mixtures to the depth giving a similar oxygen partial pressure. Carbon dioxide reduces the vasoconstrictive effects of oxygen and is probably a synergistic factor in these variations.

Mechanisms

While it would seem that an increased permeability of neuronal membrane to sodium and potassium is the immediate cause of central nervous system hyperexcitability, the mechanism by which oxygen produces this effect is not fully understood. An early hypothesis proposed the oxidation of SH-coenzymes essential in brain pyruvate metabolism. Another hypothesis suggests that the cell membrane is attacked; the lipids by oxidation of the unsaturated fatty acids and the proteins by oxidation of the sulphydryl groups. The role of a gamma-aminobutyric acid (GABA) pathway as a mechanism for maintaining brain ATP levels in preventing seizures has been discussed by Wood (1975). Brain GABA levels decrease prior to a convulsion, this decrease being specific among the brain amino acids. The decrease is reversible, as is oxygen toxicity, and the administration of GABA before an exposure to oxygen is protective. However, this hypothesis does not exclude the possible role of other neurotransmitters in oxygen neurotoxicity.

Current studies suggest that exposure to very high pressures of pure oxygen may have some practical value provided the exposure remains within the latent period of onset of convulsions (Burgess, 1977) but limited experience in man suggests that the predicted anaesthetic action of oxygen (Paton, 1967) might become manifest if breathed for a minute or two at a partial pressure as low as 5 bar.

High-Pressure Neurological Syndrome

During deep diving trials in 1964 it was noticed that divers had coarse tremors, dizziness, nausea and vomiting during bounce dives to 800 ft (243 m) with a compression rate of 100 ft min^{-1} (30 m min^{-1}). Psychometric tests revealed a decrement of manual dexterity (Bennett, 1975). Explanations based upon the possibility of inner ear dysequilibrium and the heat of compression were postulated since it was predicted, from its lipid solubility and other biophysical characteristics, that the shallowest manifestations of any helium narcosis would not be apparent before 1400 ft (400 m). Brauer (1975), who had been studying the effects of very high pressure upon animals, termed this condition the "high-pressure neurological syndrome" (HPNS) and suggested that this was an effect of pressure *per se* rather than due to the dissolved tensions of helium. This was demonstrated conclusively when similar changes were observed in liquid-breathing mice at great pressures (Kylstra *et al.*, 1967).

Manifestations

Though subjective effects of HPNS in man may be noticed as shallow as 450 ft (139 m), the depth of onset of gross tremor and more serious effects is deeper, depending on other factors such as the rate of compression (Fig. 7) and the ambient temperature.

Tremors

Frequency analysis shows the tremors to have a major component between 8 and 12 Hz. While this may not affect motor performance seriously, it is important as an early warning that other more serious HPNS manifestations are imminent.

Dizziness

Nausea and dizziness with occasional vertigo can cause disturbing symptoms during compression although they tend to improve during the first few hours after arrival at depth.

Cerebral effects

Although there is considerable variation between individuals, a loss of attentiveness has been reported in many dives to depths of 400 m or greater. Except perhaps for long-term memory, tests of intellectual function are usually unaffected. Periods of "microsleep" or somnolence occur from which the subject can be easily roused.

An electroencephalogram shows little change until the HPNS becomes severe. On-line frequency analysis may help to demonstrate early changes. Commonly there is an increasing percentage of theta waves (5–8 Hz) and a decreased percentage of alpha waves (8–13 Hz). The time course of such changes in relation to compression rate have been discussed by Bennett (1975). Auditory evoked responses may show a decrement as great as 50%. Psychomotor performance tests show decrements that may be correlated with tremors rather than mental performance. Among the other possible consequences of HPNS, a resetting of the temperature control centre (Brauer, 1975) is potentially significant when related to problems of thermal balance which are to be discussed later.

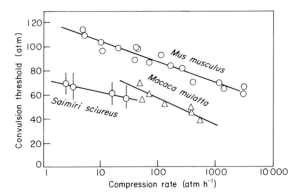

FIG. 7 Relationship between the rate of compression and the threshold pressure for convulsion in three species breathing oxy–helium (after Brauer, 1975).

Mechanisms

The fact that HPNS occurs in hydraulically compressed liquid-breathing mice is sufficient evidence to dismiss the hypothesis that HPNS is a form of helium narcosis, that it is a consequence of gas-induced osmosis in the tissues and, probably, that it is merely a manifestation of mild hypoxia. Changes in cell structure and function, membrane permeability and nerve conduction at pressures around 100 bar, have already been discussed. That the onset of HPNS is related to the compression rate suggests that a change in pressure rather the absolute pressure itself might be a factor in the development of manifestations.

Brauer (1975) suggests that the development of HPNS should be regarded as a basic change in excitability, especially pronounced along polysynaptic pathways, and a secondary development of inhibitory impulses which gain effectiveness with time. This working hypothesis would account for higher convulsion thresholds at slower compression rates but needs further elaboration to account for other HPNS features. Miller *et al.* (1973) proposed a "critical volume hypothesis" for the action of hydrostatic pressure which provides a useful explanation of the relatively rate-independant aspects of HPNS. This hypothesis also accounts for the pressure reversal of anaesthesia and for the antagonism by anaesthetic agents of pressure-induced convulsions. The hypothesis may be stated as follows: "Anaesthesia occurs when the volume of a hydrophobic region of the nerve cell is caused to expand beyond a certain critical amount by the absorption of an inert substance; an applied pressure opposes this expansion and reverses the anaesthesia". Conversely, the convulsions of HPNS are considered to occur when some hydrophobic region is compressed beyond a certain critical amount by application of pressure; absorption of an inert gas will compensate for such compression and raise the convulsion threshold pressure. Based upon the results of experiments upon animals to test this hypothesis, the addition of nitrogen to the oxy–helium breathing mixture of divers at great depth has been used in an attempt to ameliorate the manifestations of HPNS (Fig. 8). Although the results of such tests have not been wholly conclusive and have not led to a universal adoption of a mixture of helium, nitrogen and oxygen for deep diving, the response of tremor in these tests suggest that the critical volume hypothesis can be used to account for some, but not all, HPNS manifestations.

At present there is little more than general acceptance that hydrostatic pressure acts upon cell membranes and that the mechanism of some components of HPNS is also relevant to the mechanism of anaesthesia. Anaesthesia and pressure nullify each other not necessarily by acting at the same membrane site, but by modification of the integrated electrical viability of the nerve at different sites (Bennett *et al.*, 1975).

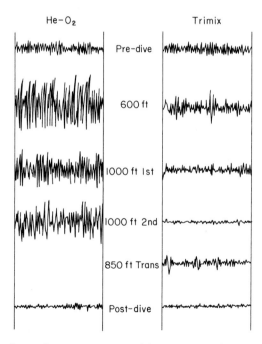

FIG. 8   Comparison of posture tremor of finger during 27 min compression to 100 feet (31 bar) when breathing oxy–helium and when breathing oxy–nitrogen–helium (trimix with 18% nitrogen). The trimix was discontinued at 850 ft during decompression and the subject returned to oxy–helium (after Bennett, 1975).

Compression Arthralgia

The onset of an ill-defined pain in one or more synovial joints during the compression appears to have no serious consequences. No association with subsequent osteonecrosis has yet been demonstrated. The aetiology is obscure and it has been suggested that when the partial pressure of inert gas in the diver's blood temporarily exceeds that in some tissues during compression, dissolved gas-concentration gradients may cause fluid to shift by osmotic action from a tissue such as articular cartilage to the relatively well-perfused adjacent ends of long bones (Bradley and Vorosmarti, 1974).

Thermal Balance

The normal processes of thermoregulation may become severely stressed when man is under water. Not only does the cold sea act as an infinite heat sink during immersion but the high thermal capacity and conductivity of the oxy–helium breathed at raised environmental pressure provides another route for heat loss. In this discussion it will be assumed that the hypothalamic centre which regulates temperature is itself working normally but there remains a possibility at very great depths that central temperature regulation may be off-set by HPNS.

The normal thermal equivalent of metabolic activity in a resting subject is about 60 watts m^{-2} of body surface area. When the balance between heat generation and heat dissipation is disturbed, the body's total heat stores will be increased or depleted and deep body temperature will be altered, possibly beyond the range 35–39 °C and into the region of thermal stress. Even within these limits of deep body temperature, thermal stress may exist in an area such as the face or hand if inadequately protected from temperature extremes. However, in diving the most serious consequence of thermal imbalance is the serious effect of relatively moderate hypothermia upon the diver's ability to respond efficiently to the other stresses present in his environment.

Immersion

With special emphasis upon surface swimming, the problems related to heat loss in cold water have been reviewed elsewhere (Keatinge, 1969). Because water conducts heat at a much greater rate than air, its temperature must be within the range 32–33 °C for it to seem neither cold nor hot.

For an unclothed man to stay in the water for many hours without discomfort, its temperature must be in the range 34–36 °C.

On immersion or submersion, cold water causes a drop in skin temperature and a vasoconstriction in the cutaneous circulation. This provides an insulative layer for the body shell which is a function of the peripheral blood flow rate. In a resting man the cooled shell becomes larger while the relatively warm core becomes smaller. An increase in muscle tension leading to shivering may be sufficient to maintain thermal balance and prevent a fall in temperature. Exercise or swimming in the water may not generate as much heat as is lost by the associated movement of the skin in the water. A dangerous and uncontrollable hyperventilation can follow the shock of sudden cold immersion. Cardiac irregularities and other manifestations of the clinical state of hypothermia may later occur. A detailed consideration of these factors and of the role of subcutaneous fat (Pugh and Edholm, 1955) is available in Webb (1975).

Protection against such consequences can largely be provided by suitable clothing, usually either a wet-suit made of closed-cell foamed neoprene, the efficiency of which diminishes with depth due to compression of the gas in the trapped cells, or a dry-suit under which, being totally waterproof, sufficient clothing can be worn to retain warmth and in which the air space can be maintained during descent by adding more compressed gas.

Helium

Helium has a high thermal conductivity and thermal capacity, increased further in proportion to its increased density at depth. In the dry environment of a compression chamber at around 20–30 bar in which the atmosphere is oxy–helium, man can tolerate a thermal comfort zone with a bandwidth of only 2 °C. The environment is highly convective and, even when men feel thermally comfortable, there is evidence that there is nevertheless an imbalance. The higher the environmental pressure, the warmer the oxy–helium must be and the narrower the comfort zone. Thus, in a single chamber those who are mildly active may feel too hot while those at rest may feel too cold.

In the water below 50 m (6 bar) supplementary heat to the diver's suit is recommended and can be achieved by means of electric heating or by the supply of hot water. In these circumstances the greatest heat loss is from the respiratory tract (Fig. 9) and at depths greater than 150 m (16 bar) heating of the inspiratory breathing gas is recommended. If cold oxy–helium is breathed at pressures around 30 bar, shivering begins almost immediately and copious secretions from the upper respiratory tract make it difficult for the diver to retain his mouthpiece (Hoke et al., 1975).

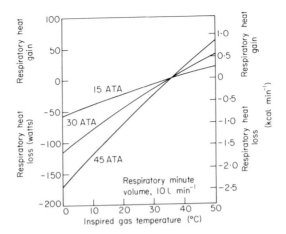

FIG. 9   Respiratory heat loss or gain when breathing normoxic oxy–helium at 15, 30 and 45 bar for a minute volume of 10 litres (after Webb, 1975).

Hyperthermia

A need for heavy physical work by the diver while wearing his suit in the bell and making preparations to dive can lead to a minor degree of heat stress. It has been postulated that the response of the peripheral circulation to this heat stress, combined with the known shifts in the vascular distribution of blood on entering the water, can lead to loss of consciousness (Lemaire and Murphy, 1977). More seriously, but very rarely, too high an environmental temperature in a compression chamber can, because of the body's inability to lose heat by any other means, lead to hyperthermia. For instance, a chamber temperature exceeding 43 °C at 20 bar can be rapidly fatal.

Weight Loss

In the course of a saturation dive most divers lose weight despite eating large amounts of food. These losses are too high to be explained simply by a fluid shift which is associated with a sustained daily diuresis attributable to suppression of antidiuretic hormone as a result of the suppression of insensible water loss. A complete caloric balance study in a 16-day saturation dive at 18.6 bar showed that food intake was high, about 3500 kcal per day, food absorption was a normal 90% of intake and energy expenditure, while elevated by 15% over sea-level control values, was about 1000 kcal per man per day less than net food intake. The five men lost an average of 0.8 kg of body fat in the 30 days of the experiment. The discrepancy between intake and expenditure has not yet been identified using conventional techniques (Webb *et al.*, 1977).

Performance

The environment

Any study of the mental and physical performance of the diver under water must largely reflect the unique nature of the environment in which he has to work. This includes a number of specific factors such as nitrogen narcosis and the high-pressure neurological syndrome, which are discussed separately.

The water itself is cold, viscous and imposes a relative weightessness upon the immersed body. The neutral buoyancy of a diver in the water requires him to use senses other than proprioception for orientation. Poor visibility due to turbidity can further diminish the diver's sense of position and, for the inexperienced, may provide some psychological stress. Both wet-suits and dry-suits impede movement and the viscosity of water impedes all movements to the extent that what is simple at the surface becomes slow and tiring below it. Anxiety can also interfere with performance; an experienced diver's attention will become narrowed to the tasks in hand while an inexperienced diver will be preoccupied with the danger. The need to concentrate on the immediate task can simultaneously reduce the level of attention required for the diver to become aware of a danger from some other source. Reviews of these and other aspects of human performance underwater, summarized below, are available in Adolfson and Berghage (1974), Bachrach (1975) and Shilling *et al.* (1976).

Cold

Studies of cognitive efficency in cold water reveal only a slight impairment but memory, however, can be significantly impaired. Cold also directly affects tactile sensibility and thus manual dexterity.

Pressure

The decrement of performance associated with increasing pressures is a consequence of many possible factors. Nitrogen narcosis, possibly exacerbated by carbon dioxide retention, has been discussed already. The direct effects of pressure upon the air-containing spaces, which have also been discussed, can lead to pain and disturbances of balance which will indirectly affect performance. The high-pressure neurological syndrome (HPNS) has both time-related and time-independent components, and either can lead to serious impairments of performance. Similarly, diminished pulmonary function associated with a need for constant awareness of the correct functioning of breathing apparatus during a dive, leads to a reduction of working efficency underwater which may be depth related.

Perception

Although the diver must be trained to work in dark and turbid water, his ability to see three-dimensionally, though reduced, is a great asset when completing certain tasks. The indices of refraction of light travelling from water, through glass to the air space in front of the eye, produce visual distortion with a virtual image at approximately 75% of the actual distance. However, learning processes and water clarity will also play a role in this judgement. The differential absorption of colour by water leads to the disappearance of red and orange perception below some 10 m, yellow and green being lost around 30 m. If natural colour is to be perceived at depth, an artificial light source is required. The visual fields are limited to a tunnel-like vision which can be attributed primarily to the necessary face mask.

Water-borne sound impinges upon the whole submerged head and it may be assumed that underwater hearing is by bone conduction. Hearing thresholds are little diminished at depth, except when a protective hood is worn, but location of the source of an underwater sound is difficult, primarily because the binaural clues are absent.

Acoustic energy can, as at the surface, provide a hazard to hearing. The sources of hazardous noise include underwater explosions and pile-driving, high-energy sonar and the circulation of breathing gas within the diver's helmet. Even out of the water, the diver is liable to be exposed to excessive noise, for instance when compressed gas is being admitted to a chamber.

The transmission of sound through water is very efficient and though hearing is impaired, when judged by a surface standard, the diver is very much aware of every sound in his relatively silent environment.

Speech

Speech is distorted by a number of factors which often act together. When breathing compressed air in a dry chamber the voice progressively takes on a nasal quality with increasing depth. Oxy–helium gas mixtures affect the voice, even at normal atmospheric pressure. Owing to the increased velocity of sound in helium, the normal resonant frequencies of the vocal tract are increased, giving the impression of a "Donald Duck" effect. Further speech distortion is caused by the need for a diver to breathe either through a mouthpiece or from the gas within a helmet. Besides changing the vocal resonances, the breathing equipment also provides significant additional background noise. The solution to these problems is largely within existing technology.

Performance evaluation

The psychometric tests which have been used in measuring diver performance are many and varied. Cognitive tests, usually computational are used under conditions where nitrogen narcosis is important, while perceptual–sensory and psychomotor tests of central nervous system status are more appropriate at pressures greater than 20 bar. There is much variation in the many separate test procedures, though some standardization of technique has been agreed between several laboratories. Besides these tests of central nervous system function there is a need to test total systems output by simulating realistically the tasks that the diver must perform. An extensive discussion of this approach is also available in the references already cited.

References

Adolfson, J.A. and Berghage, T.E. (1974). "Perception and Performance Under Water", Wiley-Interscience, New York.

Bachrach, A.J. (1975). In "The Physiology and Medicine of Diving and Compressed Air Work" (Eds P.B. Bennett and D.H. Elliott), 1–10. Baillière Tindall, London.

Barnard, E.E.P., Hanson, R. de G., Reid, B.J. and Williams, J. (1973). *Försvarsmedicin* **9**, 496–501.

Bean, J.W. and Smith, C.W. (1953). *Am. J. Physiol.* **172**, 169–174.

Behnke, A.R. (1942). *Med. Clin. N. Am.* **26**, 1213–1237.

Behnke, A.R. (1969). In "The Physiology and Medicine of Diving and Compressed Air Work", (Eds P.B. Bennett and D.H. Elliott), 226–251. Baillière, Tindall and Cassell, London.

Behnke, A.R. and Yarbrough, O. (1938). *U.S. Nav. Med. Bull.* **36**, 542–558.

Behnke, A.R. Thomson, R. and Motley, E. (1935). *Am. J. Physiol.* **112**, 554–558.

Bennett, P.B. (1975). In "The Physiology and Medicine of Diving and Compressed Air Work", (Eds P.B. Bennett and D.H. Elliott), 248–263. Baillière Tindall, London.

Bennett, P.B., Simon, S. and Katz, Y. (1975). *In* "Molecular Mechanisms of Anaesthesia", (Ed. B.R. Fink), 367–403. Raven Press, New York.

Bert, P. (1878). "La Pression Barometrique", English translation by M.A. anf F.A. Hitchcock, (1943). College Book Company, Columbus, Ohio.

Boycott, A.E., Damant, G.C.C. and Haldane, J.S. (1908). *J. Hyg. (Camb.)* **8**, 342–443.

Boyle, R. (1670). *Phil. Trans, R. Soc.* **5**, 2011–2031 and 2035–2056.

Bradley, M.E. and Vorosmarti, J. (1974). *Undersea Biomed. Res.* **1**, 151–167.

Bradley, M.E., Vorosmarti, J., Anthonisen, N.R. and Lineaweaver, P.G. (1971). *In* "Underwater Physiology, Proceedings of the Fourth Symposium", (Ed. C.J. Lambertsen), 325–337. Academic Press, New York and London.

Brauer R.W. (1975). *In* "The Physiology and Medicine of Diving and Compressed Air Work", (Eds P.B. Bennett and D.H. Elliott), 231–247. Baillière Tindall, London.

Bühlmann A.A. (1975). *In* "The Physiology and Medicine of Diving and Compressed Air Work", (Eds P.B. Bennett and D.H. Elliott), 348–365. Baillière Tindall, London.

Burgess, D. (1977). Personal communication.

Chouteau, J. (1972). *In* "Third International Conference on Hyperbaric and Underwater Physiology" (Ed. X. Fructus), 8–12. Doin, Paris.

Clark, J.M. and Lambertsen, C.J. (1971a). *J. Appl. Physiol.* **30**, 739–752.

Clark, J.M. and Lambertsen, C.J. (1971b). *Pharmcology Rev.* **23**, 37–133.

Colebatch, H.J.H., Smith, M.M. and Ng, C.K.Y. (1976). *Resp. Physiol.* **26**, 55–64.

Coles, R.R.A. (1964). *J.R. Nav Med Serv.* **50**, 23–29.

Cousteau, J. and Dumas, F. (1953). "The Silent World", Harper and Row, New York.

Cumming, G., Crank, J., Horsfield, K. and Parker, I. (1966). *Resp. Physiol.* **9**, 158–174.

Craig, A.B. (1961). *J. Am. Med. Assn* **176**, 255–258.

Dahlback, G.O. and Lundgren, C.E.G. (1973). *Försvarsmedicin* **9**, 247–250.

Davis, R.H. (1962). "Deep Diving and Submarine Operations", 7th edn, Siebe Gorman, London.

Deroanne, R., Dujardin, J., Lamy, M., Marechal, R., Petit, J.M. and Pirnay, F. (1973). *Försvarsmedicin* **9**, 352–356.

Donald, K. (1947a). *Br. Med. J.* **4506**, 667–672.

Donald, K. (1947b). *Br. Med. J.* **4507**, 712–717.

Donald, K.W. and Davidson, W.M. (1954). *J. Appl. Physiol.* **7**, 31–37.

Dugan, J. (1956). "Man Under the Sea", Harper and Row, New York.

Edmonds, C. (1973). *In* "Otological Aspects of Diving", (Eds C. Edmonds, P, Freeman, R. Thomas, J. Tonkin and F.A. Blackwood), 55–96, Australasian Medical Publishing, New South Wales.

Elliott, D.H. and Hallenbeck, J.M. (1975). *In* "The Physiology and Medicine of Diving and Compressed Air Work", (Eds P.B. Bennett and D.H. Elliott), 435–455. Baillière Tindall, London.

Evans, A., Barnard, E.E.P. and Walder, D.N. (1972). *Aerospace Med.* **43**, 1095–1096.

Fagraeus, L. and Linnarsson, D. (1973). *Försvarsmedicin* **9**, 275–278.

Fagraeus, L., Karlsson, J., Linnarsson, D. and Saltin, B. (1973). *Acta Physiol. Scand.* **87**, 411–421.

Fenn, W.O. (1969). *In* "The Physiology and Medicine of Diving and Compressed-air Work", (Eds P.B. Bennett, and D.H. Elliott), 36–57. Baillière, Tindall and Cassell, London.

Florio, J.T., Butt, W.S. and Morrison, J.B. (1979). *In* "Proceedings of the 6th International Conference on Hyperbaric Medicine", (Ed. G. Smith), 382–384. Aberdeen University Press, Aberdeen.

Goodman, M.W. (1963). *In* "Proceedings of the Second Symposium on Underwater Physiology", (Eds C.J. Lambertsen and L.J. Greenbaum), 245–246. National Acadamy of Sciences, National Research Council, Washington, D.C.

Haldane, J.S. and Priestley, J.G. (1935). "Respiration", Yale University Press, Newhaven.

Hempleman, H.V. (1975). *In* "The Physiology and Medicine of Diving and Compresses Air Work", (Eds P.B. Bennett and D.H. Elliott), 331–347. Baillière, Tindall, London.

Hesser, C.M., Adolfson, J. and Fagraeus, L. (1971). *Aerospace Med.* **42**, 163–168.

Hills B.A. (1975). *In* "The Physiology and Medicine of Diving and Compressed Air Work", (Eds P.B. Bennett and D.H. Elliott), 366–391. Baillière Tindall, London.

Hoke, B., Jackson, D.L., Alexander, J.M. and Flynn, E.T. (1975). *In* "Underwater Physiology, Proceedings of the Fifth Symposium", (Ed. C.J. Lambertsen), 725–740,

Hong, S.K. (1976). *In* "Diving Medicine", (Ed. R.H. Strauss), 269–286. Grune and Stratton, New York.

Johnson, L.R. and Van Liew, H.D. (1974). *J. Appl. Physiol.* **36**, 91–97.

Junod, T. (1835). *Ann. Gen. Med.* **9**, 157.

Keatinge, W.R. (1969). "Survival in Cold Water", Blackwell, Oxford.

Kety S.S. and Schmidt, C.F. (1948). *J. Clin Invest.* **27**, 484–492.

Kidd, D.J. and Elliott, D.H. (1975). *In* "Physiology and Medicine of Diving and Compressed Air Work", (Eds P.B. Bennett and D.H. Elliott, 487–494. Baillière Tindall, London.

Kylstra, J.A. (1975). *In* "The Physiology and Medicine of Diving and Compressed Air Work", (Eds P.B. Bennett and D.H. Elliott), 155–165. Baillière Tindall, London.

Kylstra, J.A., Natz, R., Crowe, J., Wagner, W. and Saltzman, H.A. (1967). *Science* (*NY*) **158**, 793–794.

Lambertsen, C.J. (1965). *In* "Handbook of Physiology", Section III, Vol. 2 (Eds W.O. Fenn and H. Rahn), 1027–1046. American Physiological Society, Washington, D.C.

Lambertsen, C.J. and Idicula, J. (1975). *J. Appl. Physiol.* **39**, 434–443.

Lanphier, E.H. (1954). *In* "Oxygen Consumption in Underwater Swimming", Report 14–54, US Navy Experimental Diving Unit, Washington, D.C.

Lanphier, E.H. (1963). *In* "Proceedings of the 2nd Symposium of Underwater Physiology", (Eds C.J. Lambertsen and L.J. Greenbaum), 124–133. National Academy of Sciences and National Research Council, Washington, D.C.

Lanphier, E.H. (1972). *In* "The Effects of Pressure on Organisms", (Eds M. Sleigh and A. Macdonald) pp. 379–394. Cambridge University Press, London.

Lanphier, E.H. (1975). *In* "The Physiology and Medicine of Diving and Compressed Air Work", (Eds P.B. Bennett and D.H. Elliott, eds), 102–154, Baillière Tindall, London.

Lanphier, E.H. and Rahn, H. (1963). *J. Appl. Physiol.* **18**, 478–482.

Lemaire, C. and Murphy, E.L. (1977). *Aviat. Environ. Med.* **48**, 146–148.

Lever, M.J., Miller, K.W., Paton, W.D.M. and Smith, E.B. (1971). *Nature* (*London*)

231, 368–371.

Lundgren, C.E.G. (1965). *Br. Med. J.* **2**, 511–513.

Lundgren, C.E.G. (1973). *Försvarsmedicin* **9**, 406–409.

Lundgren, C.E.G. and Örnhagen, H.C. (1976). *In* "Underwater Physiology", (Ed. C.J. Lambertsen), 397–404. Fedn. American Soc. Exp. Biol., Bethescia.

Macdonald, A.G. (1975). *In* "The Physiology and Medicine of Diving and Compressed Air Work", (Eds P.B. Bennett and D.H. Elliott), Baillière Tindall, London.

Macdonald, A.G. (1977). *Rev. Méd. Aéronant. Spat. Méd. Subaqu. Hyperbare* **16**, 277–279.

Maio, D.A. and Farhi, L.E. (1967). *J. Appl. Physiol.* **23**, 687–693.

Mead, J., Turner, J.M., Macklem, P.T. and Little, J.B. (1967). *J. Appl. Physiol.* **22**, 95–108.

Miller, J.N. and Winsborough, M. (1973). *Försvarsmedicin* **9**, 321–331.

Miller, J.N., Wangensteen, O.D. and Lanphier, E.H. (1972). *In* "Proceedings of the 3rd International Conference on Hyperbaric and Underwater Physiology", (Ed. X. Fructus), 118–123. Doin, Paris.

Miller, K.W., Paton, W.D.M., Smith, R.A. and Smith, E.B. (1973). *Molec. Pharmacol.* **9**, 131–143.

Morrison, J.B., Butt, W.S., Florio, J.T. and Mayo, I.C. (1973). *Aerospace Med.* **44**, 1120–1129.

Ornhagen, H.C. and Lundgren, C.E.G. (1976). *In* "Underwater Physiology", (Ed. C.J. Lambertsen, 399–404. Fedn. of American Societies of Experimental Biology, Bethesda.

Paton, W.D.M. (1967). *Br. J. Pharm. Chemother.* **29**, 350.

Paton, W.D.M. and Sand, A. (1947) *J. Physiol.* **106**, 119–138.

Penzias, W. and Goodman, M.W. (1973). "Man Beneath the Sea" Wiley-Interscience, New York.

Philp, R.B. (1974). *Undersea Biomed. Res.* **1**, 117–150.

Philp, R.B., Inwood, M.J. and Warren, B.A. (1972). *Aerospace Med.* **43**, 946–953.

Poulton, E.C., Catton, M.J. and Carpenter, A. (1964). *Br. J. Ind. Med.* **21**, 242–245.

Pugh, L.G.C. and Edholm, O.G. (1955). *Lancet* **ii**, 761–768.

Rahn, H. (1965). *In* "Physiology of Breath-hold Diving and the Ama of Japan", (Eds H. Rahn and T. Yokoyama), 113–138. National Academy of Science–National Research Council, Washington, D.C.

Rahn, H. and Yokoyama, T. (1965). (Eds) "Physiology of Breath-hold Diving and the Ama of Japan", National Academy of Science–National Research Council, Washington, D.C.

Regnard, P. (1891). "Recherches Experimentales Sur Les Conditions Physiques de la Vie dans les Eaux", Mason, Paris.

Rohrer, F. (1915). *Pflugers Arch. Ges. Physiol.* **162**, 225–299.

Roth, S.H., Smith, R.A. and Paton, W.D.M. (1976). *In* "Underwater Physiology", (Ed. C.J. Lambertsen), 421–430. Fedn. of American Societies of Experimental Biology, Bethesda.

Saltzman, H.A., Salzano, J.E., Blenkarn, G.D. and Kylstra, J.A. (1971). *J. Appl. Physiol.* **30**, 443–449.

Sayers, R., Yant, W. and Hildebrand, J. (1925) "Possibilities in the Use of Helium-Oxygen Mixtures as a Mitigation of Caisson Disease", US Bureau of Mines, Washington, D.C.

Schaefer, K.E. (1975). *In* "Physiology and Medicine of Diving in Compressed Air

Work" (Eds P.B. Bennett and D.H. Elliott), 185–206. Baillière Tindall, London.

Schaefer, K.E., McNulty, S.P., Carey, C.R. and Liebow, A.A. (1958). *J. Appl. Physiol.* **13**, 15–29.

Shields, T.G. (1977). *Rev. Méd. Aéronaut. Spat. Méd. Subaqu. Hyperbare* **16**, 251–253.

Shilling, C.W., Werts, M.F. and Schandelmeier, N.R. (1976). "The Underwater Handbook", Wiley, New York.

Smith, J.L. (1899). *J. Physiol.* **24**, 19–35.

Spaur, W.H. Raymond, L.W. Knott, M.M., Crothers, J.C., Braithwaite, W.R. Thalmann, E.D. and Uddin, D.F. (1977), *Undersea Biomed. Res.* **4**, 183–198. Stein, S.N. (1955). *In* "Proceedings of the First Symposium of Underwater Physiology", (Ed. L.G. Goff), 20–24. National Academy of Sciences, National Research Council, Washington, D.C.

US Navy Diving Manual (NAVSEA 0994-LP-001-9010). (1973). Best Book Binders, Carson, California.

Vorosmarti, J. and Lanphier, E.H. (1975). *In* "Underwater Physiology Proceedings of the 5th Symposium" (Ed. C.J. Lambertsen), 101–108. Federation of American Societies for Experimental Biology, Bethesda.

Webb, P. (1975) *In* "The Physiology and Medicine of Diving and Compressed Air Work", (Eds P.B. Bennett and D.H. Elliott), 285–306. Baillière Tindall, London.

Webb, P., Troutman, S.J., Frattali, V., Dressendorfer, R.H., Dwyer, J., Moor, T.O., Morlock, J.F., Smith, R.M. Ohta, Y. and Hong S.K. (1977). *Undersea Biomed. Res.* **4**, 221–246.

Wood, J.D. (1975) *In* "The Physiology and Medicine of Diving and Compressed Air Work", (Eds P.B. Bennett and D.H. Elliott), 166–184. Baillière Tindall, London.

Workman, R.D. and Bornmann, R.C. *In* "The Physiology and Medicine of Diving and Compressed Air Work", (Eds P.B. Bennett and D.H. Elliott), 307–330. Baillière Tindall, London.

Wright, W.B. (1972) "Use of the University of Pennsylvania Institute for Environmental Medicine Procedures for the Calculation of Cumulative Pulmonary Oxygen Toxicity", report 2-72. US Navy Kexperimental Diving Unit, Panama City, Florida.

Wright, W.B., Peterson, R. and Lambertsen, C.J. (1972). "Pulmonary Mechanical Functions in Man Breaphing Dense Gas Mixtures at Great Depths". Report 14–72 US Navy, Experimental Diving Unit, Panama City, Florida.

Zinkowski, N.B. (1975). "Commercial Oil-field Diving". Cornell Maritime Press, Cambridge, Maryland.

CHAPTER 7

Locomotor and Postural Physiology

F.J. Imms

Introduction

In man the upright posture, which is partly adopted by sub-human primates, reaches its most complete form. This posture depends upon the support of the body by the bony skeleton, the use of muscle tone to stabilize the skeleton, and the control of this tone by the nervous system. A further function of bone, muscle and nerve is the initiation and control of movement, for man, like almost all other animals, moves purposefully around his environment. The upright posture allows specialization of limb movement which is not present in other quadrupeds. The hind limbs (lower limbs in man) have remained as organs of locomotion; the fore (upper) limbs are little used in locomotion and have been adapted for performing many skilled tasks.

In this chapter, outlines will be given of the physiology of the musculo–skeletal and the nervous systems as they are involved in posture and movement in the human. Muscular contraction requires energy, and accounts will be given of the energy costs both of maintaining postures and of locomotion. Finally, the effects of immobilization on the body will be discussed.

The Skeleton

Bone

The bony skeleton gives both support and shape to the body as well as affording protection to vital organs such as the brain and heart. Bone is a living tissue and all of its components are undergoing continual removal and replacement; consequently it requires an adequate blood supply and if, following fracture, a fragment has no blood supply then this fragment may necrose.

Bone contains both organic and inorganic material. The organic component consists of collagen fibres which lie in a ground substance similar to that of other tissues. These fibres are formed by *osteoblasts*, which become trapped within the bone where they are termed *osteocytes*. The tissue derives its strength from the deposition of minute crystals of complex salts on these fibres. The skeleton contains 99% of the total calcium in the body, 88% of the phosphate and 35% of the sodium. The ions in bone crystals are in equilibrium with those in body fluids. *Osteoclasts* are giant cells associated with resorption of bone. For details of both organic and inorganic components of bone, and of the structural organization of the tissue, the reader is referred to the monograph of Fourman and Royer (1968) and to standard text-books of histology.

The formation and resorption of bone are influenced by several factors. At birth parts of the skeleton are cartilaginous, and calcification of this cartilage to form bone occurs during childhood or at puberty. Bone loses calcium if the body is immobilized, and recalcifies when activity is resumed (see p. 394). Dietary deficiency of calcium is unlikely, but Vitamin D is necessary for absorption of calcium from the intestine and lack of this vitamin leads to *osteomalacia* (rickets), a condition in which the organic matrix is under-mineralized. The equilibrium between calcium in the body fluids and in bone is also under hormonal control. Two hormones with specific actions on calcium and phosphate metabolism are parathormone, which increases removal of calcium from bone, and thyrocalcitonin which facilitates deposition of mineral into bone. Growth hormone, thyroxine, oestrogens and androgens all favour calcification: glucocorticoids, particularly when excessive quantities are administered therapeutically, cause *osteoporosis*, a condition in which both organic and inorganic components of bone are deficient.

Joints

Bones articulate with each other at joints. At some joints little movement is possible, at others there is a wide range of movement. Ligaments are bands of connective tissue which are attached to the bones forming the joint and add strength and stability to it. The simplest articulation is a fibrous linkage between bones, as occurs at the sutures in the vault of the skull. Little or no movement is possible at a cartilagenous joint, in which the two bony surfaces are covered with hyaline cartilage and separated from each other by a disc of fibrocartilage (Fig. 1(a)). Such articulations occur between the vertebrae.

Synovial joints are more complex and allow greater degrees of movement (Fig. 1(b)). Such joints occur in the limbs. Articular cartilage covers the opposing surfaces of the two bones between which movement takes place, and the two ends are joined by a fibrous joint capsule lined by synovial membrane. When movement takes place the tissues of the joint capsule stretch, causing the synovial membrane to slide on itself and also to slide over the other intra-articular structures; movement also takes place between cartilage and cartilage. The remarkable lubrication characteristic of joints has been found to be due to the combination of the rheological properties of synovial fluid and the surface of the articular cartilage.

The movement of synovial membrane is lubricated by one constituent of synovial fluid, hyaluronate macromolecules (molec. wt 1×10^{-6}) straight chain polymers of N-acetylglucosamine and glucopomide, and these large molecules stick to the synovium in a layer and keep the moving surfaces apart. The resistance to motion caused by soft tissue stretching and friction

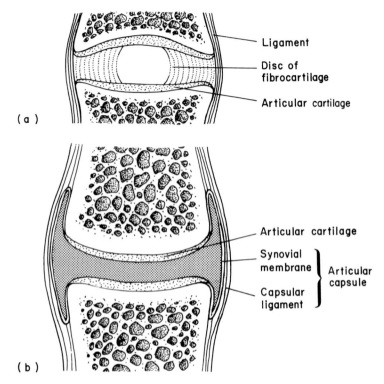

FIG. 1 (a) Section through a cartilaginous joint. (b) Section through a simple synovial joint (from Johnston and Whillis, 1949; and Warwick and Williams, 1973).

is far greater than that due to cartilage rubbing on cartilage and may be primarily responsible for clinical joint stiffness when it occurs.

The bearing surfaces provided by articular cartilage are lubricated by a system which is independent of hyaluronate. A glycoprotein fraction of synovial fluid, quite free from hyaluronate, confers on a simple buffer solution a lubricating advantage equal to that of whole synovial fluid. The action of this glycoprotein on cartilage is similar to that of hyaluronate on synovium. It sticks to the surface forming a thin layer of molecules which keep the surfaces apart—boundary lubrication. In addition, the cartilage surfaces are kept apart by a film of fluid made up of joint fluid and interstitial fluid which is "squeezed" from the articular cartilage itself when it is loaded. The effect of this film is improved further by two characteristics of cartilage: the undulations of the surface trap pools of fluid; the elasticity of the cartilage which depends upon the movement of fluid into and out of cartilage allows the cartilage to yield so that a fluid film is maintained.

This motion at the joint increases the fluid content of the articular cartilage, which is important for the role it fulfils in absorbing the energy transmitted through the joint by impact loading. Motion also improves the supply of synovial fluid and thus assists in the nutrition of the articular cartilage.

Skeletal Muscle

Skeletal muscle has two major functions in the body. First, by maintaining a state of partial contraction often known as "muscle tone", it adds stability to the skeleton and allows the adoption of sitting and standing postures (see p. 365). Secondly, muscle contracts to perform work either of a dynamic type such as walking or running, during which muscles are alternately contracted and relaxed, or of a static type such as pushing, pulling, or holding a weight, during which there is prolonged isometric contraction (see Chapter 2, by E.J. Bassey and P.H. Fentem).

Receptors in Muscles and Joints

The carefully controlled maintenance of muscle tone to allow the keeping of a posture, and the contraction of muscle during exercise requires not only motor stimulation of the muscle, but also reliable feedback signals conveying information on the position of limbs in space, the angles subtended in joints and the extent of muscle contraction (i.e. muscle tone) or stretch. There are various receptors which furnish this information. For an authoritative review of their structure and function, the reader is referred to the monograph by Matthews (1972). Although the functions of these receptors have been determined mainly in animals, the more limited experiments performed in man suggest that there are no marked species differences.

Golgi tendon organs

These receptors lie at musculo–tendinous junctions and signal changes of tension occurring at this region (Fig. 2(a)). They are slowly adapting and hence do not decrease the rate of firing when tension is sustained (Fig. 3(a)). They were originally believed to be receptors of a protective nature and to fire only when a disproportionately dangerous stretch tension

(a)

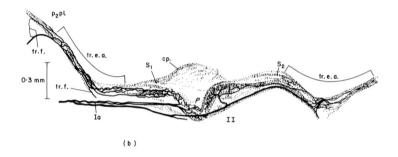

(b)

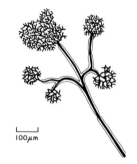

└─────┘
100μm

(c)

FIG. 2   Receptors in skeletal muscle and in joints. (a) Golgi tendon organ as seen by Cajal: *a*, tendon; *b*, large afferent fibre breaking up into sensory terminals; *c*, muscle fibres. (b) A cat muscle spindle stained with silver. The intrafusal fibres extended considerably outside the present picture, but received no further innervation. P, primary ending; S, secondary ending; cp, capsule; tr e a, trail endings; p_2pl, a p_2 plate ending; Ia and II, groups Ia and II afferent fibres respectively; tr f, motor fibres terminating as trail endings. (c) The Ruffini type ending found in joint capsules. A medium-sized afferent fibre breaks up into several separate terminal ramifications (from Matthews, 1972).

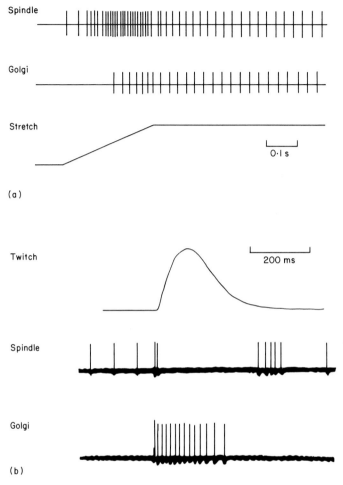

FIG. 3 Recordings from afferent nerves from limbs. (a) The contrasting sensitivities to a dynamic stimulus (stretching of a spindle ending and a Golgi tendon organ. (b) The contrasting responses of a spindle ending and a tendon organ during a twitch contraction (from Matthews, 1972).

was applied to muscle, but more recently they have been shown to respond more sensitively to increased tension during muscle contraction, and this may be their physiological response (Fig. 3(b)).

Muscle spindles

These are long, thin receptors with a complex structure which are present in muscle. They may be up to 10 mm in length and lie parallel with the muscle fibres. They respond to stretch of the muscle by increasing their rate of firing (Fig. 3(a)), whereas during muscle contraction the rate of firing is decreased (Fig. 3(b)).

Within the spindle there are two types of receptor, each with its own nerve fibre, and small striated muscle fibres known as intrafusal fibres which run the length of the spindle. The primary sensory endings are derived from a single large nerve fibre, and surround all the intrafusal muscle fibres at the equator in a spiral arrangement; hence the common name annulo-spiral endings. The secondary or "flowerspray" endings are on either side of the primary endings and discharge via a medium-sized nerve fibre.

The intrafusal muscle fibres are also of two types known as nuclear "chain" and nuclear "bag" fibres respectively. The "bag" fibres have their nuclei concentrated towards the equator of the spindle, whereas the "chain" fibres have them dispersed along their length. The secondary sensory endings are associated mainly with the chain fibres.

The intrafusal muscle fibres are innervated by small (γ efferent) fibres from small motor neurones (fusimotor) in the anterior horn of the spinal cord. The motor neurones terminate in two forms. On nuclear bag muscle fibres the motor neurones terminate as end-plates similar to those in extrafusal skeletal muscle, whereas on the chain fibres they are more diffuse and are often described as "trail" endings. The γ efferent neurones conduct at 10–15 m s^{-1} compared with motor neurones innervating extrafusal fibres which conduct at up to 120 m s^{-1}.

The primary receptor of the spindle discharges along fast nerve fibres, often termed group Ia, whereas the secondary receptors discharge along slower fibres termed group II. The Golgi tendon organs discharge through intermediary group Ib neurones.

Single fibre recording has proved invaluable in determining the functioning of spindles and tendon organs. Spindles which lie in parallel with muscle are sensitive to stretch and often exhibit tonic discharge when muscle is relaxed. Conversely, the tendon organs which are in series with muscle appear to be stimulated by contraction of muscle. Thus, during a twitch contraction, spindle activity is reduced whereas tendon organ activity is increased.

More detailed studies have revealed that the primary and secondary endings respond differently to stretch stimuli. Thus the primaries respond maximally during the application of a stretch stimulus with a reduction of firing whilst the stretch is maintained, whereas the secondaries continue to fire during the maintenance of the stretch (Fig. 4(a)). In addition, the

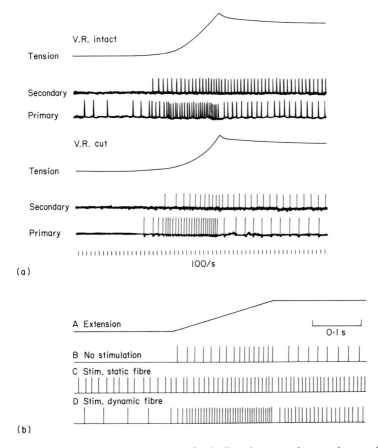

FIG. 4   (a) The contrasting responses of spindle primary and secondary endings to a rapidly applied stretch (approximately 14 mm at 70 mm s^{-1}) shown both in the presence and absence of fusimotor activities. The endings lay in the soleus msucle of the same decerebrate cat; their responses were recorded both before and after cutting the ventral roots to eliminate fusimotor activity. (b) The different effects of stimulation of static and dynamic fusimotor fibres. The response of a primary ending to stretching is shown both in the presence and absence of stimulation of a single fusimotor fibre of either kind (from Matthews, 1972).

primary endings respond with greater increase of frequency of firing if the muscle tendon is tapped as the muscle is subjected to sinusoidal stretch.

The rate of firing of both primary and secondary spindle receptors is influenced by γ efferent stimulation and contraction of intrafusal muscle fibres. The latter appear to be of two types: *dynamic* fibres appear to increase the rate of firing of a primary ending during the application of a stretch, whereas *static* fibres increase its response to maintained tension (Fig. 4(b)). The static fusimotor fibres appear to increase the sensitivity of secondary receptors, but have no influence on the primaries.

Free nerve endings

Free nerve endings are present in muscle, in fascia and around joints. They discharge through non-medullated or small-medullated nerve fibres and may be associated with sensations of pain and pressure.

Paciniform corpuscles

Paciniform corpuscles, which are sensitive to mechanical deformation and pressure, are found in fascial sheets between muscles and in loose connective tissue around joints.

Ruffini endings

These endings lie in the capsule of the joint (Fig. 2(c)) and each is believed to fire specifically at different positions of the joints. Joint position may also be signalled by Golgi tendon organs lying in ligaments associated with the joint.

The Stretch Reflex

A tap on a tendon leads to a rapid reflex contraction of the muscle, usually described as a jerk. The receptor responsible is the spindle primary, and the time interval for transmission of the signals through the spinal cord suggests that the reflex is monosynaptic, the Ia afferent fibre synapsing directly with the motor neurones in the anterior horn of the spinal cord. Neither spindle secondaries nor Golgi tendon organs are stimulated by the transient stretch induced by a tendon tap.

The stretch reflex may additionally be elicited by a maintained stretch of the muscle, giving rise to the tonic stretch reflex in which the muscle remains contracted for the duration of the stretch. The role of spindle secondaries in producing a stretch reflex has not been unequivocally demonstrated, but from their physiological actions such a role may be presumed in response to tonic stretching. Stimulation of Ib afferents from Golgi tendon organs leads to reflex inhibition of the effects of spindle stimulation in the tonic reflex situation.

Studies of the stretch reflex can reveal some of the basics of nervous integration within the body. Stimulation of stretch receptors causes not only contraction of the same muscle but also usually contraction of synergists, i.e. muscles acting at the same joint and having similar actions. At the same time, the antagonists, i.e. muscles having the opposite action at the joint, are inhibited through a disynaptic pathway. Weak reflex effects also occur in the opposite limb with inhibition of the equivalent muscle and possibly excitation of the antagonist. Thus stretching of the patellar tendon causes contraction of the ipsilateral quadriceps and relaxation of the knee flexors, weak inhibition of the contralateral quadriceps and weak stimulation of the flexors.

By contrast, the classical studies of Sherrington (1906) showed that painful stimulation of the foot caused reflex withdrawal of the ipsilateral limb, and powerful extension of the contralateral limb, i.e. contraction of ipsilateral knee flexors with relaxation of the extensors combined with contraction of the contralateral extensors and relaxation of flexors. Thus, if a person treads on a sharp object and reflexly raises the foot from the ground, a fall is prevented by the opposite leg becoming a more stable prop.

The sensitivity of muscle to stretch may be increased by fusimotor stimulation leading to contraction of intrafusal muscle fibres. In this way centres within the central nervous system, acting via the γ efferent neurones, may alter the threshold stretch needed to stimulate the spindle and the rate of firing of spindles at either a fixed length or at any particular rate of change of length.

The testing of stretch reflexes forms an important part of the clinical examination of the central nervous system. The presence of a reflex indicates that the pathway of the reflex is intact, i.e. receptors, afferent fibres, connections within the spinal cord, efferent fibre, end-plate and muscle. The commonly used tendon jerks and the spinal segments through which they relay are shown in Table 1. The character of the reflex response gives additional information. Thus increased fusimotor activity leads to reflex contractions of large amplitude and sometimes to a series of jerks in response to a single tap stimulus, a phenomenon known as clonus.

TABLE 1   Muscle tendon jerks which
may be tested clinically

Muscle	Spinal segment
Masseter	Cranial V
Biceps	C5–6
Triceps	C6–7
Supinator	C5–6
Quadriceps	L2–3–4
Gastrocnemius/soleus	S1–2

Muscle Tone

A muscle may be divided functionally into a number of motor units, each composed of a single motor neurone in the anterior horn of the spinal cord and the muscle fibres which it supplies. Discharge of the motor neurone causes contraction of the muscle fibres of the motor unit. The motor neurone may be activated either by impulses from the brain via the pyramidal tracts or reflexly by stretch, pain or other receptors. Individual motor units obey the *all or none* law of contraction, but a state of partial contraction for the muscle as a whole may exist because only a proportion of the motor units are contracting at any given time. Such a state of partial contraction of a muscle is known as "muscle tone".

The generation of muscle tone is a function of the stretch reflex. Stretch of the spindle receptors within the muscle leads to monosynaptic reflex contraction of the muscle via the α motor neurone. The role of the stretch reflex in the maintenance of tone may be demonstrated by the reduction of tone when the reflex is interrupted by section of the dorsal (sensory) root or by differential blocking of the γ efferent fibres in the ventral root.

The stretch reflex also affords a mechanism for adjustment of tone. Increased fusimotor stimulation causes contraction of intrafusal muscle fibres and thus increases the rate of firing of the spindle stretch receptors at any particular length of the muscle. This increased afferent firing reflexly activates α motor neurones and hence increases muscle tone.

The regulation of tone by fusimotor neurones is controlled by descending connections from higher centres. Some of these are facilitatory and increase muscle tone, others are inhibitory and lead to its reduction. In animals, high spinal section causes generalized flaccidity of muscle, suggesting that normally there is an excess of facilitatory descending fibres to the fusimotor neurones. On the other hand, decerebration, which implies removal of the forebrain and, in practice, part of the midbrain, causes marked increases of tone, particularly of the extensor muscles, a condition often known as decerebrate rigidity. This observation suggests that important inhibitory centres are destroyed by decerebration. These are in the basal ganglia and cerebral cortex. Important facilitatory centres remaining after decerebration are in the bulboreticular area, and activate the fusimotor neurones via the reticulospinal, tectospinal, rubrospinal and vestibulospinal tracts.

The Maintenance of Posture

In most departments of anatomy there is an articulated human skeleton which exhibits an upright posture because it hangs from a hook by a ring inserted into the skull. If the ring is unhooked the skeleton collapses to the floor, despite the fact that the bones have been assembled in the correct manner and are held together by artificial ligaments and by screws. The living body, by contrast, is a stable structure able to maintain an upright or other posture because the various components of the skeleton are maintained in their relative positions by the tone generated in skeletal muscle. This muscle tone depends upon an integrative action of the nervous system, and in situations such as general anaesthesia and fainting, when cerebral function is reduced, then tone is lost and collapse occurs.

For example, the tibia and fibula articulate with the talus to form the ankle joint. The joint is enclosed within a fibrous capsule which, since it is attached to all bones constituting the joint, forms a loose unstable support. Much firmer support is given by the thicker ligaments which cross the joint and resist the effects of traction on the joint (Fig. 5). Thus, if the foot is inverted the ligaments on the lateral aspect become stretched and prevent the movement being carried to an extreme. Maintenance of a standing posture depends upon keeping the tibia and fibula in a vertical plane at right angles to the foot. Such fixation is obtained by the tone present in the skeletal muscles acting across the ankle joint (Fig. 6). In particular, there is a balance of tone between the plantar flexors of the ankle (the soleus and gastrocnemius) attached to the calcaneum, and the dorsiflexors (tibialis anterior, peroneus longus, extensor digitorum longus) attached to the upper surface of the foot. Less importantly, lateral support is given by a balance between the muscles causing inversion and eversion at the joint.

Just as the skeleton supported only by joint capsules and ligaments is an unstable structure, so a skeleton held in place by a constant level of muscle tone would also be relatively unstable. Each joint would be held in a fixed position and a rigid posture obtained, but since the body is a tall structure standing on a narrow base, slight tilting will result in the perpendicular

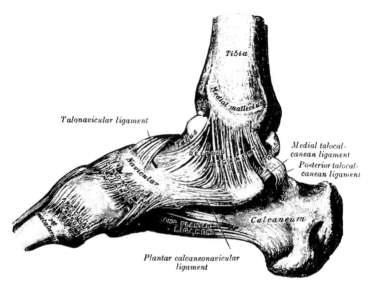

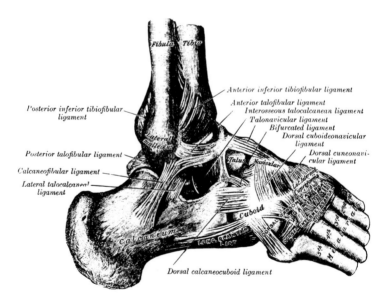

FIG. 5   Stabilization of the ankle joint by ligaments viewed from (top) medial aspect; (bottom) lateral aspect (from Warwick and Williams, 1973).

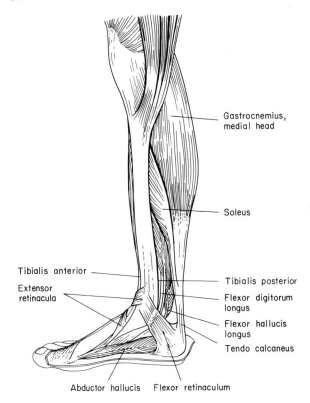

Gastrocnemius, medial head

Soleus

Tibialis anterior

Extensor retinacula

Tibialis posterior

Flexor digitorum longus

Flexor hallucis longus

Tendo calcaneus

Abductor hallucis   Flexor retinaculum

FIG. 6   The muscles acting across the ankle joint and thus giving stability to the leg (from Warwick and Williams, 1973).

from the centre of gravity falling outside the base. In these circumstances the body would fall. Such losses of equilibrium do not occur frequently because rapid adjustments of tone are made to ensure that any swaying motion has only a small excursion and that external forces on the body are counteracted.

A further method by which posture may be maintained in response to external forces is by movement of the limbs. For instance, if the body is pushed to the left by pressure on the right side of the trunk, then the left foot is moved (reflexly) a short distance laterally. This widens the base on which the body is standing, thus ensuring that a perpendicular from the centre of gravity falls within the base.

Postural Reflexes

We have seen in the previous section that maintenance of posture depends upon changes of muscle tone or movement of limbs. For such changes to be effective they must be both accurately regulated, to give the appropriate correction, and rapid. This is achieved by receptors which sense changes in body position and influence tone or induce movement through extrapyramidal pathways and the γ efferent neurones.

The reflexes involved in maintenance of posture act together in such a well-coordinated manner in the normal healthy subject that they are difficult to elicit. Disturbances of equilibrium may activate several reflex pathways, and therefore individual reflexes can only be demonstrated in animals in which other mechanisms have been eliminated or in patients in whom some mechanisms have been destroyed by disease.

Reflexes which are involved in maintaining posture (see Roberts, 1978) are described as follows.

The positive supporting reaction

Pressure on the sole of the foot produces simulataneous contraction of both extensors and flexors of the limb, so converting it into a solid pillar.

Stretch reflexes

The stretching of antigravity muscles, e.g. quadriceps femoris, leads to a reflex shortening of the muscle. Thus any tendency to bend the knee during standing will cause reflex straightening of the knee by contraction of the quadriceps. Forward flexion of the trunk will stretch the extensors of the spine, leading to reflex shortening of these muscles and of the extensors of the leg on the trunk, to restore the trunk to an upright position.

Vestibular reflexes

The vestibular apparatus is enclosed within the temporal bone and with the cochlea constitutes the inner ear (Fig. 7). The vestibular apparatus consists of two saccular compartments, the utricle and the saccule, which detect changes of position of the head in relation to the earth's gravitational field and three semicircular canals which detect angular motion of the head. The apparatus contains endolymph and is separated from the protecting bone by perilymph.

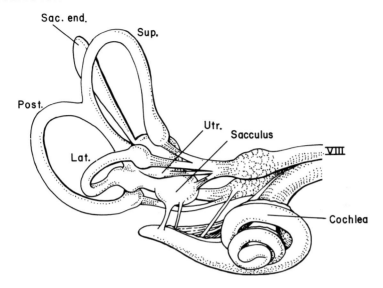

FIG. 7 Drawing of the membranous labyrinth of the right side. Sac. end = sacculus endolymphaticus; Sup. = superior, Post = posterior; Lat = lateral semicircular canals, Utr = utricle, VIII = 8th cranial nerve (from Roberts, 1978).

On each semicircular canal is a dilatation, the ampulla, which contains a projecting ridge or crista. The ridge is covered by hair cells contained in a mass of gelatinous material (the cupula). At the base of the hair cells are branches of the vestibular nerve (Fig. 8(a)). The cupula forms a curtain across the semicircular canal and is deflected by movement of fluid in the canal. This generates nerve impulses (Fig. 8(b)).

The macula, which is the equivalent of the crista in the utricle and saccule, contains chalky particles within its cupula. The pull of gravity on these particles leads to stretch of the hair cells and the generation of nerve impulses. Tilting of the head to one side leads to increased discharge from the saccule of that side and reduced discharge from the opposite saccule. Ventral or dorsal flexion of the neck both increase the nervous discharge from the utricle, which is minimal when the head is erect. The mechanisms are sensitive to a tilt of as little as 2.5 degrees.

The complex vestibular nuclei are in the pons and medulla. Vestibular information is widely relayed within the nervous system; there are connections to the cerebral cortex, giving rise to the conscious sensations of balance and motion, with the oculomotor nuclei to influence eye movements, to the cerebellum, reticular formation and other brain stem nuclei, and to the spinal cord via the vestibular spinal tracts. There appear to be

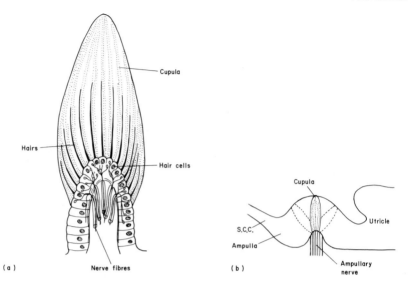

FIG. 8   Structure and function of semicircular canal. (a) Structure of crista of ampulla of semicircular canal. (b) Mode of stimulation of semicircular canals. The cupula, situated on the top of the crista, completely blocks the ampulla of the membranous canal. The cupula is caused to swing by movements of the endolymph (from Keele and Neil, 1971).

both direct and indirect pathways by which the vestibular apparatus may modify muscle tone and consequently help to maintain posture.

Ocular reflexes

It is a common observation that if we attempt to stand still with our eyes closed, after a few seconds we begin to sway. If we stand with our eyes open less sway occurs. Furthermore, patients who have vestibular damage are unable to stand with their eyes closed. It appears that the body uses visual information for regulation of posture, and this may be of considerable importance when vestibular mechanisms are deranged.

Tonic neck reflexes

These reflexes arise from receptors in the ligaments of joints between the cervical vertebrae. They can only be demonstrated convincingly in animals in whom bilateral extirpation of vestibular apparatus has been carried out, or in humans who have suffered cerebral damage. Ventriflexion of the head causes flexion of the arms and extension of the legs, whereas dorsiflexion of the head causes the arms to extend and the legs to flex. Turning of the head to the right causes extension of the right arm and leg, whilst the left limbs are flexed.

Testing of Postural Function

The ability to stand and remain still depends upon adequate function of receptor mechanisms including the vestibular apparatus and the eyes, coordinating centres of which the cerebellum is the most important, and of course motor pathways and muscle. The efficiency of postural mechanisms may be determined by measurement of the amount of sway of the body which occurs when a person attempts to stand still.

The whole basis of maintenance of posture is the correction of small errors as soon as they occur. Patients with damage to postural mechanisms will clearly sway rather more. Wright (1971) has designed a simple ataxia-meter which measures the amount of angular sway during standing (Fig. 9). It consists of a mast to which the patient is attached. The angular movements of the mast are summed by a ratchet and counter mechanism.

Sway increases with age (Fig. 10) because of aging of the central nervous system, which may often be secondary to changes in the cardiovascular

FIG. 9   The Wright ataxiameter.

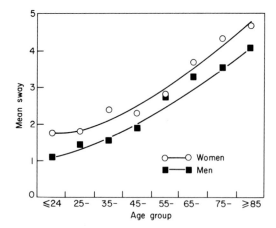

FIG. 10   The increase of sway which occurs with age (from Overstall *et al.*, 1977).

system. This impairment of the body's ability to correct rapidly small losses of postural equilibrium may contribute to the increased incidence of falling in the elderly.

Testing of Vestibular Function

More specific testing of vestibular function may be performed using a rotating chair, or by placing cool water in the external auditory meatus. These are tests for the functioning of the semicircular canals. Although the utricle and saccule are more important for maintaining a standing posture, it is not possible to test their functioning.

De Barany chair

The subject is seated, with the head forward, on a chair which can be rotated rapidly. When rotation begins the endolymph in the horizontal canal, because of its inertia, spins more slowly than the wall of the canal and effectively flows in a direction opposite to that of spin. The cupula is thus deflected in the direction opposite to that of the rotation. This sensation is interpreted as spinning in the direction in which the chair had been rotated. Sudden stopping of the chair leaves the endolymph spinning in the direction of the previous rotation, and deflection of the cupula in this direction. The subject now feels that he is spinning in a direction opposite to that of the rotation, and if asked to rise from the chair and walk in a straight line he will veer in the direction of the spin. Observation of the eyes shows side to side movements (nystagmus) with a slow component in the direction of rotation and the fast component in the opposite direction.

Caloric tests

The chair test checks vestibular function on both sides. Unilateral testing may be carried out by placing cool water in the external auditory meatus. This will set up a slow convection current in the horizontal canal and give a sensation of rotation.

Cardiovascular Adaptations to Changes of Posture

Most classical studies on the control of the circulation have been made on anaesthetized quadrupeds lying supine. In this chapter it is more important to consider the control of the circulation in conscious man and, in particular, the effects of transition from a lying to an upright posture. The maintenance of an adequate cerebral circulation during postural change is of paramount importance.

Standing up causes a transient fall in arterial blood pressure associated with an increase in heart rate (Ewing et al., 1978). It is likely that when we stand, blood pools in veins below the heart, leading to a sudden reduction of cardiac output and causing a fall in arterial blood pressure. This transient hypotension leads, via baroreceptor reflexes, to increases of peripheral arterial resistance and of heart rate. The rapidity of this reflex is illustrated in Fig. 11, which shows a reduction of interbeat interval of some 25% by the fifth heart beat after standing. A further important factor in restoring blood pressure is an increase of venomotor tone which improves venous return thereby restoring cardiac output. Restoration of blood pressure is followed by a slowing of the heart to a rate only slightly faster than that during recumbancy.

The cerebral circulation has a poor sympathetic innervation; generalised sympathetic activation as occurs on standing will cause vasoconstriction in the skin, gastro-intestinal tract and kidneys, and hence increase the relative blood flow to the brain. To restore blood flow in other tissues to the levels prior to standing, cardiac output must be restored to its previous level.

A further factor which helps to maintain venous return during standing is

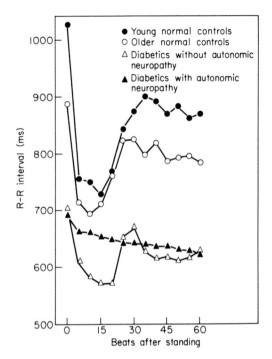

Fig. 11   The effects on the heart rate of suddenly rising from a lying to a standing position. The heart rate was measured from an ECG tracing and expressed as the interval between successive R waves (from Ewing *et al.*, 1978).

the pumping action of skeletal muscle on the veins, which have valves allowing blood to flow only from the periphery towards the heart. Compression of a segment of vein by the contraction of an adjacent muscle forces blood from that segment through the valve into the next more proximal segment.

Adaptation of the circulation may fail to occur if the autonomic nervous system is diseased (Fig. 11), or if it has been blocked with autonomic blocking agents which are frequently used in the treatment of hypertension. In these circumstances, standing causes a more marked reduction of blood pressure and the person feels dizzy. This condition is referred to as "postural hypotension". A similar phenomenon may be experienced by a healthy person if he rises quickly from a hot bath; skin blood flow is high, and is not rapidly reduced by sympathetic activation.

We may stand for prolonged periods providing that there is sufficient muscular activity to maintain venous return. Standing without rhythmical muscle activity, particularly in warm weather, may lead to a gradual decline in blood pressure. If this continues, despite strong sympathetic activation, a more drastic reflex response occurs. Blood pressure is suddenly lowered by massive dilatation of vessels in skeletal muscle, and the heart is slowed by vagal impulses. This sudden fall in pressure causes cerebral ischaemia so that the person collapses and loses consciousness. Once the body is in a horizontal position venous return is increased and as blood pressure rises the victim regains consciousness.

The ability of the autonomic nervous system to regulate arterial blood pressure, when rising to a standing position from lying or sitting, may conveniently be assessed by applying a negative pressure to the lower part of the body when the subject is lying. This procedure causes pooling of blood in the lower limbs similar to that occurring on standing. Patients suffering from autonomic insufficiency may show marked falls in blood pressure under these circumstances.

Voluntary Movement

Movement of the body depends upon contraction of skeletal muscle. Stimulation of contraction is induced by discharge of the α motor neurones in the spinal cord. These neurones may be stimulated either by the corticospinal tracts or reflexly from receptors in muscles, joints or skin. Even when the body is at rest most muscles are in the state of partial contraction known as muscle tone (see p. 364). A voluntary movement can then be considered as a large transient increase of tone. If we consider a movement such as flexion of the arm at the elbow joint which is brought about by contraction of the biceps, then clearly the antagonist muscle, the triceps, must at the same time be relaxed.

The initiation of voluntary movements is by discharge of the cells of the contralateral motor cortex, the impulses travelling along the pyramidal tracts to the anterior horn cells. It has long been known that movements of each part of the body have areas of representation in the cortex. Those parts which have a relative fineness of movement, for instance the hand, have a larger area of representation than those areas, e.g. trunk, for which movement patterns are much coarser.

With increasing knowledge of functions of the extrapyramidal system and of the γ efferent neurones, it has become progressively obvious that this system is involved in the control of voluntary movement by the higher centres (Matthews, 1972). Stimulation of γ efferent neurones leads to contraction of intrafusal muscle fibres and increases the sensitivity of the stretch receptors. This causes muscle contraction via the stretch reflex, and may be a better method of regulating the extent of muscle contraction than direct stimulation of the α motor neurones. If the desired extent of muscle shortening is set by intrafusal stimulation, then reflex contraction of the muscle occurs until the desired shortening is obtained. In this way the effects of different loads on the magnitude of contraction will be overcome, since contraction of the muscle will continue until the desired amount of shortening has been obtained.

Besides feedback information from muscle spindles on the degree of shortening of a muscle during movement, the performance of skilled and complex acts requires detailed information of the position of the moving part. This is derived by continuous feedback from Golgi tendon organs and joint position sense receptors. Afferent information passes along the spino-cerebellar tracts and dorsal columns to the cerebellum, which plays an important role in the control of voluntary muscle contraction.

Voluntary motor acts are initiated from the motor cortex by activation of both α and γ motor neurones. At the same time, information on the desired movement is transmitted from the cortex to the cerebellum. Sensory information from the moving parts reaches the cerebellum which compares the "intentions" of the cortex with the "performance" of the moving parts. The "error" between these two is calculated so that necessary correction can be made. In addition to sensory feedback from the limbs, visual information plays an important role in the control of movement. We are able to perform skilled movements of the hands more effectively with our eyes open than with them closed. In patients with cerebellar disease or damage, movements which are clumsily performed with the eyes closed are more refined when the eyes are open.

Repetitive movements such as walking or running are probably initiated by the cerebral cortex, but are continued by brain-stem and spinal centres (Miller and Scott, 1977). It is of interest that both decerebrate and spinal cats are able to perform stepping movements, providing they are adequately supported. Once initiated, walking and running will continue until conscious decisions are made to change the pattern or speed of locomotion, or to alter direction. Reflex adjustments to patterns may be made to compensate for changing circumstances, such as unevenness of the terrain.

Walking

Walking, like many other repetitive activities such as chewing, cycling or knitting, is thoroughly familiar to us, and yet how many of us could describe the movements in anatomical terms? We may recognise a gait as being normal or abnormal, but are unable to state clearly our reasons for

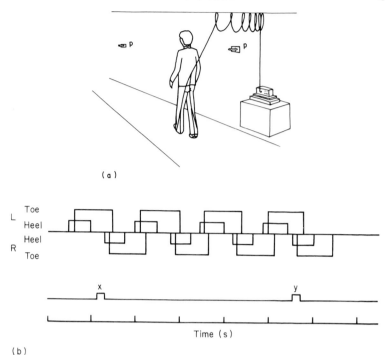

FIG. 12 (a) Apparatus used to assess time and distance factors in walking. The subject walks along a metal track wearing shoes with metal foil contacts on the heels and soles. During a passage the subject interferes with light beams crossing the track at right angles to energize photocells (p) (from Imms and MacDonald, 1978). (b) Record obtained during walking on metal track. From above downwards: contacts of L sole, L heel, R heel, R toe; outputs of photocells (x, y); time (from Imms, 1977).

this belief. If the biomechanics and physiology of normal walking are to be deciphered, and abnormalities occurring in pathological gaits understood, objective methods of representing walking must be adopted (Grieve, 1969).

Simple information regarding the speed of walking, stride length and frequency of stepping may be obtained, using apparatus similar to that illustrated in Fig. 12. It consists of a metal track on which the subject walks in shoes with metal contacts covering the heels and soles. Placement of a heel or sole on to the track completes a circuit and the durations of heel and sole contacts with the track during walking are recorded on ultraviolet paper. Beside the track are two photocells, a fixed distance apart, which are activated by light from lamps on the opposite side of the track. The subject breaks these beams during walking and hence the velocity of walking may be calculated.

The time taken for a stride (a complete walking cycle) may be measured from any fixed point on the trace, e.g. left heel strike to left heel strike. Stride length is then calculated as the product of stride time and velocity; the reciprocal of the stride time gives the frequency of stepping. The relative lengths of the support and swing phases for each limb during the walking cycle may be assessed, as may the duration of the double support time when both feet are on the ground. The symmetry of the gait may be examined by comparison of the support times and swing times for left and right limbs. The duration of the half-pace time for the left leg moving forward may be taken as the interval between right heel strike and left heel strike, and compared with that when the right leg is advancing. Using this apparatus the length of individual half-paces cannot be made, but this can be achieved using a track which is composed of multiple narrow slats on which progress is recorded by the shorting of adjacent slats by the metal contacts on the shoes.

The speed of walking may be altered by varying both stride length and step frequency. At the same time, stride length depends on the stature of the subject; at any given speed a person of short stature will take more strides than a tall person. A gait can only be completely described when measured over a range of speeds and in relation to stature. When both stride length, speed and other variables are expressed in terms of stature, there is little variation between the gaits of normal subjects (Fig. 13).

To investigate the angular changes occurring in the limbs during walking, methods of "freezing" the action are needed. This may be achieved by cinephotography followed by measurement of limb segment positions from individual frames so as to built up a composite picture of the movements of the limbs during a stride (Fig. 14). Alternatively, these measurements may be made by chronocyclography (Fig. 15). The subject wears black tights, with a luminous white stripe along the long axes of the thigh, calf and foot, and walks past a still camera containing slow film. A disc rotating in front of the camera interrupts the exposure of the film at a predetermined rate. In this way, the changes occurring during a complete walking cycle may be seen on a single photograph.

Even from such attempts to slow and pin-point the changes occurring during a walking cycle, it is conceptually difficult to understand the movements of the limbs. The relationships of the segments of limbs may be understood better if the position of, for instance, the thigh is plotted against that of the lower leg. Figure 16 shows the angle of hip extension or flexion plotted against the angle of flexion of the knee. A characteristic graph is obtained for normal subjects, and if the gait is abnormal a very different curve is produced. Imagine the "curve" obtained if the person has a knee joint fixed in full extension! The time required to produce such curves may

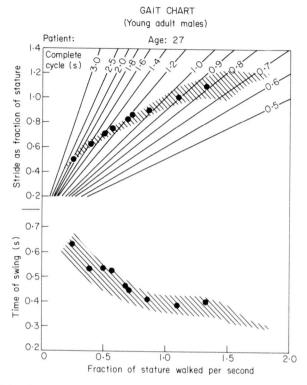

FIG. 13 Gait chart for young male subjects relating the speed of walking (fraction of stature walked per second) with the stride length (fraction of stature) and time of swing. The shaded areas contain 65 observations made on six normal young adults and the closed circles refer to measurements on one of these subjects (from Grieve, 1969).

be reduced by using a polarized light goniometer. Sensors to polarized light are attached on either the trunk and the thigh or on the thigh and the calf (Mitchelson, 1973). The subject then walks at right angles to a beam of polarized light and the outputs of the sensors are fed to an $x–y$ recorder to produce an angular plot.

The aim of walking is to impart forward motion to the body. This is achieved by the horizontal component of the force developed during pushing off with the foot, e.g. frames 1, 2 and 3 (Fig. 14). During this time the body's centre of gravity is falling, since the angle between the legs is increasing. Forward motion of the body is slowed when the opposite leg, which has been swinging, makes contact with the ground (frame 4) and takes over the support of the body. As the supporting limb passes through a vertical position the body's centre of gravity is again raised. The trunk is

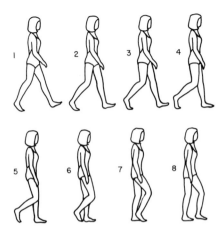

FIG. 14   A normal gait at moderate speed traced from cine film (from Wells, 1955).

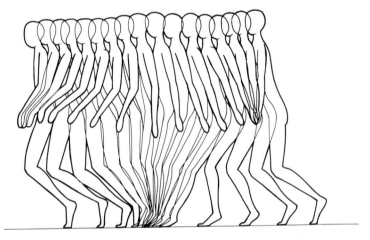

FIG. 15   Chronocyclograph of a young girl walking fast (20 exposures per second) (from Grieve, 1969).

undergoing vertical and horizontal oscillations and both the lower and upper limbs are making swinging or pendular movements. These movements are brought about by muscular contraction which implies expenditure of energy. Since walking is a well-coordinated function in which efficient transformations of energy occur, the total energy cost is relatively low; during elevation of a limb it gains potential energy which is converted to kinetic energy during the downward swing. As the limb swings beyond

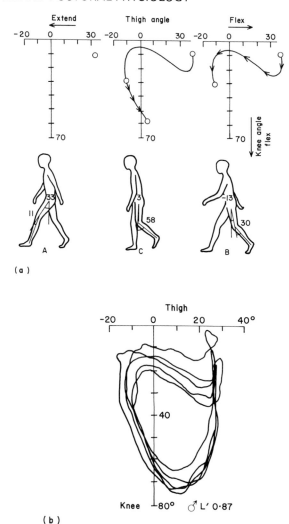

FIG. 16 (a) Construction of a thigh–knee angle chart from cine film taken during walking. (b) Angle charts at moderate speeds and strides (from Grieve, 1969).

the vertical and begins further upward movement it again acquires potential energy. This topic is considered further in the section on the energy cost of posture and locomotion (see p. 382).

The roles of individual muscles during walking may be examined by the use of electromyography. For instance, gluteus maximus, whose main action is extension of the hip joint, begins contraction just prior to heel strike and reaches a maximum as the supporting limb approaches the vertical. A secondary burst of activity may occur to steady the pelvis during propulsion by the opposite leg.

Abnormalities of gait may be caused by injuries to the lower limb and by arthritis of the joints. If a patient experiences pain in a limb he will generally "favour" that limb by reducing the duration of support phase and increasing the swing phase. Reduced muscle power due to muscular dystrophy or secondary to motor nerve injury will induce abnormal walking. Furthermore, loss of position sense in the limbs will lead to a characteristic stamping gait, as usually portrayed in characterizations of King Henry VIII.

The frequency of abnormalities of the gait increases with age, due mainly to the increasing incidence of both arthritis and neurological disease; and may be determined using many of the techniques described above. Alternatively, they may conveniently be assessed by analysis of films or videotapes of the subject walking and carrying out simple tasks such as rising from a chair, climbing and descending stairs and turning around.

Running

Running differs from walking in two important respects: first, there is no double support phase during which both feet are on the ground, and secondly there is a period during which neither foot is on the ground and the body "floats" forward. As a result of this floating the stride length is considerably longer than during walking.

The Energy Cost of Posture and Locomotion

Since the maintenance of posture and locomotion both require contraction of skeletal muscle, energy must be expended whilst we stand and when we walk or run. The increases of energy expenditure above the basal level when we sit or stand are small (Table 2).

The body's energy expenditure during walking on the level, both on smooth ground and on a treadmill, increases curvilinearly with velocity (Fig. 17). Conveniently, the relationship between energy expenditure and the sqaure of the velocity is linear (Fig. 18) and the energy expenditure, E (ml O_2 kg^{-1} min^{-1}) during walking is given by the equation

TABLE 2   The energy cost of maintaining posture

	Cal min^{-1}	$\dot{V}_{O_2}$ (ml min^{-1})	$\dot{V}_{O_2}$ (ml kg^{-1} min^{-1})
Lying—basal conditions	1.28	260	3.71
—rest	1.45	290	4.29
Sitting	1.60	330	4.71
Standing	1.80	370	5.29

(Data from Passmore and Durnin, 1955.)

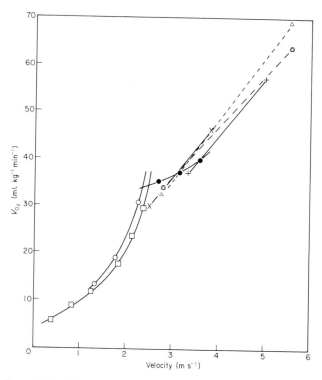

FIG. 17   The relationships between oxygen consumption and velocity of movement during walking and running on the level. Data from Wyndham *et al.* (1971): ○, walking on treadmill; ●, running on treadmill. Passmore and Durnin (1955): □, walking indoors and outdoors. Bransford and Howley (1977): running by ×, unfit and +, fit subjects. Pugh (1970): △, Olympic standard athletes on treadmill and ⊿, Olympic standard athletes on track.

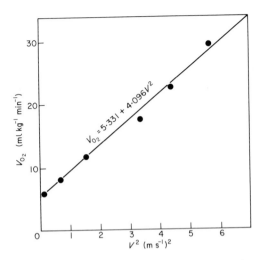

FIG. 18 The linear relationship between oxygen consumption during walking and velocity squared (from data of Passmore and Durnin, 1955).

$$E = 5.331 + 4.096V^2 \qquad (1)$$

where V is the velocity of walking in m s^{-1}. The fixed component of the energy expenditure is similar to the energy required for standing. From this equation may be calculated the energy cost (E_{km}) of walking a kilometer at a particular velocity.

The time taken to walk 1 km at V m s$^{-1} = 1000/60V$ min. Energy expended in this time

$$E_{km} = E \times \frac{1000}{60V} \qquad (2)$$

E may be derived from (1) and substituted into (2) to give

$$E_{km} = \frac{1000}{60V}(5.331 + 4.096V^2) \qquad (3)$$

which simplifies to

$$E_{km} = \frac{88.85}{V} + 68.27V \qquad (4)$$

A plot of E_{km} against V shows that the most economical velocity for walking on the flat is 1.16 m s^{-1} (Fig. 19) but over a normal range of walking speeds, say 1–2 m s^{-1} ($2\frac{1}{4}$–$4\frac{1}{2}$ mph), the energy required for walking a set distance is relatively constant. Walking either very slowly or quickly, however, considerably increases the energy requirements.

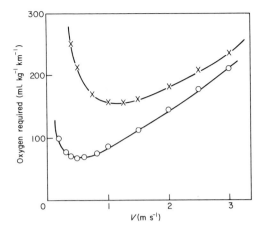

FIG. 19 The volumes of oxygen required for walking 1 kilometre at different speeds. ×———× total oxygen consumption; ○———○ net (i.e. total minus resting) oxygen consumption. (Calculated from data of Passmore and Durnin, 1955.)

The *net* energy expenditure during walking a kilometre may be calculated by subtracting the energy which the body would use at rest

$$E_{km} \text{ net} = \frac{1000}{60V}(5.331 + 4.096V^2) - R\frac{1000}{60V} \tag{5}$$

where R is the resting energy expenditure. The most economical speed of walking in these terms is 0.5 m s^{-1}.

The energy expenditure during running on the level is linearly related to the velocity (Fig. 18). Our experience that it is easier to progress by walking at slow speeds and by running at faster speeds is borne out by consideration of the energy expenditure involved in walking and running. Upward extension of the curve for walking demonstrates that at velocities exceeding 2.5 m s^{-1} the energy required to walk at a given speed is greater than that for running; downward extension of the lines for running shows that at lower velocities walking is the more efficient mode of progression.

It is interesting to consider further the demand for energy during level walking (Dean, 1965). Since no work is performed against gravity the total energy change by the body is zero. During each step energy must be expended to: raise the centre of gravity of the body; accelerate the legs and arms during swing phases; accelerate the body during horizontal and sideways oscillations.

In Table 3 are shown estimates of each of these changes in energy expenditure. Since each of these processes is reversible, transformations of

TABLE 3 Calculated component energy expenditures (kcal min^{-1}) during walking. Their sum is greater than the observed energy expenditure of walking (subject 70 kg).

	Velocity (m s^{-1})		
	0.5	1.0	1.5
Leg swinging	0.091	0.54	1.49
Arm swinging	0.003	0.03	0.120
Progressional oscillation	0.12	0.48	1.07
Sideways oscillation	0.01	0.005	0.01
Vertical motion	0.084	0.32	0.64
Total calculated energy expenditure (kcal min^{-1})	0.308	1.375	3.330
Observed oxygen uptake (ml min^{-1})	445	660	1018
Net oxygen uptake	145	360	718
Calorific equivalent (kcal min^{-1})	0.70	1.75	3.49
Mechanical work equivalent assuming efficiency $= 21\%$ (kcal min^{-1})	0.15	0.37	0.73

After Dean (1965).

TABLE 4 Energy expenditure of horizontal walking at 1.5 m s^{-1} under normal and abnormal conditions

	V_{O_2} (ml kg^{-1}min^{-1})
Treadmill	14.9
Asphalt road	16.7
Grass	18.5
Ploughed field	22.7
With crutches	21.5
Shortening of one leg	28.0
—with made up shoe	22.0

(Data from Passmore and Durnin, 1955; Imms et al., 1976)

potential energy into kinetic energy, and vice versa, will occur. Therefore, raising the centre of gravity of the body increases the body's potential energy which is then reduced when the centre of gravity is lowered. Similarly, the kinetic energy of the leg is increased by acceleration of the limb during the first part of its swing and reduced by deceleration in the second part. The muscles involved in walking are doing both positive (concentric) and negative (eccentric) work.

The calculated energy requirements for walking (Table 3) exceed the net mechanical work done by the body. This latter quantity is derived by

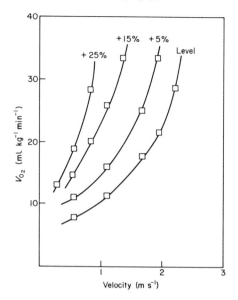

FIG. 20 Oxygen consumption during uphill walking (from Passmore and Durnin, 1955).

calculating the net energy expenditure in terms of oxygen consumption, converting this to energy units and assuming an efficiency for muscular work of 21%. It may therefore be assumed that marked interconversion of energies occurs during walking. If we imagine that the leg swings as a pendulum, then at the beginning of its swing it possesses potential energy, which is converted to kinetic energy which becomes maximal at the midpoint of swing. The pendulum now decelerates as it begins to move upwards and kinetic energy is reconverted into potential energy. Kinetic energy of forward movement may also be converted to potential energy by giving vertical lift to the body, rather as a pole vaulter uses a pole to convert his kinetic energy on the runway into potential energy.

The efficient interconversion of energies during walking can occur only if the speed is constant and the gait pattern symmetrical. Walking on uneven surfaces, such as a ploughed field, interferes with these patterns and increases the metabolic energy required for walking (Table 4). Energy expenditure is also increased in patients with leg injuries who are walking with the aid of crutches. They take strides of unequal lengths and the duration of the support phase with the injured leg is shorter than with the uninjured. A patient who had 6 cm of shortening of one leg following a femoral fracture had a gait which was clearly abnormal, and his energy expenditure during walking was raised. Raising the shoe on his injured leg improved his gait and lowered his energy expenditure.

During uphill walking or running, energy expenditure increases in proportion to the gradient (Fig. 20). In these circumstances, the body is performing work against gravity. The efficiency of this lift work may be calculated by subtracting the energy expenditure during level walking from that during grade walking, and relating this to the work performed against gravity.

For example A man walks up a 15% gradient at a velocity of 1 m s^{-1}. His oxygen uptake is 23.5 ml kg^{-1} min^{-1}; on the level it is 9.4 ml kg^{-1} min^{-1}.

Net oxygen uptake = (23.5 − 9.4) ml kg^{-1} min^{-1}
Net energy expenditure = 0.069 kcal min^{-1}
Work performed on 1 kg in 1 min = (mass) × (vertical lift)
$$= 1 \times 60 \times 0.15$$
$$= 9 \text{ kg m min}^{-1}$$
$$= 0.021 \text{ kcal min}^{-1}$$

$$\text{Efficiency} = \frac{0.021}{0.069} = 100 = 30.4\%$$

During downhill running or walking the energy expended at any velocity is less than during level walking (Fig. 21; Davies *et al.*, 1974). The energy expenditure is least when walking or running down a 10–15% slope, and on steeper negative gradients the energy expenditure approaches that of walking or running on the level.

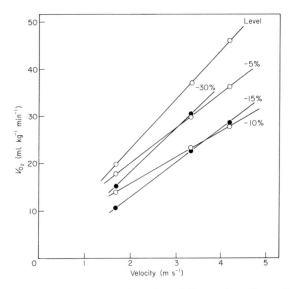

FIG. 21 Oxygen consumption during downhill running (from Davies *et al.*, 1974).

Reduced Physical Activity

Reduction of activity reverses the training processes and results in a higher resting pulse rate, reduction of V_{O_2} max muscular atrophy and reduction of strength, and an increase in the proportion of adipose tissue in the body. Such reductions of activity may occur when participation in sport ceases, at retirement from manual work, or on promotion from manual to supervisory employment. Failure to reduce calorie intake at such times results in an increase of body weight. The physiological changes associated with "detraining" have so far attracted the physiologist less than those occurring during training.

Bed rest

Bed rest is a convenient way to investigate sudden and drastic reductions of activity. Studies on healthy young men confined to bed for two to three weeks have demonstrated marked detraining of the cardiovascular system (Saltin *et al.*, 1968; Stremel *et al.*, 1976). For instance, resting heart rate increased by about half a beat per minute each day and heart volume was reduced by 10–15%. Body weight was unchanged, which may have accounted for the lack of changes in basal metabolic rate and resting cardiac output. However, there were reductions of up to 15% in the total blood volume and plasma volume.

Cardiac detraining was further demonstrated by the failure of the circulation to meet the demands of tilting and of muscular exercise. Prior to bed rest, sudden tilting of the body from the horizontal to 68° feet down raised the pulse rate by 13 beats min^{-1}, whilst systolic blood pressure was unchanged. On the day following bed rest, tilting increased the pulse rate by 36 beats, and systolic blood pressure fell by 14 mm Hg (Taylor *et al.*, 1949). In seven studies of patients confined to bed for between 10 and 20 days, reductions of aerobic capacity ranging from 5.2 to 26.4% have been described (see Stremel, 1976).

Bed rest and injury

The physiological effects of confinement to bed, following surgery or trauma, have also been investigated. Surgical trauma appears to accelerate cardiovascular deconditioning. In patients who had undergone abdominal surgery, the aerobic capacity was reduced to 80% of the pre-operative value after being confined to bed for only three days (Carswell, 1975). In patients who had undergone menisectomy the heart rate at an oxygen consumption of 1.26 1 min^{-1} was greater (145 beats min^{-1}) four days after surgery than after 14 days in bed (127) (Bassey *et al.*, 1973).

Metabolic responses to trauma

Trauma such as limb fracture, surgery or burns affects oxidative metabolism, protein anabolism and catabolism, electrolyte and water balance and removal of calcium from bone.

Cuthbertson (see Stoner, 1970) has described the changes in energy metabolism in three phases (Fig. 22). Immediately following injury the body's energy expenditure is lowered. This is described as the *ebb* phase and may persist for up to 72 hours depending on the severity of the trauma. This reduction of metabolism is not due to shortage of metabolic fuels, since there is often hyperglycaemia and elevated plasma free fatty acid levels. Animal experiments have shown that in a thermoneutral environment energy expenditure is normal, but the animal does not raise its metabolic rate when in a cooler environment. The "ebb" phase may then be the result of a temporary loss of thermoregulation. If the injury were to prove fatal, then at any time a further lowering of metabolic rate may occur, which is generally associated with circulatory shock and reduced perfusion of the tissues. In this phase, known as necrobiosis, severe acidosis is present because of both anaerobic metabolism and failure of pH regulation.

If the trauma is not fatal the ebb phase gives way to the *flow* phase, in which oxidative metabolism is increased. The extent and duration of this phase depend upon the nature and severity of the trauma as well as on the pre-injury nutrition of the subject. A major fracture may raise metabolic rate by 10–20% for up to three weeks, whereas a simple elective procedure, such as a mastectomy, will have little effect.

The flow phase is associated with loss of weight and increased excretion of nitrogen, potassium, phosphate and sulphate, suggestive of protein catabolism. Since food intake may be reduced in the early days following injury, this weight loss may be due in part to utilization of protein and fat as sources of energy. Studies on specific proteins such as plasma albumen have shown marked increases of rates of both catabolism and anabolism, and it is tempting to suggest that the purpose of increased protein catabolism is to make available essential amino acids for repair of damaged tissues. However, since little catabolism takes place in those previously on low protein diets, this view seems less than tenable. Protein catabolism is not the result of increased secretion of adreno–cortical hormones, since it occurs in adrenalectomized animals and in humans who are maintained on constant doses of glucocorticoids.

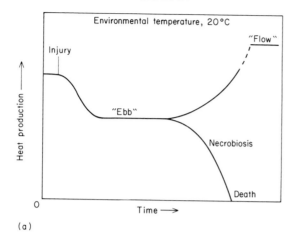

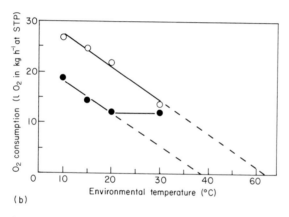

FIG. 22   Metabolic responses to injury. (a) Diagram illustrating the changes in the rate of heat production during different stages of the response to injury in the rat; (b) The relation between oxygen consumption and environmental tempera- ture of (●) rats subjected to four hours of bilateral hind limb ischaemia and (○) uninjured controls (from Stoner, 1970).

Important changes in electrolyte and water metabolism occur following injury. In many cases an important factor here is blood loss associated with the injury, but marked changes occur even in the absence of blood loss. There are reductions of total volume of extracellular fluid and of the plasma volume, because sodium and water enter cells due to altered permeability of cell membranes. Similar cellular changes occur during hypoxia, and it is tempting to speculate that sodium entry into cells may be the result of the breakdown of the sodium pump as a result of oxygen lack due to under-perfusion of tissues. Plasma volume may be further reduced by a shift of sodium and water from the vascular department into the interstitial space as a result of inflammation. Reduction of plasma volume leads to increased secretion of aldosterone and antidiuretic hormone, resulting in retention of salt and water by the kidney. A few days after injury these processes are reversed and there may be a brisk diuresis, which adds to the weight loss associated with increased oxidative metabolism.

Since marked changes in volumes of the body fluid compartments occur following injury, it is perhaps not surprising that deterioration of cardiovascular function occurs.

Long-term effects of injury

Apart from the acute changes in metabolism which follow trauma, long-term changes may occur as a result of immobilization or under-activity. For instance, a patient who has suffered a fracture of the tibia and fibula will have the leg encased in plaster of Paris splint extending from the groin to the foot; this not only immobilizes the injured limb but restricts activity of the body as a whole.

The plaster splint prevents movement of the knee and ankle joints and the patient can only contract the limb muscles isometrically. When the plaster is removed the total limb volume is reduced by up to 15%, with smaller reductions occurring in the uninjured limb. The reduction of limb volume is due to muscular atrophy, which is to some extent offset by an increased fat content. When the patient begins once more to use the limb, the muscle mass gradually increases and the proportion of adipose tissue decreases (Davies and Sargeant, 1975).

Immobilization of a limb also results in functional changes. The reduced muscle mass lowers the aerobic capacity and muscle strength of the limb. These may be further reduced by stiffness of joints and by pain produced during exertion. Stiffness, weakness and pain combine to produce abnormalities of the gait.

In the weeks following removal of the splint, the joints are mobilised and use of the limb results in increase of muscle mass, of aerobic capacity and muscle strength and improvement of the symmetry of the gait (Imms, 1977). At the same time, physical fitness of the body as a whole is increased. The recovery process may be helped by physical treatments to mobilize joints and by exercise therapy which aims to speed muscular hypertrophy and the recovery of function.

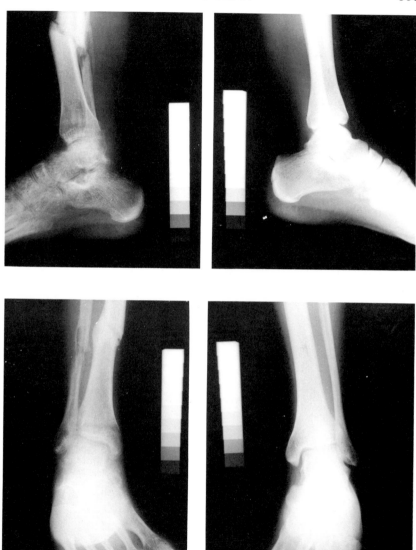

FIG. 23 Osteoporosis following fractures of the right tibia and fibula which were treated by immobilizing the limb in a plaster of Paris (POP) splint. Lateral and antero–posterior radiographs were taken one month after removal of POP splint. Each exposure includes a stepped aluminium wedge for calibration of the intensity of exposure of the film.

Immobilization and the skeleton

Bed rest produces a generalized loss of calcium from the skeleton, with calcium excretion increasing from 200 mg per day to over 400 mg per day. This calcium loss can be prevented by short daily periods of standing. Since calcium loss occurs during space travel during which the astronauts are active, it would appear that standing in a gravitational field was an important factor in maintaining bone structure. A further factor responsible for maintaining calcification of bone is muscular activity, since tilting and rippling of the bed prevents calcium loss in bed-fast patients providing that they are not paralysed (Fourman and Royer, 1968).

If a limb is immobilized, calcium is lost from the bones in that limb. This is illustrated in Fig. 23, which shows X-rays of the tibia and ankle of a patient who had a fracture of the tibial shaft. The radiotranslucency of the bones of the injured limb is less than in the uninjured limb, demonstrating calcium loss as a result of injury and immobilization.

References

Bassey, E.J., Bennett, T., Birmingham, A.T., Fentem, P.H., Fitton, D. and Goldsmith, R. (1973). *Cardiovasc. Res.* **7**, 588–592.

Bransford, D.R. and Howley, E.T. (1977). *Med. Sci. Sports* **9**, 41–44.

Carswell, Sarah (1975). *J. Physiol.* **251**, 42P.

Davies, C.T.M. and Sargeant, A.J. (1975). *Ann. Human. Biol.* **2**, 327–337.

Davies, C.T.M., Sargeant, A.J. and Smith, B. (1974). *Eur. J. Appl. Physiol.* **32**, 187–194.

Dean, G.A. (1965). *Ergonomics* **8**, 31–41.

Ewing, D.T., Campbell, I.W., Murray, A., Nielson, J.M.M. and Clarke, B. (1978). *Br. Med. J.* **1**, 145–147.

Fourman, P. and Royer, P. (1968). "Calcium Metabolism and the Bone", Blackwell, Oxford.

Grieve, D.W. (1969). *Physiotherapy* **56**, 542–460.

Imms, F.J. (1977). Physiological changes occurring during rehabilitation from injury. *Army Personnel Research Committee, Med. Res. Coun. Lond., Report No. APRC 77/4.*

Imms, F.J. and MacDonald, I.C. (1978). *Scand. J. Rehab. Med.* **10**, 193–199.

Imms, F.J., MacDonald, I.C. and Prestidge, S.P. (1976). *Scand. J. Rehab. Med.* **8**, 1–9.

Johnston, T.B. and Whillis, J. (1949). (Eds). "Gray's Anatomy", (30th edn). Longmans, Green and Co., London.

Keele, C.A. and Neil, E. (1971). (Eds). Samson Wright's "*Applied Physiology*" (12th edn). Oxford University Press, Oxford.

Matthews, P.B.C. (1972). "Mammalian Muscle Receptors and their Central

Actions", Edward Arnold, London. Monographs of the Physiological Society No. 23.

Miller, S. and Scott, P.D. (1977). *Exp. Brain Res.* **30**, 387–403.

Mitchelson, D.L. (1973). *In* "Medicine and Sport", Vol. 8. Biomechanics, III, (Ed. E. Jokl), 181–184. Karger, Basel.

Overstall, P.W., Exton-Smith, A.N., Imms, F.J. and Johnson, A.L. (1977). *Br. Med. J.* **1**, 261–264.

Passmore, R. and Durnin, J.V.G.A. (1955). *Physiol. Rev.* **35**, 801–840.

Pugh, L.G.C.E. (1970). *J. Physiol.* **207**, 823–835.

Roberts, T.D.M. (1978). "Neurophysiology of Postural Mechanisms", (2nd Edn). Butterworth, London.

Saltin, B., Blomqvist, G., Mitchell, J.H., Johnson, R.L. Jnr, Wildenthal, K. and Chapman, C.B. (1968). Circulation, **38**, *Suppl.*, **7**, 1–78.

Sherrington, C.S. (1906). "Integrative Activity of the Nervous System", (new edn, 1947). Cambridge University Press, London.

Stoner, H.B. (1970). *J. Clin. Path.*, 23, Suppl. (R. Coll. Pathal.), **4**, 47–55.

Stremel, R.W., Convertino, V.A., Bernauer, E.M., and Greenleaf, J.E. (1976). *J. Appl. Physiol.* **41**, 905–909.

Taylor, H.L. Henschel, A., Brozek, J. and Keys, A. (1949). *J. Appl. Physiol.* **2**: 223–239.

Warwick, R. and Williams, R.L. (1973). (Eds). "Gray's Anatomy", (35th edn), Churchill Livingstone, Edinburgh.

Wells, K.F. and Luttgens, K. (1976). *Kinesiology* (6th edn). Saunders, Philadelphia.

Wright B.M. (1971). *J. Physiol.* **218**, 27–28P.

Wyndham, C.H., Strydom, N.B., Van Graan, C.H., Van Rensberg, A.J., Rogers, G.G., Greyson, J.S. and Van Der Walt, W.H. (1971). *S.A. Med. J.* **45**, 50–53.

The Physiology of Stress
Part I—Emotion

S.M. Hilton

Introductory

It mat seem hardly proper for a British physiologist to admit the existence of emotion and even more outrageous to include it as a fit subject for study and discussion. Yet worthy investigators have made it their life's work, and even Sherington performed experiments to test a theory of emotion. It is, naturally enough, in the study of the functional significance of emotional expression, as well as of the central nervous mechanisms underlying this form of behaviour, that the biologist, and especially the physiologist, has had a major role to play. In this context "emotional expression" does not refer simply to the patterns of activation of the facial muscles common to so many mammalian species, as documented by Darwin (1872), nor even to the widespread activation of the skeletal musculature that characterizes such overt reactions as flight or attack; it extends to the various visceral and hormonal components of these complex responses and the elucidation of their biological role. Everyone who writes on this subject has to express his debt to Cannon (1929) for his book "Bodily Changes in Pain, Hunger, Fear and Rage", which is a landmark in the field and still one of the most stimulating accounts of the subject. It no longer needs emphasizing that the reactions are adaptive, in so far as they lead to the preservation of the organism. Cannon put this idea on a firm foundation by showing how all the bodily changes then known to occur as components of these reactions could be viewed within this framework. The experimental basis for his conclusions was provided by the detailed studies on cats carried out mainly by Cannon himself and his many co-workers, but their relevance to the human subject also, in health and disease, was comprehensively documented. That the emergency reactions do indeed involve the activation of the sympathetic nervous system and the release of catecholamines has in recent years been demonstrated for the human subject exposed to situations of "stress" and anxiety (Mason, 1972; Leshner, 1977; Levi, 1972). Likewise, the present account will have to rest largely on the results of animal experiments, in the confidence that the emergent patterns of reaction will have traces or even stronger reflections in the similar conditions of man. This assumption has been justified at whatever point it has been tested.

In this account we deal not with the alimentary and sexual reactions, whose biological implications are self-evident, but with the emergency reactions—as they have been known after Cannon—whose functional significance is most easily appreciated by reference to animal behaviour. Thus, there is no difficulty in understanding the value of the reactions of flight or attack for preservation of the individual animals, and the fascination for the physiologist then comes with the attempt to unravel the complicated stories, first of the direction and extent of the multitudinous visceral and hormonal changes occurring at different stages of these reactions and secondly of the interconnections of these changes and the various ways in which they contribute to the maximum efficiency of the organism during the emergency.

Anatomical Substratum

The regions of the central nervous system essential to the integration and performance of these emergency reactions have been established in several mammalian species, and the work so far carried out on man, though necessarily more meagre, points to the same general anatomical pattern.

Nervous structures in hypothalamus and midbrain subserving defence reactions

It was early recognized, after the pioneer work of Goltz (1892) with decorticate dogs, that the cerebral hemispheres are not essential for defence reactions. Cannon and Britton (1925) emphasized the significance of diencephalic structures for these responses, for which they coined the term "sham rage". The most extensive and detailed use of cerebral ablation, by Bard and his collaborators (Bard, 1928; Bard and Rioch, 1937; Bard and Macht, 1958), led finally to the conclusion that a large part of the brain stem plays a rôle in the mediation of most features of these responses, but that the hypothalamus must be intact for the behaviour to be well organized and of anything like normal intensity. The responses of these chronic decorticate or decerebrate preparations are elicited by mild cutaneous stimulation. Even when the level of decerebration is as caudal as at the beginning of the pons, pulling the skin of the back is sufficient of elicit most features of the "rage" reaction. (Keller, 1932), and Bard and Macht (1958) obtained some features of the reaction, in response to electrical stimulation of the tail and to a loud noise, when the line of section was so far caudal as to remove the rostral part of the pons.

More precise demarcation of the brain-stem regions integrating these reactions came with the utilization of localized electrical stimulation by means of implanted electrodes. Hess (1949) and his collaborators, in work carried out over many years, charted the regions from which various characteristic patterns of behaviour could be elicited in cats. Responses that began as alerting and culminated, if stimulation was sufficiently intense, in flight or attack were obtained most readily from that part of the hypothalamus most medial ventral and lateral to the fornix (Hess and Brügger, 1943). The fully fledged responses were indistinguishable from those that would be called "fear" or "rage" in an animal responding to a natural environmental stimulus, but Hess and Brügger (1943) preferred to use the collective term "defence reaction" (*Abwehrreaktion*) for them. The excitable region for this reaction in the hypothalamus, as located by them, is illustrated in Fig. 1. Their results indicated that the caudal part of the

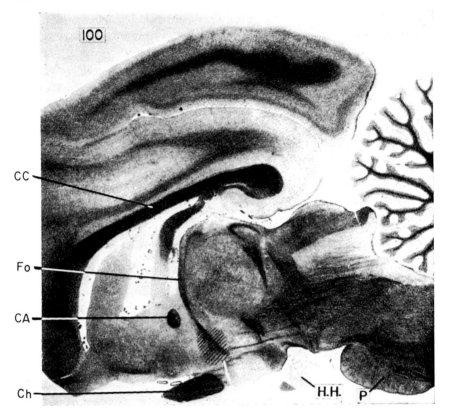

FIG. 1   Paramedian sagittal section of cat's brain. Hatched area indicates perifornical region in hypothalamus from which defence reactions were most readily
elicited. CA, anterior commissure; CC, corpus callosum; Ch, optic chiasma; Fo,
descending column of fornix; HH, posterior hypothalamus; P, pons. (From Hess
and Brügger, 1943.)

hypothalamic region connected with the central grey matter at the midbrain
level. Abrahams *et al.* (1960b) found the region to be even larger, extending
laterally at the midbrain level to occupy an area of the tegmentum ventral
to the superior colliculi. Their investigation had begun from an interest in
the sympathetic vasodilator nerve supply to the skeletal muscles, which was
known not to be involved in any of the homoeostatic circulatory reflexes,
but could be activated by electrical stimulation within the hypothalamus
and dorsal midbrain (Uvnäs, 1954; Lindgren, 1955). There proved to be
such a remarkable identity of the regions, in the hypothalamus, central grey
matter and midbrain tegmentum, from which defence reactions were
elicited in the conscious animal, and the active vasodilatation provoked in

skeletal muscle of the same animal under anaesthetic, that this vasodilatation was virtually the best objective index of the reaction itself. It thus proved an invaluable aid to mapping the exact borders of the excitable regions for the whole defence reaction, because of the large amount of information that could be gained from exploratory experiments on anaesthetised animals with stereotactically orientated micro-electrodes. In a later investigation using the same methodological approach, a caudal extension was found running dorsally in the reticular formation of the pons and medulla (Coote *et al.*, 1973). The location of the regions thus mapped is shown diagrammatically in Fig. 2.

Similar experiments on monkeys, though rather less detailed so far, have provided evidence of the same general topographical organization (Masserman, 1943; Lilly, 1958, 1960; Schamm *et al.*, 1971). Some results of electrical stimulation in the medial hypothalamus of human subjects have been reported (Heath and Mickle, 1960; Sem-Jacobsen and Torkildsen, 1960). Mild stimulation produced feelings variously recorded as restlessness, anxiety, depression, fright and horror; stronger stimulation in the posterior hypothalamus produced "rage" reactions. Almost everyone who has carried out such experiments on animals has been impelled to conclude that electrical stimulation of these regions of the brain-stem, though clearly the stimulus is abnormal, must produce changes within the brain (and thence throughout the organism) that are hardly distinguishable from those due to natural stimulation. It is beyond the scope of this chapter to discuss

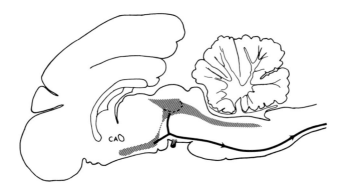

FIG. 2   Diagrammatic paramedian sagittal section of cat's brain (see Fig. 1). Hatched areas represent regions in hypothalamus, mesencephalic tegmentum, central grey matter and medulla which integrate defence reactions; solid line indicates location of efferent pathway for cardiovascular pattern of response and other visceral components. CA, anterior commissure. (Modified from Hilton, 1975.)

the interesting neurophysiological implications of this conclusion, but it seems proper at least to emphasize that, by use of some of the simplest techniques known to physiology, we can begin to explore the mechanisms responsible for phenomena that have often been thought the preserve of the psychologist or even the philosopher.

It is already clear that the function of these regions of the brain stem is not solely motor. Indeed, the ease with which stereotyped responses are elicited by natural stimuli in the various decerebrate preparations led Abrahams *et al.* (1960b) to conclude that these regions are reflex centres in the usually accepted sense. Cannon (1929) had already discussed the emergency reactions in these terms. It seemed curious therefore that interest in this concept should have waned at a time when so much evidence had accumulated for the existence of the sensory connections appropriate to this reflex function. As was pointed out by Abrahams *et al.* (1962), the discovery of sensory pathways impinging on the brain-stem—the afferent collateral system of Starzl *et al.* (1951)—had only been discussed since that time in relation to the concept of the reticular activating system. Abrahams *et al.* (1962) confirmed the findings of earlier workers and extended them a little, to show that potentials could be evoked in all parts of the integrative centre for the defence reaction, in the hypothalamus and midbrain, as response to cutaneous, auditory and visual stimuli. As they concluded, connections undoubtedly exist that would enable these regions to function as a reflex centre for the defence reaction.

Role of the limbic system

So simple a concept, like that of the spinal reflex, has to be qualified; for the central nervous system acts as a whole, and the connections to and from the telencephalon undoubtedly contribute to the emotional reactions of the intact organism. There is as yet no convincing evidence that these reactions can be initiated from specific areas of the neocortex (Hilton *et al.*, 1979). Even the cortex of the frontal lobe, which Fulton (1949) saw as the primary autonomic centre of the forebrain, would probably not be allotted such a role today, although it does contain areas which can modulate defence reactions (Timms, 1977). There now seems, however, to be no doubt of the prime significance of the limbic system. This system comprises the structures surrounding the hilus of the cerebral hemispheres, together called the great limbic lobe by Broca (1878), which are almost identical in all mammals. They include cortex, which had developed phylogenetically in association with the olfactory apparatus, and hypothalamus; as pointed out long ago by Elliot Smith (1919), this was a system that could integrate the various sensory impressions and might impart to them the quality of emotion. This concept was revived by Papez (1937), who gave it a much more definitive form. On the basis of clinical and experimental evidence, he proposed the view that the hypothalamus, anterior thalamic nuclei, cingulate gyrus and hippocampus, through their interconnections, form a circuit that can be excited from the neocortex, through the hippocampus, as well as through the hypothalamus, which "is accessible to both visceral and somatic impression". Subsequent work has given strong support to this ingenious hypothesis, the region of the limbic system most directly connected with the hypothalamus and midbrain proving to be in the amygdala, a group of nuclei subjacent to the hippocampus, within the temporal lobe. Gastaut et al. (1951) first showed that electrical stimulation of the amygdala of the cat produced defence reactions, and this has since been confirmed many times. According to Hilton and Zbrozyna (1963), the full reaction, including active vasodilatation in skeletal muscle, is produced by stimulation in the nucleus basalis amygdalae, pars medialis, and the efferent pathway from here to the hypothalamus takes a direct route along a narrow ventral band, as had been suggested by Ursin and Kaada (1960). The stria terminalis, which connects the anterior hypothalamus with the amygdala and had already been reported to contain fibres initiating defence reactions in the cat (Gastaut et al., 1951; Fernandez de Molina and Hunsperger, 1959), was found to be afferent to the amygdala, thus perhaps completing a circuit with positive feedback (Hilton and Zbrożyna, 1963). In addition, the latero–basal nucleus of the amygdala has an inhibitory influence on defence reactions (Fonberg, 1963). Again, there is similar evidence implicating these regions of the limbic system in higher mammals. Indeed, some of the earliest experimental material was obtained by Klüver and Bucy (1939), who showed complete absence of "fear" and "rage" reactions in macaques after bilateral excision of the temporal lobes. The few studies incorporating electrical stimulation in the human amygdala have yielded similar reactions, including increases in blood pressure and heart rate (Heath et al., 1955; Chapman, 1960: Sem-Jacobsen, as quoted by Bovard, 1962). In these ways a firm edifice is being built. The evidence shows how certain specific regions of the brain became specialized early in vertebrate phylogeny for the mobilization of all the resources of the organism in response to the signals of dangerous situations. The basic central nervous apparatus persists in man, with hardly any further elaboration, and the reactions themselves vary little in detail from one species to another. Even in man they are frequent occurrences in his daily life, though usually the occurrence is a mild one, and we have learnt, more or less successfully, to inhibit our movements of flight or attack. Nevertheless, so far as can be judged, the reactions of the internal organs persist, for the most part unchanged, and these may thus have become the most important changes.

Cardiovascular Response

When Cannon was carrying out his investigations it was already known that "excitement" produces sizeable increases in arterial blood pressure in man. He was interested to know how much these increases might improve the efficiency of fatigued muscle, and the early experiments by his colleagues and himself showed how, in the experimental animal, neuromuscular transmission could be facilitated, and contraction itself augmented, when the arterial blood pressure was raised. He speculated that a raised pressure might be essential to the muscular activities likely to accompany excitement and pain. From the known effects of adrenaline, a decrease in volume of the spleen, kidneys and intestines and an increase in limb volume, he argued that sympathetic discharges would probably exert the same differential effect during these responses, so that blood would be driven out of the viscera and into the skeletal muscles, which have to meet the urgent demands of struggle or escape. The small amount of information then available led him to suggest further that the blood supply to the brain and lungs would not be reduced during these reactions, whereas the coronary flow would probably even increase. He concluded: "This shifting of the blood so that there is an assured adequate supply to structures essential for the preservation of the individual may reasonably be interpreted as a fact of prime biological significance."

Uniqueness of the muscle vasodilatation

The results of subsequent investigations have gradually revealed the remarkable accuracy of Cannon's insight. Vasodilator nerve fibres in the sympathetic innervation of skeletal muscle were, in fact, described some ten years later (Burn, 1938). These special fibres were then shown not to be activated in baroreceptor (that is, depressor) reflexes (Folkow and Uvnäs, 1948), but it was only with the observations of Eliasson et al. (1951) that electrical stimulation in the hypothalamus became known to produce a large muscle vasodilatation through this innervation. They remarked that this muscle vasodilatation was accompanied by vasoconstriction in the skin and intestine, tachycardia, contraction of the spleen, pupillary dilatation and retraction of the nictitating membranes. They suggested that the vasodilator nerve fibres were being activated as part of the patterned response characteristic of the emergency reactions of "fear" and "rage". When Abrahams *et al.* (1960b) demarcated those parts of the hypothalamus and midbrain from which the muscle vasodilatation is evoked, they showed that these were indeed the regions responsible for the reactions of alerting, flight and attack in the conscious animal. As might be expected, the vasodilatation was also obtained as part of the so-called "pseud affective" (or sham rage) reflex in decerebrate cats in which the hypothalamus had been spared. It seems clear that the vasodilatation is an integral part of the defence reaction; this is worth emphasizing because, so far as we know it present, it is unique to this reaction and hence the best single index of its occurrence. By means of a technique in conscious animals for recording the changes of femoral venous blood temperature occurring during such vaso-dilator responses, it has been shown that the responses occur during defence reactions elicited either by hypothalamic stimulation or by natural environ-mental stimuli (Abrahams *et al.*, 1964; Bolme *et al.*, 1967). It seems particularly significant that the vascular response is fully developed during the early alerting stage of the behavioural reaction, when the only outward signs are pupillary dilatation, pricking of the ears and an increase in rate of respiration (Abrahams *et al.*, 1964).

Response considered as a preparatory reflex

As already indicated, when the muscle vasodilatation is elicited in anaesthetized animals by electrical stimulation within the brain-stem region integrating the defence reaction, it is but one component of a complex pattern of cardiovascular adjustment. In addition to the features noted above, the contractile force of the heart and the cardiac output increase (Rosen, 1961a), and there is also evidence of widespread venoconstriction (Folkow *et al.*, 1961; Hilton, 1965). Clearly, the cardiovascular reserves are being mobilised. A greatly increased cardiac output is being directed chiefly to the skeletal musculature, and the experiments on conscious animals suggest that this reaction occurs sufficiently early for the organism to be fully prepared within a few seconds of receiving a sudden stimulus to meet the immediate demands of widespread muscular activity, as in flight or attack. Abrahams *et al.* (1964) therefore suggested that this pattern of cardiovascular response should be classed as a preparatory reflex. The significance of such a reflex for survival of the individual is easily under-stood, yet it must be emphasized that, in order to mobilize the resources of the organism, such a preparatory reflex establishes for the time being a new state of internal equilibrium that represents a radical departure from the status quo. Mechanisms ordinarily operating to maintain the constancy of the internal milieu will therefore have to suffer a temporary abrogation.

This point is underlined by considering one further feature of the cardio-vascular response. While it is developing, the pulse pressure and heart rate are increasing simultaneously. From such observations as these, which have been made many times, Bard (1960) was led to conclude that baroreceptor reflexes must be inhibited. This was tested (Hilton, 1963; 1965) by observing the well-known depressor effect of raising the pressure in a blind sac preparation of the carotid sinus on one side and showing that this effect is strongly inhibited and may be unobtainable during stimulation in the hypothalamic region for the defence reaction. The reflex bradycardia is similarly inhibited. That this is a powerful inhibition within the central nervous system has been confirmed by a recent study in which activity in cardiac and renal sympathetic nerves was recorded in addition to heart-rate, blood pressure and regional blood flows (Coote et al., 1979). Thus, the hypothalamus imposes its own pattern of reaction and over-rides the homeostatic reflex organized at a lower (medullary) level of the neuraxis. This suggests the potency of such a basic reaction not only for promoting the continued existence of the individual by helping it deal with an emergency, but also, if the reaction should be prolonged, for embarrassing the organism by weakening some of its defences and depleting its reserves. Gantt (1960) and his co-workers have carried out many experiments showing that the tachycardia and hypertension produced by painful stimuli in dogs can be readily conditioned and that these conditioned responses are remarkably stable. Experiments carried out on conditioned muscle vaso-dilator responses point in the same direction (Abrahams et al., 1964; Martin et al., 1876).

A similar cardiovascular response, including the active muscle vasodila-tation, is provoked in anaesthetized dogs on stimulation of the same regions of the brain stem as those from which the response is elicited in the cat (Eliasson et al., 1952; Lindgren and Uvnäs, 1953; Lindgren, 1955). The cardiac responses have been obtained in conscious dogs on stimulation of points in the posterior hypothalamus (Smith et al., 1960) and when they are subjected to the stimulus of a sudden shower of water (Charlier et al., 1962). The last-named investigators also reported that the arterial blood pressure rose, though total peripheral resistance was unchanged or fell. It is not surprising, therefore, that the appropriate cardiac responses in dogs trained to run on a treadmill are recorded when the experimenter is observed to move his hand to the treadmill switch (Rushmer et al., 1960).

Brod et al., (1959) were the first to point out that a similar overall pattern of response occurs also in man. It is not necessary to document the statement that "anxiety" in man causes a rise in arterial blood pressure and heart rate. Besides this, the muscle blood flow increases (Wilkins and Eichna, 1941; Golenhofen and Hildebrandt, 1957), owing mainly to vaso-dilatation (Brod *et al.*, 1959). This vasodilatation has since been shown to be produced by atropine-sensitive nerve fibres of the sympathetic outflow (Blair *et al.*, 1959), although circulating adrenaline may make a significant contribution in some subjects (Barcroft *et al.*, 1960). Subjects suffering from chronic anxiety states show all these cardiovascular features of the defence response, including a large increase in muscle blood flow (Kelly and Walter, 1968). Moreover, a return of muscle blood flow towards normal may be the best early indication of successful therapy. The cardiac output is increased and the total peripheral resistance reduced (Stead *et al.*, 1945; Hickham *et al.*, 1948; Brod *et al.*, 1959).

This pattern of cardiovascular response in man, as in animals, is elicited by a variety of stimuli, which may become numerous during an individual lifetime because of the ease with which conditioning occurs. Even without conditioning, there is a further phenomenon resulting from repeated stimu-lation that may have a special significance. Usually, when a novel stimulus is repeated, a waning, or habituation, of the response is seen. In the case of the defence reaction, the converse can occur, and an augmentation, or sensitization, is observed (Martin *et al.*, 1976). Moreover, different com-ponents of responses may be affected differentially. As the response is such a frequent occurrence in daily life, there is much room for speculation about the role it plays in health and disease. We need to learn as much as possible about it in all its details, both in further studies on man, so as to discover how the pattern may differ from that found in animals, and by simul-taneously extending the studies on animals to complete the picture of the whole cardiovascular response. For instance, hardly anything is known as yet of the changes occurring in the profoundly important triad of organs—brain, heart and lungs. These are all regions in which the possibilities for central nervous control of the circulation have not appeared very impres-sive. Nevertheless, electrical stimulation of those parts of the hypothalamus and midbrain that integrate the defence reaction leads, amongst all the other changes, to widespread activation of the brain, including the cerebral cortex (Starzl *et al.*, 1951; French *et al.*, 1952). Since activity in central nervous structures leads to a local functional vasodilatation (Penfield and Jasper, 1955), we would expect widespread cerebral vasodilatation to be a feature of the whole response (see Chapter 9). In fact, Geiger and Sigg (1955) have reported that stimulation for 5 s in the hypothalamus, just anterior and lateral to the mamillary bodies, causes a 25–100% increase in cerebral blood flow lasting for 5–6 min. Some contribution to this increase will be made by the rise in arterial blood pressure, and this will also promote some increase in coronary blood flow. The increased heart rate and contractile force of the heart as well will add to the increase in coronary flow, quite apart from the vexed question of whether there is in addition a sympathetic vasodilator nerve supply to the coronary vessels that could be activated during such responses as the defence reactions. There is only a little information about the changes occurring in the pulmonary vascular bed during these reactions. A decreased compliance of the pulmonary arteries has been demonstrated (Szidon and Fishman, 1971), which will contribute to the mobilization of blood reservoirs; but the rate of flow is greatly increased and it would be important to investigate whether the pulmonary vasomotor innervation could in some way help to provide the most favourable conditions for gaseous exchange.

Chemoceptor inflow and response during exercise

The whole pattern of response has been discussed so far as a preparatory one, hence anticipatory of the actions that immediately follow. This is emphasized by the ease with which it may be conditioned. Nevertheless, we may reasonably ask whether any means exists for maintaining the response during muscular exertion if this should indeed be provoked. Needless to say, Cannon also took this consideration into account. He pointed to the evidence for release of adrenaline after mild asphyxiation in experimental animals as indicating a possible mechanism for producing effects during vigorous exertions that are similar to those "produced in pain and excitement". The general consensus today would not support the suggestion that asphyxia was a likely accompaniment of even the most severe physical exertion, but we may follow up Cannon's main line of reasoning by taking into account a reflex whose existence was not known to him, the chemoceptor reflex.

Hilton and Joels (1965) tried to test whether there is any interaction between the preparatory defence reflex and the chemoceptor reflex. In their experiments the integrative region for the defence reaction in the hypothalamus was stimulated electrically, as before, and the chemoceptor response was elicited by close arterial injection of cyanide into the region of the carotid bifurcation. The effects on respiration as well as those on arterial blood pressure were observed. It is well known that cyanide, by stimulating the chemoceptor endings, produces a reflex hypertnoea and rise in arterial blood pressure. When the injection was preceded by hypothalamic stimulation the reflex response was frequently facilitated. It was never inhibited, as the baroreceptor reflex had been found to be. This facilitation suggests that chemoceptor afferent inflow may, on its own, activate the very regions of autonomic control within the brain-stem that are brought into play by hypothalamic stimulation. Indeed, the possibility that the chemoceptor inflow has functional connections with the regions integrating the defence reactions has been indicated by the experiments of Bizzi et al. (1961). They showed that, in cats decerebrated at a high level with the diencephalon left intact, pseudaffective responses are readily provoked on excitation of the carotid chemoceptors by lobeline injected close arterially or on letting the animal breathe a mixture low in oxygen ($5-12\% \ O_2$ in N_2). It has been known since Bard's detailed description of the reactions of decerebrate cats that pseudaffective responses are abnormally easily elicited, even, for instance, by gentle stroking (Bard, 1928), but these results do suggest that in the normal animal the chemoceptor inflow can cause some degree of activation of the defence reaction, just as can any mild form of cutaneous stimulation. It has now become possible to study such responses in lightly anaesthetized animals, due to the introduction of the use of the steroid mixture, althesin (Glaxo), in animal experiments (Timms, 1976); for, under this anaesthetic, unlike those conventionally used (Chloralose, urethane, barbiturates), hypothalamic and midbrain structures can be activated synaptically without any apparent block or distortion of transmission. Chemoreceptor stimulation under althesin elicits the typical pattern of response of the defence reaction (Marshall, 1977) identical to the pattern produced during naturally elicited alerting in the conscious animal (Caraffa–Braga, Granata and Pinotti, 1973).

This conclusion has several implications. The one most relevant to the present discussion is the possibility that the chemoceptor drive may maintain the visceral components of the defence reaction, if the "anticipated" motor behaviour does occur. If this were so, it might be expected that oxygen lack would be a more effective respiratory stimulus during exercise than at rest, a fact indeed already well established in man.

A further point might be of great importance to physiology. Abrahams et al. (1960b) found that the defence reaction, in common with other reactions integrated at the diencephalic level, could not be obtained as a reflex response in anaesthetized animals. Of the autonomic responses they were observing, only the rise in arterial blood pressure and pupillary dilation remained, even on stimulation of the whole sciatic nerve. The muscle vasodilatation so characteristic of the reaction was never obtained. This vasodilatation was elicited, however, together with the various other manifestations of the whole reaction, as part of the pseudaffective reflex in the high decerebrate cat, when the volatile anaesthetic had been blown off. Thus, in so far as the chemoceptor afferent input is excitor to this high-level reflex, its reflex effects will also be modified by conventional anaesthetics. As more experiments are carried out on this reflex response in preparations under such anaesthetic agents as althesin, we may find that the threshold levels of stimulation are much smaller than have been thought up to the present. The results of experiments carried out so far on anaesthetized animals might turn out to be more serious artefacts than has hitherto been realised.

Respiratory Changes

Rapid and deep respiration is an obvious feature of the response to pain and emotional excitement. This is also the response obtained on electrical stimulation of the midbrain and posterior hypothalamic regions integrating the defence reaction in the cat (Hess, 1949). Tachypnoea, rather than hyperpnoea, is the usual response to stimulation in the perifornical area of the hypothalamic region, and this particular pattern of respiration has been related to the hissing that is a prominent feature of the reaction in the cat. But the combination of tachypnoea and hyperpnoea, being characteristic of muscular exertion, in which condition its utility seems self-evident, led Cannon to two interesting speculations. One is that the important effect of these changes, as part of a defence reaction, is to produce an anticipatory reduction of blood p_{CO_2}. Some findings of Douglas and Haldane (1909), quoted in this connection, showed that forced breathing for a few minutes would relieve the subsequent respiratory distress of severe muscular exertion. All later work has gone to show that p_{CO_2} during effort remains remarkably constant, except in heavy work, when it falls; it is thus difficult to understand why this first speculation has received so little attention, particularly since it would point to a possible site of the higher centre in the brain involved in the respiratory adjustments of muscular exercise.

Having accepted the desirability of minimising the respiratory distress of muscular exertion, Cannon proceeded to consider the contribution that might be made by dilatation of the bronchioles. The air flow could certainly be relatively impeded by bronchiolar constriction, as in asthma, and adrenaline, which is discharged during strong defence reactions, was known to dilate bronchioles; so, he suggested, might not such dilatation be one of the mechanisms that help to prevent a healthy man becoming "winded" during severe exertion?

Gastrointestinal Secretion and Contraction

Changes in activity of the gastrointestinal tract have been discussed less in relation to short-term effects seen during defence reactions than in connection with the abnormalities of function that may develop in chronic emotional states, especially in man. Though there must be a relationship between these sets of changes, this has provided little incentive for a study of the short-term effects. Perhaps they have in any event seemed rather uninteresting, since most have appeared to be inhibitory. Indeed, there is little to add to the conclusions documented by Cannon (1929) that pain, like fear and anxiety, will lead in animals and men to drying up of salivary, gastric and possibly even pancreatic secretion and to inhibition of peristaltic movement throughout the gastrointestinal tract.

From more recent studies of their patient, Tom, with a gastric fistula, Wolf and Wolff (1943) thought they could distinguish between the effect of anger and anxiety, which increased gastric motility, acid secretion and vascularity, and fear, which diminished them. It has likewise been reported that the former states lead to hyperactivity of the colon and the latter to quiescence (Grace, 1950). But such distinctions between emotional states are difficult to make with precision, and opposite effects have been reported, for instance by Crider and Walker (1948). If we knew the effects on patterns of activity of the gastrointestinal tract that follow electrical stimulation of those parts of the hypothalamus of established biological function, a firmer basis might be provided for criticism of Cannon's views. It has been shown that increases in tone and motility of the stomach, duodenum, jejunum, colon and rectum, but not of the ileum, can be elicited by stimulation in the tuberal region of the hypothalamus (Strom and Uvnäs, 1950), but the precise location of the regions was not established. Moreover, in such investigations the whole pattern of visceral response needs to be studied. Eliasson (1954), for instance, has shown that gastric motility is inhibited on stimulation of some parts of the hypothalamus and midbrain and excited from others, but it is uncertain what conclusions may consequently be drawn about the regulation of gastrointestinal function. Increased gastric activity occurs during vomiting and can hence be related to a general inhibition of digestive activity. Not surprisingly, Cannon (1929) discusses the evidence that vomiting may occur as a feature of a strong defence reaction, and this could provide an interpretation of various findings by Wolf and Wolff (1943).

Release of Adrenaline

Effects on cardiovascular system

In 1925 Cannon and Britton showed that, during pseudaffective reactions in a decorticate cat, the adrenal medullary hormones are released in sufficient amounts to speed the rate of the denervated heart. Not only have these observations been confirmed, but also they were soon matched by others showing release of the hormones on hypothalamic stimulation of anaesthetized cats (Houssay and Molinelli, 1925; Magoun et al., 1937). Grant et al., (1958) measured the amounts of adrenaline and noradrenaline released on stimulation of the hypothalamic region from which the vasodilator fibres to skeletal muscle were activated. The mean values for the output of adrenaline from one adrenal gland, at rest and during stimulation, were 0.06 and 0.32 μg kg^{-1} min^{-1}, respectively, and for noradrenaline were 0.19 and 0.35 μg kg^{-1} min^{-1}. The figures for the release during stimulation correspond closely to those given by von Euler and Folkow (1953) and Celander (1954) for the release in prolonged asphyxia, which is probably the most powerful natural stimulus. Much the same amounts were released on stimulation in the region of the vasodilator pathway in the medulla oblongata, and finally it was shown that this release would make no contribution to the increase in muscle blood flow unless the electrical stimulation of the brain stem was intense (Lindgren et al., 1959a and b). They appear also to have no significant effect on the rate of the innervated heart, though they will make a contribution to the increase in contractile force (Rosen, 1961b). Some results of Folkow, Johansson and Mellander (1961) obtained in experiments on the capacity vessels of the hindquarters of the cat would indicate that they are hardly affected by the amounts of noradrenaline released during defence reactions, but there is no corresponding information about adrenaline. Thus, from what is known so far of the cardiovascular effects of these catecholamines in the cat, the amounts released during defence reactions would only be expected to contribute significantly to the increase in contractile force of the heart. In man, the relative contributions of hormone and nervous system may well be different. Certainly, the muscle vasodilatation is sometimes produced largely by circulating adrenaline (Barcroft et al., 1960), and it would be most interesting to know if the amounts released could also have an important effect on venous tone. That this is possible is shown by observations of venoconstriction in man produced by small intravenous doses of adrenaline (Sharpey–Schafer and Ginsberg, 1962).

Effects on fatigue of muscles and nerves

In animal experiments adrenaline in adequate doses has been found an effective antidote to muscular fatigue. It has long been known to overcome fatigue of the neuromuscular junction and to antagonize the blocking action of curare, besides restoring the contractions of fatigued muscle

(Oliver and Schaefer, 1895; Gruber, 1922 a, b). Some augmentation, both of transmission and contraction, are also seen in unfatigued muscles, though naturally less markedly so (Bowman and Zaimis, 1958). These are direct effects of adrenaline and are independent of changes in blood flow. The former effect, which is due to a reversal of the failure of presynaptic conduction (Corkill and Tiegs, 1933; Krnjevic and Miledi, 1958), requires such large amounts of adrenaline for its demonstration that it could hardly be exhibited during defence ractions. Bowman and Zaimis (1958) showed that the amounts necessary to increase the tension developed by unfatigued fast muscles, such as the tibialis anterior, were also rather higher than would ever be expected physiologically. The soleus, a slow muscle in the cat, was much more sensitive, being affected by 0.06–0.5 µg kg^{-1} intravenous adrenaline. Curiously, however, the contraction time was much reduced, and less tension was developed. In man most muscles are mixtures of slow- and fast-contracting fibres; there could hardly be sufficient circulating adrenaline to increase the tone or contractile force of unfatigued skeletal muscles. Whether such an effect will be produced in fatigued muscles is still an open question. Nevertheless, the emotional stress of mental arithmetic in man has been found to prolong the maintenance of muscle performance, the effect being sensitive to atropine and not dependent on the blood flow (Berdina et al., 1972; Vinogradova et al., 1974). A similar cholinergic effect on prolonged isometric contraction can be demonstrated in cats during hypothalamic stimulation (Berdina et al., 1978); and this has been interpreted as due to a direct cholinergic effect on the metabolism of muscle fibres, indicated by a previously observed activation of phosphorylase a, which is also atropine sensitive (Berdina and Rodionov, 1976).

The actions of adrenaline on the contractile force of the heart and skeletal muscle may be compared with its stimulating effect on the activity of the cerebral cortex, first described by Bonvallet et al. (1954), who believed that this effect was primarily due to a direct stimulating action on the mesencephalic reticular formation. Some workers have shared this view, others have not. In particular, Baust et al. (1963) hold that the rise in arterial blood pressure caused by adrenaline is the stimulus acting on the cells of the reticular formation. But all agree that circulating adrenaline can cause alerting and effect awareness. Until now, however, the doses of adrenaline used (usually 5 µg kg^{-1} in the cat) have been much higher than those to be expected even during a strong defence reaction, so that the physiological significance of the phenomenon is still uncertain. The mere fact, however, that adrenaline can lead in excitable tissues to such ubiquitous effects, which will ameliorate and may ward off fatigue, remains strongly suggestive of a supporting role for this hormone during emotional excitement.

Effects on metabolism

Behind this idea, which was one that Cannon advanced most forcefully, lay the knowledge that adrenaline exerts a powerful effect on carbohydrate metabolism: liver glycogen is mobilized (Vosburgh and Richards, 1903), and this so readily that adrenaline leads to hyperglycaemia in the cat, for instance, when infused at a rate of $0.05\ \mu g\ kg^{-1}$ per min (Cori $et\ al.$, 1935) and similarly in man when $0.025\ \mu g\ kg^{-1}\ min^{-1}$ are infused (Cori and Buchwald, 1930). Anaerobic glycolysis is stimulated in skeletal muscle, with the output of large amounts of lactate (Cori, 1925). These basic findings have stood the test of time, as has the conclusion that fat forms the major fuel for the increased metabolic rate induced by adrenaline (Cori and Cori, 1928). The use of free fatty acids (FFA) as a fuel during muscular contraction is stimulated by adrenaline; and there is an increased level of plasma adrenaline during muscular exercise (Rennie, 1972).

One other effect of adrenaline, however, is worth mentioning, that is, its ability to induce hyperkalaemia. It was established by D'Silva (1936) and Merenzi and Gerschman (1937) that most of the potassium comes from the liver and that, during the resulting hyperkalaemia, the uptake of potassium by skeletal muscle increases (Marenzi and Gerschman, 1936). The evidence for this movement of potassium was reviewed by Fenn (1940), who also pointed out that the potassium released from muscle during contraction is probably taken up rapidly by liver. He suggested that a physiological role of the potassium movement induced by adrenaline might be to replenish the muscle loss during a burst of activity and to maintain relatively normal potassium distribution during prolonged musclar exercise.

A further point of interest comes from the fact that moderate rises of blood potassium cause a vasodilatation in skeletal muscle, possibly suffi- cient to make a contribution to functional vasodilatation in this tissue (Dawes, 1941; Kjellmer, 1961; for recent review, see Hilton $et\ al.$, 1978). It has been known for some time that the generalized muscle vasodilatation that occurs in man in response to infusions of adrenaline is due, in fact, not to the catecholamine itself, but to some as yet unidentified vasodilator substance, or substances, produced by the infused material (Barcroft and Swan, 1953). It has been suggested elsewhere that plasma kinin formation might be responsible for this effect (Hilton, 1960), but perhaps potassium plays a role here, too, in which event the muscle vasodilatation produced during defence reactions in man, in so far as it results from circulating adrenaline, may truly anticipate the vasodilatation of exercise itself.

Effects on blood-clotting

Since the early experiments of Vosburgh and Richards (1903), it has been known that adrenaline injected into the animal shortens the clotting time of blood. Cannon (1929) was already able to draw on evidence that this is a phenomenon occurring under natural conditions, such as strenuous exercise, when adrenaline is liberated (Hartman, 1927; Mills et al., 1928). In his own experiments secretion of adrenaline elicited by splanchnic nerve stimulation, or reflexly by sciatic nerve stimulation, was as effective as injected adrenaline in reducing the clotting time. As little as 1 μg kg^{-1} halved the clotting time in 20 min; it took 30–40 min to return to normal. If an intact animal was enraged, larger and more prolonged effects were obtained. Even today the mechanism of this action of adrenaline is not understood, but a provocative footnote has been added by observations that after a subcutaneous injection of adrenaline there is a great increase in the fibrinolytic activity of plasma, which is maintained for an hour or more (Biggs et al., 1947). Similar changes occur after severe muscular exercise. It would probably be wiser to postpone speculation about the functional significance of these changes until more is known as to how they are produced and the relationship, if any, between them.

Catecholamine release in man in stressful situation

Mason (1972) has reviewed the literature up to 1968; most of the evidence of catecholamine release was based on the measurement of urinary levels. Although such evidence is unsatisfactory, since urine samples will consist of urine secreted over a period of at least one hour and will bear an uncertain relationship to blood levels, and the quantity in the urine (of the order of 5% of that secreted by the adrenal gland) will depend upon the degree of breakdown which is not necessarily constant, in spite of these and other problems it could be concluded from a wealth of experimental findings that the catecholamines were released in urine as a result of exposure to stressful situations and in the subsequent emotional responses.

In recent years, plasma levels of adrenaline and noradrenaline have been measured with increasing accuracy. Although the very act of venesection can raise adrenaline plasma levels, the use of indwelling catheters has made it possible to follow rapid changes in level. Levi and his colleagues (1968, 1972) have carried out numerous investigations involving a wide variety of stimuli, and have followed changes in both adrenaline and noradrenaline levels. Substantial and rapid increases in both catecholamines are associated with anger, fear, apprehension before examinations or prior to surgical operations, as well as with general excitement.

Pituitary Hormones

Adrenocorticotrophic hormone (ACTH)

The association between several biological stimuli and release of ACTH has been documented many times in recent years, and it should now suffice to state that ACTH release would be expected as part of the defence reaction. The conclusion has not previously been put in this simple form, despite clear statements such as that of Sayers (1937) to the effect that neural mechanisms involving the brain-stem reticular formation play a major role. Harris (1960) has pointed out that the hypothalamus appears to act as an integrating mechanism where effects of afferent impulses, emotional states and humoral agents interact, and he suggested that there may be a close link between the patterns of endocrine activity and emotional behaviour integrated in this region of the brain. But no reference was made to the body of information accumulated so far about defence reactions and their central nervous control.

To select a few of the most relevant findings, it may be noted, first, that in monkeys and men the pituitary–adrenocortical system, judged by plasma 17-hydroxycorticoid levels, is remarkably sensitive to stimuli leading to distress or arousal. Mason (1959), for instance, observed striking elevations when monkeys were simply moved into a new room or cage, as well as in conditioning experiments in which the stimulus had become a "threat" that an electric shock would follow. Similar increases were reported for man, arising acutely in an aircrew on an 11 000 mile non-stop flight and as a more persistent change in patients during the first week after thoracotomy. These increases, which represented a doubling of the normal level, were of the same order as those reported in previous investigations of patients in anxiety states (Hamburg et al., 1958). Hodges et al. (1962) report similar increases in the blood of students taken just after sitting for the second M.B. (London) oral examination. They also found the blood levels of ACTH to have increased to 6–9 mu ml^{-1} (6–9 μg ml^{-1}) and pointed out that these levels are as high as those found in patients with Cushing's disease.

It has been known since the pioneer experiments of de Groot and Harris (1950) on unanaesthetized rabbits that electrical stimulation of the hypothalamus activates the pituitary–adrenocortical system and that bilateral lesions in the hypothalamus would abolish the lymphopenic response to emotional excitation. Most workers have put the site, both of effective stimulation and of effective lesions, in the posterior tuberal region of the hypothalamus (Harris, 1960). Mason (1958), working on monkeys, elicited an increase in plasma 17-hydroxysteroids on stimulation of the periventricular region of the anterior hypothalamus. It may not seem surprising that the results of lesion experiments do not give a clear localization; stimulation experiments could give more accurate indications, but no one has yet performed the heroic labour of using precise stimulation techniques in an attempt to map the whole brain-stem region from which ACTH secretion is elicited. The present information, most notably that provided by Porter (1954) on the basis of lymphopenic responses, would be consistent with a considerable overlap of the ACTH-releasing area with that integrating the defence reaction. Relevant to this is the observation that plasma 17-hydroxycorticoid levels are raised in the conscious monkey as much by electrical stimulation within the amygdaloid complex as within the hypothalamus (Mason, 1959). Results of lesions and stimulation experiments implicating the median eminence (Harris, 1960) may be interpreted as an indication that the final common path lies in this region.

One basic question remains, even after these many investigations: What is the physiological role of the adrenocortical hormone so readily released? An answer (Beck and McGarry, 1962) was that its role in carbohydrate metabolism is a minor one and that it is chiefly concerned with the distribution of body water and electrolytes, maintenance of blood pressure and glomerular filtration rate and the renal regulation of water excretion. Nevertheless, as the authors also state, the influence of cortisol in accelerating gluconecogenesis is well established, and there is also evidence that the hyperclycaemia it causes is partly due to inhibition of pyruvate catabolism. This could mean that the products of anaerobic glycolysis in muscle would be available for resynthesis of glucose. Thus, the release of ACTH will co-operate in some of its final effects with adrenaline to ensure the supply of adequate fuel for intense muscular activity. It may be recalled that there is much evidence to suggest that adrenaline facilitates the release of ACTH. The release of ACTH and cortisol in stressful situations is now well established. The earlier evidence was based on the excretion of 17-hydroxycorticosteroid in the urine. Again, as in the case of catecholamines, such evidence was not entirely satisfactory but the many reports of increases of 17-OHCS could leave little doubt that corticosteroids had been released in conditions of stress. Plasma levels have also been shown to be increased in conditions of fear and anxiety (for review see Leshner, 1977). It would appear that cortisol release accompanies stressful situations in which there is a strong emotional impact.

Vasopressin (Antidiuretic hormone, ADH)

It has long been known that operative procedures or muscular exercise will reduce the rate of urinary secretion in man. Rydin and Verney (1938) showed the antidiuretic effect in dogs of forced running on a treadmill, an unpleasant noise or weak electrical stimulation of the skin. This effect was shown to be independent of adrenaline release and the sympathetic innervation of the kidneys. It is due to a reflex excitation of the neurohypophysis (O'Connor and Verney, 1942; O'Connor, 1946). Harris (1955) emphasised that the stimuli used by these and other workers will all have elicited some degree of what is loosely called "emotional stress". Even the effect of sudden exercise cannot be due to the running itself, for the antidiuretic effect was found to wear off as the exercise was repeated (Rydin and Verney, 1938). Circulatory collapse in man, produced by sudden movement on a tilt-table from the horizontal to the vertical position, evokes an antidiuretic effect, probably by release of vasopressin (Brun et al., 1946), and the hormone has been demonstrated in the blood of human subjects during a faint due to venesection (Noble and Taylor, 1953). Since the faint itself seems to initiate the release of ADH, this has been taken as additional evidence that the hormone is liberated during so-called "stress". But what exactly is the "stress" in this experimental situation? More specifically, what is the stimulus and does it act by initiating the defence reaction? A fall of blood pressure on its own, for example, can lead to ADH release in anaesthetised animals (Ginsburg and Brown, 1957). Recently, an attempt has been made to test whether ADH is released during defence reactions in the cat (Bisset et al., 1963). The suggestion had already been made that pressor reactions obtained in encéphale isolé preparations, on electrical stimulation in the midbrain reticular formation, were due to ADH (Sharpless and Rothballer, 1961). Bisset et al. (1963) assayed the amounts of ADH released into the external jugular venous blood in anaesthetized cats on discrete electrical stimulation within the hypothalamus. They found restricted regions from which hormone release was elicited, but did not obtain any when stimulating the zone from which defence reactions are most readily elicited. It is possible that the anaesthetic had blocked the pathway from the integrative centre for the defence reaction to the supraoptic nucleus, but until this is established we have no firm evidence that ADH release is a necessary component of the defence reaction itself.

Conclusion

There is little left to add to what has been written in the preceding sections about the functional significance in emergency of the defence reaction and its various components. The diverse visceral and hormonal features seem best regarded as the preparatory stage of a reflex, which, in civilised man, may be expressed in this way alone. Curiously enough, Cannon regarded these reactions none the less as homoeostatic, for he was emphasising the long-term view; in the short run, though, they are the very antithesis, since they establish a new equilibrium only made possible by an interruption of short-term homoeostatic mechanisms. This was illustrated in dealing with the pattern of cardiovascular response by reference to the inhibition of the baroreceptor reflex, which would otherwise interfere. Some of the consequences of release of ACTH and adrenaline also are incompatible with short-term homeostasis. The powerful effects of such hormones can greatly disturb the internal milieu, particularly if no muscular exertion occurs. Their potentially harmful effect in prolonged or chronic reactions should not need to be reiterated. Thus, though responses of the organism leading to changes of the internal milieu may still be physiological, they entail unavoidable risks.

This does not detract from the imaginative hypotheses that Cannon deployed and to which reference has frequently been made here. As for the bodily responses themselves, it need hardly be said that there are still uncertainties about matters of fact, as well as of significance, and that these are multiplied with the increasing range of bodily responses to be discussed. The responses included under the heading of the defence reaction are no doubt much more complicated than can be imagined even today, but this is not to deny the growing certainly about the validity of the theoretical approach, which carries with it the forceful reminder that physiology has deep roots in biology.

Most of the bodily responses fall into place regardless of subtle psychological distinctions, and from the physiologist's point of view there has seemed little need to distinguish between "fear" and "rage" or their frequent counterparts in experimental psychology. "flight" and "attack". Nor would such distinctions help to decide the equivocal significance of such components as changes in blood-clotting or the role of ADH; on the contrary, in this branch of biology, physiology may at present be of greater assistance than psychology. It seems fitting to conclude a contemporary account of this subject with the words that Cannon used to introduce the first edition of his book, almost 70 year ago. "Fear, rage, and pain, and the pangs of hunger are all primitive experiences which human beings share with the lower animals. These experiences are properly classed as among the most powerful that determine the action of men and beasts. A knowledge of the conditions which attend these experiences, therefore, is of general and fundamental importance in the interpretation of behaviour."

References

Abrahams, V.C., Hilton, S.M. and Malcolm, J.L. (1962). *J. Physiol.* **164**, 1.

Abrahams, V.C., Hilton, S.M. and Zbrozyna, A.W. (1960a). *J. Physiol.* **152**, 54*P*.

Abrahams, V.C., Hilton, S.M. and Zbrozyna, A.W. (1960b). *J. Physiol.* **154**, 491.

Abrahams, V.C., Hilton, S.M. and Zbrozyna, A.W. (1964). *J. Physiol.* **171**, 189.

Barcroft, H., Brod, J., Hejl, Z., Hirsjärvi, E.A. and Kitchin, A.H. (1960). *Clin. Sci.* **19**, 577.

Barcroft, H. and Swan, H.J.C. (1953). "Sympathetic Control of Human Blood Vessels", Edward Arnold, London.

Bard, P. (1928). *Am. J. Physiol.* **84**, 490.

Bard, P. (1960). *Physiol. Rev.* **40**, Suppl. 4, 3.

Bard, P. and Macht, M.B. (1958). *In* "Neurological Basis of Behaviour", 55. Churchill, London.

Bard, P. and Rioch, D. McK. (1937). *Johns Hopkins Hosp. Bull.* **60**, 65.

Baust, W., Niemczyk, H. and Vieth, J. (1963). *Electroencph. Clin. Neurophysiol.* **15**, 63.

Beck, J.C. and McGarry. E.E. (1962). *Br. Med. Bull.* **18**, 134.

Berdina, N.A., Kolenko, O.L., Kotz, Y.M., Kuznetzov, S.P., Rodionov, I.M., Savtchenko, A.P. and Thorevsky, V.I. (1972). *Circ. Res.* **30**, 642.

Berdina, N.A., Kotz, Y.M., Rodionov, I.M., Thorevsky, V.I., Vinogradova, O.L. and Vishnevetskaya, M.A. (1978). *Pflügers Arch.* **378**, 149.

Berdina, N.A. and Rodionov, I.M. (1976). *Pflügers Arch.* **367**, 37.

Biggs, R., MacFarlane, R.G. and Pilling, J. (1947). *Lancet*, **i**, 402.

Bisset, G.W., Hilton, S.M. and Poisner, A.J. (1963). *J. Physiol.* **169**, 40*P*.

Bizzi, E., Libretti, A., Malliani, A. and Zanchetti, A. (1961). *Am. J. Physiol.* **200**, 923.

Blair, D.A., Glover, W.E., Greenfield, A.D.M. and Roddie, I.C. (1959). *J. Physiol.* **148**, 633.

Bolme, P., Ngai, S.H., Uvnäs, B. and Wallenberg, L.R. (1967). *Acta Physiol. Scand.* **70**, 334.

Bonvallet, M., Dell, P. and Hiebel, G. (1954). *Electroenceph. Clin. Neurophysiol.* **6**, 119.

Bovard, E.W. (1962). *Perspect. Biol. Med.* **6**, 116.

Bowman, W.C. and Zaimis, E. (1958). *J. Physiol*, **144**, 92.

Broca, P. (1878). *Rev. Anthrop.* **3** ser. **I**, 385.

Brod, J., Fencl, V., Hejl, Z. and Jurka, J. (1959). *Clin. Sci.* **18**, 269.

Brun, C., Knudsen, E.O.E. and Raaschon, F. (1946). *J. Clin. Invest.* **25**, 568.

Burn, J.H. (1938). *Physiol. Rev.* **18**, 137.

Cannon, W.B. (1929). "Bodily Changes in Pain, Hunger, Fear and Rage", 2nd edn. Appleton, New York.

Cannon, W.B. and Britton, S.W. (1925). *Am. J. Physiol.* **72**, 283.

Caraffa–Braga, E., Granata, L. and Pinotti, O. (1973). *Pflügers Arch.* **339**, 203.

Celander, O. (1954). *Acta Physiol. Scand.* **32**, Suppl. 116.

Chapman, W.P. (1960). *In* "Electrical Studies on the Unanaesthetized Brain", (Eds E.R. Ramey and D.S. O'Doherty), 334. Paul B. Hoeber, New York.

Charlier, A., Guz, A., Keatinge, W.R. and Wilcken, D. (1962). *J. Physiol.* **164**, 17*P*.

Coote, J.H., Hilton, S.M. and Perez–Gonzalez, J.F. (1979). *J. Physiol.* **288**, 549.

Coote, J.H., Hilton, S.M. and Zbrozyna, A.W. (1973). *J. Physiol.* **299**, 257.

Cori, C.F. (1925). *J. Biol. Chem.* **63**, 253.

Cori, C.F. and Buchwald, K.W. (1930). *Am. J. Physiol.* **95**, 71.

Cori, C.F. and Cori, G.T. (1928). *J. Biol. Chem.* **79**, 309.

Cori, C.F., Fisher, R.E. and Cori, G.T. (1935). *Am. J. Physiol.* **114**, 53.

Corkill, A.B. and Tiegs, O.W. (1933). *J. Physiol.* **78**, 161.

Crider, R.M. and Walker, S.M. (1948). *Arch. Surg. Chicago*, **57**, 1.

Darwin, C. (1872). "The Expression of the Emotions in Man and Animals", John Murray, London.

Dawes, G.S. (1941). *J. Physiol.* **99**, 224.

Douglas, C.G. and Haldane, J.S. (1909). *J. Physiol.* **39**, 1.

D'Silva, J.L. (1936). *J. Physiol.* **86**, 219.

Eliasson, S. (1954). *Acta Physiol. Scand.* **30**, 199.

Eliasson, S., Folkow, B., Lindgren, P. and Uvnäs, B (1951). *Acta Physiol. Scand.* **23**, 333.

Euler, U.S. von and Folkow, B. (1953). *Arch. Exp. Pathol. Phammakol.* **219**, 242.

Fenn, W.O. (1940). *Physiol. Rev.* **20**, 377.

Fernandez de Molina, A. and Hunsperger, R.W. (1959). *J. Physiol.* **145**, 251.

Folkow, B., Johansson, B. and Mellander, S. (1961). *Acta Physiol. Scand.* **53**, 99.

Folkow, B., Mellander, S. and Öberg, B. (1961). *Acta Physiol. Scand.* **53**, 7.

Folkow, B. and Uvnäs, B. (1948). *Acta Physiol. Scand.* **15**, 389.

Fonberg, E. (1963). *Acta Biol. Exp.* **23**, 171.

French, J.D., Von Amerongen, F.K. and Magoun, H.W. (1952). *Arch. Neurol. Psychiatr. Chicago* **68**, 577.

Fulton, J.F. (1949). "Physiology of the Nervous System", Oxford University Press, New York and Oxford.

Gannt, W.H. (1960). *Physiol. Rev.* **40**, Suppl. 4, 266.

Gastaut, H., Vigouroux, R., Corriol, J. and Badier, M. (1951). *J. Physiol. Path. Gen.* **43**, 470.

Geiger, A. and Sigg, E. (1955). *Trans. Am. Neurol. Ass* **80**, 127.

Ginsburg, M. and Brown, L.M. (1957). *In* "The Neurohypophysis", (Ed. H. Heller), 109. Butterworth, London.

Golenhofen, K. and Hildebrandt, G. (1957). *Pflüg. Arch. Ges. Physiol.* **263**, 637.

Goltz, F.M. (1892). *Pflüg. Arch. Ges. Physiol.* **51**, 570.

Grace, W.A. (1950). *Proc. Ass. Res. Nerv. Dis.* **29**, 679.

Grant, R., Lindgren, P., Rosen, A. and Uvnäs, B. (1958). *Acta Physiol. Scand.* **43**, 135.

Groot, J. de and Harris, G.W. (1950). *J. Physiol.* **111**, 335.

Gruber, C.M. (1922a). *Am. J. Physiol.* **61**, 475.

Gruber, C.M. (1922b). *Am. J. Physiol.* **62**, 438.

Hamburg, D.A., Sabshin, M.A., Board, F.A., Grinker, R.R., Korchin, S.J., Basowitz, H., Heath, H. and Persky, H. (1958). *Arch. Neurol. Psychiatr. Chicago*, **79**, 415.

Harris, G.W. (1955). "Neural Control of the Pituitary Gland", Edward Arnold, London.

Harris, G.W. (1960). *In* "Handbook of Physiology", Section I, Vol. II, (Ed. J. Field), 1007. Williams and Wilkins, Baltimore.

Hartman, F.A. (1927). *Am. J. Physiol.* **80**, 716.

Heath, R.G. and Mickle, W.A. (1960). *In* "Electrical Studies on the Unanaesthetized Brain", (Eds E.R. Ramey and D.S. O'Doherty), 214. Paul B. Hoeber, New York.

Heath, R.G., Monroe, R.R. and Mickle, W.A. (1955). *Am. J. Psychiatr.* **111**, 862.
Hess, W.R. (1949). *Das Zwischenhirn.* Schwabe, Basel.
Hess, W.R. and Brügger, M. (1942). *Helv. Physiol. Acta* **1**, 33.
Hickam, J.B., Cargill, W.H. and Golden, A. (1948). *J. Clin. Invest.* **27**, 290.
Hilton, S.M. (1960). *In* "Polypeptides which Affect Smooth Muscles and Blood Vessels", (Ed. M. Schachter), 258. Pergamon Oxford.
Hilton, S.M. (1963). *J. Physiol.* **165**, 56P.
Hilton, S.M. (1965). *In* "The Physiology of Human Survival", (Eds O.G. Edholm and A.L. Bacharach), 353–376. Academic Press, London and New York.
Hilton, S.M. (1975). *Brain Res.* **87**, 213.
Hilton, S.M., Hudlická, O. and Marshall, J.M. (1978). Possible mediators of functional hyperaemia in skeletal muscles. *J. Physiol.* **282**, 131.
Hilton, S.M. and Joels, N. (1965). *J. Physiol.* **176**, 20P.
Hilton, S.M., Spyer, K.M. and Timms, R.J. (1979). *J. Physiol.* **287**, 545.
Hilton, S.M. and Zbrozyna, A.W. (1963). *J. Physiol.* **165**, 160.
Hodges, J.R., Jones, M.T. and Stockham, M.A. (1962). *Nature (London)* **193**, 1187.
Houssay, B.A. and Molinelli, E.A. (1925). *C.R. Soc. Biol., Paris*, **93**, 1454.
Keller, A.D. (1932). *Am. J. Physiol.* **100**, 576.
Kelly, D.H.W. and Walter, C.J.S. (1968). *Br. J. Psychiatr.* **114**, 611.
Kjellmer, I. (1961). *Med. Exp.* **5**, 56.
Klüver, H. and Bucy, P.C. (1939). *Arch. Neurol. Psychiatr. Chicago*, **42**, 979.
Krnjevic, K. and Miledi, R. (1958). *J. Physiol.* **141**, 291.
Leshner, A.I. (1977). *In* "Emotion" (Eds D.K. Caudland, J.P. Fell, E. Keen, A.L. Leshner, R.M. Tarpy and R. Plucheck), Brooks/Cole Publishing Co, Monterey, CA.
Levi, L. (1968). *In* "Endocrinology and Human Behaviour", (Ed. R.P. Michael), Oxford University Press, Oxford.
Levi. L. (1972). (Ed.) *Acta Med. Scand.* **199**, Suppl. 258.
Lilly, J.C. (1958). *In* "Reticular Formation of the Brain", (Eds H.H. Jasper, L.D. Proctor, R.S. Knighton, W.C. Noshay, and R.T. Costello), 705. Churchill, London.
Lilly, J.C. (1960). *In* "Electrical Studies on the Unanaesthetized Brain", (Ed. E.R. Ramey and D.S. O'Doherty), 70. Paul B. Hoeber, New York.
Lindgren, P. (1955). *Acta Physiol. Scand.* **35**, Suppl. 121.
Lindgren, P., Rosen, A. and Uvnäs, B. (1959a). *Acta Physiol. Scand.* **47**, 233.
Lindgren, P., Rosen, A. and Uvnäs, B. (1959b). *Acta Physiol. Scand.* **74**, 243.
Lindgren, P. and Uvnäs, B. (1953). *Circulation Res.* **1**, 479.
MacLean, P.D. (1954). *In* "Recent Developments in Psychosomatic Medicine", (Eds E.D. Wittkown and R.A. Cleghorn), 101. Pitman, London.
Magoun, H.W., Ranson, S.W. and Hetherington, A. (1937). *Am. J. Physiol.* **119**, 615.
Marenzi, A.D. and Gerschman, R. (1936). *Rev. Soc. Argent. Biol..* **12**, 424.
Marenzi, A.D. and Gerschman, R. (1937). *C.R. Soc. Biol., Paris*, **124**, 382.
Marshall, J.M. (1977). *J. Physiol.* **266**, 48P.
Martin, J., Sutherland, C.J. and Zbrozyna, A.W. (1976). *Pflügers Arch.* **365**, 37.
Mason, J.W. (1958). *J. Appl. Physiol.* **12**, 130.
Mason, J.W. (1959). *Recent Progr. Hormone Res.* **15**, 345.
Mason, J.W. (1972). *In* "Handbook of Psychophysiology", (Eds N.S. Greenfield and R.A. Sternback), Holt, Rinehard Winston, New York.

Masserman, J.H. (1943). "Behavior and Neurosis", Chicago University Press, Chicago.
Mills, C.A., Necheles, H. and Chu, M.K. (1928). *Chin. J. Physiol.* **2**, 219.
Noble, R.L. and Taylor, N.B.G. (1953). *J. Physiol.* **122**, 220.
O'Connor, W.J. (1946). *Quart. J. Exp. Physiol.* **33**, 149.
O'Connor, W.J. & Verney, E.B. (1942). *Quart. J. Exp. Physiol.* **31**, 393.
Oliver, G and Schaefer, E.A. (1895). *J. Physiol.* **18**, 230.
Papez, J.W. (1937). *Arch. Neurol. Psychiatr. Chicago*, **38**, 725.
Penfield, W. and Jasper, H. (1955). "Epilepsy and the Functional Anatomy of the Brain", Churchill, London.
Porter, R.W. (1954). *Recent Progr. Hormone Res.* **10**, 1.
Rosen, A. (1961a). *Acta Physiol. Scand.* **52**, 291.
Rosen, A. (1961b). *Acta Physiol. Scand.* **53**, 270.
Rushmer, R.F., Smith, O.A. and Lasher, E.P. (1960). *Physiol. Rev.* **40**, Suppl. 4, 27.
Rydin, H. and Verney, E.B. (1938). *Quart. J. Exp. Physiol.* **27**, 343.
Sayers, G. (1947). *In* "Hormones in Blood", (Eds G.E.W. Wolstenholme and E.C.P. Miller), 138 Ciba Foundation Colloquia on Endocrinology, II, Churchill, London.
Schramm, L.P. Honig, C.R. and Bignall, K.E. (1971). *J. Physiol.* **221**, 768.
Sem–Jacobsen, C.W. and Torkildsen, A. (1960). *In* "Electrical Studies on the Unanaesthetized Brain", (Edited E.R. Ramey and D.S. O'Doherty), 275. Paul B. Hoeber, New York.
Sharpey–Schafer, E.P. and Ginsberg, J. (1962). *Lancet*, **ii**, 1337.
Sharpless, S.K. and Rothballer, A.B. (1961). *Am. J. Physiol.* **200**, 909.
Smith, G. Elliot (1919). *Br. Med. J.* **i**, 758.
Smith O.A., Jabbur, S.J., Rushmer, R.F. and Lasher, E.P. (1960). *Physiol. Rev.***40**, Suppl. 4, 136.
Starzl, T.E., Taylor, C.W. and Magoun, H.W. (1951). *J. Neurophysiol.*
Stead, E.A., Warren, J.V., Merrill, A.J. and Brannon, E.S. (1945). *J. Clin. Invest.* **24**, 326.
Ström, G. and Uvnäs, B. (1950). *Acta Physiol. Scand.* **21**, 90.
Szidon, J.P. and Fishman, A.P. (1971). *Am. J.Physiol.* **220**, 364.
Timms, R.J. (1976). *J. Physiol.* **256**, 71.
Timms, R.J. (1977). *J. Physiol.* **266**, 98.
Ursin, H. and Kaada, B.R. (1960). *Exp. Neurol.* **2**, 109.
Uvnäs, B. (1954). *Physiol. Rev.* **34**, 608.
Vinogradova, O.L., Kotz, Y.M., Rodionov, I.M., Thorevsky, V.I. and Shestakova, L.N. (1974). *Fiziol. Zh. SSSR*, **60**, 321.
Vosburgh, C.H. and Richards A.N. (1903). *Am. J. Physiol.* **9**, 35.
Wilkins, R.W. and Eichna, L.W. (1941). *Johns Hopk. Hosp. Bull.* **68**, 425.
Wolf, S. and Wolff, H. (1943). "Human Gastric Function", Oxford University Press, New York and Oxford.

The Physiology of Stress Part II—The Coping Mechanism

Ivor H. Mills

Introduction

If any animal is to cope with changes in its environment, it must have the necessary mechanisms, not only to record what is going on in the environment but also to give some indication of its importance in terms of any threat to the well-being or survival of the animal. Once the severity of the threat has been appreciated, the response in the animal must be adjusted so that it is appropriate to the situation. A completely novel event should produce widespread alerting effects because the dangers and subsequent processes stemming from the event will be unknown to the animal.

The threats in the immediate world of an animal may be perceived in a wide variety of ways which will be related to the sensory systems with which it is endowed. The response must be appropriate to the conditions: sensory systems must be alerted to assess more accurately the degree of threat, autonomic systems must be stimulated in appropriate ways and muscular activity must be prepared for and brought to a high degree of efficiency.

The system must be capable of responding in the most efficient way according to the perceived degree of threat. Assessment of lack of threat is as important as that of high degrees of challenge, so that the animal will not be brought repeatedly into a state of maximal readiness when the changes are not necessary. The response to the outside environment must be inter-linked with internal physiological needs for food, water, reproduction, care of the young, shelter, etc. This internal state may require different responses at one time from those at other times. The monitoring system must, therefore, be capable of altering the type of response in keeping with changed physiological needs.

From time to time environmental changes may be relatively large and perhaps permanent. In determining the types of response under the circum-stances it must be possible for the effector systems to be re-set at a new point so that account can be taken of the permanency of the change. This amounts to adaptation to the altered environment and the monitoring system will have to operate about a new set point.

In man some challenges may result from his ability to think and perhaps generate new conditions for himself and to which he must be able either to adapt, if they are relatively permanent, or with which he must cope if they are transitory. As in all gregarious species, some challenges will arise from the activities of other members of the species. Any internal demands that he generates by his thinking may become challenges which require great coping skills if they impinge upon or are frustrated by the other members of the society.

Oft-repeated or severe continuing challenges, whether self-generated or only external, or both, may make such demands upon the mechanisms involved in coping as to interfere with normal living activity. The responses to the challenges could interrupt internal processes so that coping fails to take place. In the long term this would mean that adaptation to the new environment could not occur and breakdown in the internal or social aspects of life would take place.

The Reticular Formation of the Brain

Many of the features indicated above as required of a system which enables an animal to react optimally to changes in its environment have been related to the reticular formation of the brain.

The early studies by Moruzzi and Magoun (1949), Starzl *et al.* (1951) and Magoun (1952) demonstrated that stimulation of the reticular formation led to activation of the electroencephalogram (EEG) to produce changes resembling the aroused state. This led to their describing the reticular formation as the "ascending reticular activating system". Although this anatomical structure is known to have descending as well as ascending components, the changes in the electrical state of the cortex of the brain proved a useful guide to alterations in the structure and function of the reticular formation.

The reticular formation begins in the brain stem as a continuation of the mass of interneurones of the spinal cord. It extends as a widespread pathway through the tegmentum of the brain stem (Fig. 1). It is diffuse and difficult to delineate accurately. The system originates in the medulla oblongata and pons, the longest axons arising from the medial, magnocellular part of the tegmentum. The system travels more laterally at the region of the isthmus and at the decussations of the superior cerebellar peduncles it is displaced into the dorsal tegmental region. At the ventral thalamic nucleus the reticular system splits into a pathway in the subthalamic region and a smaller component passing into the intralaminar nucleus of the thalamus. Connections are also maintained with the periaqueductal grey substance and the superior colliculus.

In the medial regions of the caudal part of the brain stem the projections arise which connect the reticular formation to the hypothalamus and the limbic system. There are reciprocal connections to these regions which have a controlling influence over endocrine and autonomic functions.

The axons in the reticular formation are of variable length (see Figs 2 and 3). The main sensory tracts, such as the spinothalamic tract, have numerous collaterals which extend at right angles into the reticular formation to produce complex interconnections. These branches link up several different types of sensory input and arise from different anatomical regions, including auditory and visual sense organs. Although some of the neurones in the reticular formation have long axons, the rest of the system is essentially polysynaptic with numerous, branching interconnections at all levels. Clearly such a system is not attempting to transmit localized or specific information but is collating impulses from various sites.

The extension into the limbic system is of great importance (see Chapter

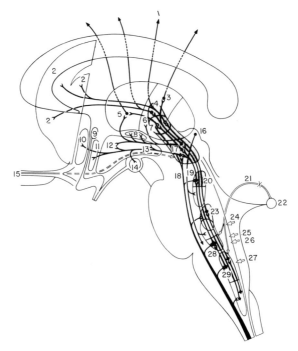

FIG. 1   Ascending fibre systems of the reticular formation. 1, Neocortex. 2, Corpus striatum. 3, Postero–lateral nucleus. 4, Thalamic intralaminar nuclei. 5, Ventro–lateral nucleus (Antero–ventral nucleus). 6, Postero–ventral nucleus. 7, Centromedial nucleus. 8, Zona incerta. 9, Anterior commissure. 10, Septal nuclei. 11, Pre-optic nucleus. 12, Lateral hypothalamic area. 13, Ventral Tegmentum. 14, Lateral nucleus of mammillary body. 15, Olfactory tract. 16, Superior colliculus. 17, Mesencephalic reticular formation. 18, Spinoreticular tract. 19, Tecto-bulbar and spinal tracts. 20, Reticular pontine nucleus (head). 21, Cerebellar uncinate fasciculus. 22, Fastigiate nucleus. 23, Reticular pontine nucleus (tail). 24, Trigeminal input. 25, Acoustic input. 26, Vestibular input. 27, Solitarius input. 28, Gigantocellular nucleus. 29, Central nucleus, medulla oblongata. (Reproduced from "The Human Central Nervous System", by R. Nieuwenhuys, J. Voogd and Chr. van Huijzen; Springer–Verlag, Berlin, Heidelberg and New York, 1978.)

8, Part I). Reticular system fibres from the mammillary peduncle pass forward in the medial forebrain bundle to lateral regions of the hypothalamus and preoptic region and to the medial nucleus of the septum (Fig. 4). This latter sends projections into the hippocampus (through the cingulum) which sends reciprocal fibres back to the medial midbrain regions. The medial forebrain bundle also sends a branch laterally to the amygdala. The hippocampus sends fibres through the fornix to the periaqueductal grey substance. Short reticular pathways extend to the tegmental region and the interpeduncular nucleus. These and longer reticular projections to the

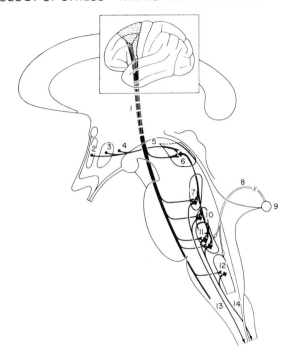

FIG. 2   Descending fibre systems of the medial reticular formation. 1, Pyramidal tract. 2, Pre-optic nucleus. 3, Anterior hypothalamic nucleus. 4, Hypothalamus. 5, Medial telencephalic fasciculus. 6, Mesencephalic reticular formation. 7, Reticular pontine nucleus (head). 8, Cerebellar uncinate fasciculus. 9, Fastigiate nucleus. 10, Reticular pontine nucleus (tail). 11, Gigantocellular nucleus. 12, Central nucleus, medulla oblongata. 13, Pontospinal tract. 14, Bulbospinal tract. (Reproduced from "The Human Central Nervous System", by R. Nieuwenhuys, J. Voogd and Chr. van Huijzen; Springer–Verlag, Berlin, Heidelberg and New York, 1978.)

medial region of the midbrain are thought to play a part in activating endocrine and autonomic mechanisms under conditions of stress (Nauta and Kuypers, 1957).

The Arousal Phenomenon

The original concept of arousal stems from the observations by Magoun and his colleagues that stimulation of the reticular formation leads to changes in the EEG of a sleeping animal to those typical of the awake state. When asleep the animal's EEG shows high-amplitude slow waves with

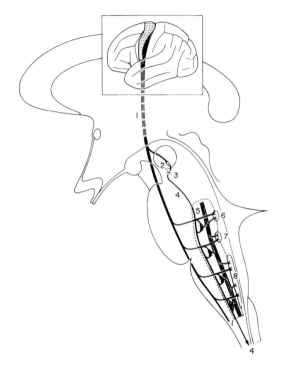

FIG. 3   Descending fibre systems of the lateral reticular formation. 1, Pyramidal tract. 2, Rubral nucleus, magnocellular part. 3, Ventral tegmental decussation. 4, Rubrospinal tract. 5, Lateral reticular formation. 6, Motor nucleus Vth nerve. 7, Nucleus VIIth nerve. 8, Nucleus VIIIth nerve. (Reproduced from "The Human Central Nervous System", by R. Nieuwenhuys, J. Voogd and Chr. van Huijzen; Springer–Verlag, Berlin, Heidelberg and New York, 1978.)

bursts of spindle activity. This so-called synchronized pattern changes relatively sharply with arousal to desynchronized fast activity of low amplitude.

Subsequently, it was shown that the arousal type of EEG does not always correlate with a behavioural aroused state and vice versa. The concept was then introduced of behavioural arousal as a separate, though usually related, entity. Lacey (1967) has introduced a third concept of arousal, namely, that associated with activity in the autonomic nervous system, which he has termed autonomic arousal.

Although all three types of arousal may be seen at the same time, relative independence of the three types of arousal can also be shown. Precision in describing in what sense the term arousal is used would help to avoid conflict of ideas. Early studies on arousal as shown by changes in the EEG were related to determination of electrical impulses in the reticular for-

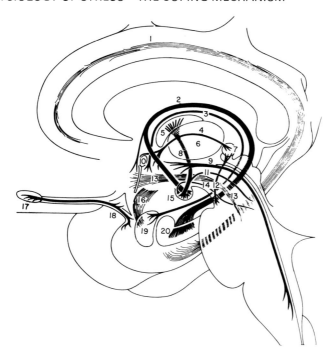

FIG. 4 The major pathways of the limbic system and the rhinencephalon. 1, Cingulum. 2, Fornix. 3, Stria terminalis. 4, Stria medulla thalami. 5, Anterior thalamic nucleus. 6, Medial thalamic nucleus. 7, Habenular nuclei. 8, Mammillo-thalamic tract. 9, Dorsal longitudinal fasciculus. 10, Anterior commissure. 11, Mammillo-tegmental tract. 12, Habenular-interpeduncular tract. 13, Medial telencephalic fasciculus. 14, Pedunculus of mammillary body. 15, Mammillary body. 16, Ansa peduncularis. 17, Olfactory bulb. 18, Lateral olfactory stria. 19, Amygdaloid body. 20, Hippocampus. (Reproduced from "The Human Central Nervous System", by R. Nieuwenhuys, J. Voogd and Chr. van Huijzen; Springer–Verlag, Berlin, Heidelberg and New York, 1978.)

mation. By picking up impulses in specific parts of the reticular formation it is possible to demonstrate that the same neurone may be activated by stimulation of more than one of the limbs. Some spontaneous neuronal activity could not be related to peripheral stimulation. Some neurones could be activated by different kinds of stimuli, e.g. by movement of hairs on one of the limbs or by an auditory stimulus.

The sequence of discharges of different neurones in the reticular system and the intervals between neuronal discharges is often dependent upon the site of stimulation and may alter with stimulation of different limbs. The pattern depends upon whether the stimulation is near to threshold or several times the threshold level. The pattern of response to repetitive stimulation suggests that the timing of early discharges may be more important in maintaining a temporal representation of the periphery than the timing of the later members of the repetitive stimuli or the total number of spikes.

Reticular neurones with a wide receptive field which respond similarly to different afferent sources may be related to generalized functions of the reticular formation. Those which respond with different temporal patterns of discharge to stimulation of different afferent sources may be related to local behavioural responses (Amassian and Walker, 1958).

Stimulation of the reticular formation

When the reticular formation in the midbrain is stimulated with impulses of 1 ms duration at 300 per second in an animal with synchronized slow-wave EEG, it produces a generalized arousal pattern all over the cortex. On cessation of stimulation an enduring potential change lasts for 1–5 s over the cortex. The changes appear simulataneously in widely separated parts of the cortex such as frontal and occipital. Very similar cortical EEG responses are obtained by repetitive stimulation of a peripheral sensory nerve. The cortical changes were not limited to the primary projection area of the stimulated nerve, but small areas of the cortex can be made to lose the potential change by local application of a weak solution of pentobarbitone to the local area.

The potential oscillations are probably related to the post-synaptic response of cerebral cortical neurones. The amplitude of the potential shift caused by stimulation of the reticular formation is always greatest in the frontal regions.

The arousal pattern in the EEG can be produced by stimulation of appropriate cortical regions. The response is exactly the same as that produced by sensory stimulation, namely, desynchronization and fast low-amplitude activity throughout the cortex. Stimulation of other cortical areas does not produce an arousal response in the EEG.

During cortical stimulation the transmission of sensory stimuli up the cord to the sensory cortex is altered. This decreases distracting stimuli when alerting stimuli are being received. Stimulation of the reticular formation decreases the impulses picked up in the cochlear nucleus which are produced in response to an auditory click. The pattern of sensory information reaching the cortex is, therefore, altered by the stimulation of the reticular formation directly or as a result of cortical stimulation. The parts of the cerebral cortex which are capable of producing arousal of the reticular formation are necessary as essential intermediates between a stimulus, such as a call, and the production of arousal changes in the EEG.

Behavioural arousal

Stimulation of the areas in the cortex which produced EEG arousal also, in unanaesthetized animals, caused behavioural arousal. With weak stimulation of the cortex the animal became alert, looked around and, on seeing no danger, returned to its random activity. With stronger stimulation in the same regions of the cortex, the animal showed signs of fear and ran to hide in a corner of the cage. With still stronger stimulation the animal raced around the cage in absolute terror trying to escape.

The corresponding states in man have been described as initial intellectual curiosity and the anticipatory set. At this alert and receptive stage the pulse rate slows. The higher stimulation would correspond to fear and is associated with autonomic arousal with increased heart rate, and the highest stimulation corresponds to terror and disorganized motor activity.

Motor activity is altered by limbic stimulation concurrently with initiation of arousal. Both facilitation and inhibition of reflex movements can be seen but both complement all the motor activity appropriate to the behavioural state that is initiated by the intensity of cortical stimulation.

Autonomic arousal is produced by stimulation of cortical areas which connect with the central cephalic brain stem. A wide variety of responses can be produced involving gastrointestinal, respiratory, vasomotor and pupillary reactions, depending upon the region of cortex stimulated (French, 1958). The defence reaction is part of the arousal mechanism and particularly involves autonomic arousal. It is described in detail in Chapter 8, Part I.

The effect of drugs on the arousal mechanism

By means of drugs it is possible to show that the arousal mechanism can be divided into three components: (1) EEG arousal; (2) behavioural arousal; and (3) autonomic arousal. Lacey (1967) has stressed that stimulation of arousal is not uniform and that there is dissociation between different components according to circumstances.

Pentobarbitone depresses or inhibits responses in the reticular formation which are evoked by stimulation of a peripheral nerve or of sensory organs such as the ear. This drug also blocks the EEG arousal produced by stimulation of the reticular formation or the thalamus.

In studies in which EEG arousal was produced in a cat by reticular stimulation with 4 V, it required stimulation of the same region by 5 or 6 V to cause behavioural arousal. The same applied to stimulation of the diffuse thalamic projection. Chlorpromazine (at 5 mg kg^{-1}) did not alter the threshold for EEG arousal by reticular stimulation but the threshold for behavioural arousal was raised to 8 V.

The threshold for EEG arousal from thalamic stimulation was slightly raised by chlorpromazine from 4 to 5 V but the threshold for behavioural arousal from thalamic stimulation was markedly raised from 6 to more than 13 volts.

Chlorpromazine may increase the filtering of sensory input and so decrease the chance of obtaining a behavioural response from thalamic stimulation. Inhibition of the response to the cochlear nucleus from a click is produced by stimulation of the reticular formation. With 8 V there was depression of the first click response but not of subsequent ones. After chlorpromazine (1 mg kg^{-1}) only 6 V were needed to suppress the response to the first click and the next one as well. Stimulation with 10 V produced a very prolonged effect, the click response taking more than 5 min to return to normal.

It is suggested (Killam and Killam, 1958) that tonic input from sensory systems is necessary for behavioural arousal, whereas it is not for EEG arousal.

The EEG is markedly altered by changes in cholinergic activity (Bradley, 1957). Physostigmine, which potentiates acetyl choline by inhibition of acetyl cholinesterase, converts the EEG to the alert, aroused pattern but has no effect on behavioural arousal in the animal which remained drowsy. However, injections of acetyl choline into the midbrain reticular formation of hibernating ground squirrels caused full awakening and increase in body temperature (Beckman et al., 1976). On the other hand, in cats, blocking the cholinergic receptors with atropine produced the high-amplitude slow waves characteristic of sleep but without causing any drowsiness in the animal. Arousing stimuli could then produce behavioural arousal without converting the EEG to the aroused pattern. If smaller doses of atropine were used, the slow wave activity was blocked by stimuli causing behavioural arousal.

These results indicate that the neurotransmission from the reticular formation to produce EEG arousal is dependent upon cholinergic transmission.

Amphetamine which is well known to cause behavioural arousal, also causes EEG arousal. Lysergic acid diethylamide had a very similar effect but in a quiet room some bursts of slow wave activity might break through.

When atropine was added to amphetamine the behavioural state was unaltered but the EEG developed slow-wave activity characteristic of sleep. Similarly, when an animal was sedated with chlorpromazine the addition of physostigmine produced EEG arousal without affecting the sedated state. This, however, has not been the experience in man. Patients who were drowsy after atropine and barbiturate anaesthesia were rapidly woken up by intravenous physostigmine (2 mg) (Hill *et al.*, 1977). Physostigmine given intravenously to normal sleeping subjects during non-REM sleep, produced REM sleep but if given during REM sleep it usually woke the subject up (Sitaram *et al.*, 1977). Coma produced by an overdose of anticholinergic tricyclic antidepressants (amitriptyline or imipramine) was rapidly reversed by physostigmine intravenously in seven patients. All became fully awake and coherent (Newton, 1975).

Although cholinergic and anticholinergic drugs differentiate between EEG arousal and behavioural arousal, the effect if not a distinct one and clearly depends upon the dose of drug or the stimulus intensity.

The effect of amphetamine on behavioural arousal is abolished by midbrain section. It appears, therefore, to act on the reticular formation in the brain-stem. However, the effect of LSD on behavioural arousal is abolished by spinal section so must depend upon the collaterals from the major afferent pathways entering the reticular formation. That the effect of LSD depends upon the input from the environment is also shown in the animal with *encéphale isolé*. It has no effect on the threshold for electrical stimulation of arousal but does facilitate arousal by auditory stimulation.

Infusion of adrenaline intravenously (5 μg kg^{-1}) produces progressive cortical activation within 15–25 seconds. It is identical with the response to repetitive reticular stimulation and lasts for 2–3 min after the injection. In animals with pre-pontine section of the brain this effect is not seen, so the adrenaline is probably acting directly on the reticular formation (Dell, 1958). Similar changes are produced by noradrenaline.

This indicates that the mechanism of stimulation of the reticular formation at the level caudad to the point of pre-pontine section must be adrenergic.

Increased levels of adrenaline not only activate the reticular system to produce an arousal type EEG but also produce intensification of vigilance and increased exploratory behaviour. The reticular stimulation produced by an alerting event in the environment may lead on to autonomic arousal and by a cholinergic mechanism stimulates increased release of adrenaline. The effect of this on the reticular system is to intensify the EEG arousal, autonomic arousal and perhaps behavioural arousal. Whether the arousal is maintained or not depends upon all the sensory input to the reticular formation, not only that which arises from exteroceptive mechanisms but also that which arises from baroceptors, chemoceptors, blood sugar levels, gut motility, etc. The extent to which cortical activity from the association areas of animals plays a part in activating arousal mechanisms is not known, but in man it certainly plays an important and sometimes a dominant role in determining not only the intensity of arousal but also its duration.

It is not possible, therefore, to get a single measure of overall arousal. A stimulus which produces EEG arousal may produce the anticipatory state in which the nervous system is set to receive incoming information. In this state the autonomic system does not show the classical arousal with increased heart rate and blood pressure and facilitated galvanic skin response (GSR) as a result of sweating. Indeed, as we have seen the reverse tends to occur in this state of expectancy. The correlation between heart rate and skin conductance or the latter and muscle potentials may be very poor (Lacey, 1967) though GSR is frequently quoted as an estimate of arousal. Lacey (1967) points out that in the anticipatory state when pulse rate slows and blood pressure does not rise, increased palmar conductance nevertheless occurs.

Stimulation of baroceptors in an isolated carotid sinus will decrease cortical activity even to the point of producing a sleep pattern EEG. Thus increased traffic along baroceptor nerves can inhibit the reticular activating system and reduce the duration of episodes of cortical motor and autonomic activity in response to a stimulus.

Limbic regulation of behavioural arousal

There are two contrasting circuits which are focussed on the amygdala. One is essentially stimulatory and the other inhibitory: both involve the relationship between the amygdala and the frontal cortex. Although drinking or eating is initiated by the hypothalamus, the duration and termination of the activity is determined by the amygdala.

Removal of the amygdala produces an animal which is usually non-aggressive and unresponsive to threat but, paradoxically, it has an elevated heart rate at rest. These animals are not capable of achieving the attentive set when they would be in readiness to respond to what comes in from the environment, a state when slow heart rate is the rule. They are in the state of being about to produce muscular effort, when a steady increase in heart rate is found (Pribram and McGuinness, 1975).

Animals without the hippocampus show a number of abnormalities. Response to stimulus tends to terminate very abruptly, just as stimulation of the reticular formation fails to maintain a potential response after stimulation ends. The maintenance of response to stimulation by the hippocampus gives it an element of perseveration after the stimulation ends.

The hippocampus has the highest concentration in the brain of histamine-sensitive adenylate cyclase: thus in response to histamine, cyclic adenosine monophosphate (cyclic AMP) is generated inside the cells (Kanof and Greengard, 1978). The precise role of these histamine (H_2) receptors is not understood at present though it is known that their stimulation depresses the firing of some neurones in the cerebral cortex (Haas and Bucher, 1975).

Cortical Excitability and Cortical Circulation

Brain blood flow was studied in early experiments by Ingvar (1957) by cannulation of the superior sagittal sinus. When arousal was stimulated an increase in blood flow was demonstrated which started a few seconds after desynchronization of the EEG took place. Passive increases in blood flow never produced an arousal response in the EEG. During the arousal state an increased utilization of oxygen also occurred.

This study indicated that increased activity of the brain was associated with increased cortical blood flow. Cerebral blood flow was subsequently measured by means of the radioactive inert gas ^{133}Xe. The gamma radiation was detected outside the skull. In initial studies the ^{133}Xe in solution in saline was injected into one carotid artery. The inert gas is taken up by the nerve tissue and then is washed out by the blood flowing through the tissue. From the partition coefficient for xenon between blood and nerve tissue the rate of blood flow per unit weight of brain can be calculated. The washout rate for the grey matter is faster than that from the white matter. The radioactive counter records a curve which can be analysed into two components which represent the blood flow through grey and white matter of the brain. Since the xenon which is carried away by the blood from the brain is almost entirely exhaled from the lungs in one passage, there is no significant alteration of uptake because of recirculation. Care has to be taken that the arterial p_{CO_2} is normal because this alters cerebral blood flow.

Studies have been carried out when the individual was at rest and when some mental activity was taking place. In the early studies very few counters were used to pick up the gamma radiation from the brain. Four counters were used by Ingvar and Risberg (1967) in one of their early studies of mental activity in which the subject in one of the determinations of cerebral blood flow was asked to repeat a series of digits backwards. Throughout the test the largest number of digits the person could correctly recite back was maintained.

In control studies at rest the flow in the frontal region was always greater than elsewhere. The mean grey matter flow was 74.8 ml per 100 g min^{-1} and the mean white matter flow was 20.9 ml per 100 g min^{-1}. During mental activity the mean flow in the grey matter increased by 5.7 ml per 100 g min^{-1}, that is by 8%.

Subsequently the technique has been much more refined and now 16 or 32 detectors are used, arranged in a pattern on one side of the skull, and a computer is used to calculate the distribution of blood flow (Wilkinson *et al.*, 1969; Risberg and Ingvar, 1973; Ingvar, 1975).

By carrying out studies before and after the mechanism to be studied, the changes in regional cerebral blood flow could be determined: these were mainly in the grey matter. Movements of the hand, talking and reading all produced alterations of flow in the expected motor regions of the brain. The frontal area was also always involved. Reading, in addition, caused increased blood flow in the visual cortex. Solving a reasoning problem (looking at a series of geometric shapes and detecting which one did not have a feature in common with the other four) increased the flow in both the pre-central (prerolandic and anterior frontal) and post-central (parietal and temporo-occipital) association cortex. When the same test was repeated, only the increased flow in the post-central association cortex remained. The increased flow in the pre-central association cortex appeared to be related to the novelty of the manoeuvres. The anterior frontal region is thought to be involved in logical and abstract thinking. The parietal and temporo-occipital regions are related to integrated and complex visual mental activities.

In subjects with normal brain states the regional blood flow has been shown to correlate well with regional cerebral metabolic rate for oxygen, which was measured with radioactive oxygen-15 produced from a cyclotron. When one hand was exercised vigorously both the blood flow and the regional metabolic rate for oxygen increased in the appropriate motor area (Raichle, 1975). These studies suggest that increased activity of the cerebral cortex is associated with increased metabolic activity and with an increase in brain blood flow in the same region. An increase in cerebral blood flow in a region of the brain may, therefore, be used as an indicator of the part of the cerebral cortex which is being activated.

Cerebral blood flow may also be measured by implanted heat clearance probes. These have been used to assess the response to noise in cats. Initially the noise produces a rise in cerebral blood flow. With repetition of the noise over time, the EEG and blood flow responses die away and eventually cease. Similarly, if a hungry cat sees a mouse and catches and eats it, there is an increased cerebral blood flow in the hypothalamus and cortex. With successive mice, the response eventually decreases and disappears (Betz, 1975).

Cerebral blood flow can also be measured with inhalation of ^{133}Xe (Obrist *et al.*, 1975), and this has the advantage that blood flow in both cerebral hemispheres can be measured but the disadvantage that it takes longer than with the technique of carotid injection. It is more difficult to associate changes with short-term events. However, it has been possible to show that changes in blood flow during attacks of migraine are similar in both hemispheres not only during the aura stage when blood flow is reduced but also in the headache phase when it is increased (Edmeads, 1977).

Bilateral regional cerebral flow rates have been made with the inhalation technique before and during verbal reasoning and spatial reasoning. At rest both hemispheres had similar blood flows with a frontal predominace. During the verbal reasoning test both hemispheres had an increase in flow from 68.9 to 79.7 on the left and from 68.8 to 78.1 ml per 100 g min^{-1} on the right. The differences, though small, were significant when tested by paired t test. In the spatial tests the rise in cerebral blood flow was less, but slightly more on the right side: on the left from 68.9 to 73.8 and on the right from 68.8 to 75.7 ml per 100 g min^{-1} (Risberg et al., 1975).

The Effects of Vasoactive Substances on Cerebral Blood Flow

Noradrenaline

There is no uniformity of view on the effect of noradrenaline on cerebral blood vessels. This may be because of variations in the procedures in different studies. In the investigation of Ekström–Jodal et al. (1974a) the radioactive krypton wash-out technique was used to study flow through cerebral grey matter prior to and during infusion of noradrenaline intravenously. They discuss their results in relation to those of other workers. With the smallest doses infused (0.03 µg kg^{-1} min^{-1}) there was no effect on cerebral blood flow but some change in mean arterial pressure was produced. Larger doses decreased cerebral blood flow with a maximum effect at about 0.4 µg kg^{-1} min^{-1}. The effect was reversed by the α-adrenergic antagonist phentolamine. During noradrenaline infusion alterations of blood pressure by withdrawal or re-infusion of blood showed excellent autoregulation of cerebral blood flow.

Serotonin (5-HT)

The subject of serotonin and its effects on the vasculature in general is dealt with at length by Garattini and Valzelli (1965). Most workers agree that serotonin constricts cerebral vessels. However, direct infusion into the common carotid artery was found to produce no change in overall flow but a decrease in cerebral blood flow and an increase in external carotid artery flow in the baboon (Deshmukh and Harper, 1971) and an increase in cerebral blood flow was found with infusion of serotonin into the common carotid artery in the dog (Swank and Hissen, 1964).

Studies by Ekström–Jodal *et al.* (1974b), using the technique of wash-out of krypton-85 to measure blood flow through cerebral grey matter, showed a consistent fall in blood flow with intravenous serotonin. Even the smallest dose ($0.8 \, \mu g \, kg^{-1} \, min^{-1}$) produced an appreciable fall in flow (about 70%) in two hours. Regardless of starting level the final flow in all their experiments was between 40 and 60 ml $kg^{-1} \, min^{-1}$. Even with hypoxic vasodilatation, serotonin decreased flow. However, alteration of blood pressure, by changes in blood volume during serotonin infusion showed excellent autoregulation.

The same workers have shown that intravenous L-tryptophane also produces cerebral vasoconstriction with a decrease in cerebral grey matter blood flow (Ekström–Jodal *et al.*, 1974c).

Dopamine

Dopamine is well known to act as a vasodilator, when given at moderate doses, in the renal circulation (McNay *et al.*, 1963), in the mesenteric vascular bed (Eble, 1964), and in the coronary circulation (Schuelke *et al.*, 1971). Its effect on the cerebral circulation was studied by von Essen (1974).

In general, small doses of dopamine tended to cause vasoconstriction and a decrease in cerebral blood flow through grey matter and somewhat larger amounts had the reverse effect. With even larger doses the cerebral blood flow was again reduced. The decreases in cerebral blood flow were abolished by the α-adrenergic antagonist phentolamine and the increases in cerebral blood flow were abolished by the specific dopamine receptor antagonist pimozide. Autoregulation of cerebral blood flow with changes in blood pressure caused by variation in blood volume, was always good and unaffected by the drugs.

The cerebral circulation has definite dopaminergic receptors mediating vasodilatation but the α-adrenergic receptors which lead to vasoconstriction must be more readily available to dopamine, since the effect of small doses of the drug was always to decrease blood flow through the grey matter.

The cerebral utilization of oxygen was not affected by any of the vasoactive drugs or by their antagonists. This suggests that the changes in blood flow do not by themselves alter neuronal function. The changes in blood flow produced by activation of specific nerve groups or pathways were always associated with a corresponding change in oxygen utilization (Raichle, 1975). It appears, therefore, that under these circumstances some metabolic factor dependent upon the greater activity of the neurones is responsible for the changes in blood flow. The neuronal transmitters either do not reach the receptors on the blood vessels or, if they do, they are not in sufficient concentration to alter blood flow.

Biogenic Amine Responses to Stress

Normal life in an open environment automatically leads to a series of challenges. As discussed in the Introduction to this chapter, these challenges are sensed in a variety of ways and the information is fed into the reticular formation of the brain. The intensity of the stimulus and the frequency of the stimuli determine the neuronal response and so the behavioural response. The point at which the challenging events can be referred to as stress is impossible to define. There is a complete spectrum of behavioural responses, from alerting and cessation of random activity to terror and a violent effort to escape.

Under a variety of stress conditions the turnover rate of noradrenaline in the brain of rats has been shown to be increased though the level of noradrenaline at any one time may be decreased (Barchas and Freedman, 1963; Bliss and Zwanziger, 1966; Corrodi et al., 1963; Maynert and Levi, 1964; Thierry et al., 1968). The stresses which have been studied include emotional, immobilization, exercise and exposure to cold. However, the uptake of ^{3}H-noradrenaline into synaptosomes of the cerebral cortex was only rarely altered by a variety of stresses which are known to be severe enough to increase adrenal steroid response (Hendley et al., 1977).

One of the most powerful stresses is electroconvulsions. Increased tyrosine hydroxylase activity was found in the hemispheres after electroconvulsion in rats in the noradrenaline synthesizing regions but not in the dopamine synthesizing regions of the brain. By blocking the conversion of dihydroxyphenylalanine to noradrenaline it was shown, after the first, third and fifth days of repeated shocks, that the turnover of noradrenaline was increased. There was no such evidence of increased turnover of 5-hydroxytryptamine or dopamine (Modigh, 1976).

In mice reared in isolation the rate of noradrenaline turnover was lower in the striatum than in animals reared in groups of 30. This was not true of the noradrenaline turnover rates in the hemispheres (Modigh, 1973). However, increased turnover of noradrenaline in the cerebral cortex was found by Korf et al. (1973) using footshock as the stimulus. They showed that the locus coeruleus is essential to mediate this. Frank and Raab (1975) found the lowest noradrenaline turnover in the basal ganglia to occur in mice housed eight or 16 together with higher values in those alone or in groups of 32 ($p < 0.01$ in both cases for males and < 0.05 for females). In earlier studies Welch and Welch (1968) had produced results similar to those of Modigh (1973). When animals reared in isolation were put together in groups for the first time intense fighting broke out. There was no difference in Modigh's studies in the turnover of 5-hydroxytryptamine between the isolation or group reared animals. During fighting the levels of noradrenaline in the brain fell but the turnover rate was increased by 50–80%. This rate was measured by the increase in precursor during inhibition of decarboxylation, and the effect was similar in noradrenaline and dopamine dominant regions of the brain (Modigh, 1973). The increased turnover as assessed in this technique reflects an increase in activity of tyrosine hydroxylase, which is the rate-limiting step in the biosynthesis of both noradrenaline and dopamine (Levitt et al., 1965; Udenfriend, 1966). It has been shown to be increased in relation to impulse flow in peripheral sympathetic nerves (Sedvall et al., 1968) although it is also affected by feed-back control from post-synaptic receptors (Kehr et al., 1972).

The lowered rate of noradrenaline turnover in isolated mice (Modigh, 1973; Welch and Welch, 1968) is compatible with decreased activation of the reticular formation as a result of the isolation. Sensory input from the presence and activity of other mice, which occurs in the grouped mice, would be absent in the isolated ones. With groups of 32, as used by Frank and Raab (1975), the density would cause crowding and would lead to more agitation and stimulation than with eight or 16 in the same-sized container. This is the conclusion of Welch quoted by Frank and Raab (1975). The sudden grouping of the isolated mice by Modigh (1973) would lead to an initial overwhelming stimulation of the reticular formation as a result of the intensely challenging, novel situation. Hence, the defence reaction would be precipitated and there would be an extreme response. The fighting would become a manifestation of the fight or flight reaction.

The fall in the level of noradrenaline in the brain after 30 minutes of violent fighting amongst the isolated mice when they were grouped was accompanied by an increased turnover of noradrenaline. These findings indicated that there was depletion of noradrenaline stores, although the concomitant increase of tyrozine hydroxylase activity implied that there was an increased production of noradrenaline.

In isolated mice allowed to fight for 150 minutes the brain concentration of dopamine is reduced as well as that of noradrenaline (Welch and Welch, 1969a). It seems, therefore, that increased dopamine neurotransmission, as well as noradrenaline activity, occurs in fighting. Fighting is also caused by stimulation of dopamine receptors by apomorphine (McKenzie, 1971) as well as by stimulation of catecholamine receptors (Randrup and Munkvard, 1969). Decreased aggressiveness was reported to be produced by inhibition of tyrosine hydroxylase with α-methyl-tyrosine (Welch and Welch, 1969b) which decreases both dopamine and noradrenaline synthesis. Modigh (1974) used this drug to block catecholamine synthesis and demonstrated that fighting produced a more rapid fall in brain catecholamines. This indicates that the stores of preformed catecholamines were being used up.

Not all investigators have distinguished between dopamine and noradrenaline when they have made studies of catecholamine metabolism in the brain. Geyer and Segal (1974) infused either dopamine or noradrenaline or saline into one lateral ventricle of rats immediately before subjecting them to footshock in pairs. They showed that noradrenaline reduced the incidence of aggressive posturing but dopamine increased it. They suggest that the balance between noradrenaline and dopamine in appropriate regions of the brain determines aggression. When noradrenergic neurones were destroyed by injections of 6-hydroxydopamine, some aggressive activity was demonstrated as a result of footshock. This was inhibited by noradrenaline infusion into the lateral ventricle but not by dopamine infusion.

Modigh (1974) repeated his studies on isolated, grouped and fighting-isolated mice using the technique of inhibition of tyrosine hydroxylase and tryptophane hydroxylase. In these studies the amount of dopamine, noradrenaline and 5-hydroxytryptamine in the whole brain was measured in the three groups. When synthesis of biogenic amines is inhibited, the rate of fall off in the concentration of the amines in the brain depends upon the rate of nervous impulse flow (Anden et al., 1966). Modigh used two inhibitors of tyrosine hydroxylase, namely, α-methyl tyrosine which does not inhibit tryptophane hydroxylase, and α-propyl dopacetamide which inhibits both hydroxylases and so the synthesis of all three transmitters.

When tyrosine hydroxylase was blocked using α-methyl tyrosine, there was a greater fall in stored noradrenaline in grouped mice than in the isolated mice. Using the same inhibitor, Modigh showed that there was a fall in the stores of brain dopamine in isolated mice put together in groups of 20, and hence fighting took place.

With both hydroxylase inhibitors there was a greater fall in stored brain noradrenaline in the fighting animals than in all other groups.

These studies confirm that the stress of fighting initiated by putting isolated mice together in a group causes a great increase in the nerve traffic in noradrenergic neurones and to a lesser extent in dopaminergic neurones in the brain.

Similar changes in adrenal gland tyrosine hydroxylase have been shown by Axelrod et al. (1970). Socially deprived mice had lower concentrations of the enzyme and mice socially stimulated by living in a series of intercommunicating boxes joined to a central feeding area had increased tyrosine hydroxylase. There were also corresponding changes in the noradrenaline and adrenaline contents of the adrenal glands.

There seems little doubt that a variety of stresses in mice and rats produce a similar increase in nervous impulses in noradrenergic neurones in the brain and that despite a greater turnover of noradrenaline the actual level of this transmitter frequently falls as a result of the stores being used up by the frequency of release at the synapses. Much of this noradrenaline is metabolized inside the brain or cerebrospinal fluid into 3-methoxy-4-hydroxy-phenylglycol sulphate and 3, 4-dihydroxy-phenylglycol sulphate. The increased production of these metabolites during footshock stress has been shown to arise from both injected ^{3}H-noradrenaline and ^{35}S sulphate (Stone, 1975).

The amount of 5-hydroxytryptamine in the brain was the same in grouped and isolated mice and did not change as a result of fighting. However, when a decarboxylase inhibitor was given the turnover of 5-hydroxytryptamine, as shown by the accumulation of 5-hydroxytrytophane, was greatly increased in the fighting animals (Modigh, 1973).

When trypophane hydroxylase was inhibited there was no evidence that fighting caused greater nerve impulse traffic in serotoninergic neurones.

Arousal and Efficiency

In the section on the arousal phenomenon in this chapter the arousal phenomenon was described and three categories were indicated: (1) EEG arousal; (2) behavioural arousal; and (3) autonomic arousal. The changes in behavioural arousal produced by stimulation of the parts of the cerebral cortex which produced EEG arousal, depended upon the intensity of stimulation. Autonomic arousal occurred only with the higher levels of stimulation of the cortex. With the highest stimulation, behaviour became chaotic and not organized.

The various ways in which arousal can be assessed do not always correlate, as Lacey (1967) has pointed out. At the point of anticipation the heart rate slows but even then skin conductance due to increased sweating may be increased (galvanic skin resistance, GSR, is lowered). As response to the expected event begins the heart beats more rapidly and more forcefully and the blood pressure is often increased. Muscles have increased blood flow and increase in tension, sometimes showing a tremor. If the aroused state persists for an hour or more an increased excretion of adrenaline in the urine can be measured (Levi, 1965). This latter study shows that the same pattern may be produced by pleasantly exciting events as well as unpleasant ones.

The studies of Levi and his colleagues indicate that the extent to which the urinary excretion of adrenaline increases in different individuals subjected to the same arousing experience varies very considerably.

A number of studies suggest that an increase in arousal level is associated with increased efficiency; however, further increases in arousal lead to a fall off in efficiency. The relationship between arousal and performance can be represented by an inverted-U relationship. This is referred to as the Yerkes–Dodson law. As Corcoran (1965) points out, the same efficiency could be found at two points on such a curve, unless the peak performance was being measured at optimum arousal. If arousal cannot be precisely measured the difficulty will be to assess whether the subject is on the rising part of the curve or at the same level of efficiency on the falling part of it. Corcoran endeavoured to assess this by using tests which should change the level of arousal of the subjects. His results indicated that the personality of the subject determined whether the increased arousal test produced an elevation of performance or a decrease. In general introverts behaved like more highly aroused subjects and extroverts as less highly aroused.

Because of these personality differences in states of arousal and response to events it is extremely difficult to get even relative estimates of arousal which apply to all subjects. Self-assessment proves to be about as accurate as any other (Poulton, 1970). In general, people are aware of increases in arousal level, especially when they are becoming more excited or more anxious, but they are less aware of a fall off in arousal when they may gradually lose efficiency. With repetition of a certain task the arousal may become gradually less but if the response becomes more automatic there may be no decrease in the standard of performance.

Constant vigilance is required of many workers or of many people under certain conditions, e.g. in industrial production or in prolonged car driving at speed. Over time all people have a progressive decrease in efficiency, which may be quite fast in some types of test, especially if great concentration is needed to spot relatively small differences which occur at infrequent intervals. Performance depends upon whether the conditions tend to maintain arousal or, as with isolation, conditions automatically lower arousal with time.

Loss of sleep has two components. One is an increased arousal mechanism and the other is the effect of fatigue. If specialist knowledge is required and perhaps someone's life is at stake, arousal may be raised despite fatigue and success be maintained. After about seven hours loss of sleep over a day or two, compensation for the fatigue effect cannot be brought about by the threat of challenge to specialist knowledge.

In tasks which demand concentration and discrimination loss of sleep has a marked effect on performance (Poulton, 1970). Under these circumstances infrequent events are more often missed and the time taken to respond to a signal becomes longer.

Noise which is sufficient in intensity to be uncomfortable interferes with responses and also increases the percentage of errors in a task demanding discrimination. However, when loss of sleep was combined with intense noise the percentage of errors was significantly less than that produced by loss of sleep alone (Poulton, 1970). It seems, therefore, that the intense noise raises the arousal level and so partly overcomes the fatigue effect of sleep deprivation.

On the other hand when various intensities of white noise were presented to rats, the curvilinear relationship between intensity of stimulus and response was not distorted by simultaneous delivery of variable foot shocks (Lind, 1976).

Similar to the effects of noise combined with sleep loss were the studies of Frankenhaeuser et al. (1974) who tested the effects of alcohol alone and combined with finger shocks on simple and complex reaction times. Alcohol generally impaired reaction times but the superimposition of finger shocks compensated, and sometimes more than compensated, for the negative effect of alcohol. In general, it has been pointed out by Poulton (1976) that stressful environments have in many cases been shown to improve performance in vigilance tests.

Starvation in animals increases activity and this has been used as an indication of arousal. When amphetamine at various dosages was administered to fed and starving rats, it was found that the effects of the drug were greatly potentiated by starvation (Campbell and Fibiger, 1971) and these effects were not altered by adrenalectomy or sympathectomy (Simpson, 1974). It was concluded that the effects were central in origin.

Profound effects on arousal have been produced by selective destruction of catecholamine neurones by injections of 6-hydroxydopamine. This leads to prolonged aphagia and adipsia, loss of ability to acquire avoidance responses and decreased exploratory activity (Smith, 1976). By microinjections of 6-hydroxydopamine the pathways of the catecholamines in the brain have been followed. Injections into the anterolateral hypothalamus produced animals which could not place their forelimbs in response to visual stimuli, yet righting and hopping reactions were normal.

In open field tests the rats with lesions in the anterolateral hypothalamus showed marked deficiencies in activity, but in the home cage they were normal. Active avoidance behaviour was never developed in rats with these lesions. Administration of amphetamine, which increases arousal by releasing central catecholamines, produced no response at all in the rats with lesions in the anterolateral hypothalamus.

The catecholaminergic system in the brain appears to play a major role in the effector system once arousal has been stimulated by appropriate sensory input from the environment. In man the arousal can be self-generated by mental activity and cardiovascular changes may be produced which resemble some of the effects of increased arousal during mental arithmetic (Ludbrook *et al.*, 1975). The increased arousal under piece work conditions is shown by adrenal cortical and medullary overactivity, as measured by the increased excretion of adrenaline, noradrenaline and corticosteroids (Timio and Gentili, 1976). It is clear that the arousal mechanism is intimately involved in the complexities of human social interactions.

References

Amassian, V.E., and Waller, H.J. (1958). *In* "Reticular Formation of the Brain", (Eds H.H. Jasper, L.D. Proctor, R.S. Knighton, W.C. Noshay, and R.T. Costello), 69. Churchill, London.

Andén, N.-E., Fuxe, K., and Hökfelt, T. (1966). *J. Pharm. Pharmacol.* **18**, 630.

Axelrod, J., Mueller, R.A., Henry, J.P., and Stephens, Patricia M. (1970). *Nature (London)* **225**, 1059.

Barchas, J.D., and Freedman, D.X. (1963). *Biochem. Pharmacol.* **12**, 1232.

Beckman, A.L., Satinoff, E., and Stanton, T.L. (1976). *Am. J. Physiol.* **230**, 368.

Betz, E. (1975). *In* "Brain Work", (Eds D.H. Ingvar, and N.A. Lassen), 366. Munksgaard, Copenhagen.

Birzis, L., and Hemingway, A. (1957). *J. Neurophysiol.* **20**, 91.

Bliss, E.L., and Zwanziger, J. (1966). *J. Psychiatr. Res.* **4**, 189.

Bradley, P.H. (1958). *In* "Reticular Formation of the Brain", (Eds H.H. Jasper, L.D. Proctor, R.S. Knighton, W.C. Noshay, and R.T. Costello), 123. Churchill, London.

Campbell, B.A., and Fibiger, H.C. (1971). *Nature (London)* **233**, 424.

Corcoran, D.W.J. (1965). *Br. J. Psychol.* **56**, 267.

Corrodi, H., Fuxe, K., and Hökfelt, T. (1963). *Life. Sci.* **7**, 107.

Dell, P.C. (1958). *In* "Reticular Formation of the Brain", (Eds H.H. Jasper, L.D. Proctor, R.S. Knighton, W.C. Noshay, and R.T. Costello), 365. Churchill, London.

Deshmukh, V.D., and Harper, A.M. (1971). *In* "Brain and Blood Flow", (Ed R.W. Ross Russell), 136. Pitman, London.

Eble, J.N. (1964). *J. Pharmacol. Exp. Ther.* **145**, 64.

Edmeads, J. (1977). *Headache* **17**, 148.

Ekström–Jodal, B., von Essen, C, and Häggendal, E. (1974a). *Acta Neurol. Scand.* **50**, 11.

Ekström–Jodal, B., von Essen, C., Häggendal, E., and Roos, B.-E. (1974b). *Acta Neurol. Scand.* **50**, 27.

Ekström–Jodal, B., von Essen, C., Häggendal, E., and Roos, B.-E. (1974c). *Acta Neurol. Scand.* **50**, 3.

Frank, K., and Raab, A. (1975). *J. Comp. Physiol.* **99**, 153.

Frankenhaeuser, M., Dunne, E., Bjurström, H., and Lundberg, U. (1974). *Psychopharmacologia (Berlin)* **38**, 271.

French, J.D. (1958). *In* "Reticular Formation of the Brain", (Eds H.H. Jasper, L.D. Proctor, R.S. Knighton, W.C. Noshay, and R.T. Costello), 491. Churchill, London.

Garattini, S., and Valzelli, L. (1965). "Serotonin and the Cardiovascular System", Elsevier, Amsterdam, London, New York.

Geyer, M.A., and Segal, D.S. (1974). *Behav. Biol.* **10**, 99.

Haas, H.L., and Bucher, U.M. (1975). *Nature (London)* **255**, 634.

Hendley, E.D., Burrows, G.H., Robinson, E.S., Heidenreich, K.A., and Bulman, C.A. (1977). *Pharmac. Biochem. Behav.* **6**, 197.

Hill, G.E., Stanley, T.H., and Sentker, C.R. (1977). *Canad. Anaesth. Soc. J.* **24**, 707.

Ingvar, D.H. (1958). *In* "Reticular Formation of the Brain", (Eds H.H. Jasper, L.D. Proctor, R.S. Knighton, W.C. Noshay, and R.T. Costello), 381. Churchill, London.

Ingvar, D.H. (1975). *In* "Brain Work", (Eds D.H. Ingvar and N.A. Lassen), 397. Munksgaard, Copenhagen.

Ingvar, D.H., and Risberg, J. (1967). *Exp. Brain Res.* **3**, 195.

Kanof, P.D., and Greengard, P. (1978). *Nature (London)* **272**, 329.

Kehr, W., Carlsson, A., Lindqvist, M., Magnusson, T., and Atack, C. (1972). *J. Pharm. Pharmacol.* **24**, 744.

Killam, K.F., and Killam, E.K. (1958). *In* "Reticular Formation of the Brain", (Eds H.H. Jasper, L.D. Proctor, R.S. Knighton, W.C. Noshay, and R.T. Costello), 111. Churchill, London.

Korf, J., Aghajanian, G.K., and Roth, R.H. (1973). *Neuropharmacology* **12**, 933.

Lacey, J.I. (1967). *In* "Psychological Stress", (Eds M.H. Appley and R. Trumbull), 14. Appleton-Century-Crofts, New York.

Levi, L. (1965). *Psychosom. Med.* **27**, 80.

Levitt, M., Spector, S., Sjoerdsma, A., and Udenfriend, S. (1965). *J. Pharmacol. Exp. Ther.* **148**, 1.

Lind, P.M. (1976). *Br. J. Psychol.* **67**, 413.

Ludbrook, J., Vincent, A., and Walsh, J.A. (1975). *Clin. Exp. Pharmacol. Physiol. Suppl.* **2**, 67.

McKenzie, G.M. (1971). *Brain Res.* **34**, 323.

McNay, J.L., McDonald, R.H. Jr., and Goldberg, L.I. (1963). *Pharmacologist* **5**, 269.

Magoun, H.W. (1952). *Res. Publ. A. Nerv. Ment. Dis.* **30**, 480.

Maynert, E.W., and Levi, R. (1964). *J. Pharmacol. Exp. Ther.* **143**, 90.

Modigh, K. (1973). *Psychopharmacologia (Berlin)* **33**, 1.

Modigh, K. (1974). *Acta Pharmacol. Toxicol.* **34**, 97.

Modigh, K. (1976). *Psychopharmacology* **49**, 179.

Moruzzi, G., and Magoun, H.W. (1949). *E.E.G. Clin. Neurophysiol.* **1**, 455.

Nauta, W.J.H., and Kuypers, H.G.J.M. (1958). *In* "Reticular Formation of the Brain", (Eds H.H. Jasper, L.D. Proctor, R.S. Knighton, W.C. Noshay, and R.T. Costello), 3. Churchill, London.

Newton, R.W. (1975). *J. Am. Med. Assn* **231**, 941.
Obrist, W.D., Thompson, H.K., Wang, H.S., and Wilkinson, W.E. (1975). *Stroke* **6**, 245.
Poulton, E.C. (1970). "Environment and Human Efficiency", Charles C. Thomas, Springfield, Ill.
Poulton, E.C. (1976). *Aviat., Space Environ. Med.* November, 1193.
Pribram, K.H., and McGuinness, D. (1975). *In* "Brain Work", (Eds D.H. Ingvar and N.A. Lassen), 428. Munksgaard, Copenhagen.
Raichle, M. (1975). *In* "Brain Work", (Eds D.H. Ingvar and N.A. Lassen), 372. Munksgaard, Copenhagen.
Randrup, A., and Munkvard, I. (1969). *In* "Aggressive Behaviour", (Eds S. Garattini, and E.B. Sigg), 228. Wiley and Excerpta Medica Foundation, New York, Amsterdam.
Risberg, J., and Ingvar, D.H. (1973). *Brain* **96**, 737.
Risberg, J., Halsey, J.H., Wills, E.L., and Wilson, E.M. (1975). *Brain* **98**, 511.
Schuelke, D.M., Mark, A.L., Schmid, P.G., and Eckstein, J.W. (1971). *J. Pharmacol. Exp. Ther.* **176**, 320.
Sedvall, G., Weise, W.K., and Kopin, I.J. (1968). *J. Pharmacol. Exp. Ther.* **159**, 274.
Simpson, L.L. (1974). *Psychopharmacologia* **38**, 279.
Sitaram, N., Mendelson, W.B., Wyatt, R.J., and Gillin, J.C. (1977). *Brain Res.* **122**, 562.
Smith, G.P. (1976). *Ann. N.Y. Acad. Sci.* **270**, 45.
Starzl, T.E., Taylor, C.W., and Magoun, H.W. (1951). *J. Neurophysiol.* **14**, 461.
Stone, E.A. (1975). *Life Sci.* **16**, 1725.
Swank, R.L., and Hissen, W. (1964). *Arch. Neurol* (*Chicago*) **10**, 468.
Thierry, A.M., Javoy, F., Glowinski, J., and Kety, S.S. (1968). *J. Pharmacol. Exp. Ther.* **163**, 163.
Timio, M., and Gentili, S. (1976). *Brit. J. Prev. Soc. Med.* **30**, 262.
Udenfriend, S. (1966). *Pharmacol. Rev.* **18**, 43.
von Essen, C. (1974). *Acta Neurol. Scand.* **50**, 39.
Welch, B.L., and Welch, A.S. (1968). *J. Pharm. Pharmacol.* **20**, 244.
Welch, B.L., and Welch, A.S. (1969a). *Commun. Behav. Biol.* **A3**, 125.
Welch, B.L., and Welch, A.S. (1969b). *In* "Proc. Int. Symp. on the Biology of Aggressive Behavior", (Eds S. Garattini, and E.B. Sigg), 188. Excerpta Medica Foundation, Amsterdam.
Wilkinson, I.M.S., Bull, J.W.D., du Boulay, G.H., Marshall, J., Ross Russell, R.W., and Symon, L. (1969). *J. Neurol. Neurosurg. Psychiat.* **32**, 367.

CHAPTER 9
Biological Rhythms

J.N. Mills,* D.S. Minors and J.M. Waterhouse

Introduction

Probably the most important principle taught by physiologists is that of homeostasis—the constancy of the body's internal environment despite fluctuations in the external environment. Although the basic concepts of "*la fixité du milieu intérieur*", as expounded by Claude Bernard and modified by Cannon, still hold, the body's internal environment is no longer considered to be "fixed" in any individual. Part of the variation can be

* Professor Mills is now deceased.

accounted for by our knowledge that homeostatic mechanisms of the body do not maintain a controlled variable exactly at the "set point". This is because the sensors cannot detect an infinitely small error and there is always a certain delay between sensing an error and correcting it; as a result, the value of the controlled variable oscillates about its "set point" But in addition to this source of variation, there is another, described later in more detail when body temperature is considered, that results from a daily variation in the "set point". This is hardly surprising, since our external environment is largely rhythmic. Indeed, ancient man must have been aware of these rhythms, as the acquisition of animal food relies upon knowledge of the rhythm of activity and inactivity of the prey and that of plant food upon awareness of the changes of the seasons.

The frequencies of physiological rhythms encompass every division of time, ranging from the alpha rhythm in the electroencephalogram which oscillates about once per 0.1 s, the cardiac rhythm with a period of about 0.8 s, the respiratory rhythm with a period of about 5 s, to a lunar rhythm in some species which are dependent upon unusually high or low tides for their security or feeding habits, and the annual reproductive rhythms shown by most species. In each of these cases, the essential characteristic of a rhythm is that it repeats itself over several, sometimes very many, cycles with greater or lesser precision. The term cannot appropriately be applied to a sequence of changes which is undergone only once by any individual, such as the single progress from the fertilized ovum to the grave. Rhythms can, however, be described in populations, even where any individual contributes only a single observation. It is thus entirely proper to speak of rhythmicity in the time of birth or death, and much fundamental work has been done upon the rhythm in the time of eclosion of insects.

The material in this chapter deals mainly with adults, but in the concluding section the different aspects of rhythms in infants, the aged and psychiatric patients are described.

The low-frequency rhythms tend to predominate in both plants and animals, and usually have a period which corresponds with that of an external environmental periodicity. The study of these rhythms has become a science in itself, termed Chronobiology. Indeed, this science has now coined its own vocabulary. A definition of some of these terms follows, but for a fuller dictionary, readers are referred to the Glossary of Chronobiology (Halberg et al., 1973 or 1977).

The dominant periodicity in our environment is the regular alternation between day and night which recurs once every 24 hours. Since our habits of sleep, rest and activity, work and leisure, eating and drinking, all largely follow a routine governed by the alternation of night and day and the social organization which results from this, it is hardly surprising that many physiological quantities oscillate with a period close to 24 hours. For a variety of reasons these rhythms have been the most extensively studied. Formerly rhythms which oscillated once per 24 hours were termed "diurnal". Through a tiresome quirk of several languages, however, the work "day" is applied both to the full period of 24 hours, and to the light portion of this period, so the word, with the adjectives derived from it, such as diurnal, is potentially ambiguous. It is therefore customary to restrict the adjective "diurnal" to the daylight portion of the cycle, contrasting it with "nocturnal". Today the term "circadian" has come into common use to refer to those rhythms with a period of around 24 hours.

It might be argued that any given circadian rhythm observed in an individual is simply a bodily response to the many persuasive external 24-hour periodicities of the solar day. For instance, the regular rhythms of high diurnal and low nocturnal deep body temperature and urine flow may simply be due to nocturnal inactivity and abstinance during sleep. It has been shown, however, in individuals screened from all time clues, either in polar regions, deep caves or in specially constructed isolation units, that many circadian rhythms can continue as self-sustained oscillations. When observed in such timeless conditions rhythms are termed "free-running". It is further observed that the period of free-running rhythms invariably deviates from an exact 24 hours; it was for this reason that the term "circadian" was originally introduced by Halberg. The strict definition of the term "circadian" is a period of between 20 and 28 hours; rhythms with periods below this range are then described as "ultradian" and those with periods above the circadian range as "infradian".

The existence of free-running rhythms indicates that there must be some, as yet unidentified, innate timing mechanism capable of controlling these rhythms and which, in timeless conditions, is incapable of exact 24-hour periodicity. Rhythms which have thus been shown to arise from within the organism are qualified by the term "endogenous", to distinguish them from "exogenous rhythms" which are entirely dependent upon an external periodicity, and which will not show continuous oscillations in a timeless environment.

The vast majority of observations upon circadian rhythms are inevitably made within the normal periodic environment of the solar day. Such conditions are called "nychthemeral" and may thus refer to a nychthemeral rhythm without prejudice as to whether it is endogenous or exogenous in origin. As will be seen, it is likely that the majority of manifested nychthemeral rhythms result from both exogenous and endogenous influences. External periodic influences also interact with the endogenous timing mechanism in another way; they are able to synchronize the endogenous oscillator to the same period, a process known as "entrainment". Thus, although a rhythm may have a period deviating from 24 hours when observed in free-running conditions, it is usually entrained by external conditions. Any environmental periodicity which is thus able to entrain the innate timing mechanism is referred to as a *"zeitgeber"* (German: time-giver) or "synchronizer".

Mention must be made of the way in which rhythms are measured and quantified. Since most or all bodily variables are subject to biological noise, it has been necessary to introduce techniques through which rhythms may be identified in the presence of such noise. The usual method is that of fitting a cosine curve. Since the shape of variation over 24 hours, however, is variable and may not be symmetrical, as indicated in Fig. 1, it would appear at first sight that the fitting of a cosine curve is a very crude tool. Where the data studied are fairly extensive, however, the fitting of a cosine curve has been a remarkably efficient tool at demonstrating the existence of a rhythm and at defining its period. The usual statistical criterion for the fitting of a cosine curve is the minimization of summed squared deviations and the criterion for rhythmicity is that the variance due to the fitted curve shall be significantly less than the residual variance. Even such clearly non-sinusoidal rhythms as in Fig. 1(iii) provide remarkably good fits by this criterion. However, a cosine curve in such instances is useless for defining the phase of the rhythm, since, with a cosine curve, the time of the peak level, the acrophase, must be equidistant from the minima on either side. Where a rhythm is symmetrical, even though grossly non-sinusoidal, as in Fig. 1(ii) the fitted cosine curve will give a good estimate of the maximum and minimum as well as of the period. Fourier's theorem states that any

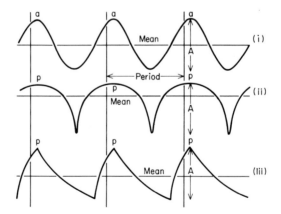

FIG. 1 Different forms of rhythm. (i) sinusoidal; (ii) symmetrical but non-sinusoidal; (iii) asymmetrical. A. amplitude, a. acrophase, p. peak. The vertical lines divide the traces up into cycles (from Conroy and Mills, 1970).

rhythm can be accurately fitted by a sufficient number of cosine curves of different periods, but the reproducibility of the shape of rhythms from one subject to another is seldom sufficient to justify such procedures. The simultaneous fitting of two cosine curves of different periods is usually reserved for circumstances when there is biological reason to suppose that two periods may be present in the data.

A major difficulty in the study of human circadian rhythms lies in our ignorance of the relevant controlling mechanisms. For instance, it is possible to demonstrate clear-cut circadian rhythms in the renal excretion of potassium and of sodium. Do these result from an inherent rhythmicity in the cells of the renal tubule? Do they reflect simply variations in the concentration of a blood-borne hormone? Do they result from variations in a nervous signal to the kidney? Or do they result from a complex integration of the various functions of the kidney, glomeruli and tubules, whereby variations in the transport characteristics of any ionic species are liable to have some effect upon the renal handling of other ionic species? Some attempt will be made in succeeding sections to sketch in what little we know of the causal nexus between the different rhythms described, and to provide tentative answers to these questions.

Endocrines

The rhythmicity of endocrine organs is most commonly studied by measuring the concentration of hormone in the plasma, but this reflects a comparable rhythm in production only if the rate of removal is not a major variable. Removal may be by destruction, by excretion, or by tissue uptake. Since, however, the plasma concentration is generally taken to be a useful index of the functional effect, rhythms become of interest even if they do not exactly parallel the rhythm in production. Episodic secretory bursts are quite common among endocrines, for example growth hormone, adreno-corticotrophic hormone (ACTH) and prolactin, so that the demonstration of a circadian rhythm is usually in combination with such ultradian components. For this reason, and because of the short halflife of many hormones, frequent blood sampling may be necessary to define the circadian rhythm with any precision; this limits the practicability when working on small animals in whom the blood loss would be substantial. Current methods of immunoassay, however, usually require only very small volumes of blood, so this is not a major difficulty in man. The problem of frequent venepuncture can be obviated by the use of an indwelling catheter, and with sampling apparatus that is portable and can be strapped to the subject, a reasonable amount of movement is possible while taking measurements throughout the 24 hours. There still remain two strategies, both of which have been employed. One may either collect discrete samples at fairly frequent intervals, which have been as brief as every 10 minutes through 24 hours, or one can collect continuously and thus obtain an integrated concentration over the period of time occupied by the collection of each distinct sample. The former technique indicates that the pituitary hormones and cortisol, which are those which have been most extensively studied, are released in small discrete packets, each release causing a momentary peak in plasma concentration which is followed by an exponential decline reflecting the normal processes of removal. The connection between the timing of these bursts and other factors such as the stages of sleep will be considered later; so too will the possible relationship between ultradian components and the lower frequency circadian rhythms. Circadian rhythmicity thus consists in the uneven but reproducible distribution of these episodes throughout the day.

Urinary excretion has also been used as evidence of rhythmicity, the substances usually studied being metabolites such as glucuronide conjugates. This yields an integrated value limited in frequency only by the interval over which sensibly complete bladder emptying can be secured. It cannot be used during sleep, however, without either waking the subject or sampling from an indwelling catheter. There is also a delay, not always precisely known, between secretion or plasma concentration and excretion in the urine. Further, the precise method of urinary handling is seldom known, so that variations in excretion could depend in part upon variations in renal rather than in endocrine function. With modern refinements in blood sampling and analysis, inferences from urine analysis are now seldom necessary and they are chiefly useful in defining annual or other rhythms of very long periods, when many days of sampling are required, and will not be further discussed.

In this section, the adenohypophysis and other endocrines under its control will be considered first, and then the pineal. The other hormones will be dealt with later when discussing their metabolic effects or the target organs upon which they act.

Useful reviews are by Weitzman (1976) on adenohypophyseal rhythms in general, by Weitzman and others (1975) upon hormone rhythms in relation to sleep, by Krieger (1975) upon adrenocortical rhythms and by Brownstein (1975) upon the pineal.

Adrenal cortex and ACTH

Of all the endocrine rhythms, that in the adrenal production of glucocorticoids has been the most extensively studied. It will be considered in some detail since it also serves as an example for many aspects of circadian rhythmicity.

When cortisol was originally studied by measuring its concentration in plasma in samples collected every three or four hours, high values were found early in the morning, falling to minimal values about bed-time; and when samples were collected by indwelling catheter it was found that the highest levels were often slightly before waking. Although in a diurnal animal such as man, the peak values are in the morning, in a nocturnal animal such as the rat, they are in the evening. Cortisol is the first hormone for which it was shown that the rhythm is complex, consisting of a series of sharp peaks with subsequent decline, the circadian pattern resulting from the aggregation of most of these peaks in the morning. It has been demonstrated that the rhythm in plasma concentration of cortisol does in fact result from rhythmic secretory activity of the gland and not, for instance, from a rhythm in metabolic disappearance. The rhythm has a large endogenous component, since it does not immediately change its phase if the subject alters the phase of his habits by eight or 12 hours. Indeed, it is one of the rhythms most resistant to adaptation and it may be at least a week after a time shift before the rhythm is adjusted to the new time. There is, however, an exogenous influence, in that under free-running conditions there is a sharp fall in cortisol concentration in plasma whenever the subject goes to sleep. Other influences, particularly stress, may cause increased secretion at any time of day and thus disturb the normal rhythm. There is satisfactory evidence that the adrenal rhythm is controlled by ACTH, whose plasma concentration shows a corresponding rhythm, and that this in turn is controlled by a rhythmic production of corticotrophin releasing factor (CRF), which is probably neurally controlled. There is evidence in man, as well as in other species, that a cholinergic pathway is involved, and an adequate dose of a vagolytic drug can inhibit the morning rise of plasma cortisol if, but only if, it is given between the hours of 16.00 and 24.00 (Ferrari et al., 1977). A failure of exact correspondence between secretory episodes of ACTH and of cortisol may mean simply that the adrenals respond to the integrated concentration of ACTH, clustered episodes of ACTH liberation leading to a single burst of cortisol; this again may be true of other endocrines stimulated by trophic pituitary hormones.

It has been remarked that the time when most cortisol is produced during sleep coincides with the time when REM is most frequent. More careful analysis suggests that there is no detailed correspondence between the two processes and their coincidence is probably indicative of no causal relationship. Even so, because of the functional relationship between the hypothalamus and pituitary, it is likely that many hormones that are influenced by the pituitary will be affected by changing neural activity. An obvious example is the effect of stress upon cortisol secretion.

The cortisol rhythm is not dependent upon direct negative feedback; when the adrenals fail to produce their hormone, as a result of removal from rats or of Addison's disease in man, the rhythmic production of ACTH continues, though at a higher level in the absence of the usual negative feedback; and after hypophysectomy in rats, the rhythm in CRF content of the median eminence persists. Observations on humans with naturally occurring central nervous lesions, or upon rats with artificially inflicted lesions, concur in suggesting that a region of the brain around the hypothalamic region is necessary for the manifestation of this rhythm, and the suprachiasmatic nucleus has been specifically implicated. The possible significance of this area of the brain will be outlined later (see page 483).

Gonadotrophins and gonads

Early in puberty, when luteinising hormone and follicle stimulating hormone are first produced in substantial amounts, they are secreted mainly during sleep, so that throughout puberty a nychthemeral rhythm can be observed. This is in part endogenous but largely sleep dependent. If subjects shift their sleeping habits by 12 hours, peaks of plasma concentration both at the former and at the new time of sleep can be observed, but adaptation to such a time shift is very rapid. The established pubertal rhythm in plasma LH is accompanied, in boys, by a similar rhythm in testosterone, although in early puberty the first appearance of a raised nocturnal LH is not associated with a similar rise in testosterone concentration. This suggests that the development of rhythmicity in testosterone secretion results from an earlier development of the pituitary rhythmicity, and also that the established rhythm is not dependent on a negative feedback from raised plasma testosterone concentration.

In girls also, when substantial production of FSH and LH begins at puberty the highest levels are found at night, but the highest levels of plasma oestradiol are in the early afternoon. To explain this anomaly it has been conjectured that either the ovary responds much more slowly than the testis to LH, or that nocturnal oestrogen production is suppressed by prolactin.

By adulthood the diurnal secretion of gonadotrophins in males has increased towards the nocturnal level, and thus there is no longer a clear-cut nychthemeral rhythm. Published plots of plasma testosterone against time of day do not suggest any very reproducible rhythm, but some observations suggest that the pattern is more reproducible in a single individual than between different subjects. Authors who have made extensive observations upon small numbers of subjects have demonstrated statistically significant mean rhythms. In one study the highest levels were during sleep and the concentration fell on waking, whether at the normal hour or during the night, but in another study the highest levels were in the morning or early afternoon. An annual rhythm also has been claimed, and the light–dark cycle may be an important stimulus. In view of the effect of thought or anticipation of sexual activity upon the production of testosterone, it is possible that the rhythm in testosterone production might reflect also some characteristic of the social habits of the subjects studied. On the other hand, peak testosterone and sexual activity levels do not coincide.

For oestradiol much less evidence is available than for testosterone. Its plasma concentration can undergo very wide variations in the course of 24 hours but they are not consistent between different women. There also appear to be substantial differences in the nychthemeral pattern of oestradiol and of progesterone at different phases of the menstrual cycle.

A circadian rhythm has been described in the concentration of oestriol in maternal blood late in pregnancy, the values in the afternoon being below those in the morning or evening. This steroid is mainly of foetoplacental origin, so this rhythmicity is unlikely to be related to the rhythms so far discussed in the adult. It opens up the possibility that nychthemeral rhythmicity is present before birth and thus not dependent upon prior exposure to alternating light and darkness; but it is also possible that the observed rhythms have been imposed upon the fetus by some aspect of the maternal environment.

Growth hormone

The plasma concentration of growth hormone, though affected by food intake, maintains a clear rhythm in the fasting subject. Concentration is highest in the earlier hours of sleep, when most slow-wave sleep (stages 3 and 4) is observed, and the rhythm is very largely dependent upon the time of sleep, being rapidly reversed in phase if the time of sleeping is shifted by 12 hours. In a particularly interesting study in a blind subject whose rhythm of sleep and wakefulness failed to conform to the ordinary 24-hour day, but free-ran with a period of 24.9 hours, the association of growth hormone secretion with slow-wave sleep was observed both when the subject slept according to his natural 24.9-hour period, and when he attempted to conform to the conventional 24-hour day. Further evidence associating slow-wave sleep and hormone secretion comes from a study of aged subjects who have less stage 4 sleep and are said not to show a nocturnal peak of the hormone. Evidence against there being a simple causal relationship includes, first, the poor correlation between the effect of drugs upon growth hormone secretion and upon the various sleep stages, and secondly, the observations that later during normal sleep the secretory episodes are less closely correlated with episodes of slow-wave sleep; also there is very little correlation on the very unusual schedule of sleeping for one hour in every three.

Prolactin

This, like growth hormone, attains higher concentrations in the blood during the night than during the day; this pattern may also be entirely sleep related, since the rhythm shifts immediately and completely with three-, six- and 12-hour shifts of sleep onset. Unlike the early peaks of growth hormone, the peaks are often higher towards the end of sleep, and there appears to be no close relationship to any sleep stage. The concentration declines rapidly when the subject wakes, whether spontaneously or whether he is woken prematurely, but most subjects have a secondary peak in the afternoon. In late pregnancy the circadian rhythm is lost and the concentration throughout the 24 hours is higher than the peak level attained towards the end of sleep in non-pregnant women.

The prolactin rhythm has been demonstrated as early as the second or third year of life. It is particularly susceptible to stress, even to circumstances as trivial as venepuncture, or a first experience of hospital life.

Thyrotrophin and thyroid hormone

The most commonly recorded pattern of plasma thyroid stimulating hormone is a high level shortly before the beginning of sleep, which subsequently declines, though others have recorded the highest values near the end of sleep. The evening rise occurs at much the same hour if sleep is postponed by twelve hours leading to a reversal of the normal sleep/waking habits; it is therefore clearly not sleep-dependent and would appear to be endogenous. Any direct effect of sleep is much less clear. A failure of the normal decline, if subjects stayed awake, has been interpreted on the supposition that sleep exerts an inhibitory influence, but, conversely, a rise of plasma TSH has sometimes been observed in subjects taking a short sleep at abnormal times of day. It is possible that some of these recorded differences result from different methods of study, some workers collecting frequent small samples, while others collected integrated samples over sizeable intervals of time. There are also contradictory accounts of a greater or lesser sensitivity of the pituitary production of TSH to an identical dose of thyroid releasing hormone given in the morning or evening.

Rhythms of thyroid hormone are even less clearly defined than those of TSH, and they reflect not only variations in production but also in disposal, which is increased by meals, partly, but not wholly, as a result of biliary excretion. The commonest finding is that there is no rhythm in free thyroxine but a rhythm in the total thyroxine, with a maximum in the afternoon, dependent upon a rhythm in the concentration of the binding protein which in turn is posture dependent.

Melanocyte stimulating hormone

Melanocyte stimulating hormone has been less studied than other pituitary hormones; the concentration in most men is said to be higher early in the morning than in the evening, as with cortisol, but the pattern varies widely between individuals. In the rat, however, its rhythm is reported to be almost reversed in phase when compared with the corticosterone rhythm.

Pineal

Rhythms in the pineal have been studied in glands removed at various times of the nycthemeron from small rodents and birds, and a rhythm of very large amplitude has been consistently found in the activity of N-acetyltransferase, which appears to be the rate-limiting factor in the synthesis of melatonin. Both this enzyme and melatonin attain maximal concentration in the gland during the dark phase, both in nocturnal species such as rats and in diurnal species such as sparrows. The development of a highly specific radioimmunassay has permitted the demonstration of a similar rhythm of melatonin in human plasma.

It should be noted that, in humans, the concentration rises before subjects retire to sleep in the dark, and falls before they rise in the light; the rhythm is thus not a passive consequence of changes in posture or illumination. The rhythm has also been shown to persist in groups of subjects remaining awake for 72 hours and repeating their pattern of meals and activity every three hours, so that the major influences with a period of 24 hours were removed.

Recent work indicates that the pineal rhythm is secondary to a rhythmic sympathetic input from the suprachiasmatic nucleus acting through the superior cervical ganglion.

In birds, the melatonin rhythm contributes to the daily rhythm of activity and sleep, but the function of melatonin in mammals is largely unknown. There is some evidence that it is involved in the longer term in photoperiodic responses to changing day length, such as the responses of the reproductive organs and changing fur colour. The alternation of winter and summer thus imposes an annual rhythm. Annual rhythms in some variables, however, have been recorded in animals screened from such seasonal alterations in length of day or of temperature.

Kidney

It is usual to sleep for some eight hours without being aroused by the need to pass urine, but, during the day, such a long period seldom elapses before the need is felt. It has indeed long been realized that the rate of urine flow by night is considerably below the mean rate over the course of the 24 hours. With the advent of flame photometers it was also discovered that the nocturnal rates of excretion of sodium and of potassium as well as of chloride are substantially less by night than by day and that the rhythms of these electrolytes approximately parallel one another and the rhythm in urine flow. With increasing automation in urine analysis, similar rhythms have been shown to be present in the great majority of urinary constituents. A summary of these rhythms in urinary excretion in healthy young subjects following a fairly regular nychthemeral existence is given by Minors et al. (1976). For those who want an account of renal rhythms under a wider range of conditions, a review by Wessen (1964) is available.

Sodium, potassium, chloride

The characteristic pattern for sodium, potassium and chloride consists of a low excretion during the night, rising as soon as the subject wakes and reaching a peak slightly before or commonly up to a few hours after mid-day. When comparison is made between the different rhythms in a large number of subjects, it is found that, although the timing of chloride and sodium excretory patterns are closely parallel, the chloride leads the sodium by about one and a half hours. Potassium excretion does not parallel sodium quite so closely and the rhythm is approximately two and a half hours ahead in phase. In some subjects a second excretory peak is seen fairly regularly in the late evening shortly before bed-time.

Flow, calcium, urate, creatinine

Urine flow and excretion of calcium and urate follow a very similar pattern to sodium but with rather less regularity. As with other electrolytes there are often two excretory peaks. Creatinine excretion has often been considered as so constant that it is used as a means of checking the completeness of urine collections, whether as a check on the adequacy of bladder emptying in short-term observations or on the retention of the whole of a 24-hour sample during experiments performed under hospital conditions. However, if the excretory pattern is observed over many days in sequence, it is usually found that this is rather lower during the night than during the day, and if the mean of a substantial number of subjects is taken, there is a significant rhythm which can be fitted by sine curve with a peak at much the same time as that of sodium excretion. It is quite likely that this reflects a rhythm in glomerular filtration rate, as has been shown in a limited number of subjects when blood samples have also been collected and the glomerular filtration rate has been assessed from the clearance of endogenous creatinine or, more accurately, from that of infused inulin. The amplitude seems to vary substantially between individuals, however, so that in some it is difficult to detect a significant rhythm.

Phosphate, hydrion

The pattern of phosphate excretion differs markedly from that of other urinary constituents and it is one of the few which is excreted in larger amounts during sleep than during the waking day. The excretion commonly falls abruptly on waking, with minimum values about midday, so that it cannot be well described by a sinusoidal rhythm.

The urine by night is commonly highly acid, whereas by midday it may be slightly alkaline. If the acid components are analysed in detail, it is found that total excretion of hydrion, assessed as the sum of titratable acid and ammonium, is high by night and falls to near zero in the morning when there may be quite substantial excretion of bicarbonate. It is usually found that hydrion excretion is an almost perfect mirror image of the excretion of potassium.

Other constituents

A circadian rhythm in excretion of magnesium has also been claimed, as indeed for trace elements such as copper and for urinary excretion of lead and of mercury in workers exposed to these metals in the course of industrial processes. Of major organic constituents, urea and uric acid have both been found to be lower by night than by day.

One obvious practical implication of all these findings is that in metabolic studies, or indeed in any clinical condition in which rates of urinary excretion might be of diagnostic importance, random samples are of little, if any, value and care should be taken to obtain a complete 24-hour sample. This point will be elaborated later.

Factors influencing renal rhythms

Not only because the formation of urine is such a complex process, but also because the kidney has a homeostatic, as well as an excretory, role to perform, many factors affect urine production. Moreover, most urinary constituents are obviously directly affected by the nychthemeral pattern of behaviour. Some of these factors will now be considered.

Homeostasis

For most inorganic and many organic substances, renal excretion makes an important contribution to homeostasis and therefore must be determined in part by the presence of plethora or deficiency within the body. Consequently, ingestion of most inorganic compounds leads to an increase in their renal excretion, but the time scale of such homeostasis varies very widely between constituents. At the one extreme, increased water intake leads to increased urinary excretion within half an hour or so and ingestion of potassium salts can lead to an almost equally rapid response, although homeostasis after ingestion of sodium or chloride is considerably slower.

Meals

Most inorganic constituents are consumed with meals; however, the presence of the rest of the food may delay absorption, so that the timing of meals is not necessarily a major influence upon circadian rhythms. If one wishes to demonstrate the endogenous nature of rhythms, it is usual to control carefully the rhythmic intake of food by distributing this in frequent small meals spaced uniformly throughout the 24 hours and at as short intervals as is convenient or acceptable to the subjects. Such a procedure has been used in demonstrating that the rhythm of most major urinary constituents is in fact endogenous, or at least not dependent upon meals. Also, the assumption of a uniform posture, either recumbent or sedentary throughout the 24 hours, has been used in such studies. There is remarkably little systematic observation of the effects of normal meals upon renal function. One study in which meals were taken at normal times resulted in the peak of urea excretion occurring in the afternoon whereas, if the same amount of food was taken about 18.00 hours, maximal urea excretion was delayed by several hours.

Sleep

Sleep at any hour has generally been observed to be accompanied by a fall of deep body temperature and an increase in alveolar p_{CO_2} and hence may contribute to both these rhythms. A systematic study of the effects of sleep at all hours of the 24 (Mills *et al.*, 1978a) has shown that it is accompanied consistently by a diminished body temperature, a lower potassium excretion and a higher excretion of phosphate. The changes are, however, small and do not contribute much to the ordinarily observed rhythms, which are present whether one confines one's values to those obtained from waking or to those obtained from sleeping subjects.

Mechanisms by which the nycthemeron affects the kidney

The means whereby these different renal rhythms are controlled is very largely unknown. Clearly, the amount of a substance excreted into the urine will be determined by the amount filtered at the glomeruli, in turn dependent upon both the plasma level and the glomerulat filtration rate, and the amount reabsorbed or secreted by the rest of the nephron

Plasma concentrations and glomerular filtration rate

Variations in plasma concentration of phosphate and of potassium may play some part in inducing the appropriate variations in renal excretion, but they do not appear to be by themselves a sufficient cause.

For other substances, enough data of sufficient accuracy are often not available. The small amplitude rhythm of glomerular filtration rate has already been discussed, but an assessment of its importance has not been systematically carried out.

Tubule function

A large factor in determining urinary excretion must be the reabsorptive and secretory activities of the tubule. A number of hormones and other factors is known to modify these processes.

Variations in plasma cortisol concentration appear to be a major factor in inducing the renal potassium excretory rhythm. Moore Ede *et al.* (1976) have shown that in squirrel monkeys, in the absence of rhythmic plasma cortisol levels, the kidneys show an excretory rhythm for potassium whose period departs substantially from 24 hours. They suggest that this indicates an intrinsic rhythmicity in the renal tubular cells, so that a controlling factor (cortisol) adjusts a rhythmicity spontaneously present in the kidney rather than imposes an external rhythm upon an arrhythmic system.

Under nychthemeral habits, the plasma concentration of aldosterone is largely determined by postural changes; it rises with plasma renin concentration with ambulance during the day, and falls during recumbency at night. Earlier workers stated that there was no endogenous rhythm, the levels remaining fairly uniform in subjects who remained recumbent throughout the 24 hours, but more recent investigations have shown that there is in such subjects a small rhythm which is approximately in phase with the rhythm of cortisol. Perhaps it is due to the same cause, namely the rhythm in ACTH production. In more prolonged recumbency, however and particularly in subjects with a low sodium intake in whom both aldosterone and renin concentrations in the plasma are high, it is claimed that a similar nocturnal elevation of plasma aldosterone is paralleled by, and perhaps caused by, a rhythm in renin production. This rhythm was absent in a small group of subjects with transplanted, and hence denervated, kidneys.

The observation that nocturnal parathyroid hormone levels are higher than diurnal values might go some way to explain the tendencies for calcium excretion to be less, and phosphate excretion to be more, at night.

The small rise in p_{CO_2} at night, coupled with the more important decreased potassium secretion by the distal parts of the tubule at this time, are believed to offer an explanation of the acid urine produced nocturnally.

In conditions of controlled and restricted water intake, the solute load, comprised mainly of sodium, potassium and chloride, may be an important factor in determining the rate of renal water excretion, although when water intake is varied, variations in ADH production are a more probable influence. There are reports that ADH levels are higher by night than by day; but evidence against any rhythm also exists. In view of the many factors that are known to influence ADH secretion, this uncertainty is hardly surprising. Indeed, so many factors have been claimed to influence renal function in general that it would seem that a full understanding of the basis for renal rhythm is still very distant.

Adaptation to altered time schedules

Since urinary rhythms are easy to measure objectively they have been used by many workers in trying to assess the adaptation of rhythms to altered time schedules, such as to life on a day of abnormal length or, of more immediate practical interest, to sudden phase shifts as are experienced by circumglobal travellers. The alteration in the phase of the excretory rhythm, most commonly of sodium, potassium or flow, or often of body temperature, is generally used as an indication of how far the body's internal rhythms have adapted to the new time and will be considered more fully in a later section. Since subjects after such a time shift are usually following nychthemeral habits in the new phase in accordance with those around them, the inferences drawn would be valid only if the rhythms were wholly endogenous. Such is unlikely to be the case, and it is entirely possible that much of the adaptation commonly claimed in such circumstances reflects solely the direct external rhythmic influences of meals, sleep and posture. Figure 2 shows such an instance where, on the first day after a westward eight-hour time shift, the potassium excretory rhythm had maintained the former timing whereas the sodium rhythm had adopted the new timing. Consequently, instead of being in phase with one another, they were exactly out of phase. The most likely explanation is that the potassium rhythm had a much stronger endogenous component than the sodium rhythm. This latter was primarily affected by some factor in the new phase of external influences, though the relevant factor or factors, whether posture or meals, are not known. Recent studies suggest that the endogenous rhythmic processes adapt considerably more slowly than the various published claims would suggest. Further comment relating to this dissociation of sodium and potassium rhythms will be made later.

Temperature

Owing to the relative ease of measurement and the clinical application, the circadian rhythm of body temperature has probably been more exten-

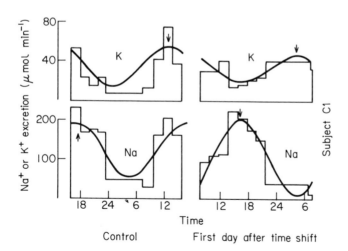

FIG. 2   Subject C1, potassium and sodium excretion on a single control day and on the first day after time-zone shift. The cosine curves of control days have been fitted to the three consecutive control days, those for experimental days to the actual day represented. Arrows indicate the acrophase (from Elliott *et al.*, 1972).

sively measured than any other. In most subjects, temperature on waking is lower than temperature on retiring, though in a minority this difference is reversed. After waking, the temperature usually rises fairly rapidly and thereafter it may either remain at a fairly steady plateau level throughout much of the day or continue to climb slowly to a maximum in the late afternoon or evening. When observations continue throughout the 24 hours the time of the maximum is most commonly in the late afternoon and of the minimum in the small hours of the morning. When the rhythm is defined by the fitting of a cosine curve the acrophase most frequently lies between 14.00 and 20.00. The rhythm shows remarkable day-to-day reproducibility in an individual, though the exact pattern varies between different individuals.

In that minority of subjects in whom the temperature on rising is above the bed-time temperature, the phasing tends to be a little earlier than usual, so that the morning rise has started before waking and the evening fall proceeds for some time before bed-time. When subjects are classified into introverts or extroverts, or into morning ("lark") or evening ("owl") types, it is found that these differences show some correlation with the diurnal course of their body temperature, the morning types and introverts tending to have the earlier temperature rise (Blake, 1971).

Origin of temperature rhythms

The temperature rhythm is one of the most stable, with a large endogenous component, and is least affected by exogenous influences, continuing little altered if subjects remain recumbent and awake throughout the 24 hours. When measurements are made of skin temperature or rate of heat loss from the extremities, it is found that both have a rhythm substantially out of phase with the rhythm of deep body temperature. Indeed, it appears that the major cause of the evening fall is a substantial increase in heat loss largely due to increased blood flow through the extremities. Attempts to interfere with these rhythms show how resistant they are. The rhythms are not affected by changes in ambient temperature even though these must initially alter the heat loss from the subject. If one attempts to raise the body temperature in the small hours of the morning, when it is normally minimal, the subject starts sweating vigorously and, if the sweat is allowed to evaporate, temperature is effectively reduced. More generally, the deep body temperature at which sweating begins has been shown to vary circadianly, being in phase with the normal temperature rhythm. The simplest interpretation of all the observations is that the circadian rhythmicity resides in the "set point" of the hypothalamic "thermostat" and this regulates body temperature around a higher level in the afternoon and a lower level in the small hours of the morning. How close this "thermostat" is to the "internal clock" will be discussed later.

Factors affecting temperature rhythms

The temperature rhythms can be modified by exogenous influences such as exercise or sleep.

Sleep can reduce body temperature. If body temperature is measured in subjects sleeping at unusual hours, it is almost invariably lower than when they were awake at that hour. The sleeping subject still shows a circadian rhythm, with an acrophase and amplitude similar to those observed on a nychthemeral existence (Mills *et al.*, 1978a).

Adaptation to time changes

The phase of the temperature rhythm is not immediately altered if one changes one's time of sleeping and of activity. This is in accord with the view that it has a large endogenous component. Many observations have been obtained upon night workers and people working on other abnormal shifts. The observed rhythms seldom correspond with any precision to a diurnal rhythm shifted by an amount appropriate to the change in time of work, for example, but the change in routine is more complex than this anyway. Thus, most night workers, instead of alternating work, leisure, sleep as do the majority of day workers, alternate instead work, sleep, leisure. The detailed operation of these differences has never been fully explored, but the result is that the rhythm is often described as being partially adapted, since, even though the higher values correspond to the time of work, they may be falling during the working hours, instead of rising as on normal schedule. Though very little work has been done upon this, it is generally believed that the rhythm of sleep–wakefulness is important as a synchronizer in humans. Perception of light is another external influence which may contribute to the process of adaptation, since many workers have found rhythms that were of abnormal phase or of rather low amplitude in blind subjects. Similar results are reported for people indigenous to high latitudes. Further comment upon possible synchronizers will be made later.

Practical importance

The practical importance of the temperature rhythm is still uncertain. There is a sizeable body of evidence suggesting that it is causally connected with performance rhythms, at least on a variety of tests that can readily be to applied repetitively, though how far these correlate with performance in real industrial situations is still uncertain. Further consideration will be given later.

For a review of temperature rhythms, see Conroy and Mills (1970).

Cardiovascular and Respiratory Systems

There is a review of cardiovascular rhythms by Smolensky et al., (1976), and of cardiovascular, haemopoietic and respiratory rhythms in Conroy and Mills (1970).

Cardiovascular system

Under nychthemeral conditions, variations of metabolic rate are very largely a reflection of rest and exercise, with sleep comprising the time of most complete rest. Since changes in metabolic rate readily affect the cardiovascular and respiratory systems, rhythms in these systems with a strong exogenous component are bound to exist. Those who have attempted to measure an endogenous rhythm in metabolic rate have used uniform conditions of rest and wakefulness at all hours of the 24, and have usually found a circadian rhythm of very low amplitude with lowest values in the small hours of the morning. The synchrony with the body temperature rhythm suggests that these small metabolic changes may be a reflection of the temperature rhythm, as does the observation that the amplitude of the temperature rhythm could satisfactorily account for that of metabolic rate. It is quite likely that similar considerations hold for the pulse rate, whose nychthemeral rhythm is largely a reflection of variations in metabolic activity, but which shows a rhythm under constant conditions of rest which may be caused by a direct effect of temperature upon the cardiac pacemaker.

Blood pressure too shows a rhythm with minimum values during the night and maximum values during the day. The development of miniature measuring systems which can be worn for as much as 24 hours and connected to an indwelling arterial catheter has shown that the amplitude of the nychthemeral rhythm is much larger than was formerly supposed, the blood pressure falling to surprisingly low values during non-REM sleep. A full description of the nychthemeral rhythm has also been aided by electronic advances that enable data to be continuously processed and recorded (see, for example, Wilson et al., 1977). Blood pressure and heart rate both rise during the REM episodes of sleep. This has led some workers to postulate in ultradian rhythm of about 90 minutes length for both variables, but there is conflicting evidence on this point (Wilson et al., 1977).

As is familiar, in addition to the effects of wakefulness and the different stages of sleep, both pulse rate and blood pressure are grossly influenced by a variety of other factors such as stress, metabolic demands, and postural changes. In conditions of constant rest, the rhythms persist with similar phasing, but are of smaller amplitude. The observation that in hypertensive patients, the rise in blood pressure above the value on waking can be reduced by administration of an adrenergic β-antagonist supports the view that the autonomic nervous system is an important contributor to the diurnal rise. By a similar argument, the very extensive interaction between the external environment, the activity of the autonomic nervous system and the control of the cardiovascular system will account for the large exogenous component that is present in many cardiovascular rhythms.

Eosinophil rhythm

Of the haemopoietic rhythms, that in eosinophils is best known. It has a maximum value in the early hours, a steady decline until about noon and a climb late in the evening. It is generally accepted that it is produced largely by plasma cortisol levels, for example being absent in patients with Addison's disease. It is not known to what extent, if any, the number of eosinophils is affected by the individual bursts of secretion from the adrenal cortex. There is a number of observations that indicates some exogenous influence upon the eosinophil rhythm; e.g. the morning fall in circulating cell numbers is delayed by keeping the awake subject blindfolded or in darkness.

Respiratory rhythms

During sleep, alveolar p_{CO_2} is often quite substantially raised and p_{O_2} reduced, suggesting respiration is depressed more than metabolism. Indeed, a very low amplitude rhythm in alveolar p_{CO_2} has been detected in subjects remaining recumbent and awake through the 24 hours, and probably reflects a rhythm of sleepiness and wakefulness. The activity of the respiratory centres in the medulla or their sensitivity to CO_2 is believed to be dependent in part upon descending impulses from higher nervous centres which, in turn, reflect, among other things, the degree of drowsiness or alertness. Thus the alveolar p_{CO_2} rhythm is presumably a secondary consequence of the rhythm in alertness. Similar consideration might well apply to the autonomic outflow to the cardiovascular system.

There is evidence from a number of studies that the calibre of the pulmonary airways varies circadianly, being narrowest during the night and widest soon after noon. The different measurements used which reflect this resistance are the Peak Expiratory Flow, the Forced Expiratory Volume and the Dynamic Compliance. The rhythmic changes are much the same in asthmatics as in normal subjects, although the absolute values are very different. The amplitude of these rhythms is not large, being seldom more than 10% of the mean. An adrenergic β-antagonist has been found to have its maximal effect upon airway calibre at the time when this calibre is smallest, and its minimal effect when the calibre is largest; thereby the rhythm is flattened and becomes statistically insignificant. It is thus possible that the rhythm reflects the circadian variations in adrenaline production and/or sympathetic outflow.

Of other respiratory functions, the unevenness of ventilation/perfusion ratios over the lung has been found to decrease by night, as indicated by the decrease in the difference of nitrogen tension between alveolar air and arterial blood. This change occurs promptly when subjects lie down and is probably a wholly exogenous rhythm, reflecting no more than the effect of a change in posture upon local pulmonary blood pressure.

Finally, the effect of standard degrees of submaximal exercise, performed on a bicycle ergometer, has been investigated at different times of the day, by measuring a number of respiratory and cardiovascular variables. If the exercise values were compared with resting values obtained at the same time of day, the differences were independent of the timing of the exercise (Davies and Sargeant, 1975).

Adaptation to altered time schedules

In view of the large exogenous component in most cardiovascular and respiratory rhythms, it would be expected that adaptation to altered time schedules would appear to be rapid. The limited data available support this supposition. The exception to this is the slow rate of adaptation of eosinophil rhythms; however, such an exception only serves to emphasize the importance of cortisol in this rhythm.

Metabolism, Cell Cycle

The ingestion of food produces many changes in the level and metabolism of substrates in the blood. Not only will ingestion of a meal raise plasma concentrations of substances, and so exert a direct effect upon mechanisms concerned with the cellular uptake and processing of food-stuffs, but also it will stimulate the secretion of those hormones concerned with gut activity (e.g. gastrin). Another effect of the changed plasma level will be to stimulate those hormones concerned with metabolic disposal (e.g. insulin), while inhibiting those that are of value during starvation (e.g. glucocorticoids, glucagon and adrenaline). These considerations suggest that there is a strong exogenous component in metabolic processes, and certainly the advantages of a system by which metabolism can immediately respond to changes produced by uptake or use are obvious.

The recent development of specific assays for many hormones has greatly increased out knowledge of metabolic rhythms, but most work has been performed upon animals rather than humans. In part, this must reflect the many problems associated with biopsy of adipose tissue, muscle or liver, for example.

Carbohydrates

In general, on waking after a period of some eight hours' fasting, liver and muscle glycogen levels in rats are minimal and climb throughout the period of activity. In humans too, liver glycogen levels are depleted at the end of sleep. Plasma glucose is, perhaps surprisingly, lowest about noon, but this reflects the influence of insulin, the concentration of which is maximal at about this time. The effect of meal-times is clearly shown in experiments in which food was taken entirely as a single meal, either at breakfast time or in the evening; the peak level of plasma insulin was altered accordingly, so that it occurred about three hours after the meal. By contrast, the glucagon peak in these experiments was less closely dependent upon mealtimes, occurring about four hours after the breakfast, but seven and a half hours after the evening meal. This suggests at least some endogenous component. The usual afternoon peak in plasma insulin, also, does not seem to be wholly dependent upon food. In a group of overweight women who were receiving minimal food intake, in the form of 220 kilo-calories per day distributed in four equal meals, the timing of the peak level of insulin concentration was similar to that in subjects on a normal diet. In a recent study upon rhesus monkeys, oscillations in plasma glucose were measured that were in phase with insulin, but 180° out of phase with glucagon concentrations. Such a relationship was less clear in humans, however, (Goodner et al., 1977).

The glucose tolerance curve has been shown to vary with the time of day; a standard dose of glucose at midday or in the early afternoon produces a much greater rise of plasma concentration than does the same dose in the morning. This is not only because the pancreas produces more insulin in the morning in response to a standard glucose load, but also because a given dose of insulin is more effective at this time. Variations in the body's production of, and sensitivity to, insulin are of obvious importance in optimising the treatment of diabetics. However, it is usually necessary to adjust the treatment empirically, since each patient might be producing a certain (but unknown) amount of insulin himself.

Amino acids

Another aspect of metabolism which has received considerable attention is the variation in the plasma concentration of amino acids. The total amino acid concentration in the blood rises from a minimum about 0400 to reach a peak in the late evening and then falls abruptly during the early hours of the night. The amplitude of this rhythm is not very large, however; its relation to the mean is largest for the relatively scarce amino acids tryptophane, cystine, tyrosine and methionine, and much smaller for the abundant ones such as glycine and glutamate. The plasma concentration is the result of a series of distinct processes: addition from the diet and the breakdown of body proteins; removal by tissue uptake, by protein anabolism, by transamination and by oxidation of individual amino acids. As already discussed, some of these processes will have distinct exogenous components, but the rhythm is said to survive minimal protein intake or even starvation.

An example of the operation of these different influences comes from a comparison of the rhythms of plasma phenylalanine in normal and in phenylketonuric children. In healthy children, the plasma concentrations of phenylalanine and tyrosine rise during the day and fall during the night, representing the effect of diurnal dietary intake. In phenylketonuria, the rhythm of tyrosine is similar in phase, but the rhythm in phenylalanine is reversed, the concentration falling during the day and rising by night. The major metabolic pathway which is lacking in phenylketonurics might thus be supposed to be responsible for the nocturnal fall in healthy children. On the other hand, in the phenylketonurics, protein metabolism appears to predominate during the night and to contribute phenylalanine to the plasma.

Particular interest has been attached to the rhythms in tyrosine and tryptophan owing to their function as precursors of the neurotransmitters serotonin and catecholamines. The amino acids pass across the blood–brain barrier by facilitated diffusion, but the relationships between their concentrations in the plasma and the brain, and between the precursor and transmitter concentrations inside the brain, are not simple. For example, in rats, small amounts of tryptophane within the range of physiological intake can raise the tryptophane concentration in the brain; at a time when brain serotonin concentration is initially low, this becomes raised, but it cannot be increased above its normal maximum, perhaps through limitation by the enzymes involved in the conversion from tryptophane. In contrast, ingestion of protein, rather than tryptophane alone, fails to increase brain tryptophane owing to competition by other amino acids for the transport system, the result being preferential tyrosine uptake and catecholamine synthesis. Rhythms in brain serotonin, tryptophane and tyrosine and plasma tryptophane and tyrosine concentrations have been described, but the phase relationships between them do not always suggest a strong causal network. Nevertheless, this tentative link between dietary intake and brain transmitter levels offers a possible mechanism by which meal-times could act as synchronizers.

There is an extensive literature upon nychthemeral variations in a large number of liver enzymes, most particularly of tyrosine aminotransferase and of tryptophane 2, 3-deoxygenase, and their induction by influences arising from meals. Such information is not available in man, and so it is impossible to build up an adequate picture of the processes underlying amino acid rhythms in humans.

A review of plasma amino acid rhythms is given by Feigin *et al.* (1967).

Fats

There is little evidence of circadian rhythmicity of plasma phospholipids and cholesterol. By contrast, serum triglyceride and lipoprotein concentrations show circadian variations, the lowest values being on waking, with a subsequent increase until evening. The rhythm in non-esterified fatty acids is the opposite, being maximal on waking and falling throughout much of the day. The nature of these rhythms has not been studied under conditions in which the effects of a nychthemeral routine have been minimized.

Cell cycle

Some years ago it was established that the number of mitotic figures in prepuces removed at circumcision was greater at night than during the daytime. Later work upon human skin from various sites has confirmed the presence of a rhythm in mitotic frequency. All are agreed that maximum activity is at about midnight. Similar results have been obtained in other species and with many tissues that contain rapid rates of cell replacement, but in the cancerous condition the mitotic rate has been reported as being constant at a high value.

The cause of the normal rhythm is uncertain, but the inverse relationship between it and that of plasma cortisol has lead some to link the two. The observation that exercise or stress suppresses the rhythm has lead others to implicate adrenaline, and the observation, not made by all investigators, that damaged tissues lose their rhythmicity when healing is taking place, has lead others to implicate chalones. It has been suggested that these factors might interact in some way that is, as yet, ill-understood.

Mitosis is only part of a regular sequence which includes DNA duplication, cell division and messenger-RNA synthesis. Not surprisingly, rhythms in all these variables have been described. The suggestion has arisen that some aspect of gene expression might be the physical basis of the internal "clock". To date, the evidence, mainly from unicellular organisms, is sparse, and alternative hypotheses can claim similar degrees of factual support! It can always be argued that data from unicellular organisms have little relevance to multicellular animals in general and to man in particular. Whatever the merits of this argument, isolated organs, tissues and even cells of human and mammalian origin do show evidence of rhythmicity. The possibility of a hierarchy of "clocks" in a multicellular organism will be discussed later. A review of hypotheses relating to rhythms at the cellular level is given by Edmunds Jr (1976).

Higher Nervous Function

Sleep and wakefulness

One of the most obvious of human nychthemeral rhythms is the alternation of sleep and wakefulness. In all parts of the world except the polar regions, it might be supposed that this is not an inherent rhythm at all but is determined by the regular alternation of light and darkness. There are now, however, records available of between 100 and 200 human subjects who have spent considerable periods alone and isolated from all obvious indications of the alternation of day and night. They have been either down deep caves or in specially constructed isolation units. By a variety of communication systems, their times of retiring and waking have been recorded and the majority of subjects follow a fairly regular rhythm of alternating sleep and wakefulness, with a period slightly more than 24 hours, perhaps 25 hours on average. In some individuals this rhythm has been exceedingly irregular but, by various mathematical devices, it is possible to extract an underlying rhythm. It was the deviation of the rhythm from a period of exactly 24 hours that prompted the introduction of the term "circadian".

It might be supposed that one sleeps because one has been awake for a sufficiently long time to feel tired, and that the duration of sleep is also a self-limiting process, the subject waking when he has had enough. If this were an adequate explanation then one would not have to suppose any underlying rhythm. Aschoff *et al.* (1971) have produced an ingenious argument to disprove this supposition. Since there is a random element in the duration of a person's day or activity period (α), in such circumstances, on might suppose that, after an unusually long day, the subject would be more tired and would need more sleep than usual. Thus the subsequent sleep or rest period (ρ) would also be longer than the average. In other words, there would be a positive correlation, even if weak, between the duration of (ρ) and that of the preceding (α). If, however, waking were determined by some inborn rhythm, then a long day would leave less time available for the subsequent rest, and (α) would be negatively correlated with the subsequent (ρ). A negative correlation is in fact observed. The argument has been developed further. The simplest interpretation of the data is that both time of retiring and time of waking are determined by endogenous rhythms. The argument is fairly elaborate, and the interested reader is referred to the original paper by Aschoff *et al.* (1971).

The actual level of sleepiness or wakefulness can only be assessed subjectively and there is a variety of different techniques for doing this. The subject might be asked to assess the level on a five-point scale, or else to place a mark on a vertical line whose upper point represents maximal wakefulness, and lowest point, maximal sleepiness. If this is done in subjects remaining awake for 24 hours, a rhythmic alternation is usually observed, although it is superimposed upon an increasing tiredness as the duration of wakefulness increases. This was particularly well demonstrated in an experiment upon groups of Swedish soldiers who spent 72 hours awake and engaged in shooting at a simulated rifle range and following a regular three-hour cycle of work, meals, urine production and so forth. Figure 3 shows the mean of their self assessed fatigue level. It can clearly be seen that, superimposed upon the progressive increase in fatigue resulting from 72 hours of activity, there are two drops when, each morning, they started feeling distinctly less fatigued, even though they had had no sleep. It is indeed a common observation of many people who have spent 24 hours or more awake, whether for reasons of pleasure or work that, some time in the small hours of the morning, they feel a maximal need for sleep, which is distinctly less pressing as the morning wears on.

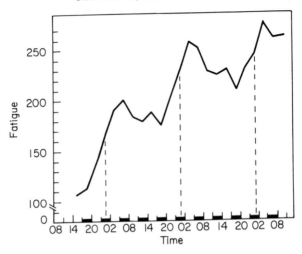

FIG. 3   Fatigue, subjectively assessed, in 63 subjects awake and active for three days (from Conroy and Mills, 1970).

Sleep stages

Sleep is a composite process which is conventionally classified from polygraphic records into five stages, designated 1, 2, 3 and 4, and the entirely different form of sleep, when rapid eye movements occur, known as REM, or paradoxical, sleep. In an ordinary night there is a rhythmic alternation between these states following a period of something like 90 minutes. This periodicity is an ultradian rhythm, and its possible relation to growth hormone release has already been described. Most of the slow-wave sleep (stages 3 and 4) comes early in the night and most of the REM towards the end (Fig. 4). If sleep is taken at other than the conventional hours, then the amounts and distribution of both slow-wave sleep and of REM may vary. There is evidence suggesting that the amount of REM follows a fairly prominent circadian rhythm. Two methods of studying this are to observe either subjects living upon a 21-hour day with seven hours of sleep and 14 hours of wakefulness, or subjests who take their eight hours of sleep in two separate four-hour portions, one always from midnight until 04.00, the other randomly positioned. On either of these schedules, sleep is taken on different occasions over all hours of the 24. If the amount of REM per hour in a seven-hour or a four-hour sleep is plotted against the time of day, it is well described by a sine curve indicating a peak of REM slightly before midday and a minimum slightly before midnight. The commonly observed progressive increase of REM during an ordinary night's sleep thus

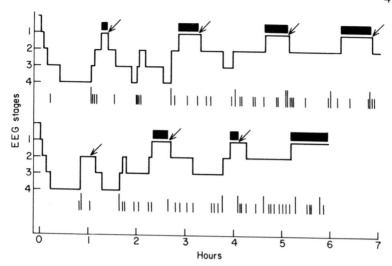

FIG. 4 Continuous plotting of EEG patterns for two representative nights. The thick bars above the EEG lines indicate periods of REMs coinciding with emergent stage 1 EEGs. The longer vertical lines at the bottom of each record indicate major body movements (changes in position); the shorter ones, minor movements. The arrows mark the successive EEG pattern cycles (from Kleitman, 1963).

apparently represents the rising phase of this sine curve. Conversely, if the amount of slow-wave sleep per hour is similarly considered, it is independent of the time of day at which the sleep is taken, but is related almost entirely to the duration of previous wakefulness. Thus, the longer the subject has been awake before a four-hour sleep, the more slow-wave sleep will he have. For a fuller account of the rhythmicity of sleep stages, see Hume and Mills (1977).

Rhythms in performance

Other aspects of higher nervous function of more obvious practical importance are much more difficult to measure objectively. Many people would claim, in contradiction to the supposition that sleep leaves one refreshed and at one's maximum ability, that they do not feel competent for any difficult mental task as soon as they get out of bed. Some individuals would claim that it is many hours before they feel at their best performance level. The difficulty in studying rhythms of performance is that most means of assessing it have a quite considerable practice effect. A large number of points, covering a number of cycles, is needed, so that, if any performance test improves substantially with practice, this change can be separated from the underlying rhythm. Partly for this reason, much attention has been given to relatively simple aspects of performance such as reaction time (for example, the time taken to extinguish a light as fast as possible), differential reaction time (for example, the time taken to extinguish a red light with the left hand, a green one with the right), or the ability to carry out relatively simple manual tasks such as sorting packs of cards.

For other tests, a number of difficulties must be overcome:

1. With a large number of examples, such as arithmetical sums, the standard must be uniform.
2. Motivation or distraction, changes in which will introduce random noise into the data, must be controlled.
3. If one wishes to measure performance throughout the 24 hours, instead of merely during the waking hours, one must guard against deterioration through lack of sleep or, rather surprisingly, of poor performance during the earlier minutes after being awakened from sleep. (Performance may indeed continue to deteriorate for at least 10 minutes after the subject has been woken).

A number of experimental designs have been used in an attempt to overcome some of these difficulties. One is to randomize the order of measurements at different times of the day, spreading them over months so that the subject is never tested twice without an interval of at least some days. Another design entails making measurements at three-hourly intervals over a period of 24 hours, and repeating this on two subsequent occasions, each time starting at a previously untested hour. In this way, every hour of the 24 will have been sampled. Further, by using a large number of subjects and changing the hour at which each sequence of eight sampling times is started, the effects of immediately being awakened from sleep, or of having been kept awake for a long time, can be removed by calculation of the mean.

All these experimental designs have given clear evidence of circadian rhythms in most of the tests that have been used, although the detailed form of the rhythm varies substantially between individuals. Some maintain a fairly steady plateau during most of the waking hours with a decline in performance during the night. Minimum performance is usually some time before the usual hour of waking, further disproving any notion that improved performance results solely from having had sufficient sleep. In tests where the subject is required to work as rapidly as possible, and yet to attempt to maintain accuracy, the circadian rhythm is usually clearest in the speed of working but is often absent in the number of errors made.

Performance and temperature

A number of lines of evidence points to a relationship between performance rhythms and temperature.

Even though there is a close parallel between the temperature rhythm and many different types of performance, this might be coincidence, many rhythms being maximal in the afternoon and minimal during the night. As has already been mentioned, there are differences between individuals as to whether their temperature on waking or during the late evening is higher. It has been found that those whose waking temperatures are the higher also have a better performance at this time than just before retiring ("larks"), and those with higher evening temperatures perform better at that time ("owls"). In studies upon subjects working on a shift system or undergoing real or simulated time-zone shifts, rates of adaptation of temperature and performance rhythms are often in parallel.

The associations described so far might be explained by a raised temperature improving performance directly or indirectly by temperature changes influencing the degree of arousal. The former possibility is suggested by the observation that raising one's temperature in the middle of the night, by getting into a warm bath, raises one's performance. The objection that such a procedure would raise one's level of arousal, and that this would account for the improvement in performance, cannot be sustained; it is found that getting into a cold bath in the middle of the day produces a similar amount of "arousal" but cools the body and decreases its performance level.

For reasons already given, most data have been obtained from "simple" tests of performance, but even these consist of more than one component. Any test involves a mixture of perception, some central nervous process, and a motor stage. Each differs in the relative weighting of the different components. In some tests, the perceptual recognition is the main contributor to variations in performance; in others, the motor response; while in others again, there is a more complex mental process involved. Some workers have found that tests with a substantial memory component follow a rather different time course from those which depend largely upon speed of working. It is a reasonable supposition that there is a number of distinct abilities which oscillate circadianly, and which contribute to performance in different proportions on the different tests, but there has, as yet, been no systematic attempt to describe performance in terms of a limited number of such variables.

Other factors influencing performance–vigilance

In addition, it is certain that factors other than temperature alone affect performance rhythms. Factors such as motivation and distraction have been mentioned already, but another variable that may be identifiable is the ability to maintain attention consistently, known as alertness or vigilance. This is usually tested by asking the subject either to pick out some particular sound from random noise, or to identify the appearance of a given pattern on a cathode ray screen. One difficulty of this form of testing is that a fairly prolonged period of observation, perhaps half an hour, is necessary for a single test.

There is fair evidence that the secretion of catecholamines is correlated with vigilance; this has not been extensively studied, and observations have been made mainly on subjects following a usual nychthemeral existence. In such circumstances, both noradrenaline and adrenaline achieve their highest levels in the blood in the morning and are low during sleep. In one study, catecholamine rhythms have been measured in a large number of subjects, continuously awake for 75 hours. Despite the fact that they were having no sleep and were screened from external indications of the alternation of day and night, their alertness and performance underwent circadian variations which were paralleled by variations in adrenaline concentration in the blood. Similar variations in noradrenaline have been claimed, but they are not so clear as those in adrenaline. The variations in adrenaline may reflect the rhythm in cortisol production, since cortisol leads to the conversion of noradrenaline into adrenaline, but the less well-defined noradrenaline rhythm presumably reflects rhythmicity in the adrenals. The problem that parallelism does not imply causality has already been mentioned: there is some evidence that the relationship between alertness and adrenaline is not purely fortuitous. Thus, "larks" are more alert and perform better in the morning, and decline towards the evening. "Owls" are the opposite in these respects, and, in addition, have a different rhythm of catecholamine production.

Since so many factors might influence brain function, it is not surprising if difficulties of interpretation or inconsistencies of data can arise. This has already been discussed in connection with "simple" performance tests, but with tasks of greater complexity, the problems will increase enormously. Thus, the well-documented result that the experience of pain, whether pathological or experimental, is felt more markedly later in the day might reflect a change in sensory threshold, alertness, motivation, or any interpretative process in the brain, or all of these. No doubt, many contradictory findings in the literature are reflections of the impossibility, or sheer impracticability, of controlling all interfering influences.

Industrial situations

Performance in real industrial situations also will be affected by so many variables other than endogenous circadian rhythms that there is only a limited number of really useful observations. Perhaps the most informative comes from the records of a Swedish gas works, where, every hour, readings had to be taken from meters, simple arithmetic carried out, and the results entered on a sheet. These tasks were all double-checked so that it was possible to identify mistakes. Over a period of 20 years, these mistakes were always most frequent at 02.00 or 03.00. Since the workers worked eight-hour shifts, if the high incidence of errors corresponded to, for instance, fatigue at the end of the shift, there would have been three, rather than one maxima of errors in the 24 hours.

Investigators of workers on shift systems have produced data that are not always consistent between variables. If the variable is influenced mainly by, for instance, cortisol or temperature, one would expect it to adapt slowly; if it is affected mainly by influences such as distraction or motivation, a more rapid rate of adaptation would be predicted. Obviously, any intermediate mixture of influences is possible.

A review of rhythms in performance is to be found in Colquhoun (1971).

The central nervous system as the site of the "master clock"

It was mentioned earlier that unicellular organisms show rhythmicity and, indeed, have been the source of most experimental material to date. It has been shown that isolated organs, tissues, and even cells from multicellular animals and plants show rhythmicity, but since, in the whole organism, all the rhythms stay in phase with one another, the concept of a hierarchy of clocks, or oscillators, has arisen. By this is understood a system whereby a "master oscillator" or "master clock" influences the other oscillators which exist at different levels throughout the body. An analogy with cardiac muscle fibres and the sino-atrial node is obvious. The degree to which these different oscillators interact, or are driven one by another, is disputed, but all accept that there must be some sort of "master clock". The problem is in locating it.

A considerable amount of work has been performed recently upon rodents, and many believe the suprachiasmatic nucleus to be the site of that master clock, since lesions of this area of the diencephalon produce arrhythmicity. Such an argument is disputed on a number of grounds. Some believe the suprachiasmatic nucleus is not itself the clock, but only part of the pathway by which it exerts its dominance. Others believe the suprachiasmatic nucleus coordinates multiple rhythms from elsewhere, since lesions produce a multiplicity of rhythms of different periods in any individual; when data from these multirhythmic individuals are combined, the pooled result will show no rhythm.

Clinical studies in man, based mainly upon neoplasms and injuries, have suggested no more than that an area in the region of the hypothalamus might be involved. Circadian firing in cells from the lateral hypothalamus and ventromedial nuclei in rats has recently been reported.

An alternative model is based upon the idea that oscillation can be produced by negative feedback systems. Such a model suffers from two types of disadvantage. First, it seems as though rhythmicity can continue in the absence of feedback inhibition; this has been described earlier for the case of cortisol and ACTH (p. 458). Secondly, circadian oscillators have an accurate period and variable amplitude; in homeostatic oscillation, it is the amplitude which is likely to be more reliable than the period.

Even if the site of the master clock can be found, its mechanism is as uncertain as ever. Models based upon data from unicellular organisms (see earlier) have been adapted to a multicellular environment, but scientific evidence for their adequacy, or otherwise, is almost completely lacking.

A review of the site and nature of the "clocks" in multicellular organisms is to be found in Edmunds (1976) and Saunders (1977).

Usefulness of Rhythms

Circadian rhythmicity is present in single cells, in plants and in most, it not all, animal phyla. Thus it has a very long evolutionary history, and must have survival value to have evolved and persisted through so much evolutionary change. Indeed, it may have evolved many times in different organisms in response to distinct needs.

Many nychthemeral rhythms have obvious survival value: for example, diurnal or nocturnal animals should be active or drowsy at the appropriate time for their pattern of activity; appetite should be greatest when the animal is active and food is available; mental powers should be higher at the time when they may be required; resistance to stresses and strains, in which adrenal cortical secretions are involved, should be greater at those hours when they are most likely to be encountered. Other rhythms have less important value, such as the low urine flow at night, enabling one to sleep undisturbed for eight hours. In the case of yet other rhythms, more than one interpretation can be placed upon their usefulness; the case of the temperature rhythm has already been considered. It is by no means so clear why an endogenous rhythmicity, rather than a direct responsiveness to the alternation of light and darkness, is functionally desirable, though one may speculate on various specific circumstances where anticipation of the changes of illumination would leave the animal better prepared for the actual time of dawn or dusk. A simple example would be the case of a nocturnal animal that needs some endogenous timing to enable it to estimate nightfall from within the safety of its burrow.

Whatever reservations we might have in attempting to interpret the evolutionary significance of rhythms, our knowledge of their existence, and, increasingly, of their underlying mechanisms, can be put to use in a variety of ways. Some of these are concerned with mitigating the effects of disrupting rhythms and will be considered later; others, concerned with clinical diagnosis and treatment, will be considered here, as follows.

Clinical diagnosis

Much clinical diagnosis depends upon quantitative measurements, with a subsequent assessment of whether these fall within or outside the range of "normality". "Normal" values of course vary, and it has been argued that the term "normal" should be abandoned in favour of "reference values". Further, the clinician is accustomed to accept upper and lower tolerance limits of normality, while the statistician would prefer a mean with its standard deviation and consequent confidence limits, which include 95% or 99% of a normally distributed population.

The wider is the range of a measurement in healthy subjects, the more difficult it is to diagnose departure from it. Thus, any means of reducing this range, by elimination of the causes of variation, is an aid to diagnosis. Simple examples in common use include differences between the sexes, and differences depending upon age or size. Some such relationships, as for instance the dependence of vital capacity upon height and/or weight, are not at all close, but they still serve to reduce the total variation in the adult population. Likewise, nychthemeral rhythms show some variation between different individuals, but, provided that they have sufficient amplitude and are sufficiently reproducible, allowance for them will narrow the range of acceptable healthy values. This has long been recognized, for example, with regard to temperature; the old assumption that a "normal" temperature in the evening is a better indication of recovery from fever than a "normal" morning temperature was, of course, based upon implicit recognition of the circadian rhythmicity of body temperature.

When the population as a whole is considered, it has been known for some time that there are rhythms both in the frequency of births and of deaths, the peak occurring in both cases in the early hours of the morning. The rhythm in mortality seems to apply to post-operative, unoperated and domiciliary cases alike and implies at what time of the 24 hours the emergency services need to be alerted. But rhythms in the individual patient also can have clinical significance and some examples are now considered.

Blood samples

When blood samples are collected for diagnostic purposes, the time of collection is of obvious importance. Some constituents, such as plasma sodium, vary very little in concentration over the 24 hours, while others, most notably the endocrines and food substances, show very large oscillations. A common clinical practice is to collect blood samples for diagnosis first thing in the morning when subjects are supposed to be in a steady "basal" condition. With many variables, this supposition is valid, but for others, it may be fallacious. Plasma phosphate, for example, often shows its largest changes in concentration soon after waking in the morning, and, if collection is spread over even as short a time as an hour, and individuals differ slightly as to the exact timing of their rhythms, a morning sample may not itself be very reliable.

The collection of a series of blood samples to define the rhythmicity of the individual is hardly practicable as a routine procedure, but a knowledge of the characteristics of a normal rhythm can be put to advantage. Perhaps the simplest example of this is in the measurement of plasma cortisol to diagnose abnormalities of the adrenal gland. The normal subject, as we have seen, shows very large variations in concentration between the high morning values and the very low ones near midnight. In Addison's disease, values are continuously low, while in Cushing's syndrome they are continuously high, although still subject to considerable oscillations. If only a single sample is to be taken to confirm or disprove diagnosis of Addison's disease, it would be best taken first thing in the morning, but, for Cushing's syndrome, late in the evening. In either of these two conditions, at least two samples, one in the morning and the other in the evening, would obviously be better still.

Urine samples

As far as urinary excretion is concerned, it is a fairly common habit in metabolic studies to ask for a complete 24-hour collection. The very large nychthemeral variations in excretion rate of many of the major constituents which have already been discussed obviously render such a procedure necessary. It is possible, however, that, in some instances, diagnosis might be aided by collecting a sample at a particular time. For example, hydroxyproline excretion has been used as an indication of collagen breakdown. Its usual rhythm may be dependent upon increased bone reabsorption during the inactivity of the night, so samples during the day might be more informative if one were searching for evidence of breakdown of collagen in other tissues. There are also instances, and they may well become more numerous with further study, where disturbance of the normal urinary rhythmicity is of diagnostic value. Perhaps the most familiar of these is the large amount of sodium and water excretion at night in cardiac failure. This is commonly described as a reversal of the usual rhythm. It is, in fact, probably a wholly exogenous effect; in the healthy subject, recumbency leads to increased excretion of both water and sodium, but this is relatively small compared with the changes involved in the usual nychthemeral rhythm. In serious cardiac failure, this postural effect is grossly exaggerated and over-rides the nychthemeral rhythm.

Malignancy

In the example just considered, an intrusive exogenous influence disturbs the usual nychthemeral rhythm. In other circumstances, the disease process itself is associated with abnormal rhythmicity. The adrenals have already been considered, and there are other examples, particularly in the field of malignancy. For example, entirely abnormal temperature rhythms may be observed over a breast containing carcinoma. The abnormal rhythm seems to take many forms, in different individuals appearing to have a period of hours, days or weeks, or to show random fluctuations. The concentrations in plasma of many hormones also have been claimed to show altered rhythmicity in malignant conditions. Unfortunately, it is not yet possible to ascribe these changes in rhythms to any particular aspect of the malignancy, but there is little doubt this this problem will become an active field for research.

In part because of the possibility of associating changes in bodily rhythms, with the early stages of cancer, Halberg (1973) has put forward the idea that individuals could regularly perform simple physiological measurements, such as pulse rate, blood pressure or temperature, upon themselves. This technique of autorhythmometry could become very powerful inasmuch as the subject would automatically have whole stretches of data, obtained in conditions of health and during his normal nychthemeral routine, with which to compare small changes that would not be described clinically as "abnormal".

Chronopharmacology and chronotherapeutics

Since most drugs act upon rhythmic physiological processes, one might not be surprised to find that the efficacy of a drug depends upon its time of administration. This field of chronopharmacology is already extensive, and is growing rapidly. It has been reviewed by Reinberg (1976). The briefest outline of this field follows.

The sensitivity of some tissues to a given plasma concentration of a drug shows a nychthemeral variation. Thus a particular dose of a drug can result in 80% mortality at one time of day, and less than 10% at another time (see Fig. 5). The variation in therapeutic effectiveness is seldom as striking as this, but it has been put to clinical use. For instance, induction of abortion by intra-amniotic prostaglandin has been found maximally effective after injection at 08.00.

In addition, the administration of a given amount of drug will not always produce the same concentration in plasma. This can be due to differences in rate of uptake into the blood, or to removal from it by metabolism or excretion. There is, for example, a rhythm in the enzyme system in the liver which oxidizes barbiturates. Therefore, the action of a fixed dose varies nychthemerally, even apart from any variation in the background sleepiness of the subject. Similarly, the rate of uptake and elimination of ethanol, the latter partly by liver metabolism, also varies nychthemerally. This is a fact

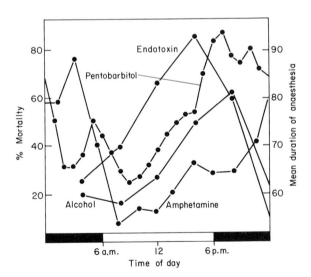

FIG. 5   The rhythmic sensitivities of rodents, maintained in light–dark cycles, to toxic and narcotic agents (from Palmer, 1976).

which can be put to good use by anyone wishing to attain a certain state of inebriation at minimal expense! The processes whereby salicylates are eliminated in the urine operate more slowly at one time of day than another, so that, at some times of day, a fixed dose of salicylate will have a more prolonged effect; this principle will be true of any drug for which the rate of elimination depends upon urinary pH.

Most drugs have multiple effects and, in many instances, there are undesirable side effects as well as the effect intended therapeutically. Considerable effort has been devoted to studying the possibility of maximizing the therapeutic, while minimizing the undesirable, effects by appropriate timing of dosage. As yet, the effort has been of relatively slight practical benefit. An example is the case of a diuretic, in which the intended natriuresis is combined with the undesirable kaliuresis. In a study upon healthy, non-hypertensive males, it was established that the ratio of potassium to sodium loss was least when the drug was administered at 08.00. But the advantage of therapy in the morning was minimal, and another study obtained results which would advocate administration at a different time. The most fruitful field here, in which some results are being achieved, is in the use of drugs where the margin between the therapeutic and toxic dose is very narrow, most notably anticarcinogenic drugs. Here there are extensive possibilities which are being actively explored (see, for example, Focan, 1976). There is evidence that many malignant tumours show rhythms quite out of phase with those of the host, a characteristic which is not only of possible use in diagnosis, but which also presents a challenge to therapeutics. It may be desirable to study the rhythm of the malignant cells as well as of the host when trying to optimize the timing of administration. It might be possible to arrange therapy so that the malignant cells will be maximally, and the host cells minimally, sensitive. It is even conceivable that one might be able to manipulate the rhythms of the patient deliberately, so as to secure the maximum asynchrony between the normal and the tumour cells. This possibility exists because, in general, malignant cells have a much more prominent endogenous, and less prominent exogenous, component to their rhythms.

Disruption of Rhythms

Man is biologically a diurnal animal, receiving his richest variety of sensory information through the eyes, and his circadian rhythms have evolved to fit him for this role. The introduction of artificial lighting, however, has enabled him to live, work and play at hours different from those for which these rhythms were evolved. In industrial societies, numbers of people are compelled by the nature of their occupation, or induced by financial reward, to work at hours at variance with their normal rhythms.

The importance of circadian rhythms to shift workers and long-distance travellers is an extensive subject which has received abundant study. In the case of those who fly large distances round the world, all the environmental rhythms of light and darkness and of the social habits of other people are shifted to approximately the same extent. The situation of these travellers can thus be described, to a very close approximation, as a simple time shift, an advance or retard of so many hours, achieved by westward or eastward flight through an equivalent number of time zones. On the other hand, the night or shift worker seldom, if ever, experiences such a simple displacement. For instance, his sequence of work, sleep and leisure is different from that of the diurnal worker, namely work, leisure and sleep. There are further differences inasmuch as the shift worker is at variance with most of society, and often attempts to conform to its hours on his days of rest, thereby greatly decreasing any opportunity of a permanent adaptation to night work.

Time-zone transitions

Most studies, whether after real flights or simulated time-zone shifts in isolation units, have followed similar protocols: there have been frequent measurement of a variable for several days before the time shift, and for several days afterwards, the fitting of sine curves to the data and the computation of the acrophase and other descriptors of the rhythms. The gradual or abrupt adaptation of these calculated values to conform to the new time-zone has commonly been referred to as "entrainment". The vast majority of results already referred to in this chapter have been obtained in this way.

It is often forgotten, however, that the rhythms observed during a customary nychthemeral existence arise both from an endogenous rhythmicity and from the operation of a variety of external influences, of which sleep, meals, and postural changes are probably the most important. After a time shift, the external influences alter their phase immediately, the endogenous ones at an uncertain rate. The evidence that human circadian rhythms are due, at least in part, to some endogenous rhythmic process has been described earlier. So too have some of the implications of the widely accepted hypothesis that different rhythms are affected to different extents by exogenous and endogenous influences. Thus, when subjects have been observed under nychthemeral conditions after a time shift, any failure of rhythms to adjust rapidly to the new phase clearly represents a persisting influence of an unentrained internal clock, but any rapid adaptation may well be the effect of a strong exogenous component of the rhythm inundating any contribution from the endogenous rhythm. By way of illustration, (1) The secretion of cortisol by the adrenal gland is one of the rhythms which is least affected by the usual external factors, although it can be substantially increased by a variety of forms of stress. Its rhythm has often been observed to take about a week to become fully adapted after a substantial time shift. (2) Owing to the ease of measurement the rhythm of body temperature has also been very extensively studied. This can be directly affected by the new time schedule, since temperature normally drops during sleep at any time of day, but the rhythm seldom shows complete adaptation under about a week, which again presumably represents the interval needed for full entrainment of the endogenous rhythm. (3) It is commonly maintained that the rhythm in cortisol secretion is a major influence upon urinary potassium, evidence for which view is that this ion, like cortisol, often takes many days to adapt to a new time zone. In contrast, the excretion of sodium, chloride, urea, urate and phosphate are more immediately entrained by time-zone shifts. These observations may thus merely represent the dominant operation of external influences.

What these external influences might be is uncertain. Social factors, light, meals, posture and sleep have all been claimed as synchronizers and reference to some of these has already been made at relevant points in this chapter. There seems to have been no systematic study of the relative influence of these synchronizers in humans, but recent work upon the squirrel monkey indicates the importance of light and feeding in this species. (Sulzman *et al.* 1977).

If the entrainment of endogenous rhythms to time shifts is to be observed, then conditions that minimize exogenous factors must be enforced, but this has seldom been done. Recently a series of experiments, designed to investigate the adaptation of the endogenous rhythms to a time-zone shift, has been performed by our group. Adaptation to an eight-hour eastward or westward shift was investigated under nychthemeral conditions. In addition, exogenous influences were minimized on two occasions, one before and the other after the shift, when the subjects underwent "Constant Routines". These consisted of keeping the subjects awake and sedentary for 24 hours, and feeding them an identical snack hourly, in order to replenish losses of water and electrolytes in the urine. The results for uric acid excretion in one subject after an eastward shift are shown (Fig. 6). Note that, on a nychthemeral existence, adaptation appears substantial, but that, on the constant routine, adaptation (presumably of the internal clock) is far less, and even in the wrong direction.

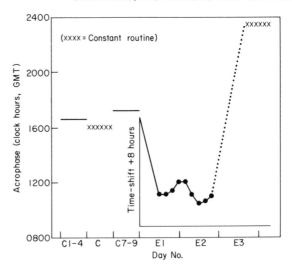

FIG. 6 Acrophases of urinary uric acid on nychthemeral and on constant routine, before (C1–9) and after (E1–3) simulated 8-h eastward shift. Control acrophases are derived from a single cosine fit to each stretch of data, covering 3–4 days. After time shift, cosines are fitted pergressively to 24-h sections with increments of 4 h. Note irregular compression of time scale on abscissa before time shift (from Mills *et al.*, 1978b).

With regard to the subjective well-being of the individual, there are some rhythms whose usefuless is such that a failure of adaptation can be a serious inconvenience. There seems to have been little systematic survey, however, of how far the malaise suffered by travellers, and loosely described as "jet lag", results specifically from the failure of these rhythms to adjust.

There are records of difficulty in adjustment of the time of sleeping. Westward travellers, as from Europe to America, may be sleepy in the evening and have little difficulty in going to sleep at night, but wake very early in the morning by American time, at a time which corresponds roughly to their time of rising at home. This disability can persist for a week. Conversely, eastward travellers may have difficulty in getting to sleep at night, but feel sleepy in the morning. This ability to sleep at unfamiliar times varies greatly between individuals, and sleep laboratories generally record that the young have little difficulty in sleeping at any hour of day or night; the difficulty more commonly affects older individuals. Failure to adjust one's appetite is again a disability which probably operates very variably with different individuals, as does failure to adjust a regular hour of bowel movement. Unadapted urinary excretory rhythms have been clearly shown to be an important difficulty encountered by travellers, who very commonly fail to sleep through eight hours during the night, owing to being awakened by a full bladder.

Mental ability is, to most travellers, of much more interest than more simply measured quantities such as the rate of excretion of sodium or chloride. Large numbers of people travel substantial distances around the world to engage in business negotiations or scholarly conferences, or to give lectures to learned societies or other bodies. Their ability to perform well at the appropriate time is a matter of considerable concern. Unfortunately, as has already been considered, the required abilities are not very readily measurable. There must be some doubt as to how far the simple tests that have been used in the study of rhythms in performance, with the possible exception of vigilance tests, reflect the kind of abilities required. It is quite likely that there are sizeable inter-individual differences, so that the rules enforced by some large business corporations, relating to the time which must elapse after travelling and before engaging in important negotiations, may be inadequate for some individuals, but excessive, and consequently uneconomic, for others.

Apart from such disabilities just described, it is commonly speculated that the asynchrony between different components of rhythms after travel may be harmful. The many rhythms, observable under ordinary nychthemeral conditions, have fairly fixed phase relationships to one another. Some of these may result from causal connections between them, some may be fortuitous, but it is often speculated that some may be of functional importance, so that disability of an unspecified nature may result when one rhythm has adapted considerably before another, and they are, therefore, completely out of phase. A very simple example of such asynchrony (Fig. 2) is in the sodium and potassium excretory rhythms of a subject on the first day after a simulated eight-hour eastward flight. These two rhythms normally peak nearly at the same time, but on this first day they were almost almost precisely out of phase, maximum potassium excretion coinciding with minimum sodium. An explanation for this asynchrony has already been offered (p. 467). It is not suggested that this necessarily has any harmful functional implications, but it is given as an instance of the kind of asynchrony which may result and which is often supposed to be harmful. Nevertheless, such an example of the differing rate of adaptation of constituents gives valuable clues as to the nature of the causal nexus between rhythms that exist in the body (See Mills, 1973). There is a full review of adaptation to time-zone shifts by Aschoff *et al.* (1975).

Night Work and Shift Work

The initial change from normal hours of work can be likened to a time-zone shift, except that, as already mentioned, the individual knows he is at variance with most people's habits. But the problem is compounded by the observations that most night workers attempt to follow a diurnal pattern on their rest days and that shift workers often change their hours of work. In effect, therefore, both groups are undergoing continual changes of routine, equivalent to a traveller forever circling the globe.

Most studies of night or shift workers have used a protocol whereby the rhythms are assessed just before, and then on the first and a later day(s) after the change in routine. In agreement with experiments upon travellers, data have indicated a wide range in the rate at which different rhythms adapt, and some rhythms (e.g. pulse, urinary uric acid) adapt more rapidly than others (e.g. plasma cortisol, deep body temperature). But, since all experiments have been performed in the presence of external influences, the adaptation, if any, of the endogenous rhythms is, of course, unknown. The possibility that the asynchrony between different rhythms, and even between the endogenous and exogenous components of a single rhythm, is associated with the malaise reported by many shift workers indicates an aspect of this field which requires further study.

It must be remembered, however, that a balance must always be struck between, on the one hand, intellectually satisfying experiments that involve complex tests which are very susceptible to interference, and, on the other hand, experimental protocols and tests that are less sophisticated, but relate to industrial practice. This reservation applies particularly to psychometric testing, and recent experiments have attempted to duplicate real situations more closely.

It has often been suggested that the frequent time shifts experienced by shift workers may be injurious to health; despite surveys on human populations, and animal experiments, there is, as yet, little evidence to support such suggestions. It is generally agreed, however, that there is a difference in the ease with which workers accept shift work, and there have been claims that, with experience, the worker adapts more easily.

Colquhoun (1971) has reviewed the field of shift work and performance measurement; Rutenfranz et al. (1977) deal with different shift systems and some associated social problems.

Infants, the Aged and Psychiatric Subjects

It has been seen how, for many aspects of rhythm research, our ignorance is at least partly caused by an inadequate amount of data. In studies upon infants, the aged and psychiatric patients, this paucity is especially marked and some explanation for this comes from the ethical problems that are involved. To separate the child from its mother or a normal environment is generally regarded as unacceptable; equally it is unacceptable to sample blood frequently from such a small individual and it is most difficult to collect urine samples reliably.

With psychiatric patients, as has been described by Sollberger (1974), it is sometimes not possible to make them understand the nature of the experiment, and the patients are often unreliable when tests must be performed or urine samples given.

With aged subjects it is undersirable to submit them to potentially traumatic experiences such as venepuncture, or to cause distress by imposing upon them a stringent experimental regimen. In addition, there is the difficulty experienced by the aged of micturating on demand.

Nevertheless, some progress in all these fields has been made, and the following outline will attempt to show that some clues pertinent to the problems of circadian rhythms in general can be obtained from these groups.

Infants and the development of rhythms

This field has been reviewed recently by Mills (1974, 1975). Most investigators have concentrated upon the pattern of sleep and wakefulness. It is agreed that there are no detectable circadian rhythms during the first week post partum, and the amount of sleep during the hours of darkness becomes greater than that during the daytime by about the third week. This difference increases steadily thereafter, but an unbroken night's sleep with no sleep during the daytime usually takes more than a year to develop.

Such results cannot be interpreted unambiguously as evidence for an endogenous rhythm since the infant is surrounded by more activity and attention during the daytime and is, where possible, left unattended at night. Thus the development of a rhythm in sleep and wakefulness might be the response to exogenous influences. Further comment upon this will be made later.

A valuable study that suggests that at least a component of the sleep–wakefulness cycle is endogenous and develops during the first weeks of life has been reported by Martin-du Pan (1970). He describes the rhythms of a child that was kept in constant light and fed on demand. At first the times of sleep were irregular, but by about the sixth week a rhythm began to emerge that seemed to have a period greater than 24 hours, that is, it was free-running. At eight weeks the rhythm showed a period closer to 24 hours, as did that of urine flow, though the phase was unusual, showing maximum flow and minimum sleep during the night. As soon as the child was placed in a nychthemeral environment (week 12) the rhythms rapidly adjusted their phase, showing maximum flow and activity in the forenoon. This study also indicates that the ability of the individaul to respond to external synchronizers develops early in infancy.

Studies of the rhythms in deep body temperature, plasma cortisol, pulse rate and urine flow all show similar stages of development, namely arrhythmicity at birth and an increasing amplitude during the first year of life. However, the rate of establishment of the rhythms varies and so gives clues as to the causal nexus between them. Thus: (1) the rhythm of urine flow develops at the same rate as that of electrolyte loss; (2) there is a similar time course between the activity rhythm and that in pulse rate; (3) the rhythm of deep body temperature develops simultaneously with that of sweating.

The plasma cortisol rhythm develops most slowly; such an observation might do no more than emphasize the comparatively small exogenous component in this rhythm in the infant as in the adult.

Hellbrügge (1974) has provided evidence that the lack of circadian rhythmicity does not mean that the new-born is arrhythmic. He has found evidence of ultradian rhythms (with a period of about four hours) in the sleep-wakefulness cycle of the full-term child in its first week; these rhythms regress as the circadian rhythms develop. Premature babies show neither type of rhythm at birth but develop first ultradian and then circadian rhythms, always slightly later than the full-term child. This delay in the establishment of rhythms in the premature child suggests that some process of maturation is involved before rhythms can be manifest.

Another factor which is closely associated with the sleep–wakefulness cycle is the demand for feeding, which recurs about every four hours. Activity has been reported to be maximum at 06.00 when the child had been without food for more than four hours. But the rhythm in food demand does not change with the same time-course as that of the emergence of the circadian rhythm of sleep and waking, and the extent of the causal relationship between the two rhythms is not known.

The recent reports that in adults ultradian rhythms exist in urine flow, capillary blood flow and various psychological functions and patterns of brain electrical activity, together with the knowledge of the episodic secretion of many hormones, all suggest that the relationship between ultradian and circadian rhythms is not as simple as has been described above. Clearly the possibility exists that circadian do not replace ultradian rhythms but rather coexist as the result of some form of interaction between ultradian components. Models describing such interactions have been used at a cellular level to produce rhythms of a longer period from rapid biochemical oscillations. But again, many more data, against which to test these models and their application to these problems, are required.

One rhythm which is an exception, in that it is well developed at birth, is that of the frequency of mitoses in prepuces removed at circumcision (see earlier). Although this might indicate an inherited endogenous rhythm, it would be mistaken to assume that the foetus lives in a constant environment. There is evidence that foetal movement is influenced by maternal movement and, in foetal lambs, rhythmicity in both respiratory movement and renal function has been reported. Further, at the end of pregnancy in the human, maternal blood oestriol levels, derived from the foeto-placental unit, show higher values on waking than in the afternoon.

Finally, occasionally there have been accounts of seemingly healthy individuals who, while living in a nychthemeral environment, showed free-running rhythms (for example Elliott *et al.*, 1970). Perhaps this indicates that the development of an endogenous rhythm and the ability to respond to external synchronizers develop separately, the latter failing in these subjects. It is possible that the developing infant must be exposed to environmental rhythmicity at some stage of maturation if he is fully to develop endogenous rhythms and the ability to respond to synchronizers. The conflicting reports of the amplitude of rhythms in inhabitants of polar regions, where the alternation of light and dark is often reduced or absent, do not enable this possibility to be accepted or rejected. However, it has been observed that the urinary rhythms of infants aged about two months disappear when studied in constant darkness, but persist in similar conditions at 12 months. These observations argue for an externally mediated rhythmicity at the younger age, and accord with the view that these endogenous rhythms are not inherited but develop is a result of continual exogenous rhythmicity. Against this view is the work of Martin-du Pan (see earlier) which is easier to interpret as a developing ability of the child to manifest its inherited potential to become rhythmic.

Such concepts are not dissimilar from those based on recent work upon the visual cortex in kittens; in this system there is evidence for a "critical period" during which it is important that the kitten experiences a normal visual environment if normal development and function of the striate cortex are to ensue.

Rhythms in the aged

This is a field which has not received much attention until recently and no general review is available. Accordingly it is necessary to give a number of specific references for the interested reader.

Not surprisingly, studies of the sleep–wakefulness rhythm in institutionalized subjects have shown in them a marked 24-h rhythmicity, but the subjects took more naps in the daytime and woke more often at night than did younger controls. Aged subjects have less stage 4 sleep and the absence of a peak in serum growth hormone secretion at night. The possible connection between stage 4 sleep and hormone secretion has already been commented upon; however, prolactin levels do not change (Murri et al., 1977).

A variety of physiological and psychological variables has been measured in subjects of 65 years or more living on regular regimes of sleep, activity and mealtimes (Lobban and Tredre, 1967; Descovich et al., 1974; Scheving et al., 1974). Most results support the view that the rhythms change with age in three ways: (1) the mean value decreases; (2) the amplitude decreases; (3) the timing of the peak becomes more variable. A partial exception to these generalizations comes from a study of a group of plasma metabolites, including lactate, pyruvate, glycerol and acetoacetate (Alberti et al., 1975); in these, there was an increase in the average concentration.

By far the most inconvenient consequence of these changes in rhythm is the necessity to pass urine at night (nocturia), often on more than one occasion. Brocklehurst (1971, 1972) has argued that, since so many rhythms diminish with age, nocturia might be no more than one manifestation of a general "cerebral senescence" that would produce, for example, an "uninhibited neurogenic bladder". This would account not only for the inability to hold large volumes of urine, a cause of nocturia, but also "stress incontinence" and difficulty in the act of micturition. Whatever the merits of this argument, it does not readily explain the increased urinary electrolyte excretion at night, and it is probably an osmotic consequence of this that produces the higher rate of nocturnal urine flow. Further, the relevance of cerebral senescence to the other rhythms that change with age is not known. An alternative explanation of nocturia—an increased response to changes in posture—has been proffered (Lindeman, 1975).

It has been found in younger subjects that there is a direct relationship between the mean value and the amplitude of a rhythm in urinary constituents (Mills and Waterhouse, 1973). If this relationship is a general, though unexplained, property of all circadian rhythms, then an explanation for the lowered mean value of rhythms in the aged would still be required. This idea of a changed mean, together with one's general impression of the process of ageing, leads to the view that a degeneration of homeostatic mechanisms underlies the changes in circadian rhythms. Presumably the variation in timing of the peak of the rhythm also could easily be incorporated into such a theory of ageing, but the difficulties attendant upon explaining circadian rhythms with reference to homeostatic mechanisms have already been described.

Psychiatric patients

The use of the word "lunatic" indicates that rhythmicity has been associated with madness since mediaeval times. More scientific approaches have endorsed rather than dispelled this belief, but of the wide range of periods that is now associated with psychiatric disorders (Sollberger, 1974), only circadian rhythms will be considered here.

There is some evidence that in both schizophrenia and manic depression there are abnormal rhythms.

The case for schizophrenia has been summarized by Mills *et al.* (1977). Even though schizophrenics are normally very passive individuals, it is commonly observed that, in institutions, they retire, rise and request their meals earlier than necessary. Measurements in 107 patients of oral temperature throughout the 24 hours indicated that the temperature was higher than that of controls on waking, and fell earlier in the evening. In a study upon two male schizophrenics placed in an Isolation Unit with no indication of the passage of time, the urinary, temperature and activity rhythms did not deviate significantly from 24 hours. However, this is shorter than the free-running rhythm reported for normal subjects, and the possibility exists that the rhythms in schizophrenia result from an internal clock that runs too quickly and must be continuously delayed by the nychthemeron. Alternatively, one may interpret the data to indicate that the period of the rhythms in schizophrenics is normal, but that, by some unknown mechanism, the peaks of their rhythms occur earlier than in controls.

Niskanen *et al.* (1976) have described raised tryptophane levels in manic depressives and reported that these levels decreased towards normal as the psychiatric disorder improved. The interpretation that was placed upon these results is that raised plasma tryptophane is a compensatory mechanism that will increase entry into the brain of this precursor, of the neurotransmitter serotonin. Certainly there is general agreement that brain serotonin levels are changed in depression. The relationship between this psychiatric disorder and changed circadian rhythms is two-fold: (1) changed plasma levels of any single amino acid will alter the pattern of brain uptake of all amino acids and the rhythmic synthesis of transmitters (see earlier); (2) a changed deep body temperature rhythm in manic depression has been reported, but the neurotransmitter basis of this, if any, has not been established.

It is not to be inferred from this work upon schizophrenia and manic depression that both disorders are caused by primary alterations of the clock(s) in these patients. Far more likely is the view that the primary disorder (possibly inadequate serotonin synthesis in manic depression) will

have widespread effects including those upon the mechanism or manifestations of the internal clock. If it is accepted that the clock is to be found in the
hypothalamic area of the brain (see earlier) then the observation that circadian rhythms are irregular in some cases of anorexia nervosa or hypothy-
roidism might well indicate a more direct link between the abnormalities.

Postscript

This brief account has indicated that circadian rhythms are important in
many branches of biology and medicine. Yet a full understanding of the
rhythms and their implications in health and disease will require more than
mere collection and mathematical descriptions of data, even though the
contribution by made the development of suitable mathematical techniques
must not be overlooked.

Many studies that have been published suffer from inadequate sampling
frequency or protocol (or both), and, in so many cases, there has been no
serious consideration of mechanisms that might account for the results or
of theoretical frameworks into which the results might be fitted. It is to be
hoped that models put forward to explain the nature of circadian rhythms,
and experiments designed specifically to test these models, will appear more
frequently in the future.

Since this chapter was written two useful textbooks dealing specifically
with human circadian rhythms have appeared. They are: "Comprehensive
Endocrinology: Endocrine Rhythms", (1979), (Ed. D.T. Krieger, Raven
Press, New York; "The Circadian System of Man. Results of Experiments
under Temporal Isolation", (1979), R.A. Wever, Springer, New York,
Heidelberg, Berlin.

In addition, two of the authors of this chapter have completed their own
textbook which should be available in 1981: "Circadian Rhythms and the
Human", (1981), D.S. Minors and J.M. Waterhouse, John Wright,
London.

Further Reading

Aschoff, J., Gerecke, U., Kureck, A., Pohl, H., Rieger, P., von Saint Paul, U. and
Wever, R. (1971). Interdependent parameters of circadian activity rhythms in
birds and man. *In* "Biochronometry", Biology and Agricultural Division
National Academy of Sciences, pp. 3–27. Washington. Nat. Acad. Sci.

Aschoff, J., Hoffman, K., Pohl, H. and Wever, R. (1975). Re-entrainment of circadian rhythms after phase-shifts of the *zeitgeber*. *Chronobiologia* **2**, 23–78.

Blake, M.J.F. (1971). Temperature and time of day. *In* "Biological Rhythms and Human Performance", (Ed. W.P. Colquhoun), 107–148. Academic Press: London and New York.

Brownstein, M.J. (1975). Minireview. The pineal gland. *Life Sci.* **16**, 1363–1374.

Colquhoun, W.P. (1971). Circadian variations in mental efficiency. *In* "Biological Rhythms and Human Performance", (Ed. W.P. Colquhoun), 39–107. Academic Press: London and New York.

Conroy, R.T.W.L. and Mills, J.N. (1970). "Human Circadian Rhythms" Churchill: London.

Davies, C.T.M. and Sargeant, A.J. (1975). Circadian variation in physiological responses to exercise on a stationary bicycle ergometer. *Br. J. Industr. Med.*, **32**, 110–114.

Edmunds, L.N. Jr. (1976). Models and mechanisms for endogenous timekeeping. *In* "An Introduction to Biological Rhythms", (Ed. J.D. Palmer), 280–361. Academic Press: New York and London.

Elliott, A.L., Mills, J.N., Minors, D.S. and Waterhouse, J.M., (1972). The effect of real and simulated time-zone shifts upon the circadian rhythms of body temperature, plasma 11-hydroxycorticosteroids and renal excretion in human subjects. *J. Physiol.* **221**, 227–257.

Feigin, R.D., Klainer, A.S. and Beisel, W.R. (1967). Circadian periodicity of blood amino acids in adult man. *Nature* (*London*), **215**, 512–514.

Ferrari, E., Bossolo, P.A., Vailati, A., Martinelli, I., Rea, R. and Nosari, I. (1977). Variations circadiennes des effets d'une substance vagolytique sur le système ACTH-secretant chez l'homme. *Annales d'Endocrinologie* (*Paris*), **38**, 203–213.

Focan, C. (1976). Circadian rhythm and chemotherapy for cancer. *Lancet* **ii**, 638–639.

Goodner, C.J. Walike, B.C., Koerker, D.J., Ensinck, J.W., Brown, A.C., Chidockel, E.W., Palmer, J. and Kalnasy, L. (1977). Insulin, glucagon and glucose exhibit sustained oscillations in fasting monkeys. *Science* **195**, 177–179.

Halberg, F. (1973). Laboratory techniques and rhythmometry. *In* "Biological Aspects of Circadian Rhythms", (Ed. J.N. Mills), 1–26. Plenum Press: London, New York.

Halberg, F., Katinas, G.S., Chiba, Y., Garcia Sainz, M., Kovacs, T.G., Künkel, H., Montalbetti, N., Reinberg, A., Scharf, R. and Simpson, H. (1973). Chronobiologic glossary of the International Society for the Study of Biologic Rhythms. *Int. J. Chronobiol.* **1**, 31–63.

Halberg, F., Carandente, F., Cornelissen, G. and Katinas, G.S. (1977). Glossary of chronobiology. *Chronobiology*, **4**, [Supplement 1].

Hume, K.I. and Mills, J.N. (1977). Rhythms of REM and slow-wave sleep in subjects living on abnormal time schedules. *Waking and Sleeping* **1**, 291–296.

Kleitman, N. (1963). "Sleep and Wakefulness", 2nd Ed. University of Chicago Press: Chicago and London.

Krieger, D.T. (1975). Rhythms of ACTH and corticosteroid secretion in health and disease, and their experimental modification. *J. Steroid Biochem.* **6**, 785–791.

Mills, J.N. (1973). Transmission processes between clock and manifestions. *In* "Biological Aspects of Circadian Rhythms", (Ed. J.N. Mills), 27–84. Plenum Press: London, New York.

Mills, J.N., Minors, D.S. and Waterhouse, J.M. (1978a). The effect of sleep upon human circadian rhythms. *Chronobiologia* **5**, 14–27.

Mills, J.N., Minors, D.S. and Waterhouse, J.M. (1978b). Entrainment to abrupt time shifts of the oscillator[s] controlling human circadian rhythms. *J. Physiol.* **285**, 455–470.

Minors, D.S., Mills, J.N. and Waterhouse, J.M. (1976). The circadian variation of the rates of excretion of urinary electrolytes and of deep body temperature. *Int. J. Chronobiol.* **4**, 1–28.

Moore-Ede, M.C. Schmelzer, W.S., Kass, D.A. and Herd, J.A., (1976). Internal organization of the circadian timing system in multicellular animals. *Federation Proc.* **35**, 2333–2338.

Reinberg, A. (1976). Advances in human chronopharmacology. *Chronobiologia* **3**, 151–166.

Rutenfranz, J., Colquhoun, W.P., Knauth, P. and Ghata, J.N. (1977). Biomedical and psychosocial aspects of shift work. *Scand. J. Work Environ. and Health* **3**, 165–182.

Saunders, D.S. (1977). Biological rhythms: anatomical location of photoreceptors and pacemakers. *In* "An Introduction to Biological Rhythms", 125–141. Blackie: Glasgow, London.

Smolensky, M.H., Tatar, S.E., Bergman, S.A., Losman, J.G., Barnard, C.N., Dasco, C.C. and Kraft, I.E. (1976). Circadian rhythmic aspects of human cardiovascular function: a review by chronobiologic statistical methods. *Chronobiologia* **3**, 337–371.

Sulzman, F.M., Fuller, C.A. and Moore-Ede, M. (1977). Environmental synchronizers of squirrel monkey circadian rhythms. *J. Appl. Physiol. (Respirat. Environ. Exercise Physiol.)*, **43**, 795–800.

Weitzman, E.D. (1976). Circadian rhythms and episodic hormone secretion in man. *Ann. Rev. Med.* **27**, 225–243.

Weitzman, E.D., Boyar, R.M., Kapen, S. and Hellman, L. (1975). The relationship of sleep and sleep stages to neuro-endocrine secretion and biological rhythms in man. *Rec. Progr. Horm. Res.* **31**, 399–466.

Wesson, L.G. (1964). Electrolyte excretion in relation to diurnal cycles of renal function. *Medicine* **43**, 547–592.

Wilson, D.M., Kripke, D.F., McClure, D.K. and Greenburg, G. (1977). Ultradian cardiac rhythms in surgical intensive case unit patients. *Psychosom. Medicine* **39**, 432–435.

References for Rhythms in Infants, the Aged and Psychiatric Subjects

Alberti, K.G.M.M., Durnharst, A. and Rowe, A.S., (1975). *Biochem. Soc. Trans.* **3**, 132–133.

Brocklehurst, J.C., Fry, J., Griffiths, L.L. and Kalton, G., (1971). *J. Am. Geriat. Soc.* **19**, 582–592.

Brocklehurst, J.C., Fry, J., Griffiths, L.L. and Kalton, G. (1972). *Age and Ageing* **1**, 41–47.

Descovich, G.C., Kühl, J.F.W., Halberg, F., Montalbetti, N., Rimondi, S. and

Ceredi, C., (1974). *Chronobiologia* 1, 163–171.
Elliott, A.L., Mills, J.N. and Waterhouse, J.M., (1970). *J. Physiol.* 212, 30–31P.
Hellbrügge, T. (1974). *In* "Chronobiology" (Eds Scheving, L.E., Halberg, F. and Pauly, J.E.), 339–341. Igaku Shoin Ltd, Tokyo.
Lindeman, R.D. (1975). *In* "The Physiology and Pathology of Human Ageing" (Eds R. Goldman and M. Rockstein), 19–38. Academic Press, New York and London.
Lobban, M.C. and Tredre, B.E. (1967). *J. Physiol.* 188, 48–49P.
Martin-du Pan, R. (1970). *La Femme et l'Enf.* 4, 23–30.
Mills, J.N. (1974). *In* "Scientific Foundations of Paediatrics", (Eds J.A. Davis and J. Dobbing), 758–772. Heinemann, London.
Mills, J.N. (1975). *Chronobiologia* 2, 363–371.
Mills, J.N. and Waterhouse, J.M. (1973). *Int. J. Chron.* 1, 73–79.
Mills, J.N., Morgan, R., Minors, D.S. and Waterhouse, J.M. (1977). *Chronobiologia* 4, 353–360.
Murri, L., Barreca, T., Gallamini, A., Massetani, R. (1977). *Chronobiologia* 4, 135.
Niskanen, P., Huttunen, M., Tammiren, T. and Jääskeläinen, J. (1976). *Br. J. Psychiatr.* 128, 67–73.
Scheving, L.E., Roig, C., Halberg, F., Pauly, J.E. and Hand, E.A. (1974). "Chronobiology", (Eds L.E. Scheving, F. Halberg and J.E. Pauly), 353–357. Igaku Shoin, Tokyo.
Sollberger, A. (1974). *In* "Chronobiology", (Eds L.E. Scheving, F. Halberg and J.E. Pauly), 515, 516. Igaku Shoin, Tokyo.

CHAPTER 10
Sleep

M.W. Johns

Basic Concepts

Fundamental rhythm of sleep and wakefulness

All mammals and many other animals lower on the evolutionary scale sleep at least once daily throughout their lives, apart from those unusual times when some animals hibernate.

Sleep is a rapidly reversible state of reduced activity and responsiveness to the environment, usually experienced at a time of day or night when, from the point of view of the particular animal's adaptation to the environment, energy might best be conserved rather than expended (Webb, 1974). The ready reversibility of sleep is one characteristic that distinguishes it from coma and hibernation, which are also states of reduced activity and reactivity.

505

Over the past 25 years major advances have been made in the detailed description of physiological changes which take place during sleep, but there is still much to be learnt about its fundamental nature. This chapter will be limited mainly to a discussion of sleep in man, although some investigations which could be carried out only in animals must be mentioned.

Electroencephalogram and electro-oculogram during sleep

Much of the sleep research carried out in modern times was made possible by an observation reported in 1929 by Hans Berger that the electroencephalogram (EEG) during sleep was markedly different from that during wakefulness. Soon afterwards it was observed that there were cycles of change in the frequency and amplitude of the EEG during a normal night's sleep, and that these changes were paralleled by variations in the apparent depth of sleep as measured by the auditory stimulation required to waken the subject at the time (Blake and Gerard, 1937). However, it was not until the early 1950s, when another important discovery was made in relation to movements of the eyes during sleep, that it became widely appreciated that sleep was not of a uniform nature during the night. Eye movements can be recorded continuously by means of the electro-oculogram or EOG. Each eye acts as a dipole because of the constant corneo–retinal potential and eye movements therefore produce changes in the potential difference between electrodes fixed to the skin at each side of the orbit. Aserinsky and Kleitman (1953) reported that at approximately 90-minute intervals the eyes of sleeping subjects flicked from side to side beneath closed eyelids for periods lasting several minutes at a time. These rapid eye movements, or REMs as they were called, were in marked contrast to the slow, wandering eye movements which occurred at most other times during sleep. When the REMs occurred the EEG resembled that recorded during drowsiness, with relatively low-amplitude waves having a higher frequency than was usual during sleep. If the subjects were woken then they usually reported having just been dreaming, whereas this seldom occurred when they were woken at other times.

These important observations led to the recognition of a distinctly different type of sleep associated with dreaming, a state which has become known as REM sleep. The distinction between REM sleep and the other type of sleep, called non-REM or NREM sleep, has been a major topic of research over the past 20 years.

Stages of sleep

At present, sleep is usually categorized into five stages—four NREM stages and REM sleep—defined on the basis of the EEG, EOG and sometimes also the electromyogram (EMG) taken from the submental muscles (Rechtschaffen and Kales, 1968).

Stage 1   The drowsy state between alert wakefulness and definite sleep. The transition from wakefulness is associated with a generalized slowing of the EEG into the theta range (2–7 cycles per sec) and loss of obvious alpha waves (8–12 cycles per sec). The conjugate, saccadic eye movements of wakefulness give way to slow, pendular eye movements which are often non-conjugate, that is, each eye moving independently of the other.

Stage 2   Spindles, K-complexes and delta waves appear in the EEG. Spindles are bursts of waves with a frequency of between 13 and 15 cycles per second lasting 0.5–1 s. A K-complex is a wave-form with a well-defined negative wave of relatively high amplitude ($>$ 100 μV) followed by a positive wave and then by a short burst of 12–14 cycles per second waves of lower amplitude. Some K-complexes occur as responses to an external stimulus such as a noise which is not loud enough to waken the subject, but most K-complexes occur spontaneously. Delta waves of low frequency (1–4 cycles per second) and high amplitude ($>$ 75 μV) also appear first during stage 2.

Stage 3   Between 20 and 50% of the EEG trace taken up by delta waves. Spindles and K-complexes are also present.

Stage 4   More than 50% of the EEG trace is taken up by delta waves. Spindles and K-complexes are less obvious than in stages 2 and 3.

REM sleep   Characterized by the concomitant appearance of a low voltage, mixed frequency EEG, as in stage 1, and of REMs recorded in the EOG. Spindles and K-complexes occur rarely, if at all.

REM sleep is sometimes referred to as paradoxical sleep because the similarity of the EEG during stage 1 and REM sleep would suggest that both stages were very light sleep. In fact, the arousal threshold is at least as high during REM sleep as that during stage 2—hence the paradox. By contrast, other NREM stages are sometimes collectively called orthodox sleep, although these terms are usually reserved for experiments with animals rather than man. Stages 3 and 4 are sometimes considered together as delta-wave or slow-wave sleep.

EEG

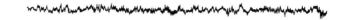

Awake

EOG

EEG

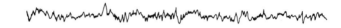

Stage 1

EOG

1 s

Stage 2

Stage 3

Stage 4

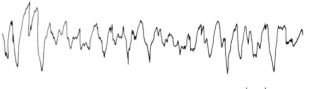

1 s

REM Sleep

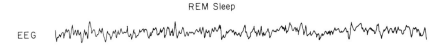

FIG. 1 Records of the electroencephalogram (EEG) and electro-oculogram (EOG) in the awake state and the various stages of sleep.

Sleep cycles

The various stages of sleep tend to recur cyclically during a night's sleep (Fig. 2). The mean interval between the beginning of successive REM sleep periods (a sleep cycle) is about 100 minutes and there are usually 3–6 sleep cycles per night. Approximately one or two hours of NREM sleep usually precede the appearance of the first REM sleep period. Sometimes, at the time when the first REM sleep period might be expected, there is instead a brief period of stage 2 sleep with K-complexes and spindles, but with REMs recorded in the EOG. The duration of REM sleep periods tends to increase during the night from a mean of about 13 minutes for the first period to between 20 and 45 minutes in subsequent periods. The amount of stages 3 and 4 sleep (delta-wave sleep) generally decreases in successive cycles. For an average night's sleep in young adults, approximately 24% of the time is spent in REM sleep, 4% in stage 1, 51% in stage 2, 10% in stage 3, and 11% in stage 4. The distribution of stages within each sleep cycle and across the whole night's sleep varies with age and with a wide range of environmental changes, psychological stress, as well as with medical and psychiatric illnesses, as described in detail by Williams et al. (1974).

Sleep and Body Systems

Cardiovascular system

The heart rate decreases by an average of about 4.5 beats per minute with sleep onset (Johns et al., 1976). There is usually a further small decrease

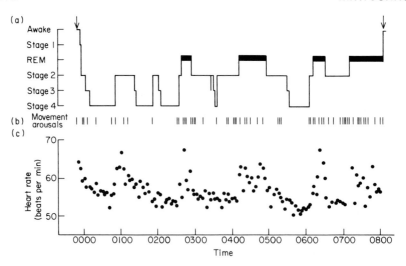

FIG. 2   (a) The distribution of sleep stages during an undisturbed night's sleep in a young man. The arrows indicate times of going to bed and getting up. (b) The intermittent occurrence of body movements during sleep. (c) The mean heart rate, measured over 20 beats at a time, varies according to the stage of sleep and the time of night. The rate is higher and more variable in REM sleep than in NREM sleep.

from stage 2 to stages 3 and 4. The heart rate usually increases during REM sleep and, in particular, becomes more variable (Fig. 2). Body movements during sleep and periods of wakefulness during the night are also associated with phasic increases in heart rate but these effects have been eliminated from the results shown in Fig. 2. However, brief episodes of bradycardia and tachycardia, lasting perhaps 10 seconds, occur spontaneously during sleep. These disappear after bilateral cervical vagotomy or methylatropine administration despite an intact sympathetic nerve supply to the heart. They are, therefore, due mainly to changes in vagal activity.

Much slower changes in heart rate take place over periods of hours during the night, regardless of the stage of sleep or wakefulness at the time. The lowest mean heart rate is likely to be found between 4 and 6 a.m. The origin of this circadian rhythm is uncertain although it is probably not simply a reflection of the diurnal variations in body temperature. The circadian rhythms of heart rate and of core temperature are often out of phase with one another by several hours on particular nights.

Arterial blood pressure decreases at the time of sleep onset and may decline to 55 mm Hg systolic and 30 mm diastolic pressure during the night in healthy adults, levels which in the waking state would be considered to be alarmingly low (Bevan *et al.*, 1969). This is because cardiac output falls without a concomitant increase in peripheral resistance. There is vasodila-tation in the skin at sleep onset and, in experimental animals, there is cutaneous vasoconstriction during REM sleep (Khatri *et al.*, 1967). This involves a redistribution of blood flow rather than a significant change in total peripheral resistance. In the cat, mesenteric and renal arterial blood vessels dilate during REM sleep when the cutaneous vessels are constricted (Mancia *et al.*, 1970). Similarly, in cats there is a marked increase in blood flow to the cerebral cortex and rhombencephalon but a decrease in the mesencephalon during REM sleep compared with the flow during NREM sleep (Reivich *et al.*, 1968). On arousal to wakefulness from slow-wave sleep there is a marked increase in blood flow to the mesencephalon and a decrease in the rhombencephalon (Baust, 1967). These variations seem to reflect regional differences in metabolism of the central nervous system in the various states of sleep and wakefulness.

Body movements, muscle tone and reflexes

Major body movements, such as flexing an arm or turning the trunk from one side to the other, are common during sleep even when, by other criteria, it is restful sleep. There are usually between 20 and 60 such movements per night. They occur mainly in stage 2 and REM sleep, very seldom during delta-wave sleep, although periods of the latter are usually terminated by a major body movement (Fig. 2).

Tonic activity in the skeletal muscles decreases during the period of relaxation in preparation for sleep and may decrease further on falling asleep. With the onset of REM sleep the remaining level of tonic activity recorded in the EMG is abolished. In addition, there is a marked reduction in the activity of spinal reflexes such as the H-reflex. During this state of hypotonia and hyporeflexia, phasic increases in skeletal muscle activity occur and these produce the rapid eye movements as well as twitching of the facial muscles and limbs (Baldridge *et al.*, 1965).

Respiratory system and oxygen consumption

The mean respiratory rate is commonly about two cycles per minute less in all stages of sleep than it is in resting wakefulness. Respiratory move-ments during sleep are more thoracic than abdominal, unlike those during wakefulness. They are usually regular in slow-wave sleep, but during stage 2 and REM sleep Cheyne–Stokes respiration is common, even in healthy subjects who may have bursts of tachypnoea and periods of apnoea which may last up to 1 minute (Webb, 1974). Oxygen saturation may decrease to about 92%, p_{CO_2} rises and blood pH falls by about 0.03 units. There is reduced sensitivity to hypercapnoea and acidaemia during normal sleep, and in patients with impaired ventilatory capacity temporary cerebral hypoxia is a frequent occurrence (Pierce et al., 1966). Oxygen consumption (metabolic rate) is about 10 per cent less during sleep than wakefulness. The lowest metabolic rates occur during delta-wave sleep and higher rates during REM sleep, although this is influenced by the frequency of body movements (Brebbia and Altschuler, 1965).

Genito-urinary system

Penile erection in males and clitoral erection in females occurs regularly during REM sleep and is unrelated to the sexual content of dreams at the time.

The rate of urine formation decreases and urine osmolarity increases during sleep. Failure of this mechanism may be at least partly responsible for nocturnal enuresis. With increasing age that proportion of the 24-hour urine output which is excreted at night increases and this commonly causes nocturia in otherwise healthy men and women.

Endocrine system

The secretion of ACTH from the anterior pituitary gland, and hence of cortisol from the adrenal cortex, is episodic rather than continuous. Bursts of secretion occur at intervals of one to two hours (Hellman et al., 1970). However, such episodes are usually inhibited during the early hours of sleep; the secretory periods then increase, reaching a maximum between about 5 a.m. and midday. During the night secretory episodes often coincide with either REM sleep or periods of arousal to wakefulness although this is not a one-to-one relationship (Alford et al., 1973a). After acute reversal of sleep and wakefulness the circadian rhythm of adreno-cortical activity takes between one and two weeks to renew its original phase relationship to sleep and wakefulness (Weitzman et al., 1968).

The diurnal rhythm of physical activity and social behaviour, especially that of sleep and wakefulness, is probably the most important factor in entraining the phase of the circadian rhythms in adrenocortical and adrenomedullary activity, even when time cues from clocks or the sun are absent (Aschoff et al., 1971).

Growth hormone secretion from the anterior pituitary gland is also episodic. For most people, a major part of the day's growth hormone secretion occurs during the first period of delta-wave sleep within an hour of sleep onset. After acute reversal of sleep and wakefulness, growth hormone secretion, unlike that of cortisol, maintains its relationship to the initial period of delta-wave sleep (Sassin *et al.*, 1969). The plasma concentrations of glucose and insulin do not change significantly during sleep. Nor have any other metabolic changes been recognized in the peripheral circulation which might explain the growth hormone secretion during sleep. There appears to be a fairly specific relationship between the neural mechanisms controlling delta-wave sleep on the one hand and those controlling growth hormone releasing factor in the hypothalamus on the other hand.

Prolactin secretion is also increased episodically soon after sleep onset. It usually attains its highest plasma concentrations later during the period of sleep and is not obviously related to particular stages of sleep (Sassin *et al.*, 1973). As with growth hormone, prolactin secretion retains its relationship to sleep onset immediately after reversal of sleep and wakefulness.

The plasma concentrations of luteinizing hormone (LH) and follicle stimulating hormone (FSH) do not show consistent differences between sleep and wakefulness, either in prepubertal children or in adults. However, at puberty the episodic secretion of LH is closely related to sleep onset, even after acute reversal of sleep and wakefulness, and in this respect it resembles GH and prolactin in its secretory pattern (Boyar *et al.*, 1972). Diurnal variations in the secretion of thyroid stimulating hormone, oestradiol and testosterone are not so closely related to sleep (Alford *et al.*, 1973b).

The overall pattern of plasma hormone concentrations, particularly those during the initial part of the sleep period, is such that anabolism should be promoted and catabolism reduced, as manifested, for example, by a relatively high rate of brain protein synthesis during sleep (Adam and Oswald, 1977).

Body temperature and sweating

With sleep onset, the skin temperature on the trunk and the limbs usually increases by between 1 and 3 °C; that is, to temperatures of 34 or 35 °C. This occurs over a few minutes as a result of cutaneous vasodilatation with a decrease in vasoconstrictor tone. As a consequence of the increased skin temperature, the rate of sweating increases in those areas of skin most concerned with thermoregulation (Satoh *et al.*, 1965). The core temperature measured in the rectum or at the tympanic membrane of the ear usually reaches its maximum for the day in the early evening. It falls during sleep, reaching a minimum of little more than 36 °C in healthy subjects, usually between about 4 and 6 a.m., after which it increases again. Most sleep, therefore, is associated with mechanisms which promote the loss of body heat at a time when the metabolic rate and heat production is already low. Nevertheless, sleep is interrupted if the degree or rate of heat loss is too high, as in very cold environments. After acute reversal of sleep and wakefulness, the normal circadian rhythm of body temperature may take several days to regain its usual relationship to behavioural state (Kleitman, 1963).

Unlike most sweat glands, those on the palms of the hands and soles of the feet are generally inhibited during sleep. This change has formed the basis of one simple method for monitoring sleep and wakefulness by continuous measurement of the electrical resistance of palmar skin (Johns, 1971).

Mental activity during sleep

The study of mental activity during sleep depends on the experimental subject's ability to recall and describe it when awake. This ability is most frequently demonstrable when the subject is woken during a REM sleep period or within a few minutes of its ending. Thereafter, recall is rapidly lost. Whether or not a subject often remembers dreams depends to some extent on his personality (Williamson et al., 1970).

When normal subjects are woken from NREM sleep they often report having had some form of mental activity which is best described as thought-like rather than dream-like, the former being more conceptual, less emotional and having lower perceptual intensity than the latter (Rechtschaffen et al., 1963). Throughout sleep, normal subjects can discriminate to some extent between "meaningful" auditory stimuli and other sounds of the same intensity (Oswald et al., 1960). A mother may wake at the first soft cries of her baby but remain asleep when a dog barks. Thus, sleep is a state of reduced responsiveness to the environment and of altered mental functioning but, nonetheless, a state in which some information processing of sensory input continues.

Functions of sleep

The ubiquity of sleep–wakefulness rhythms and the many differences between REM and NREM sleep have provided the basis for speculation about the overall needs for sleep and, lately, about the separate functions of the different kinds of sleep. Early suggestions that dreaming and REM sleep were essential for psychological wellbeing in man prompted many investigations into the nature of this requirement (Hartmann, 1973). However, as we shall see later in this chapter (p. 520), the absence of REM sleep for prolonged periods in man has not consistently led to any recognizable behavioural changes. The finding of an increased proportion of REM sleep in neonates led to speculation that REM sleep serves an ontogenetic function, organizing the developing central nervous system (Roffwarg *et al.*, 1966) and the oculomotor system in particular (Berger, 1969). So far, these are no more than working hypotheses.

The decrease in body temperature by an active process of heat loss during sleep, the inhibition of cortisol and of adrenaline secretion at a time when growth hormone, prolactin and, at puberty, also LH secretion are increased —these findings are all consistent with sleep, and particularly delta-wave sleep, being a period of restoration and anabolism after increased catabolism during wakefulness. Perhaps sleep is needed after many hours of wakefulness because of fatigue in the neural mechanisms subserving selective attention and the maintenance of psychological defences, that is, of the most highly integrated functions of the central nervous system. However, there is another important function of sleep which has often been neglected, and that is to do with the entrainment of intrinsic circadian rhythms in bodily function—for example, the rhythms of body temperature and of adrenocortical activity. After a change in these rhythms it takes several days at least to synchronize them and until this is achieved sleep may not be as refreshing as usual. To this extent, therefore, the usual amount of sleep may be a necessary but not a sufficient condition for maximal "restoration" after wakefulness. The mechanism by which sleep entrains other intrinsic rhythms, or under some circumstances fails to do so, remains to be elucidated.

Neurophysiological Basis for Sleep and Wakefulness

Reticular activating system

The classical animal experiments of Moruzzi and Magoun (1949) and of Lindsley *et al.* (1949) clearly established that the behavioural state of wakefulness and the associated activation of the cerebral cortex (manifested by low-amplitude, high-frequency EEG) were dependent upon structures in the midbrain reticular formation. However, it became important to distinguish between an ascending reticular activating system which was concerned primarily with cortical activation and another, descending reticular activating system involving the hypothalamus which was necessary but not sufficient for the maintenance of behavioural arousal (Jouvet, 1967). In the waking state the diversity and multitude of sensory inputs and humoral factors (particularly adrenaline) affect the reticular activating systems. Accordingly, one theory is that sleep is a passive phenomenon arising from a reduction in sensory input, particularly from the muscle spindles; sleep may be considered the absence of wakefulness.

Sleep an active state

It is now known that sleep, like wakefulness, is a state which is actively maintained by a complex interaction between many different parts of the central nervous system. An "active" theory of sleep was first strongly propounded by Hess (1929) who stimulated the thalamus and produced cortical synchronization (deactivation) and behavioural sleep. However, there are so many other regions, in the cerebral cortex, internal capsule, hypothalamus, cerebellum and medulla, all of which produce similar effects after stimulation, that this method is not very helpful in defining relationships between the structures directly concerned with producing sleep (Jouvet, 1969a).

The most important sleep-inducing structures are located in the lower brain-stem, particularly in the pons and medulla. These include mainly serotonergic neurons in the raphe system (Fig. 3). The extent of partial destruction of this raphe system determines the amount of NREM sleep obtained by cats (Jouvet, 1969b). By contrast, REM sleep is selectively

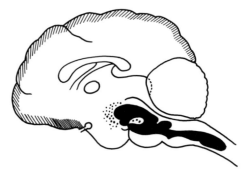

FIG. 3   A sagittal plan of a cat's brain showing the ascending reticular activating system (dots) and the sleep-inducing system of raphe nuclei (black). Decortication (removal of cross-hatched areas of cerebral cortex) suppresses slow waves but does not eliminate the oscillation between REM and NREM sleep (after Jouvet, 1969a).

suppressed by lesions in the dorso–lateral pontine tegmentum which do not impair NREM sleep. Thus, REM sleep and NREM sleep involve different structures in the central nervous system.

The cerebral cortex is necessary for the appearance of slow waves in the EEG but not for the oscillation between REM and NREM sleep. Within REM sleep, the tonic inhibition and the phasic activity also involve different neuronal systems. Neurons in the nucleus locus coeruleus, the dorsal part of the medio–lateral pontine tegmentum, trigger the descending inhibitory part of the reticular formation which blocks, both at pre- and post-synaptic levels, the discharges of spinal motor neurones and hence causes the generalized atony of REM sleep. This atony is dependent at least partly on a cholinergic mechanism because it is suppressed by atropine. By contrast, the phasic events and the cortical activation during REM sleep originate in a restricted area of the pons, the nucleus reticularis pontis caudalis (Jouvet, 1967).

Just before the disappearance of muscle tone at the start of REM sleep, spike discharges appear in the pons, soon after in the lateral geniculate nucleus, and then in the occiptal cortex and the oculomotor nucleus (Jouvet, 1967). These pontogeniculo-occipital (PGO) spikes are often related to the appearance of individual rapid eye movements and other phasic events although this is not a one-to-one relationship.

Jouvet (1969b) has suggested that a priming, serotonergic mechanism in the caudal part of the raphe system may act on cholinergic neurones which, in turn, trigger the final noradrenergic mechanisms of REM sleep located in the nucleus locus coeruleus. Recent investigations involving extracellular recordings from individual neurones in the pons have identified two groups of cells in the nucleus locus coeruleus (NLC) on the one hand, and in the gigantocellular tegmental field (FTG) on the other hand, which show reciprocal relationships in their discharge rates during both REM and NREM sleep (Hobson et al., 1975). There is a tonic increase in the discharge rate of TFG neurones during REM sleep which appears to be due to decreased inhibition from noradrenergic neurones in the N. locus coeruleus. However, serotonergic neurones in the dorsal raphe nucleus also show a decrease in firing rate during REM sleep and these cells have synaptic input to giant cells in the FTG too. The details of several possible relationships between groups of cells of this type remain to be elucidated. Nevertheless, a computer model of discharge rates in two groups of neurones having reciprocal interactions similar to those between the NLC and FTG neurones gives a clear indication that such interactions could provide the basis for the oscillation between REM and NREM sleep (McCarley and Hobson, 1975).

Population Studies

The pattern of sleep and wakefulness in individual people is subject to their voluntary control, to some extent at least. Nevertheless, most people adopt sleep habits which are fairly regular and which are arrived at by an interaction between many biological, psychosocial, occupational and environmental influences. For example, most people sleep at night rather than during the day, except when their occupation demands otherwise. A midday siesta is obtained in some communities by the appropriate organization of their daytime activities. Nevertheless, there is little to suggest that the amount of sleep usually obtained per 24 hours differs markedly between the most primitive and the most economically advanced communities. Of greater importance are the changes in sleep habits with age—changes which are presumably present in all communities.

Changes with age

There are changes in both the subjective and the objective characteristics of sleep with increasing age. Infants have more than one period of sleep each day. As the central nervous system matures their ability to remain awake throughout the day increases and by the age of 4 or 5 years most of their sleep is restricted to the night. The mean duration of sleep per 24 hours decreases from about 15 hours in the neonate to about 7.7 hours (standard deviation ± 1.5 hours) by the age of 20 years (Kleitman, 1963; Tune, 1969). There is a further small decrease until the sixth decade, after which the duration of sleep often increases again (Tune, 1969; Johns et al. 1970; Johns, 1975b). The mean delay before falling asleep, the number and duration of awakenings during the night and of naps during the day all increase progressively with age in adulthood. The amount of stage 4 sleep decreases markedly from childhood and is virtually absent in old age, perhaps reflecting a progressive degeneration of neurones in the cerebral cortex (Feinberg and Carlson, 1968). By contrast, REM sleep decreases from about 50% of sleep in babies to between 20 and 25% in adults of all ages (Williams et al., 1974).

In general, the times of going to bed and of falling asleep at night and of waking up in the morning become progressively earlier with increasing age past young adulthood (Johns et al., 1970) (Fig. 4). There may be national differences in these times. For example, the sleep period for each age group in Britain occurs about 30 minutes later than it does in Australia, although it is of the same duration (Tune, 1969). This may reflect cultural differences as well as differences in the hours of sunlight.

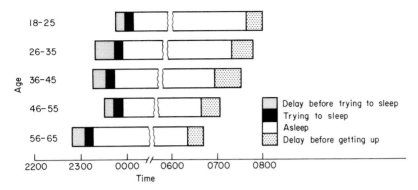

FIG. 4   Chart to illustrate sleep patterns at different ages.

Differences between sexes

The sleep habits of healthy men and women of the same age are very similar in terms of the duration and timing of sleep and the distribution of sleep stages. However, complaints of insomnia and the taking of hypnotic drugs are considerably more common among women than men of all ages (Johns *et al.*, 1971b).

Breakdown

The usual pattern of sleep and wakefulness in a particular subject may become disrupted in several ways: by the total deprivation or prevention of sleep; by selective deprivation, when particular stages of sleep are prevented and others permitted; by partial deprivation, whereby the total amount of sleep per 24 hours is curtailed; by displacement of the usual amount of sleep to earlier or later times of the day than is customary for that subject; by insomnia, or the relative inability to initiate or maintain sleep when given the opportunity to do so; by hypersomnia, in which there is a relative inability to maintain wakefulness for the normal periods.

Total sleep deprivation

When wakefulness is maintained "continuously" for several days there is an increase in drowsiness which is generally progressive from day to day but which becomes worse at the times when sleep would normally occur. Brief lapses become increasingly frequent during which there is deterioration in psychomotor performance, misperception, lability of emotional affect, regressive behaviour and perhaps hallucinations, as well as neurological signs such as bilateral ptosis and hyperactive reflexes. If the subject be sufficiently motivated his performance can be maintained at high levels for short periods at a time. Boredom and lack of incentive cause the overall level of performance to deteriorate more rapidly by increasing the frequency of lapses (Wilkinson, 1965). During these lapses, which may last only a few seconds at a time, the subjects actually have brief periods of stage 1 sleep as indicated by temporary slowing of the EEG, heart rate and respiratory movements (Naitoh *et al.*, 1971). Thus, many psychomotor, cognitive, perceptual and affective changes which result from prolonged sleep loss are actually the concomitants of the drowsy state. This may be unusually prolonged and recurrent under such circumstances, but otherwise is probably similar to the period of stage 1 which normal people experience briefly before falling asleep at night (Foulkes and Vogel, 1965). The changes are all readily reversible with recovery sleep.

Other physiological changes which are sometimes observed during total sleep deprivation, such as an increase in muscle tone and in plasma concentrations and urinary excretion of 17-hydroxycorticosteroids, are probably the results of the subjective strain involved in trying to stay awake rather than the direct effects of sleep deprivation (Kollar *et al.*, 1966).

After sleep deprivation, especially if it is prolonged, subjects fall asleep quickly and sleep for longer than usual, sometimes for 15 hours or more. They have increased amounts of stage 4 sleep and their arousal thresholds are higher than normal in all stages.

Selective deprivation of REM sleep or stage 4

After the distinction was made between REM and NREM sleep it was logical that the method of selective deprivation should have been used in an attempt to explain the separate function of these states. By arousing subjects as soon as they entered each REM sleep period it was possible to limit drastically the time spent in that state. Although there were early suggestions that this procedure might produce a reversible psychosis (Dement, 1960), none of the subsequent investigations has shown any significant changes following REM sleep deprivation in man (Dement, 1969). Nonetheless, there is usually a marked excess or rebound of REM sleep during the recovery sleep on the next night or two, and this has been taken to indicate a need for REM sleep. In more prolonged REM-deprivation experiments in animals, lasting up to 69 days, behavioural changes do occur, particularly in relation to abnormal sexual drive or hypersexuality, which disappears after recovery sleep and does not occur after selective deprivation of NREM sleep (Dement, 1969).

As with REM sleep, the selective deprivation of stage 4 in healthy adults also results in a rebound of that stage during the first nights of recovery sleep. If delta-wave sleep is prevented the amount of growth hormone secretion at night is reduced. It is possible that the restorative function of sleep is dependent upon some delta-wave sleep occurring. However, there are wide variations in the amount of delta-wave sleep usually obtained at night by healthy subjects, variations which do not relate to the subjective awareness of having had a good or a poor night's sleep (Johns, 1975a).

Partial sleep deprivation and sleep displacement

Based on an electrical analogy that the restorative function of sleep might be likened to the recharging of a battery, it would seem reasonable that the shorter the period of sleep the less the restoration. In fact, reducing the amount of sleep at night by two hours (from eight to six hours) does not much affect the ability to perform tasks next day. It is not until the amount of sleep is suddenly curtailed to less than three hours per night that there is a significant performance decrement next day (Wilkinson, 1968).

In such experiments, however, it is not possible to change the duration of sleep over a few nights without also changing the times at which sleep begins and ends. The timing of sleep has a profound effect on the phase of circadian rhythms in bodily function, some of which directly affect psycho-motor performance (Hauty, 1963; Klein *et al.*, 1968). Sudden displacement of the usual amount of sleep, so that it is obtained several hours earlier or later in the day has an effect on subsequent mood and performance which is similar to that following partial sleep deprivation (Taub and Berger, 1973).

Some apparently healthy adults regularly obtain only about three hours of sleep each night without obvious ill-effects (Jones and Oswald, 1968). They obtain about as much delta-wave sleep as normal subjects do in seven or eight hours of sleep.

Insomnia

Insomnia is a disorder which involves taking a long time to fall asleep initially, waking up frequently during the night or waking up too early in the morning. Thus, it is a highly subjective phenomenon, dependent in part on the subject's expectation of sleep which may be quite unreasonable. Each or all of these aspects of the disorder may be present in a particular subject. Insomnia does not necessarily involve much sleep deprivation, either in terms of the total amount of sleep or of the amounts of REM sleep and delta-wave sleep obtained. In some cases the complaint of insomnia may be based on psychological tension and distress which the subject would rather not experience by being awake; the complaint is really one of having "too much" wakefulness rather than "not enough" sleep (Brezniova et al., 1975).

A "poor" night's sleep is commonly one on which it takes at least 30 minutes to fall asleep, or on which sleep is interrupted frequently by spontaneous awakenings as shown in Fig. 5. Here the amount of delta-wave sleep obtained is not abnormal. Such a "poor" night's sleep is usually associated with psychological tension and often with neurotic illness. Even the minor psychological stress of sleeping in a sleep laboratory often produces a "first night effect" in healthy subjects in which they take longer to fall asleep, they wake up more frequently and have less REM sleep than on subsequent nights (Agnew et al., 1966). As might be expected, insomnia is very common among psychiatric patients in most diagnostic categories (Ward, 1968).

There is a marked increase in the prevalence of insomnia in old age, due at least in part to progressive degenerative changes within the central nervous system. This is accelerated by cerebral arteriosclerosis (Feinberg and Carlson, 1968). Other disorders of cerebral metabolism which may be associated, for example, with fever, thyrotoxicosis, or perhaps hypobaric hypoxia suffered at high altitudes (above 3000 metres) also tend to cause insomnia.

Some drugs such as caffeine and amphetamine delay the onset of tiredness and sleepiness by stimulating the reticular activating system. Other drugs such as the barbiturates cause a general, although not a uniform, depression of activity in the central nervous system and promote sleep. The latter drugs commonly interfere with the normal pattern of sleep stages and after prolonged use can be a cause rather than a cure of insomnia (Johns, 1975b).

Hypersomnia

Hypersomnia is a condition in which relatively normal sleep continues for much longer than usual. It is associated sometimes with obesity and respiratory insufficiency in the Pickwickian syndrome but can also occur in the absence of demonstrable organic disease (Kales and Kales, 1974).

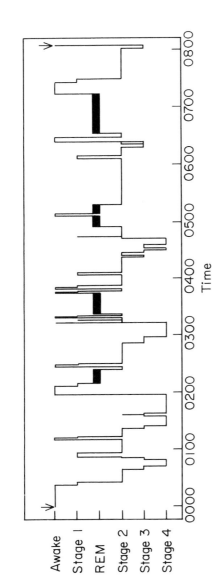

FIG. 5 The distribution of sleep stages during a "poor" night's sleep in a young man. There is a delay before sleep onset and frequent arousals to wakefulness occur during the night.

Applications

Need to understand sleep and wakefulness

An understanding of sleep and wakefulness can be of practical value in several situations which may be considered in two main categories: the impact of environmental conditions and of work schedules on the usual pattern of sleep and wakefulness; and the investigation and treatment of sleep disorders.

Work schedules and sleep

Modern man often seems to organize his activities over the 24-hour day on the basis of economic restraints and the requirements of machines rather than man. This is a special problem with shift work. It is possible to work all day and night, say, once a week, or to change from night shift to day shift every other week if motivated to do so. Many people, particularly young adults, are highly adaptable in this regard. However, some people find these changes distressing. We have seen earlier (p. 512) that circadian rhythms of body temperature and adrenocortical activity may take more than a week to adjust to acute reversal of sleep and wakefulness. Some shift workers are unable to adapt these circadian rhythms effectively after many weeks. There may be several reasons for this. Perhaps the most important reason is that they tend to revert to their normal daytime activities on their days off work each week so that they never have long enough on one routine to adapt. They may also have difficulty sleeping during the day because of noise or heat. Their poor sleep may not enable their circadian rhythms to be entrained. As a result their performance at boring jobs may be impaired in the early hours of the morning. In dealing with this problem the first requirement is to provide for quiet situations where night-shift workers can sleep during the day. This is especially true when the work demands a high level of vigilance for prolonged periods; for example, in naval personnel who scan a radar screen waiting to detect some relatively rare event as soon as it occurs. In another situation, astronauts must have a strict time schedule in the absence of appropriate periods of light and dark during space flight so that they can sleep at regular periods every 24 hours.

Treatment of sleep disorders

Sleep disorders are very common in the community. Hypno–sedative drugs which are often prescribed in the treatment of insomnia are among the most commonly prescribed of all drugs. While most of these drugs are of therapeutic value when used occasionally or for a few consecutive nights, they often present problems in long-term use (Johns, 1975b). Not the least of these problems is the tendency for many people each year to take an overdose of such drugs to attempt or commit suicide. The barbiturates have caused most trouble in this regard (Johns, 1977a). They should no longer be prescribed for insomnia now that much less toxic and more effective drugs of the benzodiazepine type are readily available. (Johns, 1975b).

Methods for assessing sleep

Each of the methods available for assessing human sleep had advantages and disadvantages which must be weighed up when choosing the most appropriate method for a particular purpose (Johns, 1971). The so-called objective methods involving recordings of the EEG, EOG and perhaps the EMG throughout sleep are very tedious and expensive if many subjects or nights are involved. However, these methods are the most accurate available and are essential if information is required about particular sleep stages. There is some evidence that the attachment of scalp and other electrodes is a continuing source of disturbance to sleep, causing more awakenings than would occur spontaneously (Johns and Doré, 1978).

By contrast, subjective reports about sleep habits in general or about sleep and wakefulness on particular days are easily and cheaply obtained from relatively large numbers of subjects. Such reports may be made as daily records in a sleep diary (Tune, 1969) or in a general sleep question-naire (Johns *et al.*, 1971a). The validity and accuracy of both types of report have been established in normal subjects although insomniacs tend to exaggerate their degree of sleep disturbance (Johns, 1977b; Johns and Doré, 1978).

References

Adam, K. and Oswald, I. (1977). *J.R. Coll. Phycns*, **11**, 376–388.
Agnew, H.W. Jr., Webb, W.B. and Williams, R.L. (1966). *Psychophysiology* **2**, 263–266.
Alford, F.P., Baker, H.W.G., Burger H.G., de Kretser, D.M., Hudson, B., Johns, M.W., Masterton, J.P., Patel, Y.C. and Rennie, G.C. (1973a). *J. Clin. Endo-crinol. Metab.* **37**, 841–847.
———. (1973b). J. Clin. Endocr. Metab., 37, 848–853.
Aserinsky, E. and Kleitman, N. (1953). *Science (N.Y.)* **118**, 273–274.

Aschoff, J., Fatranska, M., Giedke, H., Doerr, P., Stamm, D. and Wisser, H. (1971). *Science (N.Y.)* **171**, 213–215.

Baldridge, B.J., Whitman, R.M. and Kramer, M. (1965). *Psychosom. Med.* **27**, 19–26.

Baust, W. (1967). *Electroenceph. Clin. Neurophysiol.* **22**, 365–372.

Berger, H. (1929). *Arch. Psychiat. Nervenkr.* **87**, 527–570.

Berger, R. (1969). *Psychol. Rev.* **76**, 144–164.

Bevan, A.T., Honour, A.J. and Stott, F.H. (1969). *Clin Sci.* **36**, 329–344.

Blake, H. and Gerard, R.W. (1937). *Am. J. Physiol.* **119**, 692–703.

Boyar, R.M., Finkelstein, J., Roffwarg, H., Kapan, S., Weitzman, E. and Hellman, L. (1972). *New Eng. J. Med.* **287**, 582–586.

Brebbia, D.R. and Altschuler, K.Z. (1965). *Science (N.Y.)* **150**, 1621–1623.

Brezinova, V., Oswald, I. and Loudon, J. (1975). *Br. J. Psychiatr.* **126**, 439–445.

Dement, W.C. (1960). *Science (N.Y.)* **131**, 1705–1707.

Dement, W.C. (1969). *In* "Sleep. Physiology and Pathology", (Ed. A. Kales) 245–264. Lippincott, Philadelphia.

Feinberg, I. and Carlson, V.R. (1968). *Arch. Gen. Psychiat.* **18**, 239–250.

Foulkes, D. and Vogel, G. (1965). *J. Abnorm. Psychol.* **70**, 231–243.

Hartmann, E.L. (1973). "The Functions of Sleep", Yale University Press, New Haven and London.

Hauty, G.T. (1963). *Aerospace Med.* **34**, 100–105.

Hellman, L., Nakada, F., Curti, J., Weitzman, E.D., Kream. J., Roffwarg, H., Ellman, S., Fukushima, F. and Gallagher, T.F. (1970). *J. Clin. Endocrinol. Metab.* **30**, 411–422.

Hess, W.R. (1929). *Am. J. Physiol.* **90**, 386–387.

Hobson, J.A., McCarley, R.W. and Wyzinski, P.W. (1975). *Science (N.Y.)* 55–58.

Johns, M.W. (1971). *Arch. Intern. Med.* **127**, 484–492.

Johns, M.W. (1975a). *Psychol. Med.* **5**, 413–418.

Johns, M.W. (1975b). *Drugs*, **9**, 448–478.

Johns, M.W. (1977a). *Br. Med. J.* **1**, 1128–1130.

Johns, M.W. (1977b). *Ergonomics* **20**, 683–690.

Johns, M.W. and Doré, C. (1978). *Ergonomics,* **21**, 325–330.

Johns, M.W., Eagan, P., Gay, T.J.A. and Masterton, J.P. (1970). *Br. Med. J.* **2**, 504–512.

Johns, M.W., Gay, T.J.A., Goodyear, M.D.E. and Masterton, J.P. (1971a). *Br. J. Prev. Soc. Med.* **25**, 236–241.

Johns, M.W., Hepburn, M. and Goodyear, M.D.E. (1971b). *Med. J. Austral.* **2**, 1323–1327.

Johns, M.W., Thornton, C. and Doré, C. (1976). *J. Psychosom. Res.* **20**, 549–553.

Jones, H.S. and Oswald, I. (1968). *Electroenceph. Clin. Neurophysiol.* **24**, 378.

Jouvet, M. (1967). *Physiol. Rev.* **47**, 117–177.

Jouvet, M. (1969a). *In* "Sleep. Physiology and Pathology", (Ed. A.Kales), 89–100. Lippincott, Philadelphia.

Jouvet, M. (1969b). *Science (N.Y.)* **163**, 32–41.

Kales, A. and Kales, J.D. (1974). *New Eng. J. Med.* **290**, 487–499.

Khatri, I.M. and Freis, E.D. (1967). *J. Appl. Physiol.* **22**, 867–873.

Kollar, E.J., Slater, G.R., Palmer, J.O., Docter, R.F. and Mandell, A.L. (1966). *Psychosom. Med.* **28**, 101–113.

Klein, K.E., Wegmann, H.M. and Bruner, H. (1968). *Aerospace Med.* **39**, 512–518.

Kleitman, N. (1963). "Sleep and Wakefulness". 2nd edn. University of Chicago Press, Chicago and London.

Lindsley, D.B., Bowden, J. and Magoun, H.W. (1949). *Electroenceph. Clin. Neurophysiol.* **1**, 475–486.

McCarley, R.W. and Hobson, J.A. (1975). *Science (N.Y.)* **189**, 58–60.

Mancia, G., Baccelli, G., Adams, D.B. and Zanchetti, A. (1970). *Am. J. Physiol.* **220**, 1086–1093.

Moruzzi, G. and Magoun, H.W. (1949). *Electroenceph. Clin. Neurophysiol.* **1**, 455–473.

Naitoh, P., Pasnau, R.O. and Koller, E.J. (1971). *Biol. Psychiatr.* **3**, 309–320.

Oswald, I., Taylor, A.M. and Treisman, M. (1960). *Brain* **83**, 440–453.

Pierce, A.K., Jarrett, C.E., Werkele, G. and Miller, W.F. (1966). *J. Clin. Invest.* **45**, 631–636.

Rechtschaffen, A., Verdone, P. and Wheaton, J. (1963). *Arch. Gen. Psychiatr.* **9**, 536–547.

Rechtschaffen, A. and Kales, A., eds (1968). "A Manual of Standardized Terminology, Techniques and Scoring System for Sleep Stages of Human Subjects", Bulletin 204, U.S. Public Health Service, Bethesda.

Reivich, M., Isaacs, G., Evarts, E. et al (1968). *J. Neurochem.* **15**, 301-306.

Roffwarg, H.P., Muzio, J.N. and Dement, W.C. (1966). *Science (N.Y.)* **152**, 604–619.

Sassin, J.F., Parker, D.C., Mace, J.W., Gotlin, R.W., Johnson, I.C. and Rossman, L.G. (1969). *Science (N.Y.)* **165**, 513–515.

Sassin, J.F., Frantz, A.G., Kapen, S. and Weitzman, E.D. (1973). *J. Clin. Endocrinol Metab.* **37**, 436–440

Satoh, T., Ogawa, T. and Takagi, K. (1965). *Japan J. Physiol.* **15**, 523–533.

Taub, J.M. and Berger, R.J. (1973). *Psychosom. Med.* **36**, 164–173.

Tune, G.S. (1969). *Br. J. Psychol.* **60**, 431–441.

Ward, J.A. (1968). *Canad. Psychiat. Ass. J.* **13**, 249–257.

Webb, P (1974). *J. Appl. Physiol.* **37**, 899–903.

Webb, W.B. (1974). *Percept, Mot. Skills* **38**, 1023–1027.

Wilkinson, R.T. (1965). In "The Physiology of Human Survival", (Eds O.G. Edholm and A.L. Bacharach), Academic Press, London and New York.

Wilkinson, R.T. (1968). *Progr. Clin. Psychol.* **8**, 28–43.

Williams, R.L., Karacan, I. and Hursch, C.J. (1974). "Electroencephalography (EEG) of Human Sleep: Clinical Applications", Wiley, New York.

Williamson, R.W., Heckel, R.V. and Boblitt, W.E. (1970). *J. Clin. Psychol.* **26**, 300–301.

Weitzman, E.D., Goldmacher, D., Kripke, D., MacGregor, P., Kream, J. and Hellman, L. (1968). *Trans. Amer. Neurol. Assoc.* **93**, 153–157.

CHAPTER 11

Instrumentation for Physiological Measurements

D. Hill

Quantitative measurements of human physiology must rely on the availability of suitable instrumentation. Apart from those taken in hospital intensive care units and on special volunteers in research establishments, many of the measurements will be made on ordinary volunteers undertaking specific tasks. This fact inevitably focuses attention on the need for non-invasive methods. Ethical committees now take a far more rigorous look at the conditions for experiments with volunteers, and the invasive measurement of arterial blood pressure, for example, would require exceptional permission. However, the convenience of non-invasive methods must be offset against the fact that they are indirect and results thus obtained must always be subjected to a careful scrutiny. There may well be doubtful "grey areas", e.g. the indirect measurement of blood pressure when it is below 70 mm Hg. The correction factor used to align the corresponding direct and indirect methods may be dependent on physiological variables, e.g. the haematocrit. Much can be inferred by a skilled observer, regarding the pathology of the heart, from a study of electrocardiogram (ECG) wave forms, but such inferences can be wrong. The ECG does not, by itself, express the haemodynamic action of the heart in terms of pressures and volumes of blood pumped. Because it is simple to record, it is widely used and its limitations are widely understood. It is in this spirit that other non-invasive physiological measurements should be approached.

The main measurements in general use are: blood pressure and flow; body temperature, respiratory rate and volume and the composition of respired gases; blood pH and blood oxygen and carbon dioxide tensions; and heart rate together with the observation of the ECG. In special circumstances, such as in hypobaric chambers, it may be necessary to monitor the electroencephalogram (EEG) and in movement studies the activity of specific groups of muscles by means of the electromyogram (EMG). The list can be extended to include techniques such as the recording of eye movements, but this chapter will concentrate on the more common measurements.

The Measurement of Arterial Blood Pressure

Indirect measurement

Crul and Payne (1970) have described four categories of indirect blood pressure measurement involving the use of a single or double cuff placed around a limb (usually an upper arm). In relationship to the cuffs, the techniques are: (1) oscillotonometry within the lumen of the cuff; (2) auscultation of sounds under it; (3) detection of arterial wall movements under it; and (4) detection of the onset of flow below it. Pressure in the cuff is, in all cases, raised to a level above the subject's systolic pressure and the pressure is gradually deflated, preferably falling linearly with time. In the Riva–Rocci–Korotkoff technique pulsations can be detected using a stethoscope over an artery distal to the cuff. In this way, pulsations can be detected when the intra-arterial pressure just exceeds the cuff pressure.

These indirect methods cannot be used to follow rapid changes in blood pressure since the cuff has to be inflated and slowly deflated. The detection of systolic pressure is relatively simple, but the estimate of diastolic pressure depends on the disappearance or the muffling of the sounds, and this is a less certain measurement (Bordley et al., 1951; Kenner and Ganer, 1962).

A number of alternatives to the use of the stethoscope have been adopted. Geddes and Moore (1968) have demonstrated that by mounting a crystal microphone wholly within the conventional arm-encircling cuff placed anywhere on the upper arm, it was possible to detect high-intensity sounds. Pulsations can also be detected with the cuff connected to an arterial pressure transducer, and by this means the systolic and diastolic pressures can be measured with the cuff placed on the outside of the subject's jacket sleeve. This is a useful facility in the case of subjects wearing clothing during exercise studies or climatic tests. Pulsations can also be detected by observing the electrical impedance variations for a limb segment distal to the cuff. Four band electrodes are placed around the segment, the outer pair being fed with a constant current of a few milliamperes at a frequency of the order of 100 kHz, and the changes occurring in the impedance of the segment as the cuff is deflated are observed as voltage changes across the inner pair of electrodes.

An obvious next step is to automate the inflation and deflation of the cuff and the detection of the onset and subsequent cessation of the Korotkoff sounds or impedance pulsations. In a typical arrangement, a compressor is activated to inflate the cuff, often at multiples of five-minute intervals. When a pre-set pressure has been attained which is judged to exceed the subject's systolic pressure, the cuff is deflated via a linear leak. On detection of Korotkoff sounds, a moving pointer which has been moved downwards by the falling pointer of an aneroid pressure gauge connected to the cuff, is clamped to indicate the systolic pressure. The cessation of the sounds subsequently causes the locking of a second pointer at the diastolic pressure. The microphone amplifier is usually gated to be operated for only a limited period following each R-wave of the ECG when the Korotkoff sounds are expected (Lagerwerff and Luce, 1970). This reduces false triggering from extraneous noise or objects brushing against the cuff. Lywood (1967) has given a detailed description of an automated Korotkoff sound system, with pressure falling at a rate of about 2 mm Hg per second. Schneider et al. (1974) describe an automatic, portable, Korotkoff sound indirect system for recording systolic and diastolic pressures at work, whilst driving, and at home—including during sleep. In another system, based on three impedance electrodes, described by Janssen (1967), a pair of amplifiers detects the pulsating signals found between the common and proximal electrodes and the proximal and distal electrodes as the cuff deflates. Both signals are present only over the range systolic to diastolic pressure.

Fernandez et al. (1974) have reported on the use of an automatic indirect system for measuring and recording blood pressure in two limbs simultaneously in terms of systolic and diastolic pressure.

The oscillometer technique of Von Recklinghausen (1906) is well known in clinical practice. It possess a greater sensitivity and thus greater accuracy than the Korotkoff sound method. By means of a double cuff principle, it detects the sudden increase in the volume oscillations of the cuff which occur at the systolic pressure and their sudden decrease in amplitude at the diastolic pressure.

De Dobbeleer (1965) adapted Von Recklinghausen's approach to produce an automatic, indirect blood pressure recording technique for use with Parkinsonian patients having a marked degree of limb tremor. Basically, the inflation pressure is applied to the upper and the lower cuffs simultaneously, and the variation in phase occurring between the pressure pulsations in each is detected as the pressure is reduced from systolic to diastolic. The time delay arises from the inertia of the blood and the artery wall lying between the cuffs. At the diastolic pressure, the artery remains fully open throughout each cardiac cycle and there is no phase difference between the pulsations in each cuff. The lower is connected to a "systolic chamber" containing a thermistor bead, and both cuffs are connected to a similar "diastolic chamber" (Fig. 1). In each case, the jet of air coming from the cuff is directed at the thermistor. The systolic thermistor is only influenced by the fluctuations commencing in the lower one at the systolic pressure, the pulsating air stream cooling the heated thermistor and giving rise to corresponding changes in its electrical resistance. When these are detected, the systolic pointer of a panel pressure gauge is locked to indicate the systolic pressure. As the cuff pressure falls below systolic, pressure

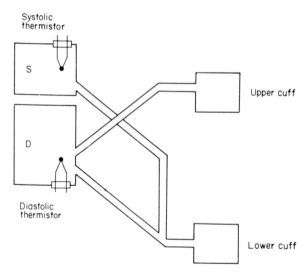

FIG. 1 Arrangement of the thermistors in De Dobbeleer's double cuff system. (Hill and Dolan, 1976).

fluctuations from the upper cuff impinge on the diastolic thermistor before those from the lower one, and the thermistor provides a fluctuating output signal. At the diastolic pressure, the fluctuations from both cuffs arrive at the diastolic chamber simultaneously, and it is arranged that the jet from the upper one is now deflected away from the thermistor by the jet from the lower, so the signal from the diastolic thermistor ceases and this is arranged to lock the diastolic pointer of the pressure gauge at the diastolic pressure. The cuff pressure is reduced at a rate of 3 mm Hg per second and the time lag between the pressure pulsations from the upper and lower cuffs is about 30 milliseconds. Crul et al. (1967) found excellent correlations of 0.989 and 0.937 respectively for systolic and diastolic pressures in 45 supine patients. These authors make the point, however, that both the double cuff (Haemotonometer) and the auscultation methods are unsatisfactory in patients with shock, hypotension, hypothermia and in young children.

These problems have led to the development of an ultrasonic technique for the indirect measurements of arterial blood pressure. Janssen's impedance rheography approach to blood pressure measurement detects the arterial wall movements, and this can also be achieved by means of a beam of ultrasound (Poppers, 1970). When a cuff is placed around the upper arm with the cuff pressure lying between systolic and diastolic, the brachial artery can open rapidly during systole, and at the diastolic pressure this rapid motion has ceased. A beam of ultrasound is aimed at the artery from a small transducer located under the cuff. The artery acts as a moving reflector and modulates the frequency of the beam reflected from it by means of the Doppler effect. When the rapid arterial wall motion is first detected, a pressure meter is locked on the systolic value and when it ceases a second meter is locked on to the diastolic value. When the cuff pressure lies below the diastolic value, only the normal arterial wall pressure pulsation is present and this slower modulation of the ultrasonic beam frequency is ignored by the detection circuitry. Sheppard et al. (1971) reported that the method tended in patients to exceed the direct intra-aortic systolic and diastolic pressures by about 0–10 mm Hg. Massie et al. (1973) and McLaughlin et al. (1971) have used it successfully in infants, and Hochberg and Saltzman (1971) and Zahed et al. (1971) with children. Hochberg and Salomon (1971), Kazamias et al. (1971), and Arno and Slate (1971) have verified the technique in adults. Gundersen and Ahlgren (1973) used several ultrasonic emitters and receivers, rather than just one pair, to obtain a wider beam and better contact with the artery.

Compact Doppler shift ultrasonic blood flow transducers are now widely used to detect blood flow in limb arteries which can be illuminated by the beam of ultrasound. By using such a device on an artery distal to the occluding cuff, the onset of blood flow at the systolic pressure can readily be detected. This technique has also been used successfully to measure the systolic pressure of neonates and Forlini (1974) used it in subjects subjected to accelerations of up to five times that of gravity. They state that conventional non-invasive methods for blood pressure measurement fail under these conditions, due to biological vibrations secondary to skeletal muscle straining manoeuvres. A somewhat similar application is that of Kopczynski (1974) who used this method to monitor the arterial blood pressure of acutely ill patients while in flight. The high noise and vibration levels of most aircraft upset other methods.

The direct measurement

As the term "direct" implies, in these techniques the pressure measurements system must be placed in contact with the blood contained in a suitable artery for measurement of the pressure in that artery. If the pressure is to be measured at the periphery, perhaps in the radial or brachial arteries, then the artery is first punctured and a short flexible plastic cannula inserted into it and connected to the measurement system. When it is necessary to measure pressures at selected sites within the cardiovascular system, as in cardiac catheterization, a flexible catheter, typically 1 m long, is employed in place of the cannula. Because an arterial puncture is associated with a small element of risk and of introducing infection to the artery, direct measurement is reserved for seriously ill patients where an indirect method of pressure measurement may be inaccurate, or where its use is called for in a diagnostic procedure such as cardiac catheterization.

Another advantage of direct pressure recording is that it is inherently continuous, so that beat-by-beat pressure variations, perhaps due to the presence of ectopic beats, can be detected. The discontinuous nature of the indirect cuff methods, plus difficulties often encountered in getting these systems to indicate consistently the systolic and diastolic pressures, means that they may be inadequate for the measurement of pressures which are changing fairly rapidly.

Pressure transducers

Electrical pressure transducers are now universally employed for the faithful reproduction of the arterial pressure waveform. Their speed of response is more than adequate to depict the dichrotic notch, the overall frequency response of the system being limited by the inertia and friction of the fluid-filled tubing linking the transducer to the patient. The electrical output signal from the transducer is fed into a suitable chart recorder and, for an approximate indication, into a panel meter. Additional circuitry can be provided to pick-off the systolic and diastolic pressures, the pulse pressure (systolic–diastolic) and the mean pressure (Sandman and Hill, 1974).

The most commonly encountered blood pressure transducers are of the strain gauge variety, either bonded or unbonded. However, some commercial physiological measurement systems still employ linear differential variable transformer type transducers or variable capacitance type transducers.

Variable capacitance transducers

At the end of World War II, capacitance transducers achieved prominence for their ability to provide a high-fidelity signal of the arterial blood pressure and good examples of this occur in Hansen's (1949) classical monograph. Basically, the transducer consists of a parallel plate capacitor with one plate forming the moveable pressure sensing diaphragm. The other plate is fixed. For small spacings of the plates a considerable change in capacity arises from small movements of the diaphragm. It can be seen that great care is needed in the choice of the materials used in the construction of capacitance manometers in order to minimize baseline drifts arising from changes in the ambient temperature. These could produce differential rates of expansion in the body of the device (Pressey, 1953), Since the diaphragm is not connected to a core or wires, capacitance transducers have a high sensitivity and natural frequency. In practice, the capacitance of the transducer usually forms part of the tuning capacitance of an oscillator and the change in applied pressure is thus converted into a corresponding frequency change. In order to avoid problems arising from changes in the capacitance of the cable connecting the transducer to the oscillator circuit, modern practice is to mount the oscillator and discriminator inside the transducer housing (Fig. 2.) In this way, the arrangement becomes a d.c.-in–d.c.-out device with the radio frequency used to energize the circuit being confined to the interior of the transducer's body. There is also the very real advantage that now the capacitance transducer can be interchanged with a d.c.-energized strain gauge transducer when required. Because of the additional circuitry built into it, the capacitance manometer is rather larger than its corresponding strain gauge counterpart. Lilley *et al.* (1947) and Hansen (1949) described in detail the use of capacitance manometers and connecting tubing for the measurement of blood pressure.

FIG. 2 Capacitance pressure transducer with self-contained oscillator and demodulation. (Courtesy of Siemens Limited). (Hill and Dolan, 1976).

The calibration of blood pressure and other pressure transducers

It is important that, before and after, and if possible during, use the pressure transducers are checked to see that the calibration is correct. The simplest method in the case of arterial blood pressure transducers is to apply a known pressure from a sphygmomanometer which is in good condition and preferably of the mercury variety. A saline manometer is often used in the case of venous pressure or airway pressure transducers. The position of the diaphragm in the transducer's housing is usually denoted by a mark on the outside of the transducer and this should be aligned with the liquid reservoir surface of the calibrating manometer.

Pressure transducer preamplifiers are usually provided with some form of electrical calibration in the form of a push button or switch which unbalances the transducer's associated bridge circuit, thus giving rise to an output signal which is equivalent to the application of a known positive pressure, typically 20 or 100 mm Hg, applied to the transducer. This technique is extremely convenient for rapid checks of the calibration during a measurement procedure, but must not be allowed to become a permanent substitute for the periodic exposure of the transducer to known pressures as measured on a mercury or water column. Substitution of a transducer having a different sensitivity from that for which the calibration control was designed would give rise to an error.

Two three-way stopcocks are usually fitted, one to each port of the transducer's cuvette (Fig. 3), so that it can be flushed through with saline, or for the application of a calibration pressure. With the transducer mounted external to the patient, it is possible at any time quickly to check the baseline and calibration of the transducer. The transducer *in vivo* (mounted at the end of a catheter) cannot be calibrated or checked in this way, and this constraint places stringent demands on its baseline stability and calibration. These must be checked before and after use at body temperature.

A dynamic pressure calibration generator is a valuable accessory to determine the resonant frequency characteristics of various transducer/ catheter combinations, with a view to obtaining high fidelity pressure recordings or to check for the presence of small air bubbles in the system. Henry *et al.* (1967) described a pressure generator capable of applying simultaneously a static pressure (0–200 mm Hg) and a sinusoidal pressure (0–70 mm Hg) at a continuously variable frequency of 5 Hz–2000 Hz. A variable frequency pressure generator is also of value in the adjustment of a variable damping control to obtain the flattest possible pressure–frequency plot for the transducer/catheter combination.

The pressure transducer's cuvette and diaphragm and their sterilization

Pressure transducers used to measure pressures outside the body are fitted with a transparent plastic cuvette or "dome" which is approximately hemispherical in shape and has two Luer male connectors leading to its chamber, the flat side of which is either in contact with, or is formed by, the transducer's diaphragm. The Luer connection perpendicular to the diaphragm leads via a three-way stopcock to the line connecting the patient. The other connector leads via a second three-way stopcock to either a reference pressure or to a flushing solution. The dead volume of the chamber is made as small as possible and the design is such that any bubbles collecting in it can easily be flushed out. With good practice, blood should never be allowed to get back down the line into the cuvette. If, due to a mishap, this happens, then the blood should be immediately removed by flushing with heparinized saline and the cuvette later thoroughly cleaned with a biologically active detergent and left filled with a sterilizing solution if it is to be used in the near future.

Transducers should not be sterilized by autoclaving, gamma radiation or formaldehyde vapour solutions. Liquid sterilizing solutions such as Cidex or Zephiran Chloride may be used if the maker's instructions are followed

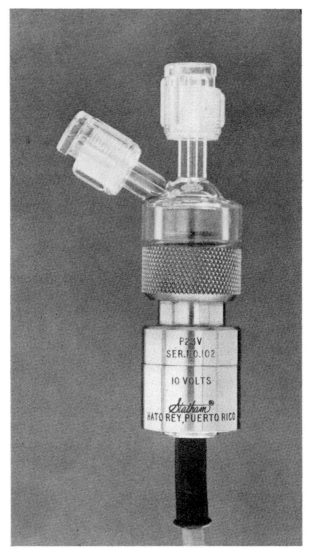

FIG. 3 Unbonded strain gauge pressure transducer having a cuvette with fittings for two three-way stopcocks. (Courtesy of Statham Instruments, Incorporated). (Hill and Dolan, 1976).

and the electrical connector is not immersed. If ethylene oxide gas is used, the transducer should be completely dry both inside and out to inhibit the formation of toxic ethylene glycol. Typically, the sterilization would be performed for one hour at 53 °C to minimize the formation of ethylene glycol in the plastic of the dome.

Electrical safety aspects of pressure transducers

In many applications an electrical pressure transducer will be connected directly to a patient's artery or vein, and sometimes to a chamber of the heart—via a column of conducting saline. Since the patient may well be earthed via another electrode, it is essential to cover all eventualities that the maximum permissible leakage current which can flow via the transducer under fault conditions should be less than the level required to produce ventricular fibrillation, with a direct connection to the heart. The requirement of the International Electrotechnical Commission is 50 μA r.m.s. at mains frequencies. The construction of the transducer must be such that the action of a surgical diathermy or a d.c. defibrillator will not cause a breakdown of its insulation. The latest models incorporate three levels of insulation. In the case of the Bell and Howell type 4-3271 transducer the primary insulation between the patient and an electrical hazard is a set of sapphire rods which support the strain gauge wires and isolate them from the metal pressure-sensing diaphragm. The second level of insulation is formed by the dual high dielectric insulators which surround the strain gauge sensing element to prevent current passing between the patient and earth. The transducer casing provides the third level of insulation. It is constructed of a non-conducting, self-extinguishing, phenylene oxide plastic. Each transducer is calibrated to provide a standard output of 50 μV cm^{-1} of Hg in order that a different transducer can be connected to a monitor without altering the sensitivity of the display. Each transducer is also calibrated so that when a 178 kΩ resistor is shunted across the positive excitation with the transducer balanced at zero pressure the output signal will be equivalent to 100 mm Hg. As an alternative to the standard cuvette (dome), the transducer can be fitted with a "Disposadome". This is a transparent dome incorporating a silicone rubber membrane which lies flush on the diaphragm and provides a sterile barrier between the patient and the transducer. Each Disposadome is pressure tested, 100% inspected and sterilized. The transducer can be sterilized using wet or dry ethylene oxide gas or cold sterilization. The use of disposable domes reduces the risk of cross-infection and also provides an electrical isolation from the transducer's diaphragm. This approach is not possible with catheter-tip transducers, but the leakage from the Millar Mikro-Tip transducer is quoted as being less than 2.5 µA at 500 V d.c.

The response of the catheter–transducer combination

When a pressure measuring system is required to reproduce faithfully a rapidly changing waveform such as that of the left ventricle with a view to deriving the maximum rate of change of pressure, it is essential that the system has an adequate frequency response, usually to at least the tenth harmonic of the fundamental frequency of the waveform. The limiting factor is not normally the transducer, but the inertia and friction introduced by the connecting column of fluid. Geddes (1970) has given the following formula for the undamped natural frequency of the combination.

$$f_0 = \frac{1.4 \times 10^3 \times d}{(V_d \times l \times p)^{\frac{1}{2}}}$$

where f_0 is in Hz, d is the diameter of the catheter or cannula in cm, V_d is the volume displacement of the transducer in mm^3 per 100 mm Hg and l is the length of the catheter, and p is the density of the fluid in g cm^{-3}. To obtain the highest possible natural frequency, the shortest, widest possible tubing should be employed. Anatomical considerations may well define the geometry of the tubing, so that the only other alternative is to use a transducer having a stiff diaphragm so that V_d is small, but of course this implies a transducer with a reduced sensitivity. A transducer for use with small-lumen catheters might have a volume displacement of only 0.01 mm^3 per 100 mm Hg in contrast to a figure of 0.8 for a high-sensitivity venous pressure transducer, the respective sensitivities being 20 and 300 microvolts per volt excitation per 10 mm Hg.

The fact that small-volume displacement transducers are available allows them to be used in conjunction with float catheters which are only 0.8 mm in external diameter. The catheter is continuously flushed with a low flow rate of saline (4 ml per hour) and can be left in place in the right heart for up to eight days without the risk of a thrombotic occlusion of the vessel used to introduce the cannula (Grandjean and Perret, 1970).

Ambulatory Monitoring of the ECG and blood pressure

The availability of miniature slow-speed cassette tape recorders capable of recording 24 hours of continuous ECG and arterial blood pressure recordings is having a major impact on the study of work situation induced stresses and the monitoring of rhythm disturbances and arrhythmias in post myocardial infarction patients and the selection and use of suitable drug therapies. Hard wired and computer systems have been designed to analyse the tapes at 60 times real time and to scan for events such as Aberrant QRS complexes, R on T beats and runs of extrasystoles. In some versions it can be arranged for the system to store "normal" beats for the patient concerned and to compare these with the series of tape-recorded beats.

Both indirect and direct arrangements are available for tape recorded blood pressures. The indirect system uses a cuff inflated from a small container of nitrogen and a microphone under the cuff to detect the Korotkoff sounds. These are superimposed on a pattern of pulses which indicate the cuff pressure. After playback and demodulation the combined tracing is viewed on a small chart recorder. The direct approach uses a pressure transducer and a float catheter which is continuously flushed with heparinized saline by a peristaltic type of pump driven by a small regulated motor. A high-resistance restrictor isolates the catheter from the pump. The whole assembly is housed in epoxy resin and worn by the patient along with the tape recorder.

Blood pressure recordings have been valuable in studying patients suffering from transient attacks of fainting, in studying the haemodynamic precursors of angina by monitoring the pulmonary artery pressure and by determining activities such as swimming and defaecation which markedly affect the blood pressure. Studies of 48 or 72 hours are becoming more commonplace and help to decide on the best pattern of work or exercise and the optimum drug regime for therapy. Spot blood pressure readings are notoriously unreliable and a continuous record during the night and daily activity is far more revealing, although more information is required, for example concerning circadian rhythm changes.

Information on the developments in this rapidly growing field can be found in the Proceedings of the International Symposia on Ambulatory Monitoring (ISAM) for 1975, 1977 and 1979. (See also *Postgraduate Medical Journal* **52**, Suppl. 7, 1976; ISAM 77, Editors F.D. Stott, E.B. Raftery, P. Sleight and L. Goulding, Academic Press London and New York, 1977; ISAM 79, Editors F.D. Stott, E.B. Raftery and L. Goulding, Academic Press, London and New York, 1979.)

The Measurement of Blood Flow

Direct measurements

Electromagnetic flow probes

Direct measurements of blood flow in man are only appliable during vascular surgery. When access to the vessel of interest is available, a cuffed electromagnetic flow probe permits both the magnitude of the blood velocity and the direction of the flow to be measured on a beat-by-beat basis. The probe operates on Faraday's principle of electromagnetic induction, i.e. when a conductor is moved in a direction perpendicular to the lines of force of a magnetic field an e.m.f. is induced in the conductor. When the magnetic field is uniform, the e.m.f. arising from the motion of the conductor is given by the product of velocity × field strength. The blood flow, the magnetic field and the induced e.m.f. are all in mutually perpendicular axes (Fig. 4). A cuffed flow probe head (Fig. 5) contains the coil and pole pieces of the electromagnet and the pick-up electrodes. The artery is passed through the slot to fit snugly within the probe. It is desirable to have a library of probes of different sizes to suit various blood vessels, as a loose fit can give rise to artefacts on the flow recording. In practice, the magnet coil of the probe is fed with an alternating or pulsed current so that an alternating flow appears at the electrodes. This is amplified and synchronously rectified. Precautions must be taken to minimize the substantial artefact signal appearing at the electrodes due to unwanted coupling from the magnet coil, and to facilitate this some designs use square or trapezoidal rather than a sinusoidal current waveform for the coil.

Flow probes are usually supplied with a calibration factor by the manufacturer, but if this factor is not used the achievement of a satisfactory

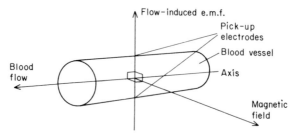

FIG. 4   The basis of operation of a transverse-field electromagnetic flowmeter.

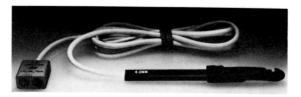

FIG. 5   A cuff-type electromagnetic flow probe head.

absolute calibration by the user requires care. The integral with respect to time of the flow tracing gives the volume of blood which has passed the probe in a given time. In one calibration arrangement, the patient's artery is clamped off on either side of the probe and 20 ml of blood from a syringe and a needle is passed through the probe and a calibration factor in terms of flow is derived from a knowledge of the integrated output from the flow probe. Strictly speaking, the flow probe output is proportional to the blood velocity and is converted to a volume flow rate on multiplying by the vessel's cross-sectional area.

Electromagnetic flowmeters are subject to sources of error arising from the non-uniformity of the electromagnetic field, the flow profile and the shunting effect of the vessel wall. A detailed description of these devices is given by Wyatt (1968; 1971). An interesting development is the extractable flow probe which can be placed around the aorta in man and subsequently withdrawn (Williams *et al.*, 1972).

The measurement of venous flow by thermal dilution

For the measurement of venous flows, a thermal dilution blood flow probe may be used (Clark, 1968). The probe contains two thermistors and is placed at the tip of a catheter which can be introduced, for example, into the veins of the lower limb. The thermal indicator substance is normal saline at room temperature, and this is injected via the catheter into the blood stream at a high velocity by means of a power-driven syringe. A local region of thermal mixing results, and provided that complete mixing occurs across the vessel cross-section, the rate of heat exchange between the saline and blood depends upon the mass flow rate of the blood. The volume flow rate can be calculated from a knowledge of the temperature of the saline and that of the mixture of indicator and blood. These are measured by the two thermistors.

Indirect measurement

Doppler shift blood flowmeters

The methods for the measurement of blood flow which have been described in this chapter so far are invasive. There is a real need for a non-invasive technique and this has led to much effort being put into the development of the Doppler shift ultrasonic blood flowmeter. A Doppler frequency shift occurs when a source of sound is moving relative to the listener. There is an apparent shift to a higher frequency as the source of sound approaches the observer and an apparent shift to a lower frequency as the source moves away from the observer. A doppler shift blood flowmeter directs a beam of 5 MHz ultrasound on the moving blood stream and the frequency is monitored of the back-scattered ultrasonic beam. The transmitted and received beams are mixed to produce an output beat frequency signal. With a 5 MHz beam frequency, the ultrasonic wavelength in tissue is approximately 0.3 mm so that a blood velocity of 3–4 cm s^{-1} produces a beat frequency of some 300–400 Hz. This simple type of Doppler instrument is not flow-direction sensitive and has been widely used for the detection of arterial wall motion for the measurement of systolic blood pressure in conjunction with a cuff and for the monitoring of relative blood flow changes in large blood vessels close to the surface, e.g. the placenta (Plass, 1964).

In a simple Doppler instrument, the factor of proportionality between the observed Doppler shift and the blood velocity involves the cosine of the angle between the ultrasonic beam and the direction of flow. Unless the beam can be aimed so that the angle is sensibly zero giving a cosine of unity then the angle must be known to obtain a velocity calibration. For arteries such as the carotids and the limb arteries which run roughly parallel to the body surface it is not possible to operate with an angle of 0°, as Hansen et al. (1976) have pointed out. Their solution is a probe carrying two transmitter and one receiver crystals. The transmitters are fed with two slightly different carrier frequencies f_{c1} and f_{c2} and give rise to a moving pattern of interference fringes. A stationary particle in the moving pattern will scatter ultrasonic energy which is amplitude modulated with $f_{c1} - f_{c2}$ (and harmonics). If the particle is moving in the same direction as the fringe pattern the modulation frequency will be increased, and reduced if the direction is opposite. In practice, because there are many different frequencies present simultaneously in the sample volume, one is dealing with the differences between two frequency spectra. To a first approximation this difference is independent of the angle between the beam of ultrasound and the blood stream and to the position of the receiver transducer. Brubakk et al. (1977) employed a direction-sensitive Doppler to diagnose valvular heart disease.

The use of range-gated and frequency modulated (McCarty and Woodcock, 1976) Doppler systems is making it possible non-invasively to visualise the internal and external surfaces of blood vessel walls and to estimate the blood volume flow rate by combining the measurement of the vessel's internal diameter with a conventional Doppler velocity measurement.

A combination of Doppler velocimetry, pressure measurements and volumetric visualizations provide a powerful alternative to more conventional investigations such as arteriography. Barnes et al. (1975) have reviewed the current and anticipated applications of Doppler ultrasound in venous, peripheral arterial and cerebrovascular diseases while Coghlan and Taylor (1976) describe four different methods for obtaining directional flow information from Doppler velocimeters.

Plethysmography

Venous occlusion plethysmography has become the classical method for measuring blood flow in the limbs of man, and the technique against which new methods are often standardised. The principle of the method is that occlusion of the circulation to an extremity, using a pressure greater than venous pressure but lower than diastolic arterial pressure, will result in blood continuing to enter the extremity from which it is unable to drain. The volume of the limb therefore increases at the same rate as the blood enters (Fig. 6) and the rate of change of volume is equivalent to the blood flow. Release of the venous occlusion allows rapid drainage from the limb and the volume returns to the preocclusion level. A series of estimates of blood flow may be made by occluding the circulation several times a minute. A convenient method is to occlude the circulation with a pressure of 60 mm Hg for 10 s with subsequent release for 7 s. If venous occlusion is maintained for longer periods, the capacity of the limb will be exceeded and the slope of the volume curve will decrease (Fig. 7).

The volume of the limb during plethysmography may be measured by placing it in a suitable rigid container. For instance, the volume of the hand may conveniently be measured by placing it in a loose surgical glove which is stuck around the wrist to a soft, thick rubber diaphragm (Fig. 8). The diaphragm is then bolted to the container. This container now has only one outlet. Changes in volume may be measured by closing this outlet and measuring changes in air pressure. It is more usual to fill the container with water at a suitable temperature and then to record the rise of water level in the chimney using either a float recorder or the change of resistance between a pair of vertical electrodes partially immersed in the water. Flow in the forearm may be similarly measured by enclosing it in a rubber sleeve attached to diaphragms at either end, which can be screwed to either end of

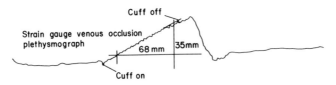

FIG. 6   Normal venous occlusion plethysmogram

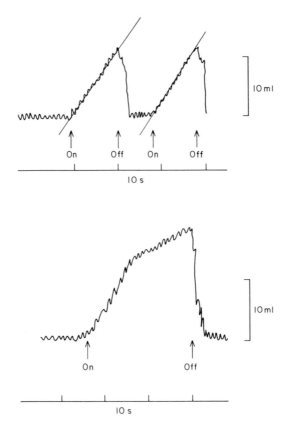

FIG. 7 (i) (Upper) Volume traces obtained during plethysmography of the hand. At ON the pressure in the collecting cuff is raised to 60 mm Hg. This is followed by an increase in the volume of the limb. At OFF the pressure is released and the volume drops rapidly to its previous level. From the slopes of the traces and the volume and time calibrations, the blood flow to the part may be calculated. (ii) (Lower) The effects of a longer period of venous occlusion.

the box. The collecting cuff is placed around the upper part of the forearm or on the upper arm. A further cuff is placed around the wrist and occluded to 200 mm Hg during the period of recording to prevent distortion of the record by venous return from the hand. Blood flows in the foot and calf may be similarly measured using suitable modified equipment.

The volumes of the forearm or the calf may alternatively be assessed from changes in circumference of the limb occurring during venous occlusion. These measurements may conveniently be made using mercury-in-rubber strain gauges (Whitney, 1953). These consist of thin soft rubber

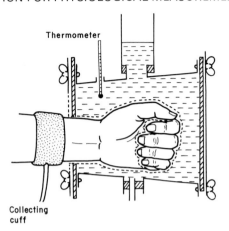

FIG. 8   Plethysmography of the hand (From Barcroft and Swan, 1953).

tubes filled with mercury which are fitted around the limb. Increase of the circumference of the limb increases the length of the mercury thread and thus increases its electrical resistance.

The recording equipment must be calibrated by introduction of known volumes of air or water into the box, or by changing the length of the strain gauge by a known amount. The slope of the volume trace may then be converted into a blood flow expressed usually as millilitres per 100 millilitres tissue per minute.

The measurement of stroke volume and cardiac output

The Fick method

The conventional techniques for the measurement of cardiac output are invasive and consist of the Fick (Fick, 1870) and indicator dilution methods. The Fick method is based upon the fact that the subject's oxygen consumption in millilitres per minute must equal the product of the cardiac output in millilitres per minute multiplied by the difference in the oxygen content of arterial and mixed venous blood. The contents can be readily measured by means of gas chromatography (Davies, 1970). The principle can also be applied to carbon dioxide output as well as oxygen uptake. Possible sources of error in the Fick method have been described by Prime and Gray (1952) and Visscher and Johnson (1953). The Fick method is often used as a standard against which other techniques for the estimation of cardiac output can be compared.

Indicator dilution methods

A small amount of an indicator substance, such as a dye or cool saline or a gamma-ray-emitting isotope such as human serum albumin labelled with iodine-131, is injected into a large vein or, preferably, via a catheter, into the right side of the heart. The bolus of indicator is passed through the right heart, the pulmonary circulation and the left heart into the systemic circulation. During the course of its passage through the heart and lungs it is assumed that the indicator becomes thoroughly mixed with the blood. In the case of the dye or cool saline, the appearance of the indicator in the peripheral circulation is usually detected by means of a densitometer or thermistor respectively and in the case of the ^{131}I by means of a collimated scintillation counter placed over the heart. If an electrical output signal is taken from the detector and displayed on a pen recorder a dilution curve is drawn out as shown in Fig. 9 when dye is the indicator.

The indicator is injected as a slug or bolus and there is an interval of several seconds (the appearance time) before the tracing rises to a peak and then decreases. The appearance time is the time for the fastest particles of dye to travel to the detector via the shortest route. In the case of dye, the curve possesses a second peak—the recirculation peak—due to dye which has passed around the circulation for a second time. The recirculation peak complicates the calculation of the cardiac output unless the speed of the detection system is fast enough to allow the curve to return to its baseline between the primary and recirculation curves. When cool saline is used, the bolus of saline will have attained body temperature before it can pass the detector for a second time so that no recirculation peak is seen with this technique.

The dye dilution cardiac output is found by means of the Stewart–Hamilton formula $\dot{Q}_{1/\text{min}} = 60 \times mg/A$ where $\dot{Q}$ is the mean cardiac output over the period of the dilution curve in litres per minute, mg is the mass of dye injected in milligrammes and A is the area under the dye curve without recirculation in mg per litre × seconds. The factor 60 gives the cardiac output in litres per minute rather than per second.

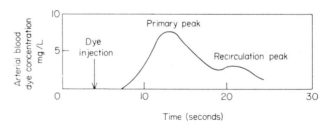

FIG. 9   A dye dilution circle (Hill and Dolan, 1976).

Indocyanine green dye which absorbs light in the 800 nm region of the spectrum—an isosbestic point for whole blood is commonly used—so that the optical absorption of the dye is unaffected by the degree of oxygenation of the blood. For an adult subject 5 mg of dye would be injected in a volume of 1 ml. The time axis of the recorder chart is scaled in seconds and the pen deflection can be calibrated in terms of the dye concentration in blood by drawing through the densitometer cuvette known concentrations of dye in blood, typically 3, 6, 9 and 12 mg per litre. These calibration mixtures are sampled via the sidearm of the three-way tap connected to the densitometer's cuvette (Fig. 10). Unless the cardiac output is low or there is marked valvular incompetence, the downslope of the dye curve prior to recirculation will be a simple exponential. This portion can be replotted to yield a straight line on semilogarithmic graph paper (Fig. 11). By projecting this line to the baseline, values can be found for the portion of the original curve which would have been present but for recirculation. These points are plotted to complete the missing lower portion of the dye curve. Using a planimeter the area under the primary curve is measured and compared.

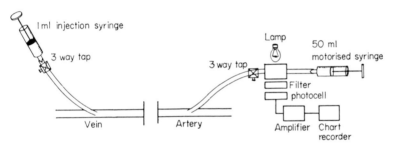

FIG. 10 Schematic diagram of the apparatus used for recording a dye curve (Hill and Dolan, 1976).

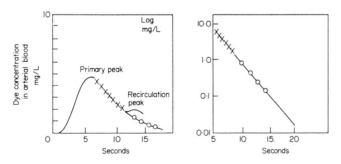

FIG. 11 The re-plotting of the downslope of a dye dilution curve on semi-logarithmic graph paper (Hill and Dolan, 1976).

with a known area constructed from the calibrations to yield a figure for the area of the dye curve in mg per litre × seconds. The subject is treated in more detail by Hill and Dolan (1976).

Thermal dilution

Another popular technique employs normal saline at room temperature as the indicator (Branthwaite and Bradley, 1968). A 4 ml injectate may be used for normal hearts, with 8 ml for hearts which are substantially dilated. Adequate mixing can be obtained by interposing only one cardiac chamber between the sites of injection and detection. Injection, using a special low heat loss catheter, may be into the right ventricle with detection by means of a thermistor located at the tip of the detection catheter in the pulmonary artery, both catheters being passed via the neck. No recirculation peak occurs since by the time the indicator has passed once round the circulation it will have attained blood temperature. The cool saline is inexpensive and can be used with repeated injections. The technique has been discussed by Lowe (1968).

The use of a radioactive indicator

Human serum albumin labelled with iodine-131 has been used for more than 20 years as an indicator in cardiac output studies (Veall and Vetter, 1958), detection being by means of a collimated scintillation counter placed over the heart. The method has the advantage of only requiring a venous injection, but yields results which are higher than those obtained with the dye or thermal dilution techniques due to the detection of radioactivity from paracardiac tissues. With suitable placing of the detector a double-peaked dilution curve is obtained, the peaks corresponding with the passage of the indicator through the right and left heart. From such a curve much information can be obtained, including the central mean transit time, the central blood volume and pulmonary blood volume (Hill et al., 1973).

Electrical impedance technique

Changes in the stroke volume as well as information on the haemo-dynamic state can be obtained using the electrical impedance technique.

A constant current of approximately 4 mA r.m.s. at 100 kHz is passed across the thorax via an outer pair of band electrodes and the resulting potential variations with each cardiac cycle are detected across the inner pair of band electrodes (Fig. 12). Figure 13 shows a typical set of normal waveforms. The value Z_0 is the standing impedance at 100 kHz between the inner electrodes and this falls by an amount ΔZ with each systole. The

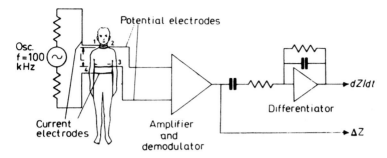

FIG. 12 Four electrode arrangement for the non-invasive measurement of cardiac output by the electrical impedance technique (Hill and Dolan, 1976).

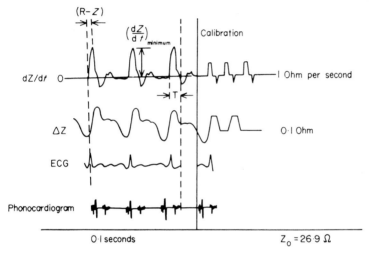

FIG. 13 A typical set of waveforms used in the monitoring of stroke volume changes by the electrical impedance method (Hill and Dolan, 1976).

maximum rate of change dS/dt_{max} of the impedance occurs during systole. By making gross assumptions such as a cylindrical thorax of uniform conductivity, Kubicek *et al.* (1966) derived a formula for the stroke volume

$$SV_{ml} = \frac{pL^2}{Z_0^2} \times \frac{dZ}{dt_{max}} \times T$$

where p is the resistivity of the patient's blood in ohm cm, L is the distance in cm between the inner band electrodes, and T is the left ventricular ejection time in seconds. When L is the shortest distance between the electrodes and the value of p for the particular blood is used, the stroke volume is likely to be within $\pm 15\%$ of the dye dilution value for patients with a normal heart action. Hill and Merrifield (1976) showed that at the 5% level there was no significant difference between the left ventricular ejection time measured with an external carotid pulse transducer and that measured from the baseline crossing of the systolic upstroke of the dZ/dt tracing and the commencement of the second heart sound or the X-point of the dZ/dt tracing. The X-point is a characteristic notch in the tracing which occurs at the time of the second heart sound in most patients. Endresen and Hill (1976) showed that normal respiration does not affect the calculated values at the 5% level for stroke volume and cardiac output measured by the electrical impedance method. Large respiratory tidal volumes can produce marked variations in the dZ/dt baseline and it may be necessary to ask the subject to hold his breath for several heart beats. If a ventilator is in use it can be stopped for a similar period. The effects of spontaneous respiration have to be tolerated with neonates and here the height of dZ/dt_{max} can be measured from some landmark on the upstroke of the systolic peak of dZ/dt for the beats of interest.

Hill et al. (1976) subjected dogs to the effects of a progressive haemorrhage and found that dZ/dt_{max} correlated well with the blood loss and the stroke work. Welham et al. (1978) showed that dZ/dt_{max} in dogs subjected to increasing blood halothane levels, correlated well with accepted indices of myocardial contractility. The measurement of dZ/dt_{max} provides a non-invasive variable which is worth considering in patient monitoring.

The measurement of Z_0 can also yield valuable diagnostic information such as the volume of intrathoracic fluid (Pomerantz et al. 1969, 1970). In one patient with a pleural effusion Z_0 rose by one ohm for each 200 ml of fluid removed. Roy et al. (1974) measured changes in Z_0 due to the effects of high altitude hypoxia, in healthy indian soldiers. The mean Z_0 at sea level was 34.6 ohm; at altitude it fell to 29.6 ohm and levelled after 10 days at 31.5 ohms. This value was similar to the mean value obtained from the permanent high altitude residents. When estimated serially in the same subject Z_0 appeared to reflect changes in the thoracic fluid volume when exposure to high-altitude hypoxia occurred.

Khan et al. (1977) have produced the following expression for Z_0 when a pair of band electrodes is located at the neck and a second pair at the level of the xiphisternum as described by Kubicek et al. (1966): $Z_0 = 4\pi \times 5.2L(1 + r^2)/r(p_1 + p_2)^2$ where L is the distance between the inner band electrodes, 5.2 ohm cm is taken as the average resistivity of the thoracic segment considered, p_1 and p_2 are the perimeters of the elliptical contour respectively at the level of the axilla and xiphisternum and r is the ratio of the major to minor axes of the elliptical segments (assumed constant). Clinically, the abnormalities due to an intrathoracic accumulation of fluid can be detected by comparing the estimated and measured values of Z_0.

The Fick (nitrous oxide) method

The advent of quadrupole mass spectrometers which can analyse respired gases on a breath-by-breath basis has encouraged investigators to look again at the Fick approach, but this time on a non-invasive basis. Nitrous oxide may be employed as a soluble gas whose uptake can be measured as the difference between the inspiratory and expiratory tidal volumes by means of an integrating pneumotachograph. Some of this uptake is due to the spread of the gas through the lung volume. Thus, a second gas—argon—is employed to determine the proportion of the gas uptake which is due to diffusion through the lung. Alveolar nitrous oxide can be calculated from a knowledge of the end-expired value. The Fick equation becomes

$$\dot{Q} = \frac{\dot{V}_{N_2O} \times 60}{F_{\bar{A}N_2O} \times \lambda_{N_2O} \times t}$$

and this can be solved on a breath-by-breath basis. The value $\dot{Q}$ is the cardiac output in litres per minute, $\dot{V}_{N_2O}$ is the nitrous oxide uptake in litres per minute, F_{AN_2O} is the alveolar nitrous oxide concentration, λ_{N_2O} is the solubility coefficient for nitrous oxide in blood (0.474 ml/ml at 37 °C), and t is the time of uptake in seconds.

Respiratory Gas Volume and Flow Rate Measurements

Gas meters

In exercise and other studies, it is frequently necessary to measure respired gas volumes with a view to calculating quantities such as the oxygen consumption and carbon dioxide output. At its simplest, when only a mean value for the tidal or minute volume is needed, some form of gas meter may be employed, sometimes with the gas first collected in a Douglas Bag. A good account of dry gas meters and the correction factors which can be applied to them is that of Adams *et al.* (1967). Reynolds (1968) has described the addition of a chopper disc to a dry gas meter to provide an electrical output signal.

Respirometers

Tidal and minute volumes are conveniently measured with a turbine-type meter such as described by Wright (1955). The mechanism is a miniature air turbine having a very low inertia. In the original mechanical version, the revolutions of the two-bladed rotor are recorded by a gear train and dial similar to that found in a watch. Gas entering the rotor chamber emerges from a series of fixed tangential slots to strike the rotor so that the meter responds to gas flows in one direction only and does not require valves. The ratio of the indicated volume to that passing through the meter is flow-rate dependent since slip of gas past the rotor will happen to a greater degree at high flow rates. In general, the Wright meter will under-read at low flows and over-read at high flows. Electronic versions of the Wright respirometer are now available in which the gear train and dial have been replaced by a light-emitting diode and phototransistor which sense the motion of the rotor. The resulting output pulses are counted to drive a meter scaled in terms of tidal and minute volume. The ranges are 200–1500 ml and 4–30 litres per minute with an accuracy of $\pm 2\%$ of the indicated reading with a response down to two litres per minute.

The pneumotachograph

A pneumotachograph is useful when individual breathing patterns have to be recorded. Essentially it consists of a resistor with a low resistance to gas flow designed so that over the stated range the flow pattern is laminar. Thus the pressure drop developed across the resistance is directly proportional to the gas flow rate. Typical resistors consist of a bundle of parallel small-bore metal tubes or a wire gauze (400 mesh) mounted between a pair of conical metal tubes (Fig. 14). Provision is usually made to heat the gauze from a low-voltage supply to prevent the condensation of water vapour from the expired air occluding the gauze. Adjacent to each side of the gauze is a series of small holes around the periphery of the cones, each set communicates with an annular pressure take-off chamber fitted with a side-arm. These are each connected via a plastic or rubber tube to one side of a sensitive differential air pressure manometer. The set of holes and the associated chambers minimize negative pressure effects which might arise at a single orifice due to the action of the gas stream. For resting adults, or during anaesthesia, a maximum flow rate of 180 litres per minute with a pressure drop of 8.5 mm of water would be suitable for a pneumotachograph while for use with exercising adults the corresponding figures would be 600 litres per minute and the same pressure drop. Fry et al. (1957) have evaluated the performance of three kinds of pneumotachograph head. For a given flow rate the pressure drop across the head will depend upon the viscosity of the gas mixture. Hobbes (1967) demonstrated that the pressure drop increases linearly with an increasing nitrous oxide concentration and

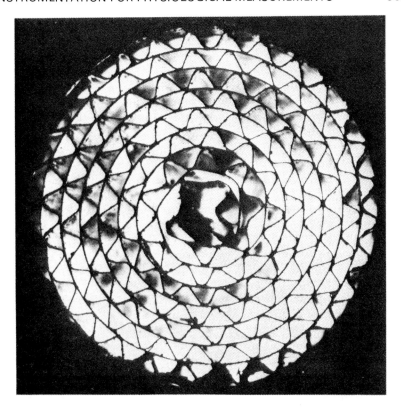

FIG. 14 Cross-section of a Fleisch pneumotachograph head (Hill and Dolan, 1976).

that the effect of saturating air at 37 °C with water vapour was to reduce the pressure drop by 1.25% in comparison with dry air at 37 °C.

When using a pneumotachograph with sharply varying gas flow waveforms such as may be produced by an automatic lung ventilator, care must be taken to see that the volumes on either side of the gauze and the transducer are equal, otherwise the pressure pulse may arrive at one side of the transducer's diaphragm before it reaches the other. This will produce a marked pressure artefact with the production of an apparently high flow rate. A changeover from a forward to a reverse flow should not give rise to any asymmetry in the calibration of the flow head. The pneumotachograph head is conveniently calibrated with a flow generator such as a vacuum cleaner and a rotameter of suitable volume flow rate capacity.

Integrating pneumotachographs

The basic pneumotachograph is normally calibrated in terms of volume flow rate (litres per minute). By feeding the output signal into an electronic integrator circuit, the volume flow rate is integrated with respect to time to yield an indication of each tidal volume. Baseline drift can be a serious problem with integrators. Any unbalance in the d.c. amplifier feeding the integrator will give rise to a d.c. level on the input to the integrator which is summed to yield a drift in the output. In some designs, the integrator is arranged to be reset to zero each time there is a baseline crossing of the pneumotachograph tracing. The integrator output then consists of a series of deflections of positive and negative polarity, the height of each corresponding with the relevant stroke volume. The incorporation of a rectifier will allow the measurement of either the inspiratory or of the expiratory tidal volumes and in conjuntion with a timer these can be summed to yield the minute volume of ventilation.

The impedance pneumograph

It has been found that a quantitative relationship exists between a subject's tidal volume and the corresponding transthoracic electrical impedance change (Baker and Hill, 1969). Baker et al. (1966) showed that over the frequency range 50–100 kHz the changes in the thoracic impedance accompanying respiration were essentially of a resistive nature and were independent of the frequency. A simple two-terminal arrangement is normally employed with a pair of disk electrodes placed bilaterally along the mid-axillary lines (Fig. 15). The impedance measured across the thorax is given by $(Z_0 + \Delta Z)$ where Z_0 is the standing impedance and ΔZ is the impedance change accompanying respiration. The electrodes are arranged to be fed with a constant current (i) of about 2 mA r.m.s. The voltage developed across the electrodes is given by $(V_0 + V_1) = i(Z_0 + \Delta Z)$, where Z is usually about 200 ohms and ΔZ is a few ohms per litre of tidal volume. After amplification the voltage appearing across the electrodes is rectified and smoothed to give d.c. components $V_{d.c.}$ and $V_{1d.c.}$. For respiratory monitoring only, an a.c. amplifier may be used and the $V_{1d.c.}$ signal variation displayed on a chart recorder. The impedance electrodes will also pick up a non-standard lead ECG if they are connected to an electrocardiograph. This ECG signal is useful for monitoring heart rate changes. Barker and Brown (1973) have described a simple two-terminal impedance pneumograph circuit.

The Z_0 reading from a two-terminal impedance pneumography may be calibrated by connecting the leads to a decade resistance box whose

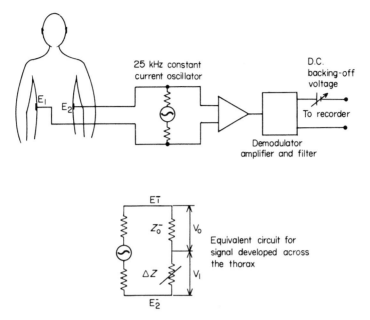

FIG. 15   Block diagram of a two-terminal impedance pneumograph (Hill and Dolan, 1976).

resistance can be altered in 1 ohm steps and adjusting the box until the recorder's deflection is the same as with the patient connected. The relationship between the tidal volume and ΔZ is different for each patient and must be determined for each individual. In the case of spontaneously breathing patients this may be done by having the patient breathe into a rapid-response spirometer. This is the preferred method, but if this is not available a respirometer can be used. For anaesthetized patients who have received a muscle-relaxant drug, a ventilator with a known range of tidal volumes may be used, or a large syringe (400 ml capacity) to inflate the patient with known volumes.

It must be remembered that the impedance pneumograph measures chest wall movements and not actual gas volumes. Thus, it is possible to record for some time an impedance pneumogram signal not far removed from normal in the presence of a respiratory obstruction. Baker and Hill (1969) found the technique convenient to use with anaesthetized patients and Pallett and Scopes (1965) have used it with neonates. Gundersen and Dahlin (1971) found that interference may be experienced at the high respiratory rates of infants, due to cardiac pulsations which are difficult to filter out without affecting the respiratory signal.

The monitoring of respiratory rate

The majority of patient monitoring systems simply make provision for the monitoring of respiratory rate; tidal and minute volume being measured as required by means of some form of respirometer. One commonly encountered technique is to tape a thermistor (thermally sensitive resistor) to the cheek of the subject so that the tidal air from one nostril flows over the thermistor bead, which should not touch the skin. During expiration, the warm expired air reduces the resistance of the thermistor relative to that obtaining during inspiration. The thermistor forms one arm of a d.c. Wheatstone Bridge circuit and the output signal from the bridge is displayed on a pen recorder. The result is a sinusoidal tracing at the breathing frequency. The thermistor can also be mounted inside a facemask, or into the airway of a patient who is intubated. Gundersen (1970) has described the use of twin thermocouples connected in series with one junction situated 5 mm for the lips and the other 5 mm below a nostril so that an output tracing is obtained whether the subject breathes through his lips or nose or both.

Other possibilities are the use of an indirect method such as a light strap or tape placed around the chest and operating a spring-loaded switch or a pneumograph consisting of a tape around the chest with each end of the tape connected to one end of a 200 mm length of corrugated rubber tube having a cork in either end and sidearms leading to a sensitive pressure transducer.

Lung function tests

The measurement of forced expiratory volume (FEV) and forced vital capacity (FVC) is made with a spirometer, which must be carefully calibrated. The apparatus should record volume to within 100 ml over a range of 10 litres: it must have a low inertia and a low resistance to gas flow. It is useful to have a timing unit incorporated.

Other pulmonary tests include the measurement of airway resistance, compliance and oxygen transfer. Fletcher and Belville (1966) used an analogue computer for the on-line calculation of pulmonary compliance and the energy cost of respiration. A mouthpiece fitted with a pneumotachograph head is used, together with a second pressure transducer to measure the difference, P, between mouth and oesophageal pressure. Then

$$\text{Work per Respiration} = \int \frac{(P\mathrm{d}V)}{\mathrm{d}t}\,\mathrm{d}t$$

A number of instruments have been designed for the measurement of oxygen consumption. The Kofranyi–Michaelis or Max Planck respirometer has been used widely for many years. It consists of a dry gas meter worn on the back; expired air passes through the meter which is designed to record flow rates of 15–50 litre min^{-1}. The subject has to wear an oral-nasal mask or a mouthpiece and noseclip. The Max Planck is easy and convenient to use, although resistance to air flow increases with rise in ventilation rates. Flows over 50 litre min^{-1} result in discomfort and cannot be tolerated for long.

The Integrating Motor Pneumotachograph, or IMP, was designed by Wolff (1961) to have a low resistance to flow, with a sampling device for expired air which could be used for relatively long periods. The IMP was excellent for field work, with the subjects engaged in a variety of activities for periods of two to three hours. However, it required skilled maintenance, and it has been superceded by the Oxylog (Humphrey and Wolff, 1977) and by the MISER (Eley et al., 1978).

The measurement of ventilation volume, collection of a representative sample of expired air and its analysis are all required to determine oxygen consumption. All these functions are combined in the Oxylog, designed by Humphrey and Wolff. It weighs 2.5 kg and can be worn in a knapsack. It is powered with rechargeable batteries with an endurance of 24 hours. There is a digital display of oxygen consumption in the preceeding minute and also a cumulative figure.

The Miniature Indicating and Sampling Electronic Respirometer, or MISER (Eley et al., 1978), consists of three parts: a photoelectric version of the Wright respirometer, a control unit including a volume counter, timer and display, and a vacuum sampler. The instrument, which is compact and robust, is powered by rechargeable cells and can run continuously for about eight hours. It weighs 600 g. Sub-samples of expired air are collected at isovolumetric intervals, hence the overall sample is representative of the total expired air.

Blood-Gas and Blood pH Analysis

Together, these topics represent an important aspect of clinical measurement instrumentation. A knowledge of the acid–base state of the patient is necessary for the management of metabolic and respiratory disorders and information concerning the oxygen and carbon dioxide tensions of the arterial blood is vital for the management of respiratory insufficiency. In current equipment, it is usual to have three electrodes for blood pH, p_{O_2} and p_{CO_2} all mounted in a single cuvette which can operate with a blood sample of 100 microlitres. The electrodes and the cuvette are thermostatted at 37 °C. Calibration of the pH electrode is by means of buffer solutions of known pH; the oxygen and carbon dioxide electrodes are calibrated with gas mixtures of known composition. The analyser frequently incorporates automatic rinsing and calibration facilities, together with checks on the correct functioning of the electrodes.

The glass pH electrode

Basically, a pH measurement system consists of the combination of a stable potential reference electrode and an indicator electrode which is placed in the solution to be measured. The indicator and reference electrodes are connected by means of a salt bridge and the potential difference appearing between them is measured with a high input resistance millivoltmeter calibrated in terms of pH units. In effect, this combination behaves as a battery whose EMF is dependent upon the pH value of the solution containing the indicator electrode.

The action of the glass type of pH electrode depends upon the fact that a potential difference appears across the interface produced when a glass semipermeable membrane is introduced between two liquid phases. The membrane allows a reversible transfer to occur of hydrogen ions, and when an equilibrium is attained across the membrane the potential is proportional to the logarithm of the ratio of the concentrations on either side of the membrane of hydrogen ions. Eisenman et al. (1966) have given a detailed account of glass pH electrodes.

Figure 16 is a schematic diagram of a pH meter. The thin-walled bulb of pH-sensitive glass contains a buffer solution, usually of pH 1, providing a known stable pH value. A chlorided silver wire dips into the buffer and connects via a high-insulation cable to the input of the pH meter. On placing the bulb in the solution whose pH is to be measured, a d.c. potential difference appears across the walls of the bulb which is proportional to the difference in pH of the buffer and the outside solution. To allow the potential difference to be measured the circuit is completed by two reversible half-cells, one of which is the silver/silver chloride wire and the other a calomel reference electrode. The latter is connected via a salt bridge to the test solution. All the electrodes and junctions are thermostatted at 37 °C, and the pH meter reads the potential difference between the electrodes in millivolts. It is of the order of tens of millivolts and varies linearly with the pH of the sample solution. The slope of about 60 mV per

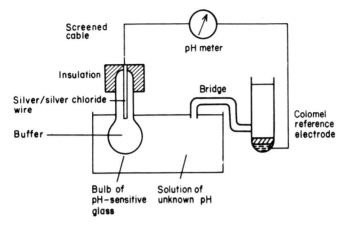

FIG. 16   Schematic diagram of a pH meter (Hill and Dolan, 1976).

pH unit depends upon the temperature of the system and varies by about 0.34% per °C. The deposition of protein material from blood will produce a gradual decrease in the slope of the electrode's calibration line, and when this becomes marked the glass should be etched according to the maker's instructions to clean it, but eventually the electrode will have to be replaced. Capillary glass electrodes (Astrup and Schroder, 1956) have the advantage that a small volume blood sample, e.g. from a finger prick, can be sucked up into the sensitive interior of the electrode. Very small pH sensitive glass microelectrodes for intracellular pH measurements have been devised (Okada and Inouye, 1976).

The calomel reference electrode has at its bottom end a plug of asbestos fibre or ceramic which provides a liquid junction between the saturated KCl solution contained inside the electrode and the sample solution. Immersed in the KCl is a cylinder of calomel (HgCl) which is connected via mercury to a platinum wire and the reference electrode terminal of the pH meter. The output from a glass electrode calomel electrode combination is normally zero when the test solution is nearly neutral (pH 7). For acid solutions the output becomes positive, whilst for negative solutions it goes negative. The internal resistance of a glass electrode is of the order of $100-1000$ MΩ so that the pH meter must have an input resistance of at least $100\,000$ MΩ to prevent it loading the pH electrode by more than 1%. A typical range for a blood pH meter is 6.000–8.000 pH units and it must be possible to detect changes of 0.001 pH units. Balance and slope controls are provided. The balance control adds a voltage in series with that from the pH electrode system and enables the output to be set to the pH value of a standard buffer solution. The slope control adjusts the gain of an amplifier to set the slope of the calibration line using two buffers. Standard buffer solutions are available from the manufacturers of blood pH systems and these are employed to calibrate the instrument. Once poured from their storage bottle, the solutions should never be returned. Rosenthal (1948) found that the temperature coefficient for blood pH measured *in vitro* is 0.0147 pH units per °C, while Adamsons *et al.* (1964) showed that the temperature coefficient depends both upon the actual pH of the blood and its metabolic state.

The p_{CO_2} electrode

The design for an electrode capable of measuring directly the partial pressure of carbon dioxide (p_{CO_2}) in the blood is that of Severinghaus and Bradley (1958) which was modified from that of Stow *et al.* (1957). Basically, the p_{CO_2} electrode is a pH-sensitive glass electrode arranged to measure the pH of a thin layer of a sodium bicarbonate solution. The bicarbonate solution is held in the pores of an inner membrane placed on the glass electrode. The inner membrane can be of cellophane and this in turn is covered with an outer membrane of polytetrafluorethylene to isolate the electrode from the blood sample with which the outer membrane is bathed. The PTFE membrane is permeable to carbon dioxide gas molecules but not to other ions which might affect the pH of the bicarbonate solution. A combination pH glass electrode and reference electrode is in contact with the bicarbonate solution. Carbon dioxide gas diffuses across the membrane in either direction depending upon the difference in partial pressures which exists across the membrane. Hydration of CO_2 in the water of the bicarbonate solution produces carbonic acid and alters the solution's pH. The output of the electrode is basically logarithmically related to the blood p_{CO_2}, a tenfold increase in p_{CO_2} is almost equivalent to a reduction of one pH unit. An amplifier with an antilogarithmic characteristic is employed to produce a final output which is linearly scaled in terms of p_{CO_2} in millimetres Hg or Pascals.

The most accurate means of calibrating a p_{CO_2} electrode is to employ blood samples which have been tonometered with gas mixtures containing known partial pressures of CO_2. In practice, known gas mixtures after humidification are simply passed over the electrode. This is convenient, but it must be remembered that with some electrodes there may be a difference in calibration depending on whether blood or gas is used. A gas mixture of 5% CO_2, 12% O_2 and 83% N_2 is often used to simulate the p_{CO_2} of normal blood and a mixture of 10% CO_2, 90% N_2 used to set the slope of the calibration line. The temperature sensitivity of the p_{CO_2} electrode is about 8% per °C so that it must also be thermostatted at 37 °C.

Coon *et al.* (1976) describe a dual function pH and p_{CO_2} sensor for use *in vivo*.

The p_{O_2} electrode

The blood pH and p_{CO_2} electrodes have much in common, but the polarographic p_{O_2} electrode for the measurement of blood oxygen tension works on a different principle. Whereas the pH and p_{CO_2} electrodes self-generate a variable potential difference, the p_{O_2} electrode is polarised with a fixed voltage and gives rise to an output current which is proportional to the blood p_{O_2}. In Fig. 17 the polarising voltage is adjusted to be in the region of 600 mV and electrolytic reduction of oxygen in the blood occurs at the platinum cathode, the polarographic cell consisting of this cathode and a silver/silver chloride reference electrode. The current is actually proportional to the oxygen available to the cathode immersed in the sample and not to the p_{O_2}. Hence if the functioning of the electrode consumes a significant amount of oxygen from the sample and this is not replenished the reading will fall. Either arrangements must be made to stir the blood mechanically or the consumption of the electrode must be minimized by using a small diameter cathode and a sensitive recording system. If the cathode is operated bare, it becomes poisoned after a period of use owing to the deposition on it of proteins from the blood. Clark (1956) solved this problem by placing both the platinum cathode and the reference electrode behind a membrane which is usually made of polytetrafluoroethylene.

A typical p_{O_2} electrode would have two ranges 0–200 and 0–2000 mm Hg with a response time (99% at 37 °C) of 30 seconds. The electrodes can be

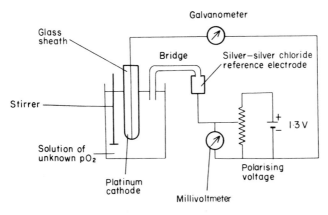

FIG. 17 Schematic diagram of a polarographic oxygen electrode (Hill and Dolan, 1976).

calibrated against the two known gas mixtures used for calibrating the p_{CO_2} electrode. One is used to set the zero and the other the gain of the instrument which is basically a current amplifier. Modern blood-gas and pH analysers such as the AVL 937C microanalyser have the pH, p_{CO_2} and p_{O_2} electrodes all mounted in a single small cuvette and can operate with samples in the 40–100 µl range, thus making it suitable for use with foetal scalp blood (Soutter et al., 1976). Automatic calibration and flushing cycles are provided and a calculator derives variables such as base excess, standard bicarbonate and oxygen saturation.

Respiratory gas measurements with the quadrupole mass spectrometer

Until the advent of mass spectrometry, it was not feasible to analyse all the respiratory gases with one instrument on a breath-by-breath basis. The earlier models of mass spectrometers were of the magnetic deflection type. The gases to be analysed had to be pre-set and the instrument tended to be bulky, but nevertheless it opened new horizons in respiratory physiology (Fowler and Hugh-Jones, 1957). The position has undergone a radical improvement with the development of the compact quadrupole mass spectrometer. This has removed the problems associated with the powerful electromagnet previously required, but now peak selection circuitry is necessary to produce signal outputs for each of the gases of interest.

The quadrupole mass spectrometer head operates at a high vacuum and is evacuated by means of a combination of a rotary oil backing pump and an air-cooled oil diffusion pump. Precautions must be taken with vacuum traps and non-return valves to prevent pump oil from contaminating the diffusion pump in the event of an unexpected mains power failure. The quadrupole mass filter consists of four parallel cylindrical metal rods mounted accurately at the corners of a square with opposite rods connected together. A radio-frequency voltage superimposed upon a d.c. voltage is placed on the rods. The potentials applied to each pair of rods are equal in magnitude, but the d.c. potentials are of opposite sign and the a.c. potentials are of opposite phase.

An ion source (Fig. 18) is located at one end of the quadrupole rod array on the axis. By means of an accelerating voltage, a beam of ions is injected through a hole in a diaphragm along the axis. The ratio of the d.c. voltage to the peak a.c. voltage is set at 0.17. There exists only one specific mass-to-charge ratio of the ions such that those ions can then travel along the axis of the rod assembly to emerge at the other end. Ions having other masses will undergo oscillations of increasing amplitude and eventually lose their charge on hitting the rods. Thus the arrangement acts as a selective mass filter. Scanning of the mass spectrum of the respiratory gas sample which is

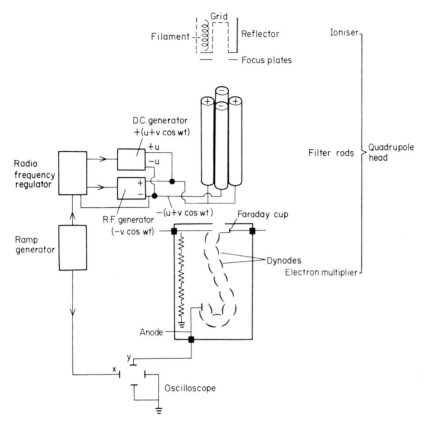

FIG. 18   Schematic diagram of a quadrupole mass spectrometer head (Hill and Dolan, 1976).

drawn into the head via a catheter is accomplished by varying the ratio of d.c. to a.c. voltage while keeping the r.f. frequency constant. This is accomplished by means of a d.c. ramp voltage. The output from the head is sampled at the various ramp voltage amplitudes which correspond with the required gases and storing the output signals on sample and hold circuits. Roboz (1968) has described a typical head as having a rod diameter of 8 mm, rod lengths of 200 mm and a field diameter of 7 mm. The radio frequency amplitude range was 0–800 V, the frequency 3–4 MHz with a power of 30 W, the frequency being stabilized to 1 part in 5×10^3 and the amplitude to 1 part in 10^5. The d.c. voltage range is 0.140 V. The selected ions are detected with a low-noise electron multiplier. The vacuum is of the order of 10^{-6} mm Hg, and spectrum can be scanned in 100 ms.

For respiratory studies, the sample is sucked down a heated stainless steel capillary tube one metre long at a rate of about 25 ml per minute by means of an auxillary vacuum pump. A typical mass range would be 2–100 mass units.

The cost of a quadrupole mass spectrometer can be justified for respiratory physiology investigations where it may be used with an integrating pneumotachograph and circuitry to calculate inspired and expired tidal volumes making allowances for the presence of water vapour. However, for respiratory gas monitoring in an intensive care unit, small bore metal tubing and a rotary input selector device to the mass spectrometer may be used to share the mass spectrometer between perhaps up to 12 patients. Blumenfeld *et al.* (1973) have described such an arrangement with time sharing of the mass spectrometer and other instrumentation under the control of a small computer. Automatic measurements were made of the heart rate, body temperature, arterial blood pressures, mean central venous pressure, respiratory oxygen and carbon dioxide waveforms and the respiratory rate.

Membrane-tipped cannulae have also been developed to enable mass spectrometers to measure continuously arterial blood-gas tensions. Brantigan *et al.* (1976) describe a Teflon® catheter for continuous blood-gas analysis *in vivo* which can be inserted into an artery via percutaneous puncture with an 18 gauge needle. The response time (63%) is 40 s for oxygen and 65 s for carbon dioxide. Other forms of cannulae have been reviewed by Seylaz *et al.* (1974).

The Measurement of Body Temperature

Some of the relevant problems are discussed in Chapter 3. It is important to realize that body temperature is a relative concept whose investigation by thermometry provides only a relative measurement. There is no one true deep body temperature since there exist internal thermal gradients between one region and another.

Thermoelectric measuring devices are convenient but are not necessarily more accurate than other, simpler equipment. However, such devices do provide for an indication of temperature remote from the site of measurement.

Body surface temperature

There are a number of complexities involved in the measurement of skin temperature, common to the difficulties of measuring any surface temperature. Thermocouples or thermistors are usually used and may be attached to the skin. A thermocouple is a circuit made from two wires of dissimilar metals in which the junctions are kept at different temperatures. The e.m.f. generated is proportional to the temperature difference between the two junctions. One of these, the reference junction, is kept at a constant temperature, conveniently in ice water at 0 °C. Using copper–constantan junctions, an e.m.f. of about 40 μV is generated per 1 °C difference between the hot and cold junctions.

Amongst the practical problems are the choice of sites to represent overall skin temperature, the control of the pressure of the sensor on the skin ensuring adequate contact without deformation, interference with local heat exchange and the effect of evaporative cooling during sweating.

Another technique is the measurement of infrared radiation from the skin, which is proportional to skin surface temperature. Infrared thermography is a non-invasive, non-contact technique for measuring the temperatures of surfaces. In recent years it has been used in medicine and physiology for recording the human skin and clothing temperatures.

The technique relies on the principle that the skin (or clothing) radiates electromagnetic energy to the surroundings, at a rate which depends on the surface temperature. The distribution of wavelengths emitted in this way shows a maximum in the infrared region of the electromagnetic spectrum. A special camera scans the surface and detects this radiation which is subsequently converted into a temperature distribution. This distribution is presented on a television screen as a thermogram where shades of grey represent particular temperatures. A refinement to this technique is to have a colour television display where particular colours are assigned to specific temperatures. The colours are calibrated by the inclusion of a standard black body reference source at a known temperature, or by direct measurement of a particular area of known colour with a thermocouple or thermistor probe.

Infrared thermography is currently used in a number of clinical conditions of which the following are examples: breast cancer screening, deep vein thrombosis, rheumatic and arthritic joints, defining the extent of burns and various vascular diseases. Many of these topics are experimental and thermography is not yet a recognized diagnostic tool. It does, however, have more obvious and immediate applications to physiology and has been used successfully to visualize the overall temperature distributions of "normal" subjects in different environmental conditions and during exercise. Its real value is that a subject may be quite free of any attachment and may exercise or perform a task while still being surveyed by the infrared camera. The thermogram presents an overall temperature picture which it is impossible to reproduce from multiple point measurements (Clark *et al.*, 1977). The sensitivity can be altered so the temperature differences represented by the various colours may be 0.2 °C to 1.0 °C.

Deep body temperature

Clinical thermometers

Mercury-in-glass clinical maximum thermometers are traditionally used for measuring deep body temperatures and have been subject to the quality specifications of the British Standards Institute since 1936. The most recent revision of this standard (BS 691, 1979) provides specifications for clinical thermometers for general use and also for subnormal range thermometers and ovulation thermometers with scale ranges of 25–40 °C and 35–38 °C respectively.

A single-use clinical thermometer which is readily disposable and bacteriologically and mechanically safer than the conventional mercury-in-glass instrument has been introduced. This thermometer incorporates an innovative approach to temperature measurement based on the differentiation of discrete temperature by the change of different mixtures of pure chemicals from the solid to liquid state at specific temperatures.

Urine temperature

Measurement of temperature in the stream of voided urine was described as long ago as 1775 by Charles Blagden. The technique has been used to record circadian body temperature rhythm and extensively in population studies on hypothermia (Fox *et al.*, 1971). The method employed in population surveys as described by Fox and his colleagues requires a screw-capped plastic bottle containing a funnel which fits into the neck of the bottle and holds the thermometer bulb in the urine flow. Rectal and urine temperatures are found to be in close agreement providing there is a urine volume greater than 100 ml.

Ear temperature

Cranial measurements of internal body temperature may be expected to be more suited to the study of human mechanisms of temperature regulation than mouth or rectal temperature observations. Such measurements were thought to provide a more valid indication of temperature changes at the central hypothalamic thermal receptors. Early devices using a thermocouple loop in contact with the tympanic membrane (Benzinger and Taylor, 1963) have largely been replaced by the thermistor bead sensor sited as deep as possible in the external auditory meatus without actually touching the ear drum (Bradbury *et al.*, 1964). The thermistor device, fitted into an ear mould (rather like a hearing aid) with the ear and head well insulated to minimize heat conduction along the body of the instrument, provides reliable aural measurement of deep body temperature under many conditions of work and external temperature. In cold ambient conditions the aural temperature reading, even with good ear lagging, is seriously affected by local cooling. Under these conditions it is necessary to measure aural canal temperature with servo-controlled heating to the outer ear so that the outer ear is heated to the same temperature as the aural canal (Keatinge and Sloan, 1975). With servo-controlled heating, aural temperature closely approximates to oesophageal temperature even in cold surroundings and follows oesophageal temperature changes with less lag than rectal temperature. In certain circumstances, for example during hyperthermia in the marathon runner when deep trunk temperature may rise to over 41 °C with no clinical signs of heat stress, the aural temperature gives a better correlation with thermoregulatory responses than oesophageal temperature (Cabanac and Caputa, 1979). This appears to be because the temperature of venous blood from the face affects the temperature of the external auditory canal in the same way that it affects central blood temperature in the cavernous sinuses.

Temperature telemetry

A number of different transducers and circuit techniques have been used for telemetering body temperature (e.g. Jacobson and Mackay, 1957; Wolff, 1961; Nagumo *et al.*, 1962; Babskii *et al.*, 1963). Thermistors have been used in some instruments, though their use does present design difficulties due to non-linear temperature coefficients.

The temperature-sensitive radio pill provides an accurate and reliable means of monitoring intestinal temperature with a minimum of discomfort to the subject. Because of their small size, radio pills can be swallowed and passed freely through the alimentary canal. The instruments have been used to study temperature variations in a large group of subjects during heat exposure (Edholm *et al.*, 1973), in temperature measurements during sleep and in hyperbaric environments during deep-sea diving activities. A typical radio pill embodies a radio transmitter operating at a frequency of 350 kHz with a frequency change of about 10 kHz per °C. Location of the radio pill in the alimentary tract during recording does present some problems since it cannot be followed continuously by X-ray radiographs. In the static subject, tracking can be accomplished by a rotating omnidirectional antenna mounted on a carriage (Jacobson and Lindberg, 1963). The carriage automatically seeks a position where the field strength is constant during the antenna rotation. Verification of the frequency stability at constant temperature indicates that the immersion of a thermistor radio pill in fluid may result in a drift corresponding to a temperature error of about 0.4 °C on the first day decreasing to 0.1 °C on subsequent days.

Temperature gradient method

Fox *et al.* (1973) reported a temperature gradient method for deep body temperature on the surface of the body. A temperature gradient will normally exist between the deep tissue and the skin temperatures. Two thermistors separated by a layer of thermal insulation (Fig. 19) sense this gradient and control a flexible, low voltage, heater situated in the probe. Using a comparator, the heater is switched on or off depending on the direction of the gradient until a thermal equilibrium is attained. The net heat flow across the insulating layer is then zero, so that the temperature gradient across the layer must be zero and the skin temperature at that point must be equal to the deep body temperature. This is indicated by the skin thermistor. Commencing with the pad at room temperature, the system takes about 15–20 min to indicate the deep body reading.

Thermocomple and thermistor probes are also used for the measurement of rectal temperature. Subcutaneous and muscle temperatures can be measured with thermocouples mounted in needles and inserted into the limbs (Claremont and Brooks, 1974).

Heat flow discs

Hatfield (1950) designed a heat flow device consisting of a semiconductor sandwiched between two thin copper discs. Measurement of heat flow was

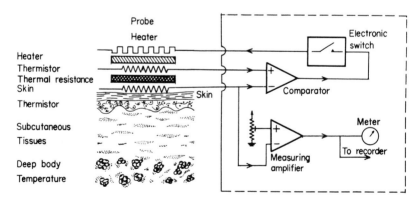

FIG. 19 Temperature gradient method for the measurement of deep body temperature.

obtained from the temperature difference between these two discs. These discs have been widely used and were convenient and effective. Unfortunately, they are no longer available owing to the difficulty of obtaining the suitable semiconductor material.

An instrument for measuring local heat transfer (Clark *et al.*, 1972) consists of a surface plate calorimeter with an electric heater sandwiched between a thin metal plate and an insulating backing, which can be attached to the skin with double-sided sticky tape. Skin and plate temperatures are measured with thermocouples. The electric power supplied to the plate is measured and when plate and skin temperatures are the same this power supply will be equal to the heat loss by convection and radiation.

Miscellaneous

The *Schlieren* technique

Schlieren is a German word meaning streak, or flaw. The technique of *Schlieren* visualization had its origins at the end of the last century in the German glass industry. Since then it has been used to identify shock waves and other aerodynamic features over models of aircraft wings, fuselages, and missiles in high-speed wind tunnels.

More recently, it has been used to study the microclimate around human subjects. In this application, the technique relies on the slight differences in refractive index that occur due to air temperature variations. The *Schlieren* system, in its undisturbed state, basically consists of a parallel bundle of light rays focussed at a particular point on a cut-off (see Fig. 20). In this state, the light rays, when viewed from behind the cut-off, appear as a particular colour (the background colour). Any disturbance to the path of the parallel light rays, for instance by the introduction of warmed air streams such as occur in the human microenvironment, cause some of these parallel light rays to be bent away from parallel. This bending is reflected from mirrors and lenses to result in a lateral shift of the focussed beam at the cut-off. This lateral shift of disturbed light rays then gives the appearance of a different colour to the disturbed parts. In this way, moving air within the microclimate can be viewed as one colour against the original background colour. The system is extremely sensitive and may be used to visualize air movements down to 1–2 cm s^{-1}. Differences in the human microclimate between standing, sitting and lying subjects can be shown.

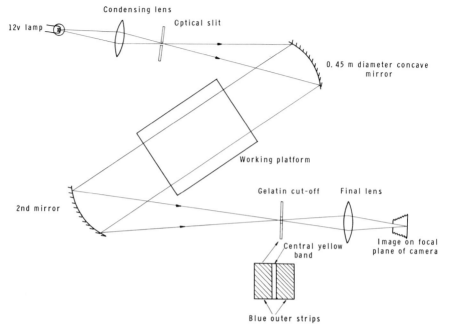

FIG. 20   *Schlieren* optical system.

The Medilog

The Medilog, a miniature portable casette recorder, was developed to provide continuous recording of physiological signals. A four-channel instrument is now available and has been used for recording intra-arterial blood pressure, ECG, respiration, EEG and EOG (Honour *et al.*, 1972). The tape recordings can be analysed using appropriately designed replay systems together with a minicomputer. Programs have been developed for processing ECG records as well as blood pressure data.

The heart rate SAMI

Wolff designed a number of socially acceptable monitoring instruments (SAMI) to meet the needs of investigators taking part in the Human Adaptability Section of the International Biological Programme. The heart rate SAMI has been the most widely used (Baker, Humphrey and Wolff, 1967). An ECG signal, after amplification and conversion into a constant charge pulse, is applied to an electrochemical integrator or E-cell. At the end of the period of recording, the E-cell is removed and reset in a replay machine which provides a digital read-out of the quantity of charge stored in the E-cell; from this the number of heart beats can be calculated. The use of the heart rate SAMI in the field has been described by Edholm (1976).

Muscle strength

The measurement of the maximal voluntary contractile force produced by a muscle (usually known as "muscle strength") can conveniently be made using strain-gauges fitted to suitably robust equipment. Important points to be observed are that the test is carried out under isometric conditions, that the position of the limbs is constant from one subject to another (e.g. the angle of knee flexion during testing of the quadriceps), that sufficient rest periods are allowed between successive maximal efforts and that the test equipment does not produce pain or discomfort sufficient to limit the effort of the subject.

Figure 21 illustrates equipment which may be used to determine the strength of the quadriceps muscle. The subject's pelvis is stabilized by a seat belt, thus giving him a firm position from which to work and at the same time preventing use of the hip flexors. The subject pulls on a strain-gauge bar by means of a strap placed around the ankle. Figure 22 illustrates equipment for measuring the strength of the hip flexors. The subject lies on

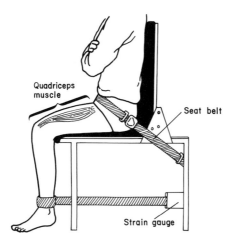

FIG. 21 Equipment for measuring the strength of the quadriceps muscle (From Edwards *et al.*, 1977).

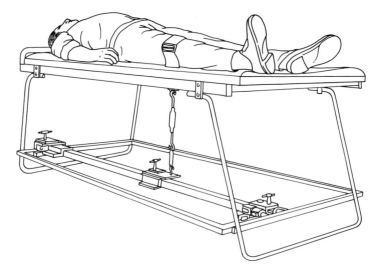

FIG. 22   Apparatus for mesurement of the strength of flexor muscles of the hip joint.

a modified bed. A belt is placed around his thigh 2 cm above the patella and passes through a slit in the bed to a strain-gauge bar attached to a sliding trolley. The strap is suitably tightened and the subject then attempts to do a straight leg raise.

Perhaps the most frequently used test of muscle strength is the measurement of the grip strength of the hand. Suitable dynamometers are small and robust and have obvious attractions for both teaching and research. However, considerable care must be taken to ensure that the equipment is constructed so that attempts to produce maximal contractions are not limited by pain.

References

Adams, A.P., Vickers, M.D.A., Munroe, J.P. and Parker, C.W. (1967). *Br. J. Anaesth.* **39**, 174–183.

Adamsons, K. Jr., Daniel, S.S., Gandy, J. and James, L.S. (1964). *J. Appl. Physiol* **19**, 897–900.

Arno, I.C. and Slate, W.G. (1971). *J. Abert Einstein Medical Centre* **19**, 5–10.

Astrup, P. and Schroder, S. (1956). *Scand. J. Lab. Clin. Invest.* **8**, 30.

Babskii, E.B., Sorin, D.M., Belousev, A.S., Shukov, U.S. and Dimonis, V.I. (1963). *Doklady Biol. Sci.* **149**, 490–492.

Baker, J.A., Humphrey, S.J.E. and Wolff, H.S. (1967). *J. Physiol.* **188**, 4–5p.

Baker L.E., Geddes, L.A. and Chaput, C.J. (1966). *J. Appl. Physiol.* **21**, 1491–1499.

Baker, L.E. and Hill, D.W. (1969). *Brit. J. Anaesth.* **41**, 2–17.

Barcroft, H. and Swan, H.J.C. (1953). "Sympathetic Control of Human Blood Vessels", Edward Arnold, London.

Barker, A. and Brown, B.H. (1973). *Med. Bil. Eng.* **11**, 352–353.

Barnes, R.W., Hokanson, D.E., Summer, D.S. and Strandness, D.E. Jr. (1975). *Proc. Soc. Photo-Optical Instr. Eng.* **72**, 65–72.

Benzinger, T.H. and Taylor, G.W. (1963). *In* "Temperature—Its Measurement and Control in Science and Industry", (Ed. J.D. Hardy), Vol. 3, 111–120. Reinhold, New York.

Blumenfeld, W., Wolff, S. and McCluggage, C. (1973). *Computers Biomed. Res.* **6**, 139–149.

Bordley, J. III, Conner, C.A.R., Hamilton, W.F., Kerr, W.J. and Wiggers, C.J. (1951). *Circulation* **4**, 503–509.

Bradbury, P.A., Fox, R.H., Goldsmith, R. and Hampton, I.F.G. (1964). *J. Physiol.* **171**, 384–396.

Branthwaite and Bradley, R.D. (1968). *J. Appl. Physiol.* **24**, 434–438.

Brantigan, J.W., Dunn, K.L. and Albo, D. (1976). *J. Appl. Physiol.***24**, 434–438.

Brubakk, A.O., Angelson, B.A.J., and Hatle, L. (1977). *Cardiovascular Res.* **11**, 461–469.

Cabanac, M. and Caputa, M. (1979). *J. Physiol.* **289**, 163–174.

Claremont, A.D. and Brooks, G.A. (1974). *Eur. J. Appl. Physiol.* **32**, 183–186.

Clark, L.C. (1956). *Trans. Am. Soc. Artif. Organs* **2**, 41.

Clark, C. (1968). *Med. Biol. Eng.* **6**, 133.

Clark, R.P., Cox, R.N., and Toy, N. (1972). *J. Physiol.***233**, 10–12

Clark, R.P., Mullan, B.J. and Pugh, L.G.C.E. (1977). *J. Physiol.* **267**, 53–62

Coghlan, B.A. and Taylor, M.G. (1976). *Ultrasound Med. Biol.* **2**, 181–188.

Coon, R.L., Lai, N.C.J. and Kampine, J.P. (1976). *J. Appl. Physiol.* **40**, 625–629.

Crul, F.J. and Payne, J.P. (1970). (Eds.) "Patient Monitoring", Exerpta Medica, Amsterdam.

Crul, F.J., Van Herwaarden, C.L. and Smulders, F.P. (1967). *Fed Proc.* **26**, 377.

Davies, D.D. (1970). *Br. J. Anaesth.* **42**, 19.

De Dobbeleer, F.D.P. (1965). *World Med. Electronics and Instrumentation, Lond.* **3**, 122.

Edholm, O.G. (1976). *In* "Man in Urban Environments," (Eds G.A. Harrison and J.B. Gibson), Oxford University Press, Oxford.

Edholm, O.G., Fox R.H. and Wolff, H.S. (1973). *Arch. Sci. Physiol.* **27**, 339–355.

Edwards, R.H.T., Young, A., Hosking, G.P. and Jones, D.A. (1977). *Clin. Sci. Molec. Med.* **52**, 283–290.

Eisenman, G., Mattack, G., Bates, R. and Friedman, S.M. (1966). "The Glass Electrode", Wiley, London.

Eley, C., Goldsmith, R., Layman, D., Tan, G.L.E., and Walker, E. (1978). *Ergonomics* **21**, 253–264.

Endresen, J and Hill, D.W. (1976). *Eur. J. Intensive Care Med.* **2**, 3–6.

Fernandez, H., Barnes, R. and McKinney, W. (1974). *Aerospace Med.* **45**, 893–894.

Fick, A. (1870). Uber die messung des blutquantums in der herzventrikeln (Translated by H.E. Hoff and H.J. Scott 1948). *New Engl. J. Med.* **239**, 122 Sitzungsh der Phys-Med., Ges. Zu Wutzburg, 36.

Fletcher G. and Belville, J.W. (1966). *J. Appl. Physiol.* **21**, 1321.

Forlini F.J. Jr. (1974). *J. Appl. Physiol.* **37**, 584–586.

Fox, R.H., Woodward, P.M., Fry, A.J., Collins, J.C. and MacDonald, I.C. (1971). *Lancet* 1, 424–427.

Fox, R.H., Solman. A.J., Isaacs, R., Fry, A.J. and MacDonald, I.C. (1973). *Clin. Sci.* 44, 81–86.

Fowler, K.T. and Hugh-Jones, P. (1957). *Br. Med. J.* 1, 1205.

Fry, D.L., Hyatt, R.E., McCall, C.B. and Mallos, A.S. (1957). *J. Appl. Physiol.* 10, 210.

Geddes, L.A. (1970) Year Book Publishers, Chicago.

Geddes, L.A. and Moore, A.G. (1968). *Med. Electron. Biol. Engng.* 6, 603.

Grandjean, T and Perret, C. (1970). *In* "Patient Monitoring" (Eds F.J. Crul and J.P. Payne), 90–95. Exerpta Medica, Amsterdam.

Gundersen J. (1970). *Scand. J. Lab. Clin.* Invest. 25, 299–301.

Gundersen, J. and Ahlgren, I. (1973). *Acta. Anaesth. Scand.* 17, 203–307.

Gundersen, J. and Dahlin, K. (1971). *Med. Bil. Eng.* 9, 541–547.

Hansen, A.T. (1949). "Pressure Measurement in the Human Organism", Technisk Forlag, Copenhagen.

Hansen, P.L., Cross, G. and Light, L.H. (1976). *In* "Clinical Blood Flow Measurement", (Ed. J. Woodcock), 28. Sector Publishing, London.

Hatfield, H.S. (1950). *J. Physiol.* 111, 10–11

Henry, W.L., Wilner, L.B. and Harrison, D.C. (1967). *J. Appl. Physiol.* 23, 1007.

Hill, D.W., Thompson, D., Valentinuzzi, M.E. and Pate, T.D. (1973). *Med. Biol. Eng.* 11, 43–54.

Hill, D.W. and Dolan, A.M. (1976). "Intensive Care Instrumentation", Academic Press, London and New York.

Hill, D.W. and Merrifield, A.J. (1976). *Acta. Anaesth. Scand.* 20, 313–320.

Hill, D.W., Mohapatra, S.N. Welham, K.C. and Stevenson, Margaret (1976). *Eur. J. Intensive Care Med.* 2, 119–124.

Hobbes, A.F.T. (1967). *Br. J. Anaesth.* 39, 899–907.

Hochberg, H.M. and Salomon, H. (1971). *Curr. Ther. Res.* 13, 129–138.

Hochberg, H.M. and Saetzman, M.B. (1971). *Curr. Ther. Tes.* 13, 482–488.

Honour, A.J., Littler, W.A., Sleight, P. and Stott, F.D. (1972). *J. Physiol.* 225, 8–9.

Humphrey, S.J.E. and Wolff, H.S. (1977). *J. Physiol.* 267, 12P.

Jacobson, B. and Lindberg, B. (1963). *Electronics* (Mar. 22), 58–60.

Jacobson, B. and MacKay, R.S. (1957). *Nature (London)* 179, 1239–1240.

Janssen, F.J. (1967). The rheographic determination of systolic and diastolic blood pressure Digest 7th. Int. Conf. Med. Biol. Engng., Stockholm, p. 221.

Kahn, M.R., Tandon, S., Guha, S.K., and Roy, S.B. (1977). *Med. Biol. Eng. Comput.* 15, 627–633.

Kazamias, T.M., Gander, M.P., Franklin, D.L. and Ross Jr. J. (1971). *J. Appl. Physiol.* 30, 585.

Keatinge, W.R. and Sloane, R.E.G. (1975). *J. Appl. Physiol.* 38, 919–921.

Kenner, T.H. and Ganer, O.H. (1962). *Pflugers Arch. Ges. Physiol.* 275, 23.

Kopczynski, H.D. (1974). *Aerospace Med.* 45, 1307–1309.

Kubicek, W.G., Karnegis, J.N., Patterson, R.P. Witsoe, D.A. and Marrson, R.G. (1966). *Aerospace Med.* 37, 1208.

Lagerwerff, J.M. and Luce, R.S. (1970). *Aerospace Med.* 41, 1157–1161.

Lilly, J.C., Legallais, V. and Cherry, R. (1947). *J. Appl. Physiol.* 18, 513.

Lowe, R.D. (1968). *In* "Blood Flow through Organs and Tissues", (Eds, W.H. Bain, A.M. Harper and W.A. MacKay), 79–89. Livingstone, Edinburgh.

Lywood, D.W. (1967). *In* "Manual of Psycho-physiological Methods", (Eds P.H. Vanables and Irene Martin), North Holland, Amsterdam.

Massie, H.L., Ziedonis, J.G. and Black, I. (1973). *Med. Instrument.* **7**, 240–244.
McCarty, K. and Woodcock, J.P. (1976). *In* "Clinical Blood Flow Measurement", (Ed. J. Woodcock), 20. Sector Publishing, p. 20. London.
McLaughlin, G.W., Kirby, R.R., Kemmerer, W.T. and Delemos, R.A. (1971). *J. Pediat.* **79**, 300–303.
Nagumo, J., Uchiyama, A., Kimoto, S., Watanuki, T., Hori, M., Ouchi, A., Kumono, M., and Watanabe, H. (1962). *I.R.E. Trans. Biomed. Electron.* **9**, 195–199.
Okada, Y., and Inouye, A. (1976). *Biophys. Struct. Mech.* **2**, 21–30.
Pallett, J. E. and Scopes, J.W. (1965). *Med. Electron. Biol Eng.* **4**, 1–14.
Plass, K.G. (1964). *I.R.E. Trans. Biomed. Electron.* **11**, 154.
Pomerantz, M., Baumgartner, R., Laurisdon, J. and Eiseman, B. (1969). *Surgery* **66**, 260–268.
Pomerantz, M. Delgado, F. and Eiseman, B. (1970). *Ann. Surg.* **171**, 686–691.
Poppers, P.J. (1970). *In* "Patient Monitoring", (Eds J.F. Crul and J.P. Payne). Exerpta Medica, Amsterdam.
Pressey D.C. (1953). *J. Sci. Instrum.* **30**, 20.
Prime, F.J. and Gray, T.C. (1952). *Curr. Res. Anesth.* **31**, 347.
Reynolds, J.A. (1968). *J. Sci. Instrum.* **1**, 433–436.
Roboz, J. (1968). "Introduction to Mass Spectrometry Instrumentation and Techniques". Interscience, New York.
Rosenthal, J.B. (1948). *J. Biol. Chem.* **173**, 25–30.
Roy, S.N., Balasubraminian, V., Khan, M.R., Kanshik, V.S., Manchandra, S.C.. and Guha, S.K. (1974). *Br. Med. J.* **3**, 77.
Sandman, A.M. and Hill, D.W. (1974). *Med. Biol. Eng.* **12**, 360–363.
Schneider, R.A., Kimmell, G.O. and Van Meter, A.P. Jr. (1974). *J. Appl. Physiol.* **37**, 776–779.
Severinghaus, J.W. and Bradley, A.F. (1958). *J. Appl. Physiol.* **13**, 515.
Seylaz, J., Pinnard, E. and Correze, J.L. (1974). *J. Appl. Physiol.* **37**, 937–941.
Sheppard, L.C., Johnson, T.S. and Kirklin, J.W. (1971). *J. Ass. Adv. Med. Instrum.* **5**, 297–301.
Soutter, W.P., Aitchison, T.C., Thorburn, J. and Sharp, F. (1976). *Br. J. Anaesth.* **48**, 1211–1218.
Stow, R.W., Baer, R.F. and Randell, B.F. (1957). *Arch. Phys. Med. Rehabil.* **38**, 646.
Veall, N. and Vetter, H. (1958). "Radioisotope Technique in Clinical Research and Diagnosis", Butterworth, London.
Visscher, M.B. and Jonson, J.A. (1953). *J. Appl. Physiol.* **5**, 635.
Von Recklinghausen, H. (1906). *Arch. Exper. Path Pharmakil* **5**, 325–504.
Welham, K.C., Mohapatra, S.N., Hill, D.W. and Stevenson, L. (1978). *Intens. Care Med.* **4**, 43–50.
Whitney, R.J. (1953). *J. Physiol. (Lond.)* **121**, 1.
Williams, B.T., Sancho-Forres, S., Clark, D.B., Abrams, L.D. and Schenk, W.C. Jr. (1972). *J. Thoracic Cardiovasc. Surg.* **63**, 917–921.
Wolff, H.S. (1961). *New Scientist* **12**, 419–421.
Wright, B.M. (1955). *J. Physiol. (Lond.)* **127**, 25p.
Wyatt, D.G. (1968v). *Phys. med. Biol.* **13**, 529.
Wyatt, D.G. (1971). *In* "IEE Medical Electronics Monographs 1–6," (Ed. B.W. Watson), 181. London Institution of Electrical Engineers.
Zahed, B., Sadove, M.S., Hatano, S. and Hsin Hsiung, Wu. (1971). *Anaesth. Analg. Curr. Res.* **50**, 699–704.

CHAPTER 12

The Nature and Control of Physiological Disturbances in Critical Illness

J. Tinker

Critical illness has many causes, all of which produce an acute, rapidly changing, and widespread disturbance of normal physiological control and function.

The workings of the various physiological systems are closely inter-related and whilst a disturbance may initially be confined to one system, the function of others is inevitably affected, and a common pattern of "multi-system failure" emerges. For instance, a severe myocardial infarction produces circulatory failure because of damage to the left ventricle, respiratory failure due to pulmonary oedema, and renal failure because blood flow to the kidneys is reduced. Alternatively, the respiratory failure of severe pneumonia can cause circulatory failure which may, in turn, lead to renal failure. Numerous other similar examples could be cited but these two serve to illustrate the functional integrity of the different systems and the chain of events that can be initiated when one, or other, of them fails.

Judged on this basis the characteristics of critical illness are, therefore, severity, multi-system involvement, acute onset and rapid rate of change.

Its management, by inference, requires frequent or continuous measurement of many physiological variables and support of certain vital functions (Fig. 1).

This chapter describes those features of acute illness which are crucial to understanding the changes in the functional state of the circulatory, respiratory and renal systems, thereby providing the basis for rational management and treatment.

Acute Circulatory Failure (Shock)

Shock is a state of low cardiac output that is insufficient to meet the oxygen demands of the tissues; it is fundamentally a disorder of blood flow.

A low cardiac output may result from damage to heart muscle with or without disturbances in heart rhythm, or from a reduction in the circulating blood volume (hypovolaemia). Accordingly, shock is classified into two categories, cardiogenic and hypovolaemic (Fig. 2). This is a practically

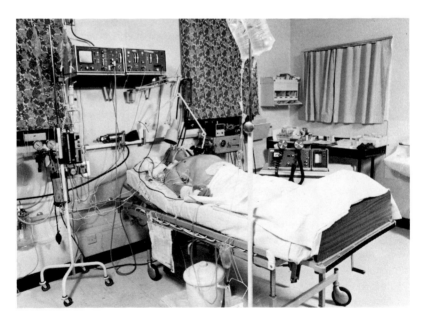

FIG.1   Intensive care medicine.

useful division but it is important to recognize that in any individual case, both factors may be operative. For example, in the form of shock commonly associated with severe bacterial infections there is a combination of both hypovolaemia and myocardial depression.

The pathophysiology of shock (Fig. 3)

Because of the reduction in cardiac output there is a reflex increase of sympathoadrenal activity with an outpouring of adrenaline and noradrenaline from the sympathetic post-ganglionic fibres and the adrenal medullae. This response increases heart rate and myocardial contractility and produces arteriolar constriction, particularly in the cutaneous and visceral

A CLASSIFICATION OF SHOCK

    I Cardiogenic: Acute myocardial infarction
                     After cardiac surgery
                     Massive pulmonary embolism
                     Pericardial tamponade

    II Hypovolaemic: Haemorrhage
                     Loss of plasma
                     Loss of extracellular fluid

FIG. 2   A classification of shock.

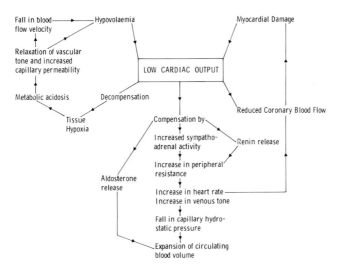

FIG. 3   The pathophysiology of shock.

circulations. The tachycardia and increased contractility improve cardiac output, and the arteriolar constriction increases peripheral resistance and restores arterial pressure towards normal. The veins are also constricted by the sympathetic influences with a decrease in their capacity which improves venous return to the heart and so cardiac output. At this stage the vascular constriction is more marked in the arterioles than the venules and, as a consequence, the capillary hydrostatic pressure falls and fluid moves by osmosis from the interstitial spaces into the circulation, increasing blood volume.

The presence of circulatory "auto-regulation" also protects against the effects of the low cardiac output. Certain tissues, for instance skin and muscle, have a passive, blood pressure-flow relationship, so that as their perfusion pressure declines they suffer a proportionate reduction in blood flow. The renal, coronary and cerebral circulations, however, are able to maintain their flows until the perfusion pressure falls to around 50 mm Hg, below which value they fall precipitously. This ability to preserve flow and compensate for changes in arterial perfusion pressure is termed "auto-regulation", and it is an intrinsic property of vascular smooth muscle independent of its nerve supply. It can be viewed as a protective mechanism whereby the body attempts to sustain the circulation to "vital" organs.

The effects of these compensatory mechanisms are boosted by a release of aldosterone from the adrenal cortex, which increases sodium reabsorption by the kidneys, and a release of renin from the juxtaglomerular apparatus, which produces an increase in the circulating levels of angiotensin, a powerful vasoconstrictor.

In spite of all the circulatory adjustments, tissue blood flow eventually decreases and the reduced oxygen supply disturbs cellular metabolism. Pyruvic acid is unable to pursue its normal pathway into Kreb's citric acid cycle and is converted into lactic acid, with a resultant decrease in ATP production. The increased quantities of lactic acid appear in the circulation, giving rise to a metabolic acidosis; this alters vascular reactivity and, despite intense sympathoadrenal activity, there is relaxation of vascular smooth muscle, more pronounced in the arterioles than in the small veins. Capillary hydrostatic pressure therefore increases, and there is transudation of fluid from the plasma into the interstitial spaces. This process is aggravated by the action of certain vasoactive substances, including histamine and plasma kinins, which increase capillary permeability.

The continuing loss of plasma from the circulation increases the viscosity of the blood and its resistance to flow. There is a fall in the velocity of flow in the microcirculation and red cells and platelets aggregate, obstructing flow to an even greater degree.

The combined effects of these later changes progressively reduce the venous return to the heart and the cardiac output continues to fall. A vicious circle is established, in which the shock process once established tends to be self-prepetuating.

Influence of changes in heart rhythm

Disturbances of heart rhythm (dysrhythmias) are frequently associated with shock or develop as a result of its haemodynamic and metabolic sequalae. Their occurrence may impair heart function and cause a further fall in cardiac output. These prejudicial effects are due to a combination of changes in heart rate, reduction in coronary blood flow, loss of atrial transport function and asynchronous ventricular contraction. The severity of the haemodynamic disturbance therefore varies with different dys-rhythmias and, especially with ventricular dysrhythmias, it can be catastrophic.

In normal individuals variations in heart rate from 40 to 150 beats per minute produce little change in cardiac output. However, in heart disease the margins for compensation are limited and high or low heart rates produce significant falls in cardiac output. An increased heart rate is a major determinant of myocardial oxygen demand and may also reduce coronary blood flow by shortening the duration of diastole. In diseased hearts the development of a tachycardia can therefore produce a situation in which oxygen demand exceeds supply and there is further myocardial depression.

The atria normally act as booster pumps to increase the volume of ventricular filling, and augment stroke volume. Loss of this function is not important in normal hearts but in abnormal situations it can impair ventricular filling.

Normal ventricular contraction is synchronized by the rapid and uniform spread of depolarization throughout the ventricular myocardium. In the presence of intraventricular conduction disturbances, however, contraction becomes asynchronous and less effective.

The clinical picture of shock

This reflects the low cardiac output and increased sympathoadrenal activity. The skin is pale, cold and sweating. There is a tachycardia, and dysrhythmias may be present; the blood pressure is low or normal. In hypovolaemic shock the venous pressure is low whilst in cardiogenic shock it is increased. The presence of oliguria indicates the reduced renal blood flow.

Measurements in patients with shock

The rationale for making particular physiological measurements is based on the nature and pathogenesis of the circulatory disturbance. Physiological supplement clinical observations, and provide an index of the severity of the disorder, indicate its rate of change, and enable treatment to be optimized.

The electrocardiogram

This continuously records the heart rate and rhythm and facilitates the rapid recognition and treatment of the various cardiac dysrhythmias.

The arterial pressure

The arterial pressure is measured routinely but has to be interpreted with care. It is the product of arterial flow and resistance, and as a result its value may only be marginally reduced when cardiac output is significantly decreased.

Although sphygmomanometer cuff measurements accurately reflect intra-arterial pressure in patients with relatively normal cardiovascular function, they are less reliable in patients with a low cardiac output and excessive peripheral vasoconstriction. In such cases the Korotkoff sounds may be difficult to hear and the radial or brachial arteries impalpable. However, direct intra-arterial measurements from a catheter in the radial or brachial artery, provide accurate recordings in such situations and also enable the arterial pressure waveform to be displayed. This is normally sharp with a rapid upstroke, a clearly defined dicrotic notch and a definite end-diastolic point. The rate of rise of the upstroke is slow if there is poor myocardial contractility, obstruction to left ventricular outflow or partial occlusion of the catheter lumen.

An arterial catheter also provides a convenient site for blood sampling in patients who require frequent blood gas estimations. The discomfort and complications of repeated arterial punctures are avoided.

The central venous pressure

The central venous pressure (CVP) is commonly measured and, provided its limitations are recognized, can be used effectively for the management of shock.

The CVP is, by definition, the pressure in the right atrium. It is a complex variable affected by the venous return to the heart, the functional state of the right ventricle, the level of venous tone, and the pressures outside the heart, both intrapericardial and intrathoracic. If the tricuspid valve is normal it equates with the right ventricular end-diastolic pressure.

To measure CVP a catheter is introduced into the right atrium via either a subclavian, internal jugular or antecubital vein. The femoral veins can be used but the risks of infection are increased and this route is not recommended. Correct placement of the catheter must be checked on a chest X-ray (Fig. 4) and ideally its tip should be located in the right atrium a distance away from the tricuspid valve. However, since the pressure in the large veins inside the chest is invariably within 1 mm Hg of the pressure in the right atrium, placement in one of these is acceptable.

From the catheter the pressure is measured continuously, using a transducer, or intermittently with a saline manometer. In all instances accurate and careful observation of a zero-reference level is mandatory; small changes in CVP (1 mm Hg) can reflect significant changes in cardiac function. Usually the mid-axillary line, in the fourth intercostal space, or the sternal angle are used for reference and, of the two, the former is preferred because it is at the anatomical level of the right atrium. It has been shown that the right atrial pressure remains almost constant in an animal rotated into any position in space, provided that the hydrostatic reference point is in the geometrical centre of the tricuspid valve.

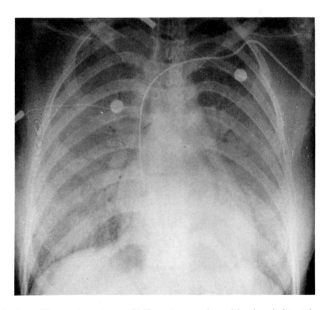

FIG. 4   A chest X-ray showing a CVP catheter placed in the right atrium.

Normal values   When measured from the mid-axillary line the normal range of CVP is 3–7 mm Hg (5–10 cm H_2O) and from the sternal angle it is 0–3 mm Hg (0–5 cm H_2O).

Clinical application of CVP measurement   Three distinct and separate features of circulatory insufficiency are related to CVP: hypovolaemia, right ventricular failure, and left ventricular failure. These can each occur in isolation but they are often combined, and hence inter-related, in a critically ill patient.

Hypovolaemia reduces CVP and the extent of the reduction gives some indication of the degree of blood volume depletion. Serial measurements are helpful and provide a guide to the rate and volume of fluid replacement.

Right ventricular failure increases CVP, but because right ventricular function is influenced by a number of factors the specific interpretation of such an increase can be misleading. In cases of cardiogenic shock the state of left ventricular function is the feature of particular interest and whilst in normal individuals there is a defined relationship between left and right atrial pressures, this no longer holds in many forms of cardiorespiratory failure. In these conditions there is no longer a predictable or consistent correlation between the functions of the two ventricles, and the use of CVP as a measurement of left ventricular function cannot be justified. A more direct measurement of the left atrial pressure, from the pulmonary capillary wedge pressure, is essential.

Many critically ill patients are mechanically ventilated and it is important to remember that the higher intrathoracic pressure present in these cases increases the CVP. A rise in intrapericardial pressure, as in pericardial tamponade, likewise increases the CVP often to a high level.

The pulmonary capillary wedge pressure

With the introduction of balloon-tipped, self-guiding flotation catheters (Swan–Ganz catheters) to measure pulmonary capillary wedge pressure (PCWP) a new era in the monitoring of critically ill patients began.

Pulmonary capillary wedge pressure refers to the pressure measured through a catheter impacted into a small branch of the pulmonary artery, in such a way that there is free communication between the catheter tip and the pulmonary venous system. It gives a valid measurement of left atrial pressure and, if the mitral valve is normal, left ventricular end-diastolic pressure. The normal range is 5–12 mm Hg.

Flotation catheters have small inflatable balloons near their tips (Fig. 5). They are inserted from either an ante-cubital, subclavian, internal jugular or femoral vein. Once in the venous system the proximal end of the catheter is attached to a pressure transducer and the catheter is then advanced as far as the large intrathoracic veins. At this point the balloon is inflated with air (1.2–1.5 ml) and the flow of blood subsequently directs the catheter tip through the right heart chambers into the pulmonary artery. Constant display of the pressure waveforms shows its location until it finally "wedges" in a branch of the pulmonary artery. On wedging, the pulmonary artery pressure contour changes to a form which resembles that of the left atrial pressure (Fig. 6). The catheters can be passed easily at the bedside and fluoroscopy is not necessary. After a measurement of the PCWP has been made the balloon is deflated until the next measurement is required, when it is simply re-inflated.

Since they were first introduced a family of balloon-tipped catheters have been developed to allow cardiac output determinations by the thermodilution method, transvenous cardiac pacing, pulmonary angiography and "His" bundle electrocardiography.

Left ventricular function It is accepted that the Frank–Starling law applies to the intact human heart, and consequently the force of ventricular contraction is dependent upon ventricular end-diastolic volume or end-diastolic wall tension. Left ventricular end-diastolic pressure is widely used as an index of end-diastolic volume and wall tension, and an increase in its value signifies either left ventricular dysfunction or hypervolaemia. For practical purposes PCWP relates to left ventricular end-diastolic pressure.

FIG. 5   The inflatable balloon at the tip of a flotation catheter.

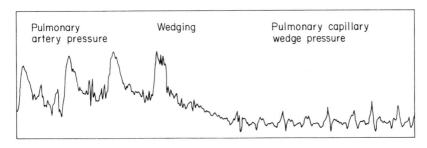

FIG. 6   Change in pressure waveform on "wedging".

With measures of both PCWP and cardiac output now readily available it is possible to construct ventricular function curves. Left ventricular stroke work (stroke index × mean arterial pressure × 0.0136) is plotted against PCWP which can be changed either by fluid administration, diuretics or vasopressors. In this way a Starling curve can be constructed and the optimum level of ventricular filling pressure determined for a particular patient (Fig. 7). Normally, as PCWP increases, there is a rapid initial rise in left ventricular stroke work which subsequently "plateaus" at higher filling pressures. With left ventricular impairment the curve is displaced downwards and to the right (Fig. 7). Improvement in performance, either in response to inotropic drugs or mechanical cardiac support, moves the curve upwards and to the left.

Arterial blood gas tension and acid–base status

Measurement of the arterial partial pressure of oxygen (p_{O_2}) and carbon dioxide (p_{CO_2}), together with pH, standard bicarbonate and base excess, is essential for quantifying the degrees of hypoxaemia and acidosis that are invariably present in acute circulatory failure. Repeat estimations are necessary to assess progress and evaluate treatment.

The serum electrolytes

The serum potassium level, in particular, has important influences on cardiac performance and irritability. Hypo- and hyperkalaemia are both

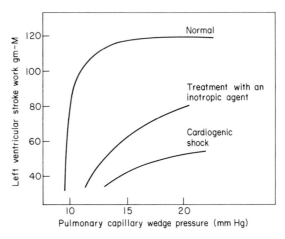

FIG. 7   Left ventricular function curves.

associated with serious disturbances of cardiac rhythm. Frequent measurements are required in patients receiving intravenous potassium and in those suffering from renal failure.

Urine volume

A falling urine volume usually indicates the reduced cardiac output and renal blood flow. Because of the risks of renal failure, urine volume is routinely measured, via in-dwelling bladder catheter, in all shocked patients.

Principles of treatment

The goals of treatment in shock are to restore a normal cardiac output and to support cardiovascular function until the underlying cause has been corrected. In all cases hypoxaemia, acid–base and electrolyte disturbances must be rectified so as to ensure an optimum "internal" environment for recovery to take place.

Hypovolaemic shock

The basis of treatment is:

1. to stop the abnormal source of fluid loss by medical and surgical means;
2. to replace the lost fluid with appropriate solutions, usually given intravenously.

The rate and volume of replacement is assessed by monitoring the numerous variables previously described. If the patient has abnormal cardiac function measurement of the PCWP is crucial for the prevention of pulmonary oedema ensuring that the left atrial pressure does not rise above that of the plasma oncotic pressure.

The type of fluid that is given relates specifically to the nature of the loss and may therefore be whole blood, plasma, saline, water, or any combination of these. If blood is the fluid of choice but is not immediately available, a number of substitutes can be used for a limited period. They include plasma protein solutions, dextrans and gelatin preparations, all collectively referred to as "plasma-expanders" because their molecular size confines them to the circulation for a relatively long time. Optimal blood flow and oxygen transport are achieved by keeping the haematocrit at around thirty per cent.

Vasopressor drugs, such as noradrenaline, will increase peripheral resistance and arterial pressure but have little place in the treatment of hypovolaemic shock where vasoconstriction is already extreme.

Cardiogenic shock

The objectives of therapy are to increase cardiac output and to decrease pulmonary venous congestion. In patients with acute myocardial infarction it is preferable if these goals can be achieved without enhancing myocardial ischaemia.

The principles are:

1. to ensure an optimum filling pressure to the left ventricle;
2. to improve myocardial contractility with inotropic drugs;
3. to reduce the afterload on the heart with vasodilator drugs;
4. to increase heart rate when bradycardia is a contributing feature;
5. to provide mechanical support for the circulation if these pharmacological measures prove inadequate.

Restoration of left ventricular filling pressure

Hypovolaemia with a low left ventricular end diastolic pressure is uncommon in cardiogenic shock but may occasionally be present as the result of diminished fluid intake, vomiting, sweating and the use of diuretics. In such cases the PCWP should be increased by careful fluid replacement to around 15 mm Hg.

Inotropic drugs

The inotropic agents are potentially capable of improving cardiac output by increasing contractility. The majority are catecholamines acting, via the $beta_1$-adrenergic receptors in the heart, to increase both contractility and heart rate. Some also act on beta-adrenergic receptors in certain vascular beds and cause vasodilatation. The ones in current use are isoprenaline, adrenaline, dopamine and dobutamine; their relative actions on heart rate, contractility and vascular resistance are shown in Fig. 8.

Isoprenaline (dose $1-8$ μg min^{-1}) acts on both $beta_1$ and $beta_2$ receptors. Cardiac output is increased and arterial pressure lowered. The vasodilatation occurs mainly in the skeletal muscle and mesenteric vascular beds and much of the increased cardiac output is diverted to these regions. Renal blood flow is not significantly altered.

	Heart rate	Myocardial contractility	Vascular resistance	
Isoprenaline	+++	+++	↓++	
Adrenaline	+++	+++	↓+↑	High doses
Dopamine	++	+++	↑	High doses
Dobutamine	+	+++	↓	

FIG. 8   Cardiovascular effects of some catecholamines.

Adrenaline (dose 10–30 μg min^{-1}) acts on beta$_1$ and beta$_2$ receptors and also affects alpha receptors in the blood vessels. With large doses the alpha action predominates and peripheral resistance increases.

Dopamine (dose 5–15 μg kg^{-1} min^{-1}) acts on beta$_1$ receptors in the heart and alpha receptors in the blood vessels. It differs from other catecholamines by acting on specific dopamine receptors in the renal arteries to produce renal vasodilatation and increase renal blood flow. It has less effect on heart rate than either isoprenaline or adrenaline.

Dobutamine (dose 2.5–10 μg kg^{-1} min^{-1}) is the most recently available inotrope. It acts on beta$_1$ receptors but has less effect on heart rate than the other three; its action on beta$_2$ and alpha receptors is relatively weak.

All the catecholamines, by increasing contractility, increase myocardial oxygen demand to a level which, in the presence of coronary artery disease, may exceed oxygen supply. Paradoxically myocardial function may then be adversely affected. Their chronotropic effects also add to the myocardial oxygen demands by inducing tachycardia.

Reducing afterload

Decreasing the resistance to left ventricular ejection enhances left ventricular stroke volume and cardiac output. It can be achieved by the use of vasodilator drugs such as sodium nitroprusside, phentolamine or nitroglycerine all of which have rapidly reversible haemodynamic actions. They produce a fall in arterial pressure (afterload) without significantly altering heart rate or contractility so that there is an overall decrease in myocardial oxygen demand. The main risk of this form of therapy is the danger of profound hypotension and reduction of coronary blood flow.

Increasing heart rate

In some patients a slow heart rate, due to sinus bradycardia or atrioventricular (heart) block contributes to the low cardiac output. It is necessary in such cases to increase the heart rate to optimal levels. Sinus bradycardia usually responds to intravenous atropine sulphate but cases of atrio-ventricular block will require ventricular pacing with an endocardial pacemaker.

Mechanical circulatory assistance

If the pharmacological measures prove inadequate, mechanical assistance of the circulation can be of value in selected cases. The two major objectives of this form of treatment are to increase the arterial pressure during diastole, so improving coronary blood flow, and to decrease left ventricular afterload during systole so reducing myocardial work and oxygen demand. Most experience has been gained with the use of intra-aortic balloon pumping devices (IABP).

An IABP consists of an inflatable balloon mounted on a catheter (Fig. 9); the balloon is placed, via the femoral artery, into the descending aorta, just distal to the left subclavian artery. It is inflated and deflated by an external power unit timed by the ECG. Inflation occurs during diastole, increasing arterial diastolic pressure, deflation is synchronized with ventricular systole, and lowers systolic arterial pressure. Essentially, the balloon pumping action augments that of the patient's own left ventricle.

Balloon pumping has been maintained, in some cases, for up to three to four weeks but in most instances, if it is to prove of value, only a few days' support is necessary. It is, however, unfortunately true that whilst IABPs produce a temporary haemodynamic improvement, the overall mortality of cardiogenic shock due to acute myocardial infarction has not been significantly altered. This is largely due to the presence of extensive and irrever-

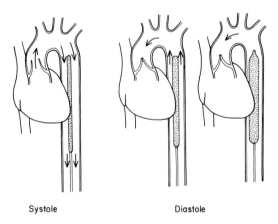

Systole   Diastole

FIG. 9   Action of an intra-aortic balloon pump.

sible myocardial damage at the time support is instituted. Earlier application of the technique, in an attempt to limit the extent of the infarction, may be rewarded by better results in the future.

Acute Respiratory Failure

Failure of gas exchange in the lungs is very common in critically ill patients, either because of primary respiratory disease or to secondary involvement of the respiratory system by disorders originating in other systems.

Definition and classification

Respiratory failure is defined, somewhat arbitrarily, in accordance with the arterial partial pressures of oxygen (p_{O_2}) and carbon dioxide (p_{CO_2}). Failure is taken to be present if, with the patient breathing room air, the p_{O_2} is less than 60 mm Hg or the p_{CO_2} is greater than 45 mm Hg.

From this definition respiratory failure can be classified into three categories:

1. Hypoxaemic (low p_{O_2}).
2. Ventilatory (high p_{CO_2}).
3. Mixed (combination of 1 and 2).

It is also important, and clinically essential, to distinguish between the acute and chronic forms of respiratory failure. The former develops quickly in a matter of days in a previously healthy respiratory system, whereas the latter progresses slowly over a period of years. Acute on-chronic respiratory failure is due to an acute respiratory insult superimposed on lungs that are already damaged.

Hypoxaemic respiratory failure

This is produced by a number of inflammatory and toxic agents affecting the lung parenchyma and pulmonary circulation (Fig. 10). In critically ill patients several of these may be present together and their effects combine to create a pattern of respiratory failure that has been labelled "The Adult Respiratory Distress Syndrome". Such a grouping is justifiable for, although separate clinical entities may be recognizable, their pathological pictures are non-specific. The common hallmarks of the pathophysiology

1. Pulmonary infection: bacteria, viruses, fungi.
2. Fluid overload and pulmonary capillary damage.
3. Oxygen toxicity
4. Obstruction of the pulmonary microcirculation:
   emboli, thrombi.

FIG. 10   Some causes of hypoxaemic respiratory failure.

are a mismatching of ventilation and perfusion and an increase in pulmonary extravascular water.

Hypoxaemia occurs because the normal relations between ventilation (V) and perfusion (Q) are disturbed; blood continues to perfuse alveoli that are no longer ventilated and an effective venous shunt is established. The magnitude of the shunt determines the degree of hypoxaemia.

The increase in pulmonary extravascular water, with interstitial and alveolar oedema, is caused either by pulmonary vascular congestion or by loss of integrity of capillary endothelium. Its presence compounds the ventilation–perfusion disturbance.

Ventilatory failure

This arises when the ventilatory forces are inadequate; the elevated p_{CO_2} signifies the hypoventilation.

Effective pulmonary ventilation requires an intact and integrated neuromuscular mechanism from the cerebral hemispheres to the chest wall and any part of this can be affected by various diseases (Fig. 11).

The combined form of respiratory failure is not uncommon and indicates involvement of both the lung and its ventilation control.

Clinical features of acute respiratory failure

The clinical manifestations of acute respiratory failure are somewhat ill defined and in the early stages can be difficult to recognize. Diagnosis and progress have both to be quantified by serial measurements of arterial oxygenation and pulmonary ventilation.

Hypoxaemia affects the central nervous system, causing hyperventilation, restlessness, confusion and, when severe, stupor and coma. The sympathetic nervous system is also activated, giving rise to tachycardia, hypertension and sweating. Cyanosis, best looked for in the mucous membranes of the mouth, is only apparent when the p_{O_2} has fallen to less than 50 mm Hg.

The features of carbon dioxide retention are even more variable. At high levels of p_{CO_2} (> 80 mm Hg) there is drowsiness, confusion and coma. With lower values there is reversal of sleep rhythm, a flapping tremor and

Central nervous system	Head injuries
	Overdose with hypnotic drugs
	High spinal cord injuries
Nerves supplying the respiratory muscles	Poliomyelitis Polyneuritis
Myo-neural junction	Myasthenia gravis
Respiratory muscles	Myopathies
Chest wall	Chest injuries

FIG. 11 Some causes of ventilatory failure.

evidence of a hyperdynamic circulation due to the vasodilator effects of carbon dioxide.

Monitoring the patient with respiratory failure

Two aspects of respiratory function require detailed observation:

1. The adequacy of oxygen exchange.
2. The adequacy of ventilation.

Both are fundamental to the diagnosis and management of respiratory failure.

The adequacy of oxygen exchange

Arterial p_{O_2} The normal value for arterial p_{O_2} is 80–95 mm Hg. Its value indicates the adequacy of arterial oxygenation, but if considered in isolation it provides little information about oxygen exchange in the alveoli. To be meaningful its value has to be compared to that of the partial pressure of oxygen in the alveoli. The difference between the two, the alveolar–arterial gradient $(A–aD_{O_2})$, is then a valid indicator of the efficiency of oxygen exchange.

Alveolar–arterial oxygen gradient The gradient can be measured at any inspired oxygen concentration, but it is most commonly determined with the patient inhaling 100% oxygen. In this situation the oxygen tension is the same in all ventilated alveoli and any gradient is a direct reflection of the degree of pulmonary shunting; i.e. the portion of cardiac output flowing past non-ventilated alveoli. It is also affected, but to a much smaller extent, by the arterio–venous oxygen content difference, the size of the cardiac output and the position of the oxyhaemoglobin dissociation curve.

The measurement obtained when the patient is breathing 100% oxygen is of great practical value but it has to be appreciated that inhaling pure oxygen can itself affect the degree of shunting by promoting absorption atelectasis in the lung. It is also impossible to deliver 100% oxygen to the alveoli unless the patient is intubated and the gradient has to be measured at the maximal inspiratory oxygen concentration possible, usually around 60 per cent.

The pulmonary shunt   Measurement of the size of the pulmonary shunt enables that proportion of the cardiac output returning to the left side of the heart unoxygenated to be quantified. Normally it is 3–5% of the total but in severe respiratory failure it can rise to over 20%. Values around 50% are recorded in extreme degrees of failure and are only compatible with life if the patient is being mechanically ventilated with high inspired oxygen concentrations.

Oxygen tension of mixed venous blood   Analysis of the arterial p_{O_2}, $(A - a)d_{O_2}$, and pulmonary shunt gives detailed information of oxygen transfer in the lung, but not of oxygen delivery to the tissues. This depends on the oxygen content of arterial blood (oxygen tension and haemoglobin content), the affinity of haemoglobin for oxygen and the rate of blood flow. Its adequacy is expressed by the oxygen tension in mixed venous blood which is normally 40 mm Hg, corresponding to a haemoglobin saturation of 75 per cent. Values of less than 40 mm Hg indicate increased oxygen extraction by the tissues and, if it is below 30 mm Hg, tissue oxygenation is probably affected.

The adequacy of ventilation

Arterial p_{CO_2} The normal level is 35–45 mm Hg and it is the prime indicator of alveolar ventilation; a normal p_{CO_2} signifies normal ventilation; a low p_{CO_2} hyperventilation; and a high p_{CO_2} hypoventilation. Its level also has an influence on arterial pH, high levels producing an acidosis and low levels an alkalosis.

Tidal volume; frequency of respiration; minute volume   These are inter-related and when they are reduced produce hypoventilation. Adjusting their relative values is an essential part of controlling mechanical ventilation.

Ratio of dead space ventilation to tidal volume   Normally this ratio is between 0.2 and 0.4; an increase occurs if there is under-perfusion of ventilated alveoli as in pulmonary embolism or shock. This leads to an increase in the physiological dead space and effective hypoventilation.

Principles of treatment

The purpose of supportive respiratory care is to correct the blood gas abnormalities and to maintain them within physiologically acceptable limits. For hypoxaemia this may entail increasing the oxygen concentration in the inspired air, intermittent positive pressure ventilation, or extracorporeal oxygenation; ventilatory failure may require mechanical ventilatory support.

Treatment of hypoxaemia

The methods used for correcting hypoxaemia are dictated by the measured level of the p_{O_2} considered in relation to the oxyhaemoglobin dissociation curve, from which three clinically noteworthy ranges of p_{O_2} can be identified (Fig. 12).

p_{O_2} *lower than 30 mm Hg*   This range is critical because the gradient of p_{O_2} from the capillaries to the cells is inadequate for oxygen transfer, particularly in the heart and brain.

p_{O_2} *between 30 and 60 mm Hg*   These values fall in the steep part of the curve where small changes in p_{O_2} are accompanied by relatively large changes in oxygen saturation and content. Any increase or decrease of p_{O_2} within this range is therefore very significant.

p_{O_2} *of 60 mm Hg or greater*   With a normal blood pH 60 mm Hg marks the point of 90% saturation and the beginning of the plateau of the curve. Above this point large increases in p_{O_2} lead to relatively small increases in saturation and content. As a consequence a p_{O_2} of 60 mm Hg is the desired goal of oxygen therapy. It is unnecessary and undesirable to give a patient oxygen in excess of that required to achieve this figure.

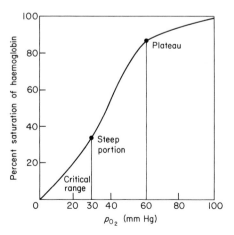

FIG. 12   The oxyhaemoglobin dissociation curve.

Increasing the oxygen concentration in inspired air

For patients who are breathing spontaneously, the inspired oxygen concentration can be varied over a wide range, from 24 to 70%, by using different types of face masks (Fig. 13).

Venturi Masks   These use the venturi principle whereby an oxygen jet draws in air and mixes it with oxygen in a certain ratio. They provide an accurate method for administering oxygen at certain specified concentrations: 24, 28, 35, 40 and 60 per cent.

Hudson and MC masks   These are two of a number of masks, with a low volume, that can provide high inspired oxygen concentration (50–60%) but in a less controlled manner than a venturi mask.

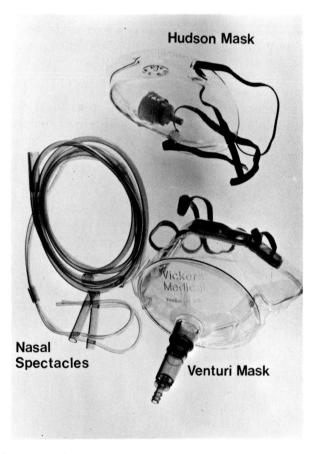

FIG. 13   Oxygen masks.

Nasal spectacles   These are useful in patients intolerant of a face mask and have the additional advantage of not requiring disconnection during eating and drinking. They give inspired oxygen concentrations ranging from 25 to 40 per cent.

Intermittent positive pressure ventilation (IPPV)

If a p_{O_2} of 60 mm Hg cannot be attained using a face mask it is very probable that the patient will require IPPV. This is achieved by intermittently inflating the lungs via a cuffed endo-tracheal or tracheostomy tube. Increasing the pressure in the airways above atmospheric pressure expands the lungs and then on its release the lungs recoil, by virtue of their elasticity, to the original volume. Respiration is therefore effected by alternately applying and removing a positive pressure by means of a ventilator (Fig. 1).

Many different types of mechanical ventilator are in use but, in principle, they can be divided into two kinds; pressure cycled and volume cycled. In a pressure-cycled machine the flow of gas from the ventilator will stop when a certain airway pressure has been achieved. With a volume-cycled machine flow continues until a specific volume is reached, the pressure required to attain this volume varying with the compliance of the lung and the resistance of the airways. For the treatment of acute respiratory failure a volume-cycled machine is preferred because it delivers a constant tidal and minute volume in spite of changes in lung pathology.

When commencing IPPV the appropriate values for tidal volume, frequency of respiration and oxygen concentration have to be set. Any necessary adjustments are subsequently made according to arterial blood gas measurements.

The reasons for the benefits of IPPV are as follows:

(i)   It improves the ventilation–perfusion ratio and decreases the pulmonary shunt by recruiting previously closed alveoli. This is presumably achieved by delivering a necessary opening pressure to those alveoli that are either collapsed or contain fluid.
(ii) It improves the distribution of gas in the lungs.
(iii) It decreases the work of breathing.

The problems of IPPV are related to the increased intrathoracic pressure that it generates. Venous return to the heart can be impaired, leading to a fall in cardiac output, and alveolar rupture can cause a pneumothorax.

If the pulmonary shunt is large IPPV with oxygen concentrations in excess of 60% may be unable to achieve a p_{O_2} of 60 mm Hg. In such instances the addition of a positive end-expiratory pressure (PEEP) can be helpful. This is applied in increments of 5 cm H_2O and produces an increase in p_{O_2} by further recruitment of atelactic alveoli and decreasing the development of alveolar oedema.

The problems associated with a PEEP arise from a further reduction in cardiac output, especially in the presence of hypovolaemia or cardiac disease, and an added risk of alveolar rupture.

Its optimum value, in terms of oxygen delivery to tissues, depends on the balance obtained between improved oxygen transfer and decreased cardiac output. If too high a level is used the fall in cardiac output may well offset the advantages of the higher p_{O_2}.

If IPPV with optimal levels of PEEP and high inspired oxygen concentrations cannot produce a p_{O_2} of 60 mm Hg, only two other measures are available.

Hypothermia Lowering the patient's temperature to 32 °C decreases tissue oxygen requirements and may increase the p_{O_2}. Measurement of mixed venous oxygen tension will indicate the benefits of this manoeuvre.

Extracorporeal membrane oxygenation The membrane artificial lung is an extracorporeal device for exchanging blood gases. Its use can support life and relieve the patient's lungs of the burden of high ventilatory pressures and high concentrations of inspired oxygen, thus, hopefully, buying time for the underlying lung pathology to heal.

The interposition of a silicone membrane between blood and respiratory gases allows effective gas exchange and reduces blood damage. Such oxygenators can therefore be used for relatively long periods of respiratory support and perfusions lasting for ten days have been carried out. The three possible routes for by-pass are arterio–venous, veno–venous and veno–arterial. The last is probably the most popular. Unfortunately, the survival record using these devices has been poor, largely because, as with IABP, patients have only been treated by this method when their respiratory disease has been extremely advanced and in some instances irreversible. Further trials and developments are needed to define its particular place in the treatment of acute hypoxaemic respiratory failure.

Treatment of ventilatory failure

Compared to hypoxaemic failure the support of inadequate ventilation is relatively simple. Intermittent positive pressure ventilation is usually required if the p_{CO_2} rises above 55 mm Hg and it is maintained until adequate normal ventilation returns. In pure ventilatory failure oxygen transfer is not a problem.

The use of drugs that purport to stimulate the respiratory centre produces little benefit and is associated with a number of unwanted side-effects.

Acute Renal Failure

Acute renal failure is characterized at its onset by a reduced glomerular filtration rate (GFR), oliguria (urine vol < 20 ml per hour), and accumulation in the body of certain substances that are normally excreted in the urine. Its occurrence in critically ill patients is usually heralded by oliguria which dominates the early stages of diagnosis and treatment.

Pathogenesis of acute renal failure

The commonest cause of acute renal failure, in the context of critical illness, is acute circulatory failure which reduces the renal blood flow and the GFR. The oliguria is, initially, a direct consequence of the low GFR coupled with a continuing tubular response to the actions of antidiuretic hormone and aldosterone. At this stage there exists a state of reduced renal perfusion with intact renal function which is reflected by the composition of the urine. The kidneys retain salt and water and maintain the ability to concentrate the urinary solute load. The urine, therefore, has a high osmolality compared to the plasma, and a low sodium concentration.

These changes are reversible if the underlying haemodynamic disturbance is corrected quickly. However, if such treatment is delayed or proves ineffective then the renal failure becomes established and there is loss of tubular function. The urine is no longer concentrated with respect to plasma and has a high sodium content. This form of established renal failure is often referred to as acute tubular necrosis, an unsatisfactory title because actual necrosis is not a histological feature.

The forms of acute renal failure caused by inflammatory diseases of the kidney and urinary tract obstruction are outside the scope of this chapter.

Biochemical changes in the plasma

Nitrogen retention

The blood urea rises by 4–6 mmol l^{-1} per day and if the patient is hypercatabolic by 10–15 mmol l^{-1} per day. The blood creatinine level also rises and the decrease in GFR is reflected by a reduced creatinine clearance.

Sodium

Hyponatraemia is common and usually indicates dilution due to a relative water excess. Hypernatraemia does occur and signifies either a water deficit or a true excess of sodium.

Potassium

Hyperkalaemia is frequent and dangerous. It is a result of reduced renal excretion of potassium aggravated by an increased release of intracellular potassium due to hypercatabolism. At serum levels of greater than 7 mmol l^{-1} it can cause cardiac arrest. Changes in the ECG are useful for assessing its severity. With moderate increases (6.0–7.0 mmol l^{-1}) the T waves are tall and symmetrically peaked, whilst above 7.0 mmol l^{-1} there may be loss of P waves and intraventricular conduction disturbances, with widening of the QRS complexes.

Hypokalaemia is less common and occurs only if there are other abnormal sites of potassium loss e.g. the gastrointestinal tract.

Bicarbonate

This is reduced, and there is a metabolic acidosis as a result of reduced excretion of hydrogen ions and an accumulation of acid metabolities from increased protein breakdown.

Clinical features

Oliguria is the presenting feature and the other symptoms and signs are manifestations of fluid and electrolyte disturbances as they affect different systems.

Respiratory

Dyspnoea is caused by pulmonary oedema and the metabolic acidosis.

Cardiovascular

Peripheral oedema and hypertension are both attributable to fluid overload.

Gastrointestinal

Anorexia, nausea, and vomiting are all common and probably result from water overload. In some patients hiccoughs are troublesome and gastrointestinal bleeding may occur in advanced failure.

Neurological

Drowsiness, confusion, muscle twitching, and coma are all late features associated with widespread metabolic disturbances.

The clinical course of established renal failure

Whatever its cause, the course of renal failure passes through three phases; the oliguric; the diuretic or polyuric; and the recovery.

Oliguric phase

This may last several days or weeks if the patient survives. Urine osmolality is low and sodium concentration is high.

Diuretic phase

This follows the oliguric phase and is marked by an increasing urine volume, up to three or four litres per day. The blood urea and creatinine levels decrease and there may be large urinary losses of potassium. The GFR remains below normal for weeks or months.

Recovery phase

Full recovery is usual. If, however, the oliguric phase persists for several weeks, a renal biopsy is necessary to define the diagnosis and to assess prognosis. If the situation is irreversible the patient may be a candidate for long-term dialysis and renal transplantation.

Principles of treatment

The first goal is to correct the low cardiac output which will, if the failure is still reversible, restore normal renal function. If, however, in spite of this and the use of large doses of diuretics, oliguria persists, the measures for treating established renal failure are instituted.

During the oliguric phase the aim is to keep the composition and volume of the body fluids within tolerable physiological limits. Initially treatment is confined to strict control of water, electrolyte and nutritional balance (conservative treatment). If these prove inadequate, dialysis is necessary.

Conservative treatment

The regime consists of:

1. *Restriction of water intake* to 400 ml per day to replace insensible water loss plus additional quantities, equal to the daily urine volume and any other losses.
2. *Restriction of sodium intake* to less than 30 mmol per day. Any abnormal losses of sodium are replaced accordingly.
3. *Restriction of potassium intake.* Hyperkalaemia is the most dangerous complication of acute renal failure and no potassium is given. If the plasma level rises to dangerous heights it can be temporarily reduced by giving glucose and insulin, which shifts the potassium into the cells, and using ion exchange resins.
4. *A high-calorie intake* of at least 2000 calories per day to minimize endogenous protein breakdown and so control urea and potassium production. Provision of a sufficient number of calories, within the restrictions imposed on fluid intake, requires the use of either 50% dextrose or a fat emulsion. Protein intake is restricted but limited quantities of amino acids are necessary to attain a positive nitrogen balance. Anabolic steroids can be given to reduce the rate of catabolism but are largely ineffective.
5. *Correction of severe acidosis* (plasma pH 7.1) with sodium bicarbonate. This can only be a temporary measure because of the inevitable sodium overloading that will result from the use of sodium bicarbonate solution.

Indications for dialysis

Failure to control the situation adequately by conservative measures points to the need for dialysis but the need for this should be realized before severe biochemical disturbances have occurred. The aim is to prevent complications rather than treat them; specific indications are a blood urea greater than 33 mmol l^{-1}, a serum creatinine greater than 900 μmol l^{-1}, uncontrollable hyperkalaemia, severe fluid overload and severe metabolic acidosis.

Forms of dialysis   Two types of dialysis procedure, peritoneal dialysis and haemodialysis can be used to treat acute renal failure. Both methods are appropriate for the majority of cases, but since peritoneal dialysis is easier to institute and manage it is used more often than haemodialysis outside specialist renal units.

Peritoneal dialysis (Fig. 1) This uses the patient's peritoneum as a dialysing membrane. Waste products, water and electrolytes move across it, between the body fluids and a specifically formulated dialysis fluid introduced into the peritoneal cavity. The direction and rate of movement of each substance is determined by the osmotic and concentration gradients present on each side of the membrane.

Dialysis is performed through a catheter introduced through the anterior abdominal wall into the peritoneal cavity. One- or two-litre volumes of selected dialysis fluid are sequentially run in and out of the peritoneum in a cyclic fashion until a satisfactory biochemical equilibrium has been established.

Choice of dialysis fluid Three kinds of peritoneal dialysis fluids are available for use in particular clinical situations:

1. an approximately iso-osmotic fluid;
2. a fluid rendered hypertonic by the addition of extra glucose;
3. a fluid with a low sodium concentration.

None of them contains potassium which can be added as the serum potassium level is brought under control. The iso-osmotic fluid is the most widely used, it avoids too-rapid shifts in water balance and hypovolaemia; the hypertonic solution is for cases who are seriously overhydrated. It removes water relatively rapidly, but can only be used for a limited period because the high glucose content makes it irritant to the peritoneum.

Repeated exchanges across the peritoneum cause a large loss of protein into the dialysis fluid, which may lead to severe hypoproteinaemia if

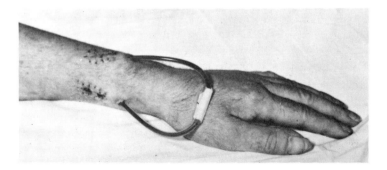

FIG. 14   An arterio–venous shunt *in situ*.

adequate nutritional replacement is not provided. Peritoneal dialysis cannot be efficiently handled if the patient has peritoneal adhesions or fistulae through the abdominal wall. Haemodialysis is then the only alternative, and it is regarded by some as the first choice in hypercatabolic patients because of its greater efficiency in controlling urea levels.

Haemodialysis This brings the patient's blood and dialysing fluid to either side of a semipermeable membrane across which necessary exchanges can occur. An arterio–venous shunt is used for access to the circulation, (Fig. 14) and blood flow is dependent on the patient's arterial pressure and may not be adequate if the patient is hypotensive.

A haemodialysis machine consists of instrumentation for delivering blood and dialysate, and the haemodialysis membrane. The most widely used membrane is cuprophane, and the large surface area required for satisfactory dialysis is achieved by arranging it in appropriate geometric configurations; flat layers, coils or hollow fibres (Fig. 15). The advantages

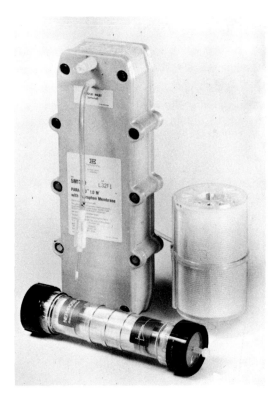

FIG. 15   Dialysers.

of the latter arrangement appear to lie in its compactness and low priming volume. Development of new membranes continues, aimed at producing membranes with better solute clearance and improved ultra filtration capabilities.

Acknowledgements

I wish to thank Miss S. Shah for typing the chapter, the medical artist and the Photographic Department of the Middlesex Hospital for their help and Dr F.D. Thompson, St Phillips Hospital, Renal Unit, for his help in preparing Figs 14 and 15.

POSTSCRIPT
The Ethics of
Human Experimentation

J.S. Weiner

At the present time there is a wide acceptance of the obligation on biomedical research workers to carry out investigations on human subjects in accordance with strict ethical codes. This liability is ultimately enforceable by law, civil or criminal, in many countries. In the UK, the Medical Research Council (MRC) many years ago issued a clear statement defining the responsibilities and liabilities that clinical and non-clinical investigators assume when they embark on research on human subjects. There is now an internationally agreed code of ethics, embodied in the "Declaration of Helsinki", adopted at the Eighteenth World Medical Assembly in June 1964, and revised by the Assembly at Tokyo in October 1975. The latter widened the scope of the code from "clinical research" to

609

"biomedical research involving human subjects". The 1976 Declaration of Helsinki is appended.

Physiologists will be aware of the explicit requirements by the American and British Physiological Societies that experiments reported in their publications must be guaranteed as complying with ethical principles conforming to those of the Declaration of Helsinki (see p. 616). Enforcement or regulation is performed in the UK by hospital and institutional ethical committees (see below).

Research workers need therefore to be fully conversant with the guidelines in the Helsinki and MRC statements, as well as with the local arrangements for securing approval and clearance of their proposed investigations.

Although the Helsinki Declaration speaks of "research" there can be no doubt that ethical responsibilities are involved in any kind of investigation requiring human subjects. These would include not only systematic experimentation but also biomedical surveys of groups, the medical examination of individuals, or the taking of blood for study even on an isolated occasion. In clinical and preventive medicine, many other procedures would be involved. In the US it is the case that investigators' responsibilities are deemed to apply to subjects exposed to possible psychological or social, as well as physical injury. Thus, even the administration of a questionnaire, or the videotaping or photography of a subject, demands conformity with the principles laid down in the accepted codes.

Before detailing the practical steps needed to safeguard both the subject and the investigator, it may be helpful to draw attention to three basic principles which underly the ethical approach to investigations on man. These are:

1. the principle of informed consent;
2. the principle of justifiable benefit or risk;
3. the inviolability of individual rights.

These principles are closely intertwined and they all raise issues of some complexity, both in theory and practice.

Informed or True Consent

"By true consent is meant consent freely given with proper understanding of the nature and consequences of what is proposed. Assumed consent or consent obtained by undue influence is valueless" (MRC Statement). Clearly a major consideration in obtaining "true consent freely given" is that full account must be taken of the relationship between the investigator and his subject. This relationship ranges from equality of standing as when scientists experiment on each other to a high degree of inequality when the potential subject is a minor.

It might be thought that when the investigator and subject are one and the same, the self-experimentor, there can be no doubt as to "consent freely given in the light of full understanding". Self-experimentation has in fact been a productive and fully accepted procedure in physiology and medicine from the days of John Hunter. Much of the history of discovery could be recounted in terms of self-experimentation, or of mutual experimentation in many famous partnerships such as that of Douglas and Haldane, or in team experimentation, as in the Harvard Fatigue Laboratory. It could, however, be argued that junior members of such research teams might feel themselves in a position of inferiority, making it difficult for them to withold consent. The example set by senior workers acting as subjects would seem on the whole to constitute a strong ethical argument for the propriety of such investigations before the present era of explicit codes of conduct. In fact, self-experimentation or team experimentation by committed and informed research workers can no more be undertaken today without approval of an ethical committee than any other arrangement between experimentor and subject. An investigator, however well informed, may need protection from his own dedication or enthusiasm. That, at least, is today's philosophy.

Wherever the subject is in a dependent relationship, safeguards are all the more essential. This is obviously the case when paid subjects are employed or where minors and children or the handicapped are involved. Equally, safeguards are necessary where medical students are required to help in demonstrations of physiological principles, even though they may be thought to be reasonably informed.

Agreement on the part of the subject must always be sought by a clear and adequate explanation of both the purposes and the conduct of the proposed procedure, but the investigator must obtain evidence that consent has been given, and this must be done by providing explanations both in writing and verbally in the presence of another person. "Written consent unaccompanied by other evidence that an explanation has been given, understood and accepted is of little value" (MRC). The explanatory statement must, of course, be acceptable to the ethical committee as well as to the subject.

Justifiable Benefit

Consent cannot be expected unless the investigator includes in his explanation of the nature and purposes of the proposed procedure, a clear indication of both the benefits and possible discomforts or risks.

The "benefits" accruing from the research or investigation can assume different degrees of certainty.

(a) Procedures of direct benefit to the individual include new or modified clinical treatment or preventive measures. Medical examination of the subject, taking of blood or other samples may be clearly intended for his benefit. It could be argued that the participation of a student in practical teaching exercises is of direct value to him.

(b) Procedures may be held to be of indirect or long-term benefit to the individual. Subjects acting as "controls" or as "placebo" recipients or participants in surveys may well provide immediate beneficial information to the community, but only indirectly to themselves.

(c) Procedures or experiments may be proposed primarily for scientific reasons, that is for a gain in knowledge, without obvious immediate or indirect benefit to the subject. Much self-experimentation or "peer" experimentation falls into this category.

The benefits expected from the investigation therefore need to be clearly appreciated and made explicit both to the subject and the ethical committee. At the same time benefits must be weighed against risk. The Declaration of Helsinki lays down: "Biomedical research involving human subjects cannot legitimately be carried out unless the importance of the objective is in proportion to the inherent risk to the subject." Even where risks and discomfort may be negligible, or almost so, as with the use of non-invasive techniques, the benefits, scientific or therapeutic, might be judged also to be so small as to be trivial. The ethical committee must be of a scientific calibre to make judgements both of risks and of benefits in relation to the worthwhileness of the proposal.

Ethical codes in the UK stipulate that medical supervision of experiments must be readily available; and that only the use of healthy volunteers will be permitted in non-clinical investigations.

Individual Rights

Ethical codes are designed to promote professional discipline, etiquette and responsibility; primarily they are concerned with the rights of the subject. Informed consent properly obtained, a high benefit or an apparent low risk will be difficult to use as defence where negligence may have occurred. Thus, ethical committees have the task of assessing the competence and experience of the investigator and the appropriateness and safety of his procedures. The conduct of experiments implies also full knowledge and adherence to the safety codes and regulations in force in the laboratory involved, over and above the precautions necessary in the investigation itself.

The inviolability of an individual's rights must be given explicit recognition by making it clear to the subject that his consent is specific to the particular experiment as described, and shall not and cannot be taken to imply consent to participate in any subsequent investigation. Above all, the subject must be made aware of his absolute right to withdraw from the investigation at any time. This is in accord with the Helsinki Code, which lays down that "The right of the research subject to safeguard his or her integrity must always be respected". The Code enjoins that "every precaution should be taken to respect the privacy of the subject and to minimize the impact of the study on the subject's physical and mutual integrity and on the personality of the subject."

Professional Discipline

It can be fairly claimed that adherence to ethical codes, such as those of Helsinki or the Medical Research Council combined with strict observance of Safety codes and regulations governing laboratory work, constitutes a substantial guide to responsible professional conduct for biomedical investigators in the particularly difficult circumstances of non-clinical investigations. But fundamentally the same ethical considerations apply to both the clinical and the non-clinical spheres. The concluding words in the MRC's statement are entirely apposite.

"The progress of medical knowledge has depended, and will continue to depend, in no small measure upon the confidence which the public has in those who carry out investigations on human subjects, be these healthy or sick. Only in so far as it is known that such investigations are submitted to the highest ethical scrutiny and self-discipline will this confidence be maintained. Mistaken, or misunderstood, investigations could do incalculable harm to medical progress. It is our collective duty as a profession to see that this does not happen and so to continue to deserve the confidence that we now enjoy."

Practical Measures

The instructions which have been in force for many years in the place of work of one of the editors (JSW at the London School of Hygiene and Tropical Medicine) provide clear guidance to investigators. (For the Code of Practice applicable to human volunteers, see p. 619. In substance they are as follows.

Before commencing any experiment using human volunteers, permission must be obtained from the Ethical Committee. Forms A, B and C should be submitted through the Head of Department and the Dean (The Head of the Institute) with the explanatory statement.

Form A is the formal application from the Head of the Department or Laboratory on behalf of the experimentor, affirming that they have both read the standing instructions as given in the School's Code of Practice.

Form B sets out

Title and aim of investigation
Period of investigation
Name(s) of principal investigator(s)
Names of other personnel
Information of subjects to be used—age, sex, source, numbers, whether expenses defrayed and other payments, etc.
Procedures to be used
Measures to ensure confidentiality
Details of other Institutions or Committees supporting the investigation.
Affirmation of method of securing subjects consent by
(1) verbal and written explanation to be given to each subject in the attached *explanatory statement*
(2) a copy of the *form of consent* to be signed by the subject on the standard form, Form C (reproduced on facing page, p. 615)

Experiments Involving Human Subjects: Form C

CONSENT TO PARTICIPATE IN AN EXPERIMENT

The experimental procedure of the investigation into:-

_____

has been explained to me in the presence of the witness to my signature and I fully understand the nature of my participation. I give my full and true consent to my participation in the procedures described in Form B.

This true consent is specific to the particular experiment named above and shall not be taken to imply my consent to participate in any subsequent investigation. I reserve the right to withdraw from this experiment at any time.

_____ signed

_____ name in
block
capitals
_____ date

Signature witnessed by:–

_____

_____ name in
block
capitals

London School of Hygiene & Tropical Medicine,
Keppel Street,
London, WCIE 7HT.

Biomedical Research: a Revised Code of Ethics

Declaration of Helsinki

Recommendations guiding medical doctors
in biomedical research involving human subjects

Adopted by the Eighteenth World Medical Assembly, Helsinki, Finland, 1964, and
revised by the Twenty-ninth World Medical Assembly, Tokyo, Japan, 1975

Introduction   It is the mission of the medical doctor to safeguard the health of the people. His or her knowledge and conscience are dedicated to the fulfilment of this mission.

The Declaration of Geneva of the World Medical Association binds the doctor with the words, "The health of my patient will be my first consideration", and the International Code of Medical Ethics declares that, "Any act or advice which could weaken physical or mental resistance of a human being may be used only in his interest."

The purpose of biomedical research involving human subjects must be to improve diagnostic, therapeutic and prophylactic procedures and the understanding of the etiology and pathogenesis of disease.

In current medical practice most diagnostic, therapeutic or prophylactic procedures involve hazards. This applies *a fortiori* to biomedical research.

Medical progress is based on research which ultimately must rest in part on experimentation involving human subjects.

In the field of biomedical research a fundamental distinction must be recognized between medical research in which the aim is essentially diagnostic or therapeutic for a patient, and medical research, the essential object of which is purely scientific and without direct diagnostic or therapeutic value to the person subjected to the research.

Special caution must must be exercised in the conduct of research which may affect the environment, and the welfare of animals used for research must be respected.

Because it is essential that the results of laboratory experiments be applied to human beings to further scientific knowledge and to help suffering humanity, the World Medical Association has prepared the following recommendations as a guide to every doctor in biomedical research involving human subjects. They should be kept under review in the future. It must be stressed that the standards as drafted are only a guide to physicians all over the world. Doctors are not relieved from criminal, civil and ethical responsibilities under the laws of their own countries.

I. Basic principles   1. Biomedical research involving human subjects must conform to generally accepted scientific principles and should be based on adequately performed laboratory and animal experimentation and on a thorough knowledge of the scientific literature.

2. The design and performance of each experimental procedure involving human subjects should be clearly formulated in an experimental protocol which should be transmitted to a specially appointed independent committee for consideration, comment and guidance.

3. Biomedical research involving human subjects should be conducted only by scientifically qualified persons and under the supervision of a clinically competent medical person. The responsibility for the human subject must always rest with a medically qualified person and never rest on the subject of the research, even though the subject has given his or her consent.

4. Biomedical research involving human subjects cannot legitimately be carried out unless the importance of the objective is in proportion to the inherent risk to the subject.

5. Every biomedical research project involving human subjects should be preceded by careful assessment of predictable risks in comparison with foreseeable benefits to the subject or to others. Concern for the interest of the subject must always prevail over the interests of science and society.

6. The right of the research subject to safeguard his or her integrity must always be respected. Every precaution should be taken to respect the privacy of the subject and to minimize the impact of the study on the subject's physical and mental integrity and on the personality of the subject.

7. Doctors should abstain from engaging in research projects involving human subjects unless they are satisfied that the hazards involved are believed to be predictable. Doctors should cease any investigation if the hazards are found to outweigh the potential benefits.

8. In publication of the results of his or her research, the doctor is obliged to preserve the accuracy of the results. Reports on experimentation not in accordance with the principles laid down in this Declaration should not be accepted for publication.

9. In any research on human beings, each potential subject must be adequately informed of the aims, methods, anticipated benefits and potential hazards of the study and the discomfort it may entail. He or she should be informed that he or she is at liberty to abstain from participation in the study and that he or she is free to withdraw his or her consent to participation at any time. The doctor should then obtain the subject's freely-given informed consent, preferably in writing.

10. When obtaining informed consent for the research project the doctor should be particularly cautious if the subject is in a dependent relationship to him or her or may consent under duress. In that case the informed

consent should be obtained by a doctor who is not engaged in the investigation and who is completely independent of this official relationship.

11. In the case of legal incompetence, informed consent should be obtained from the legal guardian in accordance with national legislation. Where physical or mental incapacity makes it impossible to obtain informed consent, or when the subject is a minor, permission from the responsible relative replaces that of the subject in accordance with national legislation.

12. The research protocol should always contain a statement of the ethical considerations involved and should indicate that the principles enunciated in the present Declaration are complied with.

II. Medical research combined with professional care (clinical research)

1. In the treatment of the sick person, the doctor must be free to use a new diagnostic and therapeutic measure, if in his or her judgement it offers hope of saving life, reestablishing health or alleviating suffering.

2. The potential benefits, hazards and discomfort of a new method should be weighed against the advantages of the best current diagnostic and therapeutic methods.

3. In any medical study, every patient—including those of a control group, if any—should be assured of the best proven diagnostic and therapeutic method.

4. The refusal of the patient to participate in a study must never interfere with the doctor–patient relationship.

5. If the doctor considers it essential not to obtain informed consent, the specific reasons for this proposal should be stated in the experimental protocol for transmission to the independent committee (**1**, **2**).

6. The doctor can combine medical research with professional care, the objective being the acquisition of new medical knowledge, only to the extent that medical research is justified by its potential diagnostic or therapeutic value for the patient.

III. Non-therapeutic biomedical research involving human subjects (non-clinical biomedical research) 1. In the purely scientific application of medical research carried out on a human being, it is the duty of the doctor to remain the protector of the life and health of that person on whom biomedical research is being carried out.

2. The subjects should be volunteers—either healthy persons or patients for whom the experimental design is not related to the patient's illness.

3. The investigator or the investigating team should discontinue the research if in his/her or their judgement it may, if continued, be harmful to the individual.

4. In research on man, the interest of science and society should never take precedence over considerations related to the well-being of the subject.

Extract from London School of Hygiene and Tropical Medicine Code of Practice

Use of human volunteers in experiments

Principles

Experiments on human volunteers must conform to generally accepted scientific principles and should be based on adequately performed laboratory and animal experimentation and on a thorough knowledge of the literature. They can be justified only if the importance of their objective is in proportion to the inherent risk to the subject. A proposal must therefore be preceded by a careful assessment of predictable risks and comparison with foreseeable benefits to the subject and/or others. Concern for the subject must always prevail over the interests of science and society. The privacy of the subject must be safeguarded and the impact of the experiment on him or her minimized.

Each potential subject must be adequately informed of the aims, methods, anticipated benefits, potential hazards of the study and the discomfort it may entail. The subject should be informed that he or she may refuse or withdraw from the study at any time. Particular care is essential if the subject is in a dependent relationship to the experimenter: in such cases consent should be obtained by a doctor not involved in the experiment and to whom the subject is not in a dependent relationship. (Based on *WHO Chronicle*, **30**, 360, (1976).)

Instructions

(i) Before commencing any experiment using human volunteers, permission must be obtained from the Ethical Committee. Forms A and B should be submitted through the Head of Department and the Dean.

(ii) Only the use of healthy volunteers will be considered. Applications for experiments on patients must be submitted to the appropriate hospital ethical committee.

(iii) If an experimenter proposes to experiment on himself, on a member of the staff of the School or on a student of the School, this must be explicitly stated in the application and safeguards prescribed against undue influence being used to persuade staff or students to volunteer,

(iv) If volunteers are to be paid or rewarded in any way, this should be stated.

(v) Medical staff involved in volunteer experiments must be personally covered by the Medical Defence Union, the Medical Protection Society or a similar body.

(vi) No proposal without designated medical supervision will be approved.

(vii) No proposal involving administration of a radioactive substance will be considered without prior approval of the use by the MRC

(viii) Witnesses of consent forms (Form C) should normally be independent of the Department in which the experiments are being undertaken.

(ix) Careful and complete records must be kept of all experiments in case legal action arises. These should record any complaint or reaction shown by the volunteer.

(x) The Committee will wish to be assured that adequate facilities (e.g. privacy for changing clothes) exist for any experiment involving volunteers.

Insurance

The School's insurers must be informed well in advance of any use of human volunteers since otherwise there will be no insurance cover. The insurers will consider each case as a separate matter and subject for insurance. They may refuse to accept the risk or may demand a high premium and/or impose conditions which are unacceptable. In that case there must be time to seek insurance cover elsewhere, such as from Lloyds underwriters. The cost of any such insurance will have to be borne by the department or research grant concerned.

Other ethical problems

Certain procedures (e.g. simple venepuncture) on human volunteers are technically not experiments. However, such subjects should invariably receive adequate explanation and should be asked to consent. Although it may not be practicable in all cases to obtain full written consent, the principles laid down above should be applied, particularly paras (v), (vi), (ix).

Ethical problems arise in testing or becoming otherwise involved in the use of materials collected from man elsewhere by possibly unethical procedures. In any case of doubt, the Head of Department and/or the Ethical Committee should be consulted.

Heads of Departments or the Dean may be consulted on any ethical problem. If necessary the Dean will refer it to the Ethical Committee.

References

Biomedical research: a revised code of ethics. (1976). *WHO Chronicle* **30**, 360–362. (Reproduced on pp. 616–618, with permission from the World Health Organization.)

Ethical Principles on the Conduct of Research with Human Subjects. (1973). American Psychological Association, Washington, D.C.

Responsibility in Investigations on Human Subjects. Report of the Medical Research Council for 1962–63 (Cmnd 2382) pages 21–25.

Author Index

The numbers in italics are those pages where References are listed in full.

Subject Index

The Principles and practice of human physiology/
editors, O.G. Edhoim, J.S. Weiner. -- London ;
New York : Academic Press, 1981.

xiii, 672 p. : ill. ; 24 cm.
Includes bibliographical references and
index.
ISBN: 0-12-231650-9

51500

1. Human physiology. I. Edholm, Otto G.
II. Weiner, Joseph Sidney, 1915-